LONG-TERM
STUDIES OF
Vertebrate
Communities

LONG-TERM
STUDIES OF
Vertebrate
Communities

Edited by

MARTIN L. CODY
JEFFREY A. SMALLWOOD
Department of Biology
University of California, Los Angeles

ACADEMIC PRESS

San Diego   London   Boston   New York   Sydney   Tokyo   Toronto

Copyright © 1996 by ACADEMIC PRESS

Academic Press, Inc.
525 B Street, Suite 1900, San Diego, California 92101-4495, USA
http://www.apnet.com

Academic Press Limited
24-28 Oval Road, London NW1 7DX, UK
http://www.hbuk.co.uk/ap/

Library of Congress Cataloging-in-Publication Data

Long-term studies of vertebrate communities / edited by Martin L.
   Cody, Jeffrey A. Smallwood.
         p.   cm.
   Includes bibliographical references (p.   ) and index.
   ISBN 0-12-178075-9 (alk. paper)
   1. Vertebrates.   2. Animal communities.   I. Cody, Martin L.,
date.   II. Smallwood, Jeffrey A.
QL605.L66   1996
596'.05247--dc20   96-12813
                                                    CIP

Printed and bound in the United Kingdom
Transferred to Digital Printing, 2011

CONTENTS

SECTION **II**

Reptile and Amphibian Communities

9 **Structure and Dynamics of an Amphibian Community:
Evidence from a 16-Year Study of a Natural Pond**

*Raymond D. Semlitsch, David E. Scott, Joseph H. K. Pechmann,
and J. Whitfield Gibbons*

SECTION **III**

Bird Communities

10 **Role of the Sibling Species in the Dynamics of the Forest-
Bird Communities in M'Passa (Northeastern Gabon)**

A. Brosset

11 **Bird Communities in the Central Rocky Mountains**

Martin L. Cody

SECTION **IV**

Mammal Communities

CONTRIBUTORS

Numbers in parentheses indicate the pages on which the authors' contributions begin.

A. Brosset (251), Muséum National d'Histoire Naturelle, URA 1183/CNRS, 91800 Brunoy, France

James H. Brown (555), Department of Biology, University of New Mexico, Albuquerque, New Mexico 87131

Martin L. Cody (1, 291), Department of Biology, University of California, Los Angeles, California 90095

Justin D. Congdon (137), University of Georgia's Savannah River Ecology Laboratory, Aiken, South Carolina 29802

James E. Diffendorfer[1] (421), Natural History Museum and Department of Systematics and Ecology, University of Kansas, Lawrence, Kansas 66045

Barry J. Fox (467), School of Biological Science, University of New South Wales, Sydney 2052, Australia

M. S. Gaines[2] (421), Natural History Museum and Department of Systematics and Ecology, University of Kansas, Lawrence, Kansas 66045

J. Whitfield Gibbons (137, 217), University of Georgia's Savannah River Ecology Laboratory, Aiken, South Carolina 29802

B. Rosemary Grant (343), Department of Ecology and Evolutionary Biology, Princeton University, Princeton, New Jersey 08544

[1]Present address: Department of Biology, University of Miami, Coral Gables, Florida 33133.
[2]Present address: Department of Biology, University of Miami, Coral Gables, Florida 33133.

Peter R. Grant (343), Department of Ecology and Evolutionary Biology, Princeton University, Princeton, New Jersey 08544

Nelson G. Hairston, Sr. (161), Department of Biology, University of North Carolina, Chapel Hill, North Carolina 27599

Charles O. Handley, Jr. (503), Division of Mammals, National Museum of Natural History, Smithsonian Institution, Washington, DC 20560

Darelyn Handley (503), Division of Mammals, National Museum of Natural History, Smithsonian Institution, Washington, DC 20560

Sally J. Holbrook (19), Department of Biological Sciences and Coastal Research Center, Marine Science Institute, University of California, Santa Barbara, California 93106

R. D. Holt (421), Natural History Museum and Department of Systematics and Ecology, University of Kansas, Lawrence, Kansas 66045

Douglas H. Johnson (391), Northern Prairie Science Center, National Biological Service, Jamestown, North Dakota 58401

Elisabeth K. V. Kalko (503), University of Tübingen, Tierphysiology, 72076 Tübingen, Germany; and Smithsonian Tropical Research Institute, Balboa, Republic of Panamá; and Division of Mammals, National Museum of Natural History, Smithsonian Institution, Washington, DC 20560

Allen Keast (49), Biology Department, Queen's University, Kingston, Ontario K7L 3N6 Canada

Joseph H. K. Pechmann (217), University of Georgia's Savannah River Ecology Laboratory, Aiken, South Carolina 29802

Eric R. Pianka (191), Department of Zoology, University of Texas, Austin, Texas 78712

Peter F. Sale (73), Department of Biological Sciences, University of Windsor, Windsor, Ontario N9B 3P4 Canada

Russell J. Schmitt (19), Department of Biological Sciences and Coastal Research Center, Marine Science Institute, University of California, Santa Barbara, California 93106

David E. Scott (217), University of Georgia's Savannah River Ecology Laboratory, Aiken, South Carolina 29802

Raymond D. Semlitsch (217), Division of Biological Sciences, University of Missouri, Columbia, Missouri 65211

N. A. Slade (421), Natural History Museum and Department of Systematics and Ecology, University of Kansas, Lawrence, Kansas 66045

Thomas J. Valone (555), Department of Biology, California State University, Northridge, California 91330

Kirk O. Winemiller (99), Department of Wildlife and Fisheries Sciences, Texas A&M University, College Station, Texas 77843

FOREWORD

To the general public, perhaps the most famous and disturbing zoological discovery resulting from a long-term field study was the detection of genocidal murder among wild chimpanzees. Jane Goodall began her chimpanzee studies at the Tanzanian site of Gombe in 1960. It gradually became apparent that chimps lived in distinct bands occupying areas largely separate from each other; but powerful behavioral mechanisms sufficient to maintain the separateness of band territories were not obvious. Compared to us murderous humans, our chimpanzee relatives apparently deserved to be idealized as relatively gentle and nonviolent.

This illusion was not dispelled until January 1974, when 7 male chimpanzees and 1 female chimpanzee of the Kasakela band were observed traveling quickly and silently southward toward chimpanzee calls coming from the territory of the Kahama band. The group suddenly came upon a Kahama male referred to as Godi, feeding in a tree. One Kasakela male grabbed Godi's leg and pinned him to the ground, while 5 other chimpanzees spent 10 minutes hitting him, biting him, and eventually throwing a large rock. The very badly wounded Godi was never observed again after the attack and is presumed to have died of his injuries.

Over the next 3 years, witnessed fatal attacks by Kasakela chimps (mainly adult males) resulted in the deaths or disappearances of 5 other Kahama chimps; 4 more Kahama chimps disappeared from unknown causes (unwitnessed attacks?); and the last 2 Kahama chimps, young females, transferred to the Kasakela band, which proceeded to occupy the former Kahama territory. At Mahale in Tanzania, where another long-term field study of chimps led by Toshisada Nishida began in 1966, no murders and only one brief nonfatal fight between adult males had been observed by 1985. However, all adult males of one of the Mahale bands disappeared, one after another, suggesting the possibility of nonwitnessed fights there, too.

Evidently, chimpanzee band territories are maintained by aggressive patrolling that in the great majority of contacts ends with mere posturing or one side fleeing, but infrequently involves fights and, even more infrequently, deaths. In retrospect, we can wonder about the rosy idealization of chimpanzees that for so long let scientists view them as gentle. But this idealization was based on real observations and lack of observations. It was not that interband genocide was observed and the observations were suppressed. Instead, it took 14 years to witness it at Gombe, and it was still unwitnessed after 19 years at Mahale.

On reflection, we should not be surprised that chimpanzee genocide proved so difficult to observe. To a chimpanzee, another chimp is a dangerous, equally matched opponent. Even the winner in a fight risks incurring disabling injuries. The few witnessed successful murderous attacks arose when a single chimpanzee of one band was surprised by many chimps of another band. Like humans who have experienced murderous violence firsthand, a chimpanzee that experiences or observes the dangers of attacking another chimp may carry the lesson for the rest of its life, without need to be reminded.

Demographic reasoning also makes it obvious that chimpanzee genocide must be rare: a frequency of murderous attacks higher than once per chimp per lifetime (around 50 years) would be incompatible with the survival of chimpanzee populations. The chances of detecting it are negligible in a 3-week field project, low in a 3-year study. Yet the omnipresent, veiled threat of murder, no matter how rarely it is carried out, is enormously significant in understanding chimpanzee behavior, distribution, life history, and demography.

Long-term field studies have transformed our understanding of murder and infanticide among animals. As recently as 1963, the distinguished behavioral biologist Konrad Lorenz could still write (in his book *On Aggression*) that animals' aggressive instincts are held in check by instinctive inhibitions against murder and that humans are unique in their murderousness (as a result of our newly acquired powers of killing with weapons no longer being restrained by our old instincts). Occasional observations or reports of animal murderousness were profoundly upsetting to scientists and were dismissed as unsubstantiated, misinterpreted, or pathological consequences of human disturbance. Since Lorenz wrote his book, however, long-term studies have now detected murder and/or infanticide by a substantial proportion of those animal species with adequate anatomical equipment for effecting it. A short list of some of the better-studied species providing examples includes African hunting dogs, cheetahs, gorillas, hyenas, Komodo monitor lizards, langur monkeys, lions, web spiders, and wolves.

Detection of murder in traditional human hunter/gatherer societies was even more difficult, and more upsetting. Many anthropologists studying the world's few remaining hunter/gatherer populations had a bigger emotional investment in idealizing "their" hunter/gatherers as gentle, nonviolent people than zoologists had in their corresponding image of gentle animals. Again, reflection makes it clear why an anthropologist's chances of observing a murder during a 3-year study of a hunter/gatherer band of 30 (15 adults, 15 children) are low: an average of 1 adult murder per 3 years would mean half of the band's adults murdered per 25-year generation, a rate incompatible with band survival. Much longer studies, and self-reports of life histories, do testify to the traditional role of human murderousness, but there are still anthropologists who have not yet gotten through the stage of regarding it as a pathological artifact of modern circumstances.

The possibility of observing important but rare events is one type of payoff from long-term studies. To an animal, an event that happens on the average once a generation or once every 10 generations would be "rare" but still potentially important for demography and for geographic range, respectively. Whether such an event is also rare to the human field observer depends on whether the animal's generation time is 1 day (bacteria), a few weeks (*Drosophila*), or 50 years (great apes). Murder is merely one example of a biologically important rare event. Others include effects of El Niño years (ENSO events) on Pacific seabirds, Galapagos finches, and Aboriginal Australian society; effects of recruitment bursts in exceptional years on Atlantic marine fishes and North American amphibia; and pulses of dispersal.

As an example of the importance (and difficulty of detection) of rare pulses of dispersal, hundreds of nonmigratory land bird species have resident breeding populations on Southwest Pacific islands that apparently must have received their bird founders overwater—because the island is a volcano that recently rose from the sea, is a coral atoll that was recently uplifted, or is sufficiently small that demographic accidents will drive any bird population there to extinction within a few decades. Nineteenth-century biologists, troubled by this puzzle posed by the existence of animal and plant populations thus "stranded" on remote oceanic islands, postulated colonization from continents via former land bridges, which (supposedly) then vanished under the sea. So many land bridges were postulated that no open sea would have remained if they had all actually existed.

In 1976, while traveling by canoe among the Solomon Islands for a month, I observed 37 of the Solomons' 141 resident land bird species flying overwater between islands. During a few days traveling by ship among the Bismarck Islands in 1972, I similarly observed 15 of the Bismarcks' 146 species flying overwater. A much longer span of observations was amassed by a missionary, Father Otto Meyer, who resided for the first 3 decades of this century on Vuatom, a 5-square-mile extinct volcanic island lying 4 miles off the much larger island of New Britain. Meyer recorded 47 species breeding on Vuatom in 1906, 4 additional species that arrived and established themselves as breeders between 1906 and 1930, and 24 species that arrived as vagrants and did not establish themselves. Some of those vagrants were observed on Vuatom only once in 24 years: for instance, one individual of the kingfisher *Ceyx websteri*, observed being chased by the similar-sized resident kingfisher *Alcedo atthis*; and a single group of the wood swallow *Artamus insignis* that arrived, fed on swarming ants for 2 days, and departed by the third day. Of the 127 bird species breeding 4 miles away on New Britain, Father Meyer never observed 55 during his decades on Vuatom. Yet some of those 55 must cross water at least every few centuries, because they are among the bird species now breeding on Long Island, a nearby island that was defaunated by a volcanic explosion in the late 17th century and has subsequently been recolonized by plants and animals. Thus, very long-term studies would be required to obtain comprehensive direct observations of overwater dispersal, a process fundamental to island biogeography.

Detection of rare events is not the sole type of payoff unique to long-term studies. Another is the detection of processes that unfold slowly compared to the duration of the usual short-term study. To me, one of the most startling ornithological experiences of my life was my return, at the age of 45, to the house in Boston, Massachusetts, where I had lived as a child. I began bird-watching there in 1944, at the age of 7, and my observations continued until I moved to California in 1966. During those years I became familiar with the resident Massachusetts birds and also read in bird books about the (to me) exotic birds common in the southeastern United States, such as bright red cardinals and that fabulous songster the mockingbird. But I never saw a single individual of those exotics during my Massachusetts childhood. My bird bibles—the 1947 edition of *Peterson's Eastern Field Guide* and Forbush's 1907 *Useful [Massachusetts] Birds and Their Protection*—didn't even mention cardinals from Massachusetts and listed mockingbirds only as "a few north to Massachusetts."

When I finally returned, after 17 years, to attend my 25th college reunion and to show my wife the house where I had grown up, I was astonished as soon as I got out of the car in front of the house. The three commonest birdsongs were now the formerly exotic cardinal and mockingbird, which had greatly extended their ranges northward due to warm climate and winter bird feeders, and the House finch, a western American species that had never even been recorded in the eastern United States until escaped cage birds established themselves around New York and then spread explosively.

I thought of what I could have learned if I had been able to make long-term observations in Boston from 1966 to 1983, while those 3 species were soaring to abundance. Which habitats did the first arrivals select and which habitats were filled only later? Did competition from the new arrivals depress populations of the original resident species? From detailed knowledge of the year when each of the new species arrived and when the original species began to decline, could one deduce which pairwise cases of interspecific competition had the strongest effect? Did the new arrivals' population densities fluctuate wildly at first, as they became adapted to the new environment, or did their breeding success start high and then decline as they filled the habitat? None of those questions could be answered by comparing Boston bird communities viewed as two snapshots, one in 1966 and the other in 1983. I had missed the opportunity to observe Nature's grand experiment in reshuffling the community.

Despite the obvious potential payoffs from long-term studies, the reasons that field biologists have tended to neglect them are equally obvious. It is difficult enough to secure foundation research support guaranteed for 3 years; hence it takes courage (and outstanding grant-getting ability) to commit oneself to a long-term study whose main payoffs are not expected to emerge until after 3 years, and which might have to be aborted before payoff because of a lapse in funding. Long-term commitment to one site brings costs of foreclosed opportunities; the same time could have been spent comparing many sites surveyed more briefly. Each approach brings unique insights, but funding problems and other practical considerations have given us more of the insights unique to brief comparative studies of many sites.

This book now addresses the balance. It reports studies by 16 scientists or teams who did have the courage to persist and who did succeed in scrounging up the necessary funds. The studies encompass all five classes of higher vertebrates. For those of us who (like myself) have concentrated on short-term studies and have had to guess at some processes responsible for what we observe, reading this book engenders feelings similar to those aroused by seeing in the newspaper the solution to yesterday's impossibly difficult crossword puzzle. Again and again, readers will find themselves saying, "So, *that* was the answer!" The authors of this book have had to sweat hard, and many words of the puzzle are not yet filled in, but they have made discoveries that could not have been made otherwise.

Jared M. Diamond
Department of Physiology
University of California,
Los Angeles, Medical School

CHAPTER 1

Introduction to Long-Term Community Ecological Studies

MARTIN L. CODY

Department of Biology, University of California, Los Angeles, California 90095

I. INTRODUCTION

A. Background

In view of the wide disparity of approaches among ecologists to their science, it is notable that considerable concordance exists in at least one respect: long-term studies are widely regarded as indispensable, and thus no elaborate case need be made for their justification. The potential for change in the populations that make up a local biota and its various assemblages and communities is an accepted fact and is compelling in its many sources. Some of the potential comes directly from variability in abiotic factors, such as climatic trends over the longer term, or in weather variations from year to year in the mid to shorter terms. Some will come indirectly from such abiotic factors, as they influence species' geographic ranges and lead to expansion or contraction; thus, at the local level some populations are favored at the expense of others as conditions for their persistence or increase are enhanced. Yet other factors for long-term change are strictly intrinsic to organisms, as they adapt and evolve and thereby occupy ecological roles in a specific area or at a given site that change or shift over time. And overall, inexorable and perhaps inevitable, is the increasing pressure on environments and habitats from human activities, providing a potential for biotic change that can be ignored only in very few circumstances. Thus, the nature, causes, and effects of variability in ecological systems can be understood only through extended study over far longer time periods than those typical of many sciences and of most other biological sciences.

Clearly, with a longer research commitment the capacity to resolve short-term phenomena is enhanced, and the possibilities for discovering and addressing those of longer term are revealed. As Gilbert White, the astute and dedicated natural historian of his southern English parish, remarked in 1779, "It is now more that forty years that I have paid some attention to the ornithology of this district, without being able to exhaust the subject; new occurrences still arise as long as any inquiries are kept alive" (White, 1877). Calls for the investment of more time, effort, and resources in long-term ecological studies have become much more persistent over the last decade or so. They appear attributable to a recognition of two rather distinct aspects of ecology, the dynamical nature of ecological systems themselves and the uses to which our understanding of the systems might be put.

First, there is the recognition that ecological processes are, innately, of long-term resolution and that change is as justifiably a quality of ecological systems as is permanence or stability. Many sorts of natural perturbation or disturbance are rare and/or aperiodic, and their effects may persist for centuries or longer. And many ecological processes, such as natural succession, the outcome of competition for limited resources, or the cycling of predator-prey systems, may take inordinately long times for documentation let alone resolution, even without persistent disturbance. Thus with an overlay of variability from abiotic factors, from long-term climatic change to shorter term variations in weather, and from inevitably pervasive anthropogenic influences, it is apparent that many ecological questions require an unusual commitment of research effort.

Second is the need for and dependence upon long-term data bases in the management of critical species, habitats, and resources. This need is often promoted by the legal aspects of species conservation and resource exploitation and becomes apparent only when the letter of such legislation is followed. Resource managers, whether of large animals in small national parks, of habitat-specialized kangaroo rats or gnatcatchers facing urban development, or of desert landscapes faced with large-scale mining concerns, inevitably conclude that their data bases are deficient and that their capacity for prediction or extrapolation is limited.

Several indicators of the importance and perceived value of long-term studies are visible. An overview of the strategic or policy aspects of conducting and funding long-term studies is given in *Long-Term Ecological Research* (SCOPE—Scientific Committee on Problems of the Environment—Volume 47, Risser, 1991), and implicit in the imperatives for monitoring our dwindling resources is the realization that such monitoring efforts will be ongoing, likely to become an essential part of environmental management in perpetuity (e.g., Spellerberg, 1991). *Long-Term Studies in Ecology* (Likens, 1989) resulted from the 2nd Cary Conference, which identified a "critical need" for commitment to a longer-term approach and adopted a statement reading, in part:

> Sustained ecological research is one of the essential approaches for developing [an] understanding [of] processes that vary over long periods of time. However, to fulfill its promise, sustained ecological research requires a new commitment on the part of both management agencies and research institutions [and] should include longer funding cycles, new sources of funding, and increased emphasis and support from academic and research institutions.

The US/LTER (Long-Term Ecological Research) program, instigated in 1977, addresses these needs (Callahan, 1991), but its aims for representative sites in major North American biomes have yet to be met.

B. Ecological Studies over Spatial versus Temporal Scales

There has been a good deal of attention in the ecological literature paid to the notion of scale (e.g., Wiens *et al.*, 1986, for an overview), which has dual spatial and temporal aspects. However, much of the development of the concept has addressed the spatial rather than the temporal concerns, leading, among other important aspects, to treatment of populations in fragmented habitats as "meta-populations" (e.g., Hastings and Wolin, 1989) and their associations in the patches as "metacommunities" (Wilson, 1992). Models that emphasize spatial structure already have a considerable history, beginning with the work of Levins (1969, 1970), Levins and Culver (1971), Horn and MacArthur (1972), and Yodzis (1978); their implications for genetics, predation (Kareiva, 1986), competition and coexistence (e.g., Chesson, 1986), diversity (e.g., Hanski *et al.*, 1993), and conservation (e.g., the MVP—Minimum Viable Population—problem, Gilpin and Soulé, 1986) are profound. Levin's (1992) perspective of these developments and their significance is particularly valuable.

A corresponding theoretical development of temporal analogs of models with spatial heterogeneity seems lacking in the literature. In the same way that a greater understanding of communities in patchy and spatially heterogeneous habitats is gained from studies with a broader, more regional perspective than from single, site-specific snapshots, so it is with communities that experience environmental heterogeneity over time. The broader perspective that comes from long-term research contributes to the understanding of how communities cope with year-to-year, and decade-to-decade, variations in conditions at a particular site.

A diagrammatic comparison of temporal versus spatial scale broadening is given in Fig. 1, but the duality of the two scales is exacerbated by their obvious qualitative differences. Logistically, a broadening of perspective in spatial scales can be achieved relatively easily, with adjustments that might take advantage of, e.g., further site replication, wider habitat gradients, a better microscope, or better satellite imagery, but there is not such a ready correction for restricted temporal scales: time alone can serve to broaden a data base in the temporal dimension. Specifically with respect to questions of community structure, stability, and replicability, there are advantages and disadvantages to broadening scales in the spatial versus temporal dimensions, the "far-flung" versus the "long-term" approach. By replicating sites within a limited time over the geographical range of a habitat (Fig. 1, left–right axis), more habitat variability is likely to be encountered (front–back axis), and the precision of intersite comparisons reduced. However, questions of convergent and parallel evolution can be addressed, as geographic counterparts may substitute in distant sites, and the effects and influences of species from adjacent (different) habitats (spillover effects, mass effects; Shmida and Ellner, 1984; Cody, 1993, 1994) can be evaluated, because throughout a habitat's range the extent and type of other habitats will likely vary. The long-term approach will lack these potential advantages, but also lack the disadvantage of establishing replicability (e.g., in terms of habitat structure) over multiple sites.

However, at a single site studied over a longer term, site stability is by no means assured; indeed, site-specific variation over time can be dramatic and, if

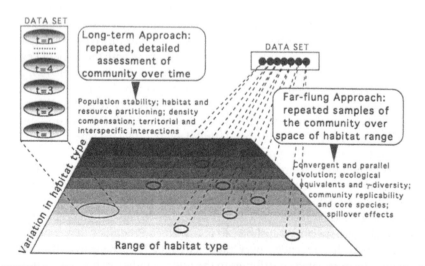

FIGURE 1. Diagrammatical representation of two different views of extended community-level studies, contrasting the far-flung approach, in which sites are replicated over space, with the long-term approach, in which a single site is studied over an extended time period. See text for discussion.

calibrated, can allow for the investigation of covarying community structure. Such attributes as density compensation, shifts in habitat use or territory preference, and changes in interspecific interactions can thus be studied as the site undergoes year-to-year variation in resource availability and community composition.

In general, the changes that occur within a site over the course of a long-term

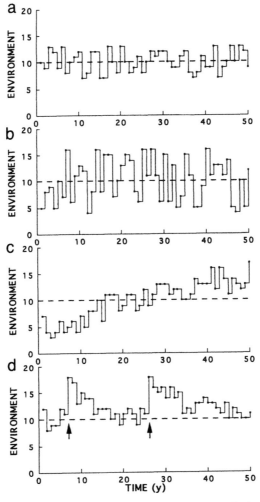

FIGURE 2. Synthetic patterns of long-term environmental variation (ordinate) over time (abscissa) at a hypothetical site of interest. Random variation (a) with a modest range and (b) with a more pronounced range among years. (c) An obvious trend for increase over time in the environmental parameter measured; (d) periodic perturbations of differing severity (at arrows) followed by asymptotic return, over time, to the original status quo.

study are among the more formidable challenges faced by this type of research. The site may change year-to-year as a consequence of, e.g., natural variability in weather patterns or may show long-term trends in response to broader scale climatic patterns or factors related to such variation, as discussed above. Figure 2 shows examples of four of the more obvious patterns in within-site variation over time, in which variation is random $y-y$ with a modest range (Fig. 2a) or with a more conspicuous range (Fig. 2b), in which a unidirectional change occurs over time throughout the study (Fig. 2c) or in which conspicuous perturbations occur periodically and from which the site gradually regains its initial (quasi-equilibrial) value (Fig. 2d). Few long-term studies at a fixed site are expected to be free from one or another of such variations or changes, and the monitoring of the site ideally, even necessarily, would encompass both the consumers (in the community of interest) and their site-based resources. Furthermore, the distinction between data sets that have basically chaotic versus nonchaotic or deterministic generators, whether in time series data of resource variations or of consumer populations that are purportedly tracking them, may be extremely subtle, as simulations have shown (Ellner and Turchin, 1995).

The extent to which the community is an open rather than a closed system will potentially affect studies at both broad spatial and temporal scales, but is more likely a handicap in long-term studies. If the community disbands and reassembles with a seasonal tempo, then factors operating beyond the study site may be important but largely unknown. For example, factors affecting overwintering survival may affect the densities of migrant birds in their breeding communities, or those influencing adult salamander survival in woodlands could constrain populations at the breeding ponds. These and other aspects of long-term studies are recurrent themes in the chapters assembled in this book.

C. Studies of Vertebrate Communities

The chapters of this book represent a diversity of approaches to long-term ecological studies, but they have in common a community rather than a populational perspective, and they address in particular communities of vertebrates. Few of the authors become embroiled in the notion of what constitutes a community; rather, each author adopts de facto his or her own working version of Charles Elton's (1927) view of community, as distinct from population, studies as those addressing "a larger fragment of nature . . . of at any rate a group of ecologically inter-related species." Some authors, notably those who perforce sample sight unseen from below the waterline of rivers and lakes or who trap at population confluences near breeding ponds, choose to talk of "assemblages" rather than "communities," but all avoid extending a debate on definition that has generally proved of no value to the field.

Our restriction of contributions to studies of vertebrates does have implications for community ecology because it is traditionally among vertebrates that we expect to find several of the properties prerequisite for a meaningful community ecology: the operation of density-dependent factors in population regulation; the involvement of larger, often longer-lived and wider-ranging animals and of generalized foraging habitats (whether herbivores or carnivores) and therefore a potential for interactions among species of a competitive nature; and the sort of "proactive" ecological life styles that come with the higher-order behavior typical of vertebrates, including perhaps acoustical, olfactory, or visual recognition. Such behavior in turn leads to habitats being evaluated and choices made, community membership or exclusion being more than a passive aggregation or a random sample, and a more discrete set of questions for ecological determination than would characterize, e.g., the plankton caught in a towed net, the insects gassed from trees, or the weedy plants found in a vacant lot.

II. CONTENTS OF THE BOOK

A. Origins

This book originated after a symposium on long-term studies of vertebrate communities was convened for the Ecological Society of America's 1993 annual meeting, held in Madison, Wisconsin. Several of the chapters in this book were first presented as symposium papers, and most symposium contributors agreed that a synthesis volume on the topic was timely and valuable. Additional topics were added as other researchers with long-term data sets were identified and recruited to the task. In some cases the chapters present a current synthesis of an ongoing study with a considerable history of publication (e.g., on Lake Opinicon fishes, Chap. 3; Galapagos finches, Chap. 12; and experimental desert rodent communities, Chap. 17). In others, the chapter might represent the first overview of a long-term research project (e.g., on prairie ducks, Chap. 13) or an attempt to relate a body of work that has been treated thus far as separate contributions (e.g., on successional small mammals in Australia, Chap. 15). All authors have addressed similar questions in as far as the specifics of their work allow: What changes are apparent over longer time periods in vertebrate communities in terms of species identities and densities? How important are such changes relative to aspects of constancy in the communities? Is community change related directly or indirectly to identifiable factors operating locally (within-site) or more distantly, with general or specific impact, or is it rather the result of the vagaries of various indeterminate or random processes? And in segregating the determinate from the indeterminate, how important is the longer time span in allowing for, first, the quantification and qualification of the changes and, second, attribution, or at least suggestion, of their causes?

B. Representation by Taxonomic, Regional, and Habitat Diversity

Although the sixteen chapters that follow are similar in their overall goals, particularly in the hope of reaching new or deeper insights into communities through long-term study, they are of course unlike in many respects. Some of the most obvious differences are listed in Table I, and foremost among these is the wide range of subject taxa. The chapters are evenly distributed over the four major vertebrate groups (fish, herpetiles, birds, and mammals), a pleasing equanimity from the taxonomic standpoint, but one implying a diversity in approaches, from sampling or census methodology to the details of foraging ecology or behavioral interactions within and among species. The relative constraints and advantages to research on different taxa have obvious implications for the type of community studies that are feasible.

Although marine fishes can be directly counted along transects or within patch reefs by divers (Chaps. 2 and 4), freshwater fish communities are studied largely by netting samples taken from murky waters (Chaps. 3 and 5); foraging habits are derived from stomach contents and deductions based on the structure of fishes' trophic equipment. Whereas most lizards are diurnal and countable, and their foraging behaviors directly measurable (Chap. 8), most salamanders are cryptic and/or nocturnal (Chaps. 6 and 7), and, along with turtles (Chap. 9),

TABLE I   Synopsis of Chapter Characteristics

By taxon:		Fish	Herpetiles		Birds		Mammals	
		1,2,3,4	5,6,7,8		9,10,11,12		13,14,15,16	
By geographic region:		North America	Central and South America			Australia	Africa	
		2,3,6,7,9, 11,13,14,17	4,5,12,16			4,8,15	5,10	
By latitude:		Temperate	→		Tropics			
		3,9,13,11,7	14,2,6,15,17,8,12,4,5,16,10					
By habitat:	Marine	Freshwater/terrestrial	Deserts	Grasslands		Scrub	Forests	
	2,4	3,5,6,7,9,13	8,17	14		11,12,15	10,16	

Note. Entries in the table are chapter numbers.

their numbers are best measured when they congregate at aquatic habitats for breeding purposes. Their ecology in the terrestrial environments, where most of their growth and survival (or otherwise) occurs, is less tractable and remains less well known. Birds are not only diurnal and active but often colorful, noisy, and obvious, and they are thus easily counted; their activities can be plotted, usually sex-specifically, and a good deal of behavior relevant to resource acquisition and partitioning is then built into an evaluation of the community (Chaps. 10–13). Like lizards, birds can be, and often are, caught and recaught, tagged, weighed, and measured, and must rival lizards (Pianka, 1986; Vitt and Pianka, 1995) as the ideal vertebrate organism for ecological studies. The case is also made, with some justification, that coral reef fishes share many of the same advantages prized by field ecologists (Sale, 1991). On the other hand, most large mammals are scarce, elusive, and/or dangerous. Small mammals have none of these qualities but are nearly invariably nocturnal, and their communities must be assessed by trapping, as in terrestrial rodents and their marsupial equivalents (Chaps. 14, 15, and 17), or netting, as in the volant bats (Chap. 16). Food and foraging habits may be studied occasionally by direct observation, e.g., in the former on moonlit nights, with attached reflectors, or via the contents of convenient cheek pouches, but in bats such information is less directly obtained, by location and location-specific auditory spectra and receiver characteristics or by examination of gut or fecal samples.

The chapters are also diverse in terms of geographic region and collectively span four continents (Table I); around one-half of the studies were conducted in North America, with the remainder reporting on research from the Neotropics, Australia, and Africa. Several chapters include a strong comparative component in which results from different regions are compared. In Chap. 4 communities of coral reef fishes from the Caribbean and Indo-West Pacific regions are contrasted, and Chap. 5 relates the structure of freshwater fish assemblages from similarly seasonally inundated habitats around Venezuelan and Zambian river systems. References are made in Chap. 15 to results on small mammal succession in post-fire vegetation in southeastern Australia and in other fire-prone habitats worldwide, and in Chap. 10 to other comprehensive work on tropical rainforest birds besides that reported from Gabon, West Africa.

A conspicuous aspect of the diversity of chapter topics is the extremely wide range of habitats in which the long-term research has been conducted. By latitude, the coverage extends from the north temperate (the lake fishes of Chap. 3) to the tropics (rainforest birds and bats, Chaps. 10 and 16; island finch communities, Chap. 12). Both aquatic and terrestrial habitats are represented. Research from aquatic habitats is exemplified by marine fish studies of both temperate (Chap. 2) and tropical (Chap. 4) reefs and by studies of freshwater habitats in temperate lake (Chap. 3) and tropical river (Chap. 5) systems. Several chapters (amphibians, Chaps. 7 and 9; turtles, Chap. 6; prairie ducks, Chap. 13) concern animals that depend on both aquatic and terrestrial habitats; for the amphibians and turtles, the aquatic habitat is prerequisite for reproduction, whereas in one component of the duck communities, it is the terrestrial habitat that provides the nesting opportunities. Desert communities are represented by Australian lizards (Chap. 8) and North American rodents (Chap. 17); grasslands to shrubland, successional or otherwise, by Rocky Mountain birds (Chap. 11), Galapagos finches (Chap. 12), Kansas rodents (Chap. 14), and small Australian mammals (Chap. 15); and forests by tropical bird and bat communities (Chaps. 10 and 16).

C. Other Variables in Long-Term Community Studies

1. Spatial Scales

Apart from variation in geographical locale, the spatial scale of the studies summarized in this volume also varies widely. Marine fish studies have used transect counts in harbors and on natural reefs and counts at isolated patch reefs (Chaps. 2 and 4). Freshwater fish studies were confined to sampling in a single lake and associated streams (Chap. 3) or broad sampling over a river system and its adjacent floodplains and marshes (Chap. 5). A local pond system and neighboring marshes were used in Michigan turtle studies (Chap. 6), broader elevational transects were employed to census Appalachian salamanders (Chap. 7), and a single pond was the focus of the amphibian community discussed in Chap. 9.

The lizard communities of Australian desert habitats (Chap. 8) have been monitored over several study plots of some hectares each within a sandplains region, and in bird community studies, data collection was focused in areally discrete plots that varied in size from 2 (tropical forest birds, Chap. 10) to 1 km² (prairie ducks, Chap. 13) to study plots of a few hectares (shrubland birds in the Rockies, Chap. 11). However, in Galapagos finches, coexisting populations are delineated by the islands on which they occur (Chap. 12). The studies of small terrestrial mammals were conducted on fixed plots from ¼ to 2 ha in size (Chaps. 14, 15, 17), but for the vagile and probably wide-ranging tropical bats (Chap. 16), trapping stations were located at fixed sites, often fruiting trees, on Barro Colorado Island in Gatun Lake, Panama.

2. Study Duration

What constitutes a "long-term" community study is of course open to question and ideally would be related to population turnover rates or the longevity of the member species. Most of the chapters report data collection periods of around two decades (mean: 20.3 years ± 5.0 SD), although some are notably longer (e.g., 30 years for Keast's lake fish studies, 25 years for prairie ducks, desert lizards, and Australian successional mammals). In temperate passerine birds and small mammals, with average life expectancies of at most a few years, a quarter-century study does seem to qualify as "long-term," although for some lower vertebrates with reduced metabolic rates and long life expectancies, such as fish and turtles, the same duration is relatively modest. The 20-year turtle study of Chap. 6, for example, is about equivalent to age at first reproduction of these species, but individuals can live at least three times this time period. The details of life expectations and replacement rates would be required to assess the adequacy of the study periods relative to community changes, and in general these details are not available. Further, important but rare events that can drastically alter community structure, such as periodic droughts and fires, might occur with a frequency much lower than every couple of decades, and recovery from such events might be slow. Thus, there is little objectivity in our classification of studies as "long-term," and in general we simply mean that lots of data were collected using extended efforts over many years.

3. Community and Species' Sizes

The communities that are the subjects of these studies are likewise variable in size and in the number of species included in the researchers' operational definitions of "community" or "assemblage." Some encompass all of the organisms within a broad taxonomic group that can be collected or sampled using a given technique (e.g., netted fishes or bats), whereas others are more narrowly focused on a suite, or guild, of species that are closer in size, ecology, resource use, and behavior (e.g., insectivorous birds or seed-eating rodents). Some indication of the variations in both community size and the size range of constituent organisms is shown in Fig. 3. Note that collectively the studies span two orders of magnitude in community sizes (±3–320) and an even wider range in body size. Even single community studies, such as those on fish, Australian lizards, or tropical bats, can include a range of species that differ in mass by at least 1.5–2 orders of magnitude.

4. Site Variation over the Duration of the Study

As pointed out above, variations are expected to occur at a study site within the duration of long-term studies, such that the conditions under which the community functions change over time. Weather shifts, newly introduced or extinct species, anthropogenic effects, etc., are some possibilities whose inci-

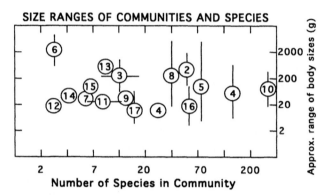

FIGURE 3. Community studies reported in this volume vary widely in the number of component species (abscissa), in the sizes of component species (ordinate), and in the size range of component species both within and among studies. Symbols within the figure refer to chapter numbers.

dence will be experienced with increasing likelihood over extended study periods. These natural or unnatural perturbations can be regarded either as a hazard or as an opportunity to learn more about community organization and resilience.

Not surprisingly, the authors of our chapters report many such incidences of changing conditions during their periods of study. Often cited are "unusual" weather phenomena, sometimes operating at broad geographical scales, such as El Niño Southern Oscillation events that affect Californian reef fishes (Chap. 2) as well as the food supplies of finches on the Galapagos Islands (Chap. 12). Precipitation patterns are shown to be particularly important to communities of species as varied as sagebrush and willows passerine birds (Chap. 11) and Chihuahuan desert rodents (Chap. 17), and they are more obviously a factor in the breeding or feeding habitats of aquatic species, such as turtles and salamanders (Chaps. 6, 7, and 9), prairie ducks (Chap. 13), and floodplain fishes (Chap. 5). Changes to habitats are not always due to abiotic factors, however; the colonization of willow shrublands by beavers in the Rocky Mountains causes a considerable change in water levels and affects foraging opportunities for several bird species, eliminating some while facilitating others (Chap. 11).

Some of the studies are conducted in habitats undergoing succession, where an assessment of vegetational change and its consequences to the vertebrate inhabitants is very much a part of the research design. Examples in this book are the post-fire successions in the Australian desert *Triodia* grassland and associated lizard community (Chap. 8) and in the *Eucalyptus* woodlands of southeastern Australia relative to small mammal succession (Chap. 15). Rodents occupying old fields in Kansas (Chap. 14) are further examples.

Anthropogenic influences are hard to avoid over the longer time periods. Human influence had much to do with introducing an alien fish, alewife (*Alosa pseudoharengus*), and weedy Eurasian milfoil (*Myriophyllum spicatum*) to Lake Opinicon (Chap. 3); surprisingly, the former had little effect on the fish community, whereas the latter was a significant contributor to habitat alteration. In the tropical forest bird study (Chap. 10), habitat clearing and road building were major factors in the incursion of second-growth plants and birds, and on Barro Colorado Island, forest recovery from partial cutting 70 years ago has lead to the reduction of bat species associated with secondary vegetation and its fruit and nectar resources.

5. Are the Communities Closed or Open Systems?

Community ecology studies are almost never conducted in (or treated as) completely closed systems, but the extent to which they are not varies dramatically with the site and with the life-history attributes of different subject taxa. At one extreme, reef fish may have widely pelagic larvae whose ambit necessarily is considered outside of the reef-centered community study. The terrestrial life of salamanders and turtles is similarly a "black box" from the viewpoint of pond-based community studies and one in which much of relevance presumably takes place, as it surely does on the migration routes and wintering grounds of birds studied on their temperate breeding grounds. Communities of the more vagile species, such as birds and bats, will presumably be more open, and community characteristics will be less completely attributable to on-site effects and influences than will be the case for more sedentary species such as lake fish, rodents, and herptiles. Using desert rodents, Brown and associates (Chap. 17) have proven that it is feasible to constrain the animals in realistically large enclosures and thereby study essentially closed, or selectively closed, communities; but naturally the opportunities for such protocols will be very limited in volant or more vagile taxa.

6. Sampling, Observation, and Experiment

As a final dimension to the research presented in this volume, mention can be made of the nature or methodology of data collection. There has been a long tradition in vertebrate community studies, particularly of terrestrial species, of observation-based study; that is, data are collected by simple observation only, without the necessity for hands-on work with the organisms. This is the case particularly with bird research, although birds can easily be caught in many habitats, banded, and weighed or otherwise manipulated. With taxa that are more easily captured, such as most lizards and small rodents, observational data are supplemented or supplanted with information derived from trapped individuals. In other taxa (e.g., salamanders and turtles), discovery is nearly tantamount to capture, and in yet others practically the only source of information is the trapping data (e.g., in fish and bat communities).

When it is logistically feasible, legal, and morally acceptable, field ecologists like to enhance observational data with those derived from field manipulations and experiments. In general, the possibilities are narrower with vertebrates (in contrast, e.g., with plants and insects) and narrower still with higher vertebrates, and in open communities of higher vertebrates the potential for conducting controlled experiments is severely limited. A useful summary of the topic, aimed at community ecologists, is well presented in Diamond's (1986) discussion.

Not surprisingly, then, the chapters here rely on field manipulation and experimentation to only a small degree, the more so as emphasis is on the long-term aspects of community constancy and change rather than on the mechanics of community structure and function. The most notable exceptions are the study on desert rodents (Chap. 17), which has a very strong component of controlled experimentation; the old-field rodent study (Chap. 14), where habitat fragmentation is experimentally manipulated; and the study in Chap. 4, where patch reefs are manipulated for settlement patterns in coral reef fishes. Further, there is a tradition of experimentation in studying the community ecology of pond vertebrates, such as fish and amphibians. Clearly, variation in data collection techniques is heavily taxon-dependent and is well reflected in the following chapters.

References

Callahan, J. T. (1991). Long-term ecological research in the United States: A federal perspective. In "Long-term Ecological Research: An International Perspective" (P. G. Risser, ed.), SCOPE, vol. 47, ICSU, Chapter 2, pp. 9–21. Wiley, New York.

Chesson, P. L. (1986). Environmental variation and the coexistence of species. In "Community Ecology" (J. M. Diamond and T. J. Case, eds.), Chapter 14, pp. 240–256. Harper & Row, New York.

Cody, M. L. (1993). Bird diversity components within and between habitats in Australia. In "Species Diversity in Ecological Communities" (R. E. Ricklefs and D. Schluter, eds.), Chapter 13, pp. 147–158. Univ. of Chicago Press, Chicago.

Cody, M. L. (1994). Mulga bird communities. I. Species composition and predictability across Australia. Aust. J. Ecol. 19, 206–219.

Diamond, J. M. (1986). Overview: Laboratory experiments, field experiments, and natural experiments. In "Community Ecology" (J. M. Diamond and T. J. Case, eds.), Chapter 1, pp. 3–22. Harper & Row, New York.

Ellner, S., and Turchin, P. (1995). Chaos in a noisy world: New methods and evidence from time-series analysis. *Am. Nat.* **145**, 343–375.

Elton, C. (1927). "Animal Ecology." Sidgwick and Jackson, London. Ltd.

Gilpin, M. A., and Soulé, M. (1986). Minimum viable populations: Processes of species extinction. *In* "Conservtion Biology" (M. Soulé, ed.), Chapter 2, pp. 19–34. Sinauer Assoc., Sunderland, MA.

Hanski, I., Kouki, J., and Halkka, A. (1993). Three explanations of the positive relationship between distribution and abundance of species. *In* "Species Diversity in Ecological Communities" (R. E. Ricklefs and D. Schluter, eds.), Chapter 10, pp. 108–116. Univ. of Chicago Press, Chicago.

Hastings, A., and Wolin, C. L. (1989). Within-patch dynamics in a metapopulation. *Ecology* **70**, 1261–66.

Horn, H. S., and MacArthur, R. H. (1972). Competition among fugitive species in a harlequin environment. *Ecology* **53**, 749–752.

Kareiva, P. (1986). Patchiness, dispersal, and species interactions: Consequences for communities of herbivorous insects. *In* "Community Ecology" (J. M. Diamond and T. J. Case, eds.), Chapter 11, pp. 192–206. Harper & Row, New York.

Levin, S. A. (1992). The problem of pattern and scale in ecology. *Ecology* **73**, 1943–1967.

Levins, R. (1969). Some demographic and genetic consequences of environmental heterogeneity for biological control. *Bull. Entomol. Soc. Am.* **15**, 237–240.

Levins, R. (1970). Extinction. *In* "Some Mathematical Problems in Biology" (M. Gerstenhaber, ed.), pp. 77–107. Am. Math. Soc., Providence, RI.

Levins, R., and Culver, D. (1971). Regional coexistence of species and competition between rare species. *Proc. Natl. Acad. Sci. U.S.A.* **68**, 1246–1248.

Likens, G., ed. (1989). "Long-term Studies in Ecology." Springer-Verlag, New York.

Pianka, E. R. (1986). "Ecology and Natural History of Desert Lizards." Princeton Univ. Press, Princeton, NJ.

Risser, P. G., ed. (1991). "Long-term Ecological Research: An International Perspective," SCOPE, Vol. 47, ICSU. Wiley, New York.

Sale, P. F., ed. (1991). "The Ecology of Fishes on Coral Reefs." Academic Press, San Diego, CA.

Shmida, A., and Ellner, S. P. (1984). Coexistence of plant species with similar niches. *Vegetatio* **58**, 29–55.

Spellerberg, I. F. (1991). "Monitoring Ecological Change." Cambridge Univ. Press, Cambridge, UK and New York.

Vitt, L. J., and Pianka, E. R., eds. (1994). "Lizard Ecology: Historical and Experimental Perspectives." Princeton Univ. Press, Princeton, NJ.

White, G. (1877). "Natural History and the Antiquities of Selborne" (Thomas Bell FRS, ed.), London (first publ. 1779).

Wiens, J. A., Addicott, J. F., Case, T. J., and Diamond, J. (1986). Overview: The importance of spatial and temporal scale in ecological investigations. *In* "Community Ecology" (J. M. Diamond and T. J. Case, eds.), Chapter 8, pp. 145–153. Harper & Row, New York.

Wilson, D. S. (1992). Complex interactions in metacommunities, with implications for biodiversity and higher levels of selection. *Ecology* **73**, 1984–2000.

Yodzis, P. (1978). "Competition for Space and the Structure of Ecological Communities." Springer-Verlag, Berlin.

Fish Communities

CHAPTER 2

On the Structure and Dynamics of Temperate Reef Fish Assemblages
Are Resources Tracked?

SALLY J. HOLBROOK AND RUSSELL J. SCHMITT

Department of Biological Sciences and Coastal Research Center, Marine Science Institute,
University of California, Santa Barbara, California 93106

I. INTRODUCTION

Understanding processes that influence community structure and population dynamics of marine fishes associated with reefs has long been a cardinal goal of marine ecological research (Sale, 1991b). Historically, competition for reef resources was viewed as a dominant process shaping local patterns of species richness, species composition, and population abundances. This view largely has been replaced over the past decade in recognition of the fact that local assemblages and most constituent populations of reef fishes are open systems whose nature and dynamics are coupled to processes that occur in other locations (Sale, 1991a; Doherty, 1991). This arises from the bipartite life histories of most marine reef organisms: typically the propagule stage (larvae or spores) disperses in the plankton away from the point of birth before settling to the reef environment occupied by older life stages (Kingsford, 1988; Kingsford et al., 1989). The linkage among local populations via exchange of reproductive output is widely believed to decouple the amount of recruitment (here defined as the appearance of newly settled young fishes) to a given local population from the production of young by adults that reside there (Sale, 1991a). This effectively removes one source of local regulation. Current models that describe dynamics of populations and structures of reef fish communities emphasize nonequilibrium processes associated with larval recruitment (e.g., Doherty, 1991; Sale, 1991a).

With respect to abundance, the sizes of local populations are viewed as a balance between external inputs (larval settlement) and subsequent mortality (Keough, 1988; Mapstone and Fowler, 1988; Roughgarden et al., 1987, 1988; Warner and Hughes, 1988; Underwood and Fairweather, 1989; Hughes, 1990; Raimondi, 1990; Doherty, 1991; Doherty and Fowler, 1994; Sale, 1991a). Temporal variation in recruitment can be great and often is attributed to oceanographic processes that transport and deliver larvae (Doherty, 1991). One widely held notion, called the recruitment limitation model, is that reef resources often may not be limiting due to an undersupply of propagules (Sale, 1977, 1980, 1991a; Victor, 1983, 1986; Wellington and Victor, 1985, 1988; Lewin, 1986; Young, 1987; Doherty and Williams, 1988; Doherty, 1991; Doherty and Fowler, 1994; Underwood and Fairweather, 1989; Sutherland, 1990; Stoner, 1990). As a result, mortality after settlement may be largely density independent, and fluctuations in density of older life stages can mirror the pattern of fluctuation in recruitment (Doherty, 1991). Such direct correlations can be obscured by variation in the density-independent mortality rates on reef-associated life stages (Warner and Hughes, 1988) and obviously will be altered by density-dependent mortality (Jones, 1991). Thus far compensatory mortality has not been detected for reef fishes with open populations, although density-dependent growth has been found relatively frequently (e.g., Jones, 1991; Forrester, 1990). Because the duration of field experiments that explore density dependence in reef fishes typically is short, it has not been possible to determine whether compensation in growth rates results in density dependence in an individual's probability of reaching adulthood. Nonetheless, the general failure to identify strong regulatory mechanisms that operate at the local reef scale reinforces the notion that the dynamics of local reef fish populations are shaped mostly by external forces and less so by local processes. Consequently, local populations of reef fishes are thought to be highly variable and relatively unaffected by local resources and events that occur on the reef.

Given the prevailing view for constituent populations, it is not surprising that current models describing attributes of local reef fish assemblages also emphasize nonequilibrial processes. Indeed, the first model that seriously challenged the resource competition perspective for reef fishes—the lottery hypothesis (Sale, 1977)—addressed aspects of community structure by positing how chance in securing a settlement location by larvae of different species could govern composition of the assemblage. Subsequently, it was recognized that strong recruitment periods can buffer local populations of perennial species (which most reef fishes are) from extinction—the storage effect (Chesson and Warner, 1981; Warner and Chesson, 1985)—and that fluctuations in recruitment could promote local coexistence of competing species with open populations (Warner and Chesson, 1985; Chesson, 1986). Of course, the existence of external sources of new young also means that local extinction of any given species of reef fish is likely to be temporary (Sale, 1991a; Holbrook et al., 1994). Further, widely fluctuating recruitment that is asynchronous among species will result in an assemblage whose structure is highly transient (Sale, 1991a). Thus, such community-level aspects as species richness, composition, and relative abundances of local reef fish assemblages are thought to be shaped primarily by recruitment history and are therefore unlikely to remain constant through time.

A central issue, of course, is how well these models describe long-term behaviors of local reef fish assemblages and populations. Despite the fundamental importance of this issue, it is not possible to resolve it because very little information exists regarding long-term trends in assemblages of marine reef fishes, particularly regarding patterns of species richness, guild structure, interrelationships among component species, or even population dynamics of individual species (Jones, 1988; Holbrook et al., 1994). Most published accounts suffer from either a short temporal framework or a limited taxonomic focus, which has resulted in a very incomplete understanding of the dynamic behavior of reef fish communities, including responses to environmental perturbation. For instance, Cowen and Bodkin (1993) sampled 45 species during a 5-year study of reef fishes at San Nicholas Island, California, whereas 65 species were sampled in Los Angeles Harbor by Allen et al. (1983) over a 1-year period.

We have studied five species of reef fishes at Santa Cruz Island, California, for 13 years and have very detailed information for two of the species (e.g., Holbrook and Schmitt, 1989; Schmitt and Holbrook, 1990a,b). Unlike data sets for many pelagic fisheries (e.g., Mysak, 1986; Hsieh et al., 1991; Ware and Thomson, 1991; Beamish and Bouillon, 1993), time series data of about a decade or less are typical for species of reef fishes, especially in temperate zones (e.g., Ebeling et al., 1980, 1990; Ebeling and Laur, 1988; Stephens, 1983; Stephens and Zerba, 1981; Stephens et al., 1984, 1994; Jones, 1988; Cowen and Bodkin, 1993; but see Holbrook et al., 1994; Davis and Halvorson, 1988). Such relatively short time frames reflect the fact that logistic constraints usually preclude long-term studies rather than any overt decision that longer studies are unnecessary. The paucity of long-term data is unfortunate since oceanographic events that potentially could influence reef fish communities can occur at intervals of a few years to decades (see following discussion), making more extensive time series necessary for adequate analysis of their effects.

While the current focus is on the open nature of most marine populations, it is the case that many species of marine reef organisms have propagule stages that do not disperse in the plankton from the parental population. Many marine invertebrates contain species that brood young and produce "walk-away" juveniles. For reef fishes, this reproductive mode is best represented by the surfperch family Embiotocidae which contains some two dozen species, all in the temperate regions of the North Pacific Ocean (Eschmeyer *et al.*, 1983). All surfperches give birth to very few fully developed juveniles that immediately occupy the adult habitat. While not a particularly speciose family, surfperches can account for a large fraction of individual fishes found on shallow, rocky reefs off the West Coast of the United States.

Our previous work on two species of surfperches in southern California has revealed a tight coupling between local reproductive output and recruitment, and the dependence of local dynamics solely on events that influence the demography of local individuals (Schmitt and Holbrook, 1990a). Thus, a nontrivial fraction of local reef fish assemblages of temperate North America is composed of species with relatively closed populations. It is not known to what extent long-term population behaviors differ between species of reef fishes with relatively open and closed local populations or whether responses to environmental change vary qualitatively between these two groups of species. Such comparisons potentially can yield great insight, but as yet no such explorations have been done.

Perhaps the most comprehensive, long-term data set for any assemblage of reef fishes is that collected at King Harbor, California, by Stephens and his associates (Stephens *et al.*, 1984, 1986). The King Harbor site now has been sampled four times per year for 20 years, with a total of 75 species of reef-associated teleost fishes observed. Of these 75 species, 9 (12%) are surfperches. These data provide an unparalleled opportunity to examine long-term trends in an assemblage of temperate reef fishes and to compare the behavior of species with relatively open and closed local populations. In the following discussion we couple exploration of temporal patterns in species richness and population abundances from this data set with a more detailed mechanistic analysis of 5 species using our 13-year body of work at Santa Cruz Island, California. In particular, we explore the responses of these local assemblages to regional environmental perturbations [i.e., El Niño Southern Oscillation (ENSO) events and changes in regional upwelling patterns] that greatly influence fluctuations in resources used by reef fishes and which affect an area much larger than that occupied by any given local population of reef fish.

The evidence presented here suggests that there are two consistent temporal patterns in the abundance trajectories of fishes, although these are unrelated to reproductive mode, of dispersing versus nondispersing young. The data also support the conclusion that long-term behaviors of local reef fish populations are shaped strongly by environmental perturbations that influence resource levels on a regional scale, independent of whether local populations are open or closed in an ecological sense. Last, for the fish assemblages examined in southern California, it appears that the local populations of all species consistently present were tracking a declining resource base over the last 2 decades.

II. METHODS AND STUDY SITES

Fish communities were studied on rocky reefs at two locations in southern California: Santa Cruz Island (35°5′ N, 119°45′ W; see Schmitt and Holbrook, 1990a,b; Holbrook et al., 1994) and the mainland coast at King Harbor (33°51′ N, 118°24′ W; see Terry and Stephens, 1976; Stephens, 1983; Holbrook et al., 1994). Reefs in each location were covered with foliose macroalgae and turf, a low-growing, debris-laden mixture of small plants and animals. Giant kelp, *Macrocystis pyrifera,* was present intermittently on a number of the reefs at Santa Cruz Island and in King Harbor.

Estimates of species occurrences and densities were made using visual counts by divers along isobathic transects, ranging from 3 to 20 m bottom depth. Multiple counts of different aged fish (usually adult, subadult, and young-of-year) were made seasonally (King Harbor) or annually during the fall months (Santa Cruz Island); sampling methodologies are described in more detail elsewhere (e.g., Holbrook et al., 1990a,b; Schmitt and Holbrook, 1990a,b; Stephens, 1983; Stephens et al., 1984). Here we present temporal trends for the entire King Harbor fish community during the time period 1973–1993 and for five members of the guild of demersal fish at Santa Cruz Island (Garibaldi, *Hypsypops rubicundus;* black surfperch, *Embiotoca jacksoni;* striped surfperch, *Embiotoca lateralis;* pile surfperch, *Damalichthys vacca;* and rubberlip surfperch, *Rhacochilus toxotes*) from 1982 to 1994. At King Harbor one common species— blacksmith (*Chromis punctipinnis*)—often forms large schools, and although present at each year of sampling, they frequently were extremely patchy in occurrence. Since they were patchily distributed, their numbers added substantial variability to estimates of overall fish density, and therefore, the species was excluded from most statistical analyses.

Throughout this chapter we take species richness to mean the number of species present in an assemblage during any calendar year. Species appeared and disappeared at King Harbor during the 20-year period. Since sampling activities were rather intense (multiple counts of fish on transects at several locations in the harbor during different seasons every year), we regard a species as absent from the assemblage (rather than just undiscovered) if it was not counted at any time during a particular year.

Detailed studies were made of reef resources used by the Santa Cruz Island guild; these included analyses of plant cover on the reef and the crustaceans contained within. Bottom cover on the reefs was characterized at the time fish were counted using a random point contact method (see Schmitt and Holbrook, 1986) to obtain estimates of the percentage cover of turf and the common species of understory macroalgae. When present, the abundance of giant kelp was estimated in band transects (Schmitt and Holbrook, 1990a). Abundance of crustaceans used by the fish as food was estimated during the fall months (a period of seasonally high crustacean abundance; see Schmitt and Holbrook, 1986) by collecting prey in randomly placed 0.1-m² quadrats. Prey-laden substrata were collected, placed in plastic bags underwater, and removed to the laboratory where prey were removed, counted, and identified. Substrata in each sample were sorted taxonomically and weighed damp to provide estimates of algal biomass. An estimate of the food base in a given year was computed from the product of the substratum cover used by fishes as foraging microhabitat and prey density.

To facilitate comparisons among species (with different absolute densities and among-year variances), we standardized the data for a species by subtracting the 20-year mean density from each yearly density estimate and dividing that by the population standard deviation. This yielded for each species a mean density of 0 and a standard deviation of 1. Simple linear regressions were calculated for the standardized density of each species against year. While this probably violates statistical assumptions, we simply used the regression technique to categorize species according to gross temporal trends in abundance. In subsequent analyses standardized abundances (averaged across species within a group), or change in standardized abundance from one year to the next, were correlated with several independent variables (i.e., intensity of ENSO, upwelling index, and food base). Published values of the Southern Oscillation Index (SOI; e.g., Norton and Crooke, 1994) were used to estimate intensity of the ENSO signal, and amount of upwelling was determined from published values of the Upwelling Index (Bakun, 1975; Norton and Crooke, 1994). Stepwise regressions also were calculated for these dependent and independent variables.

III. RESULTS AND DISCUSSION

A. Temporal Patterns in Species Richness

Overall, a total of 83 species of teleost fishes was observed at King Harbor, which represents a substantial fraction (on the order of 70%) of the teleosts recorded in the Southern California Bight (Gotshall, 1981; Eschmeyer et al., 1983). Species richness of the reef-associated fishes present in a given year varied from about 40 to 60 species (Fig. 1). The greatest species richness was found at the start of the sampling period in the mid 1970s, and the greatest

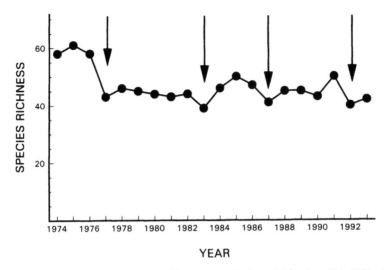

FIGURE 1.   Species richness of teleost fishes at King Harbor, California, 1974–1993. Arrows indicate periods of major environmental disturbances in the form of El Niño Southern Oscillation events.

decline between years (a net loss of nearly 20 species) occurred in 1977. Following this large decline, the total number of species present varied comparatively little and fluctuated between 40 and 50 species per year (Fig. 1; see also Holbrook *et al.*, 1994). There were four distinct declines in species richness during the 20 years of observation (arrows on Fig. 1); except for the first in 1977, each decline was followed by rapid recovery (Fig. 1). During the 20-year period, four major environmental disturbances occurred in the Southern California Bight and, together with a sharp warming in surface waters in the late 1970s (see following discussion), these four events were likely causes of the observed depressions in species richness (Stephens and Zerba, 1981; Stephens *et al.*, 1984, 1986; Holbrook *et al.*, 1994). The disturbances were ENSO phenomena, and the occurrence of each is denoted by the arrows on Fig. 1.

In the Southern California Bight, ENSO events are signaled by above-average sea-surface temperatures, reduced dissolved nutrients, elevated sea levels, an increase in the frequency and intensity of winter storms, and changes in current regimes (Mysak, 1986; Murray and Horn, 1989; Hayward, 1993). These effects can greatly alter reef resources, productivity, and even reef structures (via strong storm activity; Ebeling *et al.*, 1985). For instance, there can be dramatic reductions of nearshore kelp forests (Dayton, 1985; Tegner and Dayton, 1987; Dayton *et al.*, 1984, 1992) and decreased standing stocks offshore of phytoplankton, zooplankton, and fish (Fiedler, 1984; Fiedler *et al.*, 1986; McGowan, 1984,

1985; Smith, 1985; Hannah, 1993). Warm-water species with southern affinities become more prevalent in the Bight during El Niño periods (e.g., Cowen, 1985; Cowen and Bodkin, 1993; Mysak, 1986). Moderate to strong El Niño events have occurred on numerous occasions during the past century; the average interval between strong El Niño events is about 12.3 years, and the interval between strong and moderate events is about 5.4 years (Quinn *et al.*, 1978). El Niño events have occurred on four occasions during the last 20 years. During 1976, there was a moderate El Niño, while the 1982–1983 event was one of the strongest of the century (Mysak, 1986). Another moderate El Niño occurred in 1987, and similar effects have been apparent in the Southern California Bight intermittently from 1991 into 1994 (Hayward, 1993; Norton and Crooke, 1994).

If the ENSO phenomenon is a causal agent in depressing species richness, we might expect not only timing of the disturbances and declines in richness to coincide but also the magnitude of El Niño intensity and net decline in species numbers to be related. For the 16 years over which we have estimates of both the intensity of ENSO phenomena (indicated by the Southern Oscillation Index) in the Southern California Bight and species richness at King Harbor, the two show a very strong correlation (Fig. 2a; $r = 0.61$; $P < 0.01$); the stronger the event, the greater the net species loss. Further, net change in the number of species present from year to year was unrelated to the intensity of regional upwelling, a good proxy for primary and secondary productivity (Fig. 2b; $r = -0.04$; $P > 0.85$).

The species present in any given year reflect in part a species turnover between years, that is, a balance between previously recorded species that disappeared and the number of new species that colonized the site. Even during time periods when the assemblage was not changing substantially in species richness (i.e., non-El Niño years), about 20% of the assemblage turned over each year, which resulted in a net increase of about one species per year (Table I). The net declines in species richness that occurred during each ENSO disturbance resulted both from larger losses of species present the previous year and from smaller than usual gains of "new" species absent the previous year; on average, there was a net loss of some 6.6 species during an El Niño year (Table I). A number of species reappeared in the assemblage following each environmental disturbance, only to disappear with the next one. With respect to reproductive mode, a greater fraction of species with planktonic larvae (i.e., open populations) was lost during any El Niño event (51% of the group) than live-bearing surfperches with relatively closed populations (11%). As a result, species lost during El Niño events were predominantly those with planktonic larval stages. However, the following data suggest that the probability of going locally extinct during an El Niño event was not related directly to life history per se but rather to the degree of rarity of a local population.

It is well established that the probability that a population will go locally

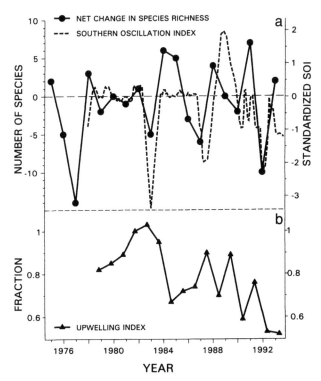

FIGURE 2. (a) Net change in species richness each year (solid circles) at King Harbor, California, 1974–1993. Also shown are the five-month running means of the standardized Southern Oscillation Index (dashed line, after Norton and Crooke, 1994) and (b) the average July–August Upwelling Index (solid triangles, after Norton and Crooke, 1994; at 30° N, 119° W as a fraction of the highest value during 1978–1993).

extinct scales inversely with population size. With respect to the reef fishes at King Harbor, of the cumulative total of 33 species that disappeared following ENSO events, over 75% had extremely small populations (≤0.05 individuals per transect when present; Fig. 3); of the 41 species that persisted throughout all

TABLE 1   Mean (SE) Number of King Harbor Species Gained or Lost during El Niño Periods at King Harbor, 1974–1993

	El Niño periods	Non-El Niño periods
Number gained	2.6 (1.1)	6.5 (0.8)
Number lost	9.2 (2.3)	5.4 (0.5)

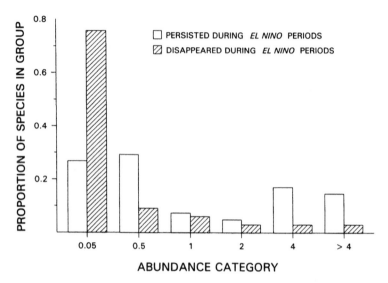

FIGURE 3. Frequency distribution of abundances (adult density per transect) of King Harbor species that persisted during El Niño periods (N = 41 species; open bars) and those that disappeared during any El Niño period (N = 33 species; shaded bars). Abundance categories are <0.05, 0.05–0.5, 0.5–1.0, 1.0–2.0, 2.0–4.0, and >4.0 fish per transect.

El Niño events, less than 30% had population sizes in this lowest abundance category (Fig. 3). Overall, the rarest species had nearly a 70% chance of disappearing during any El Niño event, compared with a 21% chance in the remaining species (densities >0.05 per transect). None of the 9 species of surfperches present at King Harbor was rare (mean number per transect, 3.5; range, 0.8–9.2), and only 1 (walleye surfperch, *Hyperprosopon argenteum*) disappeared during an ENSO event.

The 2 decades of the King Harbor data set may not be typical of historical disturbance regimes in the Southern California Bight. For instance, although several El Niño events occurred in the 20 years that preceded the onset of the King Harbor sampling, the 1957–1958 ENSO event produced a particularly strong oceanographic signal and was strongly reflected in time series data for pelagic fisheries (Mysak, 1986). Even if large changes in the King Harbor assemblage occurred as a result of the 1957–1958 event, the intervening years before the onset of sampling in 1973 should have enabled substantial recovery to occur. Although the frequent occurrence of moderate to strong ENSO events from the 1980s to the present is probably not typical of historical patterns, it is clear that the frequency and intensity of large-scale oceanographic perturbations, as well as the elapsed time between successive events, can have a substantial role in shaping patterns of species richness in a community.

There was a period from the late 1950s to the 1970s when the surface waters in the Southern California Bight were relatively cool. This extended cool-water period came to an end with the 1977 El Niño, just 4 years after the onset of sampling of fishes at King Harbor (Stephens and Zerba, 1981; Stephens et al., 1984). Surface waters rapidly warmed over the next few years and have remained elevated and relatively constant in the Bight from the early 1980s to the present. The abrupt change in thermal conditions in the Bight coincided with the initial (and greatest) decline in species richness at King Harbor (Fig. 1). Stephens and his colleagues have argued compellingly that the decline in species richness in the late 1970s reflected a disproportionately greater loss of cold-water species than of warm-water species gained (Stephens and Zerba, 1981; Stephens et al., 1984). Indeed, the King Harbor fauna of the mid 1970s contained more cold-water species than the fauna has since the mid 1980s (Stephens et al., 1984; Holbrook et al., 1994). For example, of the 14 species of cold-water rockfishes in the genus *Sebastes* that have been observed at King Harbor since 1974, 9 were present in the initial years of sampling (1974–1975), but only 3 were seen in the last 2 survey years (1992–1993). The shift in the proportion of cold- to warm-water species in the assemblage occurred rather abruptly in the late 1970s to early 1980s, and the proportional representation of these groups has remained relatively constant since then. A similar response of reef fishes to the abrupt warming in the late 1970s was also observed in other areas of the Southern California Bight (Ebeling and Hixon, 1991).

Like many assemblages, there was a substantial turnover of species of fish from year to year, independent of the effects of large-scale oceanographic perturbations or changing thermal conditions. Whereas most fish species were seen in much less than half of the samples over 20 years (Fig. 4), one group of species in the King Harbor assemblage was present in virtually every year. The *core* of the assemblage, here considered those species present in more than 17 years, represented 46% of the total species observed (34 of 74), but over 90% of the total number of individuals counted. This reinforces the fact that species that had a propensity to turn over tended to be much less common on the reef than those that persisted throughout the years. Of the 34 species in the core group, 9 (26%) were surfperches; these live-bearing fishes constituted 36% of the total number of individuals in the core group. Temporal patterns in abundances of species that form the core of the King Harbor assemblage are considered next.

B. Temporal Patterns in Population Abundances

The aggregate density of species in the core of the King Harbor assemblage remained relatively constant for most of the 20 years, fluctuating within rela-

Sally J. Holbrook and Russell J. Schmitt

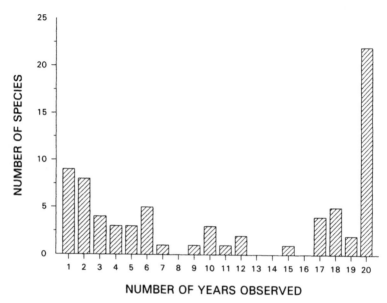

FIGURE 4. Frequency distribution of number of years that reef-associated species of teleost fish were observed during 20 years at King Harbor. The number of species observed ($N = 74$) that were present for 1–20 years is given.

tively narrow bounds (~80–120 individuals per transect), until the late 1980s when the density of fishes began a decline to about half of the long-term average (Fig. 5).

The first issue we address is whether the temporal trend in abundance of surfperches differed from that of species with planktonically dispersing larval stages. For both groups, abundances tended to decline through time, although the rate of decline differed between them (Fig. 6; the slopes are significantly different, $P < 0.001$). From this it is tempting to conclude that the temporal patterns of abundance of the fish were influenced by whether the local population was relatively open (species with dispersing larvae) or closed (live-bearing surfperches). However, differences in the slopes of the temporal trends could arise if each group differed in the consistency of temporal patterns among its members. To explore this aspect, we considered the temporal abundance patterns of each species separately. As shown in Table II, for the King Harbor species ($N = 34$) there tended to be two temporal patterns: (1) species whose abundance declined more or less linearly and significantly through time (i.e., a negative slope with years) (19 species), and (2) those whose abundance patterns were unrelated to years (13 species) or showed a positive relationship (2 species). (The number of positive slopes is about that expected by chance

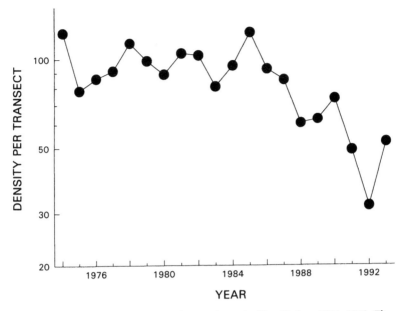

FIGURE 5. Total density (per transect) of core teleosts in King Harbor, 1974–1993. The core group contains species that were observed in ≥17 years (*N* = 34 species).

given the number of separate analyses done.) It is clear from Table II that species with dispersing larval stages showed much less consistency in their temporal abundance patterns (40% negative relationships) than the surfperches (78% negative).

For species with significant negative trends in abundance through time, there was no difference in the relationship for surfperches and species with planktonic larvae (correlation in annual abundances of the two life history groups: $r = 0.91$; $P < 0.001$). Similarly, there was no substantial difference between the two life history groups for species that did not show a significant negative relationship through time, although the correlation between the groups was weaker ($r = 0.43$; $P = 0.057$). Thus both temporal response patterns were shown by both surfperches and species with dispersing larvae, although the fractional representation in the two response clusters differed for the two life history groups. These considerations suggest that there were no substantial differences in the abundance behaviors of species with open and closed local populations.

There was a substantial and consistent difference in the abundance patterns of the two temporal response groups at King Harbor (Fig. 7). One set of species (those showing a significant negative linear trend through time) was

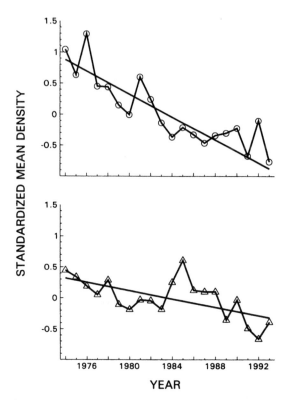

FIGURE 6.   Standardized mean adult density of surfperches (top, $N = 9$ species) and other fishes (bottom, $N = 25$ species) at King Harbor. See Table II for species lists. Grand means for each year for the two life history groups are shown. Solid lines are least-squares regressions of density against year.

most abundant in the mid 1970s, declined rapidly toward the end of that decade, remained relatively constant throughout the 1980s, and then declined steadily after 1989 (Fig. 7). Hence, the temporal trend in this group can be described as a negative sigmoid pattern. By contrast, the second group (those with a nonsignificant linear decline through time) showed a unimodal pattern of abundance through time: abundances began to rise in the late 1970s, remained high and relatively constant through the 1980s, and then also declined starting in the late 1980s (Fig. 7). Thus, the overall pattern in abundance of the core species at King Harbor (Fig. 5) consists of three phases. In the first phase (1974–1980), the decline in abundances of the negative sigmoid group was compensated by an increase in the unimodal group such that overall abundance of fish changed little (Figs. 5 and 7). The second phase (1981–

TABLE II Regressions of Standardized Fish Density against Year, King Harbor Core Species

		Temporal relationships		
		P value	Slope	r^2
Core-surfperches				
Embiotocidae				
Brachyistius frenatus	Kelp surfperch	>0.49	0.00	0
Cymatogaster aggregata	Shiner surfperch	<0.03	−0.08	0.23
Damalichthys vacca	Pile surfperch	<0.001	−0.14	0.62
		<0.002	−0.17	0.44*
Embiotoca jacksoni	Black surfperch	<0.001	−0.15	0.71
		<0.001	−0.23	0.72*
Hyperprosopon argenteum	Walleye surfperch	<0.05	−0.08	0.19
Hypsurus caryi	Rainbow surfperch	<0.001	−0.13	0.59
Micrometrus minimus	Dwarf surfperch	<0.003	−0.13	0.52
Phanerodon furcatus	White surfperch	<0.001	−0.13	0.6
Rhacochilus toxotes	Rubberlip surfperch	>0.50	0.00	0
		>0.65	−0.03	0.02*
Embiotoca lateralis	Striped surfperch	<0.001	−0.25	0.85*
Core-nonsurfperches				
Labridae				
Semicossyphus pulcher	California sheephead	>0.75	0.00	0
Oxyjulis californica	Senorita	>0.9	0.00	0
Halichoeres semicinctus	Rock wrasse	<0.001	0.12	0.44
Kyphosidae				
Girella nigricans	Opaleye	>0.20	0.00	0.08
Medialuna californiensis	Halfmoon	<0.01	−0.10	0.31
Hermosilla azurea	Zebraperch	>0.20	0.00	0.08
Pomacentridae				
Hypsypops rubicunda	Garibaldi	>0.45	0.00	0.03
		>0.95	0.00	0*
Scorpaenidae				
Scorpaena guttata	California scorpionfish	>0.1	−0.07	0.14
Sebastes serranoides	Olive rockfish	<0.02	−0.09	0.27
Serranidae				
Paralabrax clathratus	Kelp bass	>0.50	0.00	0
Paralabrax maculatofaciatus	Spotted sand bass	<0.001	−0.14	0.6
Paralabrax nebulifer	Barred sand bass	>0.20	0.00	0.09
Hexagrammidae				
Oxylebius pictus	Painted greenling	<0.001	−0.14	0.6
Cottidae				
Clinocottus spp.	Sculpin	<0.05	−0.08	0.2
Scorpaenichthys marmoratus	Cabezon	<0.03	−0.08	0.23
Haemulidae				
Anisotremus davidsonii	Sargo	>0.10	0.06	0.14
Sciaenidae				
Cheilotrema saturnum	Black croaker	<0.001	−0.14	0.64

(*continues*)

TABLE II (*Continued*)

		Temporal relationships		
		P value	Slope	r^2
Gobiidae				
Coryphopterus nicholsii	Blackeye goby	>0.45	0.00	0.03
Lythrypnus dalli	Bluebanded goby	>0.45	0.00	0.03
Clinidae				
Gibbonsia elegans	Spotted kelpfish	<0.01	−0.10	0.3
Heterostichus rostratus	Giant kelpfish	<0.15	−0.06	0.1
Malacanthidae				
Caulolatilus princeps	Ocean whitefish	<0.05	−0.08	0.2
Bothidae				
Paralichthys californicus	California halibut	>0.85	0.00	0
Pleuronectidae				
Pleuronichthys coenosus	C-O turbot	<0.001	−0.14	0.68
Pleuronichthys ritteri	Spotted turbot	<0.004	0.11	0.37

*Results for species from Santa Cruz Island.

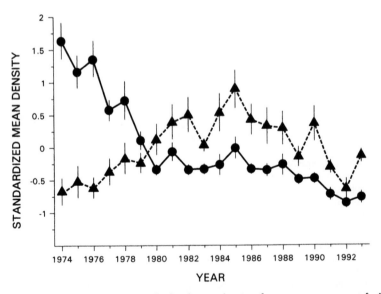

FIGURE 7. Temporal patterns in standardized mean density of two response groups of teleosts at King Harbor. Data are means ± 1 SE. Species represented by circles (*N* = 19) showed significant declines in abundances through time (Table II), while species represented by triangles (*N* = 15) did not (Table II).

1988) was a period when the abundances of both response groups remained relatively unchanged, and in the last phase (1989–1993) the abundances of both groups declined, resulting in a drop in overall abundance of fishes at King Harbor (Figs. 5 and 7). Thus, the only difference in responses of the two groups occurred during the first phase; the behaviors of both the negative sigmoid and the unimodal abundance groups were qualitatively similar since about 1981.

It is striking that the only period when the behavior of these two groups of species differed was when the surface temperature of the water in the Southern California Bight was rapidly warming. It seems plausible that this pattern is related to a difference in the responses of species to changing thermal conditions. Indeed, during this time period Ebeling and Hixon (1991) found a marked and opposite shift in the abundance of recruits of two fishes on Naples Reef in the northern region of the Bight: annual recruitment of a cold-water planktivore, blue rockfish (*Sebastes mystinus*), declined sharply between 1977 and 1982, whereas that of a warm-water planktivore, blacksmith (*Chromis punctipinnis*), increased during this period in an equally striking fashion. To explore this hypothesis for the core species in the King Harbor assemblage, we used objective criteria to classify the fishes as cold-temperate or warm-temperate species based on northern limit of geographical distribution; we defined warm-temperate species as those with northern limits south of central California and cold-temperate species as those with northern limits north of the California–Oregon border. The results (Table III) demonstrate compellingly that the nature of a species' change in abundance during the period of ocean warming was not independent of its temperature affinity; of the species whose populations declined at King Harbor between 1974 and 1980 (i.e., in the

TABLE III Number of Cold-Water and Warm-Water Affinity
Species of Reef Fishes in the King Habor Assemblage That Displayed
a Negative Sigmoid or Unimodal Pattern of Abundance through Time

| | Temporal abundance pattern | |
Temperature affinity of fish	Negative sigmoid	Unimodal
Cold-temperate	15	4
Warm-temperate	3	13

Note. The number of species in each category is given. Temperature affinity was categorized by the northern extent of the geographical distribution of a species (limits of warm-water species are south of central California; cold-water species are between the Oregon–California border and Alaska). The temperature affinity–temporal trend pattern is not statistically independent ($\chi^2 = 12.6$; $P < 0.005$).

negative sigmoid response group), 83% were cold-temperate species, whereas 77% of the species whose populations increased during this period (i.e., in the unimodal response group) were classified as warm-temperate species (Table III). As a test of the thermal affinity–abundance response hypothesis, we examined the temporal pattern in abundance of the warm-temperate planktivorous blacksmith at King Harbor, a species that did not fit the core group criterion (see Methods). The blacksmith is a temperate derivative of the tropical family Pomacentridae whose northern limit of distribution is central California (i.e., a warm-temperate species). As expected, annual densities of blacksmith were positively correlated with those in the unimodal response group ($r = 0.75$; $P < 0.001$), but were negatively correlated with those of the negative sigmoid response ($r = -0.54$; $P < 0.05$). We tested the temporal abundance patterns of five additional species at Santa Cruz Island, and in every case the observed pattern matched the predicted response. Thus, the evidence strongly suggests that the widespread and rapid shift in the thermal regime of the Bight produced abundance responses in the assemblage of reef fishes that were predictable based on whether the fish was a cold- or warm-temperate species.

Following the period of rapid change in abundance that was associated with the switch in the thermal regime, the abundances of both groups of fishes remained relatively constant throughout most of the 1980s before declining sharply toward the end of the decade (Fig. 7). To what extent can this temporal pattern of reef fishes be attributed to other large-scale oceanographic conditions in the Bight? While ocean water temperature has remained elevated but more or less constant since 1981, three ENSO events of varying intensities and durations have occurred, and the occurrence and magnitude of regional upwelling have changed, from relatively constant and strong to highly variable and weak (Fig. 2). Although the two response groups behaved similarly during these latter two phases (correlation in annual abundances across groups, 1980–1993: $r = 0.87$; $P < 0.001$), we consider them separately because of the distinctly different proportions of cold- and warm-temperate species in the two groups.

Stepwise regression models revealed that abundances of both response groups were positively associated with patterns in the intensity of upwelling and, for a given amount of upwelling, were negatively associated with the intensity of ENSO phenomena ($P < 0.05$ in all cases). These two independent variables accounted for 52 and 60%, respectively, of the overall variation in abundances of the negative sigmoid and unimodal response groups. Whereas neither of the two oceanographic processes could account for a significant amount of average annual change in the abundance of cold-temperate species (the negative sigmoid response group), 55% of the average annual change in the density of warm-temperate species (the unimodal group) was accounted for by ENSO events. In this latter group, the stronger the El Niño signal, the greater the decline in density of fish from the previous year. It may be the case that warm-temperate species are relatively more sensitive to ENSO events than cold-temperate species, although this does not fit well with the effect of El Niños on water temperature. Nonetheless, there are clear indications of upwelling and ENSO signals in the abundance patterns of reef fishes at King Harbor.

It is particularly noteworthy that the decline in the abundances of both response groups beginning in the late 1980s at King Harbor corresponds to the marked downward trend in regional upwelling in the Bight (Figs. 2 and 7). It is known that long-term trends in population abundances of certain pelagic species are associated with temporal patterns of upwelling. For example, time series analyses of Pacific hake (*Merluccius productus*), Pacific sardine (*Sardinops sagax*), and northern anchovy (*Engraulis mordax*) indicate that their long-term population trends follow long-period (40- to 60-year) oscillations in primary and secondary production, which in turn are forced by a long-period oscillation in wind-induced upwelling (Ware and Thomson, 1991). Since 1978, the highest summer upwelling in the Southern California Bight (when nutrients shaping primary productivity are otherwise at an annual low; see Bakun, 1975) occurred during 1980–1982, and the lowest levels were measured in 1992–1993 (Norton and Crooke, 1994). Since 1988, there has been an overall decline in the amount of upwelling, with exceptionally low values of the upwelling index during the 1990s (Norton and Crooke, 1994). The slope of the decline in the standardized abundance of both response groups does not differ from that of the standardized upwelling index. Thus, it appears that changes in environmental productivity, as influenced primarily by upwelling and secondarily by ENSO phenomena, may be causally related to or may generate the temporal abundance patterns of reef fishes in the Bight. Below we explore a possible mechanistic link between these large-scale environmental forcing processes and changes in the local abundances of reef fishes.

C. Are Local Populations Tracking Resources?

If the two temporal response patterns in abundance of reef fishes observed at King Harbor (Fig. 7) were related to fluctuations in such regional phenomena as upwelling, ENSO events, and thermal regimes, we might expect populations in other areas of the Southern California Bight to show temporal patterns that are qualitatively identical to those seen at King Harbor. Of the five species of reef fishes (three cold-temperate and two warm-temperate) we have examined in detail at Santa Cruz Island since 1982, four are found in the King Harbor assemblage. The four species common to both locales showed the same temporal trends in abundance at both sites. Abundances of the two cold-affinity species (black surfperch, *Embiotoca jacksoni*; pile surfperch, *Damalichthys vac-*

ca) remained relatively constant throughout most of the 1980s and then declined dramatically in the early 1990s (Fig. 8, Table II). By contrast, the
abundance patterns of the two warm-water species (Garibaldi, *Hypsypops rubi-
cunda*; rubberlip surfperch, *Rhacochilus toxotes*) were unimodal with marked
population increases from the early 1980s, constancy through the mid 1980s,
and a marked decline beginning in the early 1990s (Fig. 8, Table II). The fifth
species from Santa Cruz Island (striped surfperch, *Embiotoca lateralis*) is a
cold-temperate species that is not found in the King Harbor assemblage. If
there is an association between temperature affinity and temporal pattern of
abundance, we would predict that the abundance of striped surfperch would
trend downward in a more or less sigmoid fashion. This indeed was the pattern
displayed by striped surfperch at Santa Cruz Island (Fig. 8, Table II). The

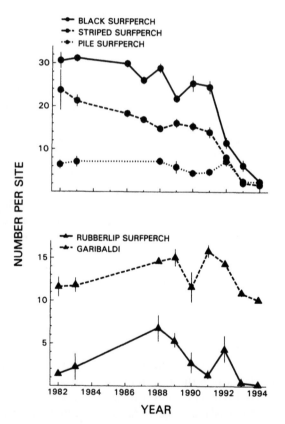

FIGURE 8. Densities of five species of reef fish in two response groups at Santa Cruz Island,
1982–1994. Shown are mean (±1 SE) adults per site (*N* = 2 sites) for three species whose
abundances declined through time (top) and two species whose abundances did not (bottom).

similarities in response patterns at these two widely separated sites within the Bight suggest that the responsible agent(s) operated on a regional rather than a local scale.

In addition to quantifying patterns of fish abundance at Santa Cruz Island, we have estimated the food base (a composite of the local abundance of foraging microhabitats and prey densities per unit of microhabitat) at 11 sites along the north coast of the island. There was some degree of variation among these sites in the extent of interannual fluctuations in the resource base, and we found that the degree of interannual variation in abundances of the five fishes matched local spatial differences in the resource base (Holbrook *et al.*, 1993, 1994; R. J. Schmitt and S. J. Holbrook, unpublished). Nonetheless, these fine-scale patterns were swamped by the broader scale temporal trends in both the resource base and the fish populations at Santa Cruz Island. For example, the overall food base has declined by around 60% at all sites over the 13-year period of observation (Fig. 9; correlation of food base with year: $r = -0.78$; $P < 0.005$).

To what extent were trends in abundances of the five species of fishes correlated with the declining resource base at Santa Cruz Island? In all cases,

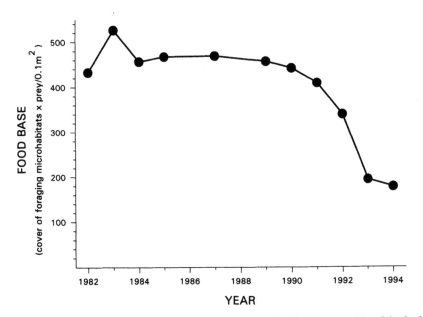

FIGURE 9. Temporal trends in the food base at Santa Cruz Island. The mean value of the food base (prey density on turf and on foliose red algae) at reefs on Santa Cruz Island, 1982–1994, is shown.

the abundance of each fish population declined in parallel with the resource base (Table IV). The correlation was highly significant for three of the species and marginally significant for the other two (Table IV). Significantly, the latter two species (Garibaldi and rubberlip surfperch) were the two warm-temperate species whose abundance patterns were unimodal; when only the period after the shift in the thermal regime (i.e., after 1983) is considered, the correlation in abundances of these fishes with levels of their food base is much improved (Garibaldi: $r = 0.86, P < 0.01$; rubberlip surfperch: $r = 0.78, P < 0.02$). These data support the conclusion that long-term, substantial declines in the local food base since the late 1980s were tracked rather well by each of the embiotocid fishes at Santa Cruz Island.

Given that populations of the fishes and the food base trended downward together, is there any evidence that the temporal pattern of the resource was influenced by large-scale oceanographic forcing processes? At Santa Cruz Island there was a very strong positive correlation between the abundance of food and regional upwelling ($r = 0.71; P < 0.01$), but no relationship between food and intensity of ENSO phenomena ($r = -0.12; P > 0.70$). It is important to note that in stepwise regression models, temporal variation in the abundance of each fish species was primarily associated with variation in the food base rather than with upwelling or El Niño events, neither of which entered into the models when food was included as an independent variable. The fact that food was the best predictor of variation in fish abundances suggests that fish were not responding to upwelling per se but rather to its effect on the resource base.

The same predictors influence changes of fish abundances at Santa Cruz Island from one year to the next. Overall, the average annual change in abundance of these fishes was positively correlated with the annual change in the resource base ($r = 0.82; P < 0.025$). Again, stepwise regression models for annual change in abundance of the fishes identified annual variation in the

TABLE IV Relationships in the Annual Abundances of Five Species of Reef Fishes at Santa Cruz Island with Annual Estimates of the Food Resource Base

Species	Correlation with food base	Relationship	P value[a]
Black surfperch	0.95	+	<0.001
Striped surfperch	0.91	+	<0.001
Pile surfperch	0.79	+	<0.01
Rubberlip surfperch	0.57	+	=0.054
Garibaldi	0.53	+	=0.069

[a]P values are for a one-tailed test of the hypothesis that fish populations track changes in the resource base.

food base as a useful predictor but not upwelling or ENSO phenomena. Annual change in the food base was positively related to upwelling ($r = 0.68$; $P < 0.05$) but not to the intensity of ENSO events ($r = -0.29$; $P > 0.4$). This reinforces the notion that fishes were responding to variation in food, which was affected primarily by upwelling, rather than to the oceanographic perturbations per se. Thus, fish populations not only tracked the long-term trends in the resource base but also variation on an annual time scale.

These considerations suggest that regional levels of upwelling during the summer influenced the amount of food available to fishes in the Southern California Bight, and they provide a mechanistic link between this large-scale environmental forcing process and temporal patterns in the abundance of reef fishes. Does this linkage make biological sense? Primary productivity in the Southern California Bight varies seasonally because during the winter the water is cold and nutrient-laden but during the summer it is warm and nutrient-poor. Summer productivity is enhanced by upwelling, which brings nutrient-rich water from deeper regions of the Bight to shallow coastal areas and, in turn, increases primary productivity in these nearshore areas. Interannual variation in the existence and intensity of upwelling, driven largely by variation in wind speed and direction, results in substantial interannual fluctuations in primary productivity in the Bight. While it is known that ENSO phenomena can reduce productivity in the Bight by bringing in warm, nutrient-poor waters, strong upwelling events can ameliorate this effect. There appears to be no correlation between the intensities of ENSO and upwelling phenomena in the Bight during the period of these observations ($r = -0.11$; $P > 0.75$). Secondary productivity depends, of course, on primary productivity, so it follows that trends in upwelling also reflect trends in the productivity of prey items harvested by reef fishes (primarily small invertebrates such as amphipod crustaceans). We believe that lower productivity of prey species consumed by these fishes is reflected in the significant reduction in prey densities through time (ANOVA of crustacean densities prior to and after 1990; $F_{5,250} = 14.09$; $P < 0.001$), independent of the amount of foraging microhabitat available to the fishes.

A second possible pathway by which variation in upwelling might influence the food base for reef fishes in the Southern California Bight is through its effect on the local abundance of foraging microhabitat. The majority of the fishes in this assemblage consume prey from benthic substrata. Most of their foraging microhabitats are species of algae, primary producers that require water-borne nutrients for growth; these are nutrients whose availability is influenced by upwelling, such that interannual variation in the biomass of red algae at Santa Cruz Island correlates with intensity of upwelling ($r = 0.64$; $P = 0.06$). Thus it appears that the influence of upwelling on the food resource of reef fishes in the Southern California Bight may be mediated through effects both on direct productivity of prey items and on the amount of microhabitat from which reef fishes harvest their food.

We have explored the demographic causes underlying the changes in populations of two species of surfperch at Santa Cruz Island. Increases in the food base resulted in a local birth rate response; increases in the adult population lagged by a year; and, by contrast, declines in the food resulted in an immediate decline in the adult populations (Schmitt and Holbrook, 1990a, unpublished).

IV. CONCLUDING REMARKS

Most current models of local community structure and population dynamics of marine reef fishes emphasize the fact that populations of most marine organisms are open, such that nonequilibrium processes can play a central role in shaping patterns (Talbot *et al.*, 1978; Doherty, 1991; Sale, 1991b). The recruitment of young to a local reef can be highly variable especially if produced elsewhere, and this has led to the widespread notion that local assemblages and their component populations of fishes are highly variable and show little predictability (Doherty and Williams, 1988; Doherty, 1991; Sale, 1991b). To date there has been little published evidence to suggest that reef fishes can track benthic resource levels (e.g., Wellington and Victor, 1985, 1988) or that older life stages on the reef are subject to compensatory (density-dependent) mortality (Shulman, 1985; see review by Jones, 1991). This lack has further undermined support for resource-based processes as major forces in shaping the observed and seemingly unpredictable temporal variability in local populations of reef fishes (Doherty, 1991).

The long-term behavior of the temperate reef fish assemblage in the Southern California Bight stands in stark contrast to the picture painted by existing models for open nonequilibrial systems. There was considerable temporal variation at the local reef scale in both species richness and population abundances; yet these local fluctuations were remarkably predictable in light of large-scale environmental forcing processes. Temporal patterns of species composition and richness at the local scale appear to be shaped largely by the regional thermal regime and ENSO phenomena which, predictably, affected the rarer species disproportionately. Abundances certainly trended with variation in the local food base in a manner consistent with population tracking of the critical resource. Variation in local resources, in turn, appears to be affected mostly by varying nutrient inputs that drive patterns of primary production in the region. The notion that environmental forcing can have a large influence on the structure and dynamics of marine populations is not new (e.g., Cushing, 1982; Sissenwine, 1984; Brodeur and Pearcy, 1992), but perhaps it is underappreciated as an explanation for local patterns of variation. Indeed, there is justifiable reason to believe that population dynamics may be qualitatively different on local and more regional (i.e., metapopulation) scales (Doherty, 1991). An emerging central issue is the spatial scale over which local populations with possible reciprocal reproductive output are affected by varying environmental conditions.

If long-term patterns in populations of reef fishes are best characterized by open nonequilibrial models, we would expect the local dynamics of species with larval stages which disperse in the plankton (i.e., having open local populations) to be qualitatively different from those of species whose local populations are relatively closed. In the temperate eastern Pacific, a sizable fraction of reef-dwelling fishes belongs to the surfperch family Embiotocidae, all of which are characterized by giving birth to few, fully developed juveniles that do not disperse widely from the parental population (i.e., they have relatively closed local populations). However, we found that local dynamics of all species were remarkably similar, regardless of whether the local population was open or closed, and all generally appear shaped by temporal patterns in productivity. The similar behavior of species with open and closed local populations suggests that the spatial scale of processes that affect productivity was large relative to the spatial scale over which the species with dispersing larvae exchange young. This would argue that local populations that are linked via the dispersal of their young are similarly affected by variation in environmental conditions that influence productivity.

The evidence presented here suggests that regional patterns of productivity can have a major influence on the long-term behavior of local reef fish assemblages. Although temperate areas such as the Southern California Bight may be subject to considerable temporal fluctuations in large-scale environmental forcing processes, there is no reason to believe that reef systems in other geographical localities do not respond similarly to variation in regional environmental conditions. The data presented here suggest that there is a critical need to reconsider the role of resource variation and to investigate its influences on a regional scale.

V. SUMMARY

Two long-term data sets of assemblages of temperate reef fish were studied to explore population and community trends in relation to environmental disturbances. The King Harbor assemblage included a total of 75 reef-associated teleost fishes and has been observed for 20 years; the Santa Cruz Island data set included five species of fish and detailed information about reef resources during a 13-year period. The assemblages responded to regional environmental perturbations such as El Niño Southern Oscillation events and changes in regional upwelling patterns that greatly influenced reef resources (especially food supply). Our analyses revealed that species of northern, cold-water affinity declined sharply during the 1970s, stayed more or less constant during the 1980s, and fell again during the early 1990s. Southern affinity species tended to increase in abundance during the 1970s into the 1980s, but since the mid-1980s these species have generally been declining as well. Similar patterns were detected at both sites. Analyses of crustacean food supply at Santa Cruz Island revealed a tremendous decline since 1990, and we conclude that local populations of these reef-associated species of fish have been tracking a declining resource base.

Acknowledgments

We thank John S. Stephens, Jr., for making his extraordinary King Harbor data set available to us. We are grateful for the many years of effort that John put into its collection, especially for his generosity in sharing it with others, and also for the support of John's work by Southern California Edison Research and Development. The help of the many divers who have assisted us underwater at Santa Cruz Island as well as in the laboratory is appreciated. B. Williamson provided technical assistance. We profited from our many discussions about this manuscript with Andy Brooks and Mark Carr. Financial support was provided by the National Science Foundation (OCE91-02191 and earlier awards).

References

Allen, L. G., Horn, M. H., Edmands, F. A., II, and Usui, C. A. (1983). Structure and seasonal dynamics of the fish assemblage in the Cabrillo Beach area of Los Angeles Harbor, California. *Bull. South. Calif. Acad. Sci.* **82**, 47–70.

Bakun, A. (1975). Daily and weekly upwelling indices, west coast of North America, 1946–71. *NOAA Tech. Rep., NMFS SSRF* **NMFS SSRF-693**, 1–114.

Beamish, R. J., and Bouillon, D. R. (1993). Pacific Salmon production trends in relation to climate. *Can. J. Fish. Aquat. Sci.* **50**, 1002–1016.

Brodeur, R. D., and Pearcy, W. G. (1992). Effects of environmental variability on trophic interactions and food web structure in a pelagic upwelling ecosystem. *Mar. Ecol.: Prog. Ser.* **84**, 101–119.

Chesson, P. L. (1986). Environmental variation and the coexistence of species. *In* "Community Ecology" (J. Diamond and T. J. Case, eds.), pp. 240–256. Harper & Row, New York.

Chesson, P. L., and Warner, R. R. (1981). Environmental variability promotes coexistence in lottery competitive systems. *Am. Nat.* **117**, 923–943.

Cowen, R. K. (1985). Large scale patterns of recruitment by the labrid, *Semicossyphus pulcher*: Causes and implications. *J. Mar. Res.* **43**, 719–742.

Cowen, R. K., and Bodkin, J. L. (1993). Annual and spatial variation of the kelp forest fish assemblage at San Nicolas Island, California. *In* "Third California Islands Symposium: Recent Advances in Research on the California Islands" (F. G. Hochberg, ed.), pp. 463–474. Santa Barbara Museum of Natural History, Santa Barbara, CA.

Cushing, D. H. (1982). "Climate and Fisheries." Academic Press, New York.

Davis, G. E., and Halvorson, W. L. (1988). "Inventory and Monitoring of Natural Resources in Channel Islands National Park, California." U.S. National Park Service, Ventura, CA.

Dayton, P. K. (1985). Ecology of kelp communities. *Annu. Rev. Ecol. Syst.* 16, 215–245.

Dayton, P. K., Currie, V., Gerrodette, T., Keller, B., Rosenthal, R., and Van Tresca, D. (1984). Patch dynamics and stability of some southern California kelp communities. *Ecol. Monogr.* 54, 253–289.

Dayton, P. K., Tegner, M. J., Parnell, P. E., and Edwards, P. B. (1992). Temporal and spatial patterns of disturbance and recovery in a kelp forest community. *Ecol. Monogr.* 62, 421–445.

Doherty, P. J. (1991). Spatial and temporal patterns in recruitment. *In* "The Ecology of Fishes on Coral Reefs" (P. F. Sale, ed.), pp. 261–292. Academic Press, San Diego, CA.

Doherty, P. J., and Fowler, T. (1994). An empirical test of recruitment limitation in a coral reef fish. *Science* 263, 935–939.

Doherty, P. J., and Williams, D. McB. (1988). The replenishment of coral reef fish populations. *Oceanogr. Mar. Biol.* 26, 487–551.

Ebeling, A. W., and Hixon, M. A. (1991). Tropical and temperate reef fishes: Comparison of community structures. *In* "The Ecology of Fishes on Coral Reefs" (P. F. Sale, ed.), pp. 509–563. Academic Press, San Diego, CA.

Ebeling, A. W., and Laur, D. R. (1988). Fish populations in kelp forests without sea otters: Effects of severe storm damage and destructive sea urchin grazing. *In* "Community Ecology of Sea Otters" (G. VanBlaricom and J. Estes, eds.), pp. 169–191. Springer-Verlag, Berlin.

Ebeling, A. W., Larson, R. J., Alevizon, W. S., and Bray, R. N. (1980). Annual variability of reef-fish assemblages in kelp forests off Santa Barbara, California. *Fish. Bull.* 78, 361–377.

Ebeling, A. W., Laur, D. R., and Rowley, R. J. (1985). Severe storm disturbances and reversal of community structure in a southern California kelp forest. *Mar. Biol.* 84, 287–294.

Ebeling, A. W., Holbrook, S. J., and Schmitt, R. J. (1990). Temporally concordant structure of a fish assemblage: Bound or determined? *Am. Nat.* 135, 63–73.

Eschmeyer, W. N., Herald, E. S., and Hammann, H. (1983). "A Field Guide to Pacific Coast Fishes of North America." Houghton Mifflin, Boston.

Fiedler, P. C. (1984). Satellite observations of the 1982–1983 El Niño along the U.S. Pacific coast. *Science* 224, 1251–1254.

Fiedler, P. C., Methot, R. D., and Hewitt, R. P. (1986). Effects of California El Niño 1982–1984 on the northern anchovy. *J. Mar. Res.* 44, 317–338.

Forrester, G. E. (1990). Factors influencing the juvenile demography of a coral reef fish. *Ecology* 71, 1666–1681.

Gotshall, D. W. (1981). "Pacific Coast Inshore Fishes." Sea Challengers, Los Osos, CA.

Hannah, R. W. (1993). Influence of environmental variation and spawning stock levels on recruitment of ocean shrimp (*Pandalus jordani*). *Can. J. Fish. Aquat. Sci.* 50, 612–622.

Hayward, T. L. (1993). Preliminary observations of the 1991–1992 El Niño in the California current. *CalCOFI Rep.* 34, 21–29.

Holbrook, S. J., and Schmitt, R. J. (1989). Resource overlap, prey dynamics, and the strength of competition. *Ecology* 70, 1943–1953.

Holbrook, S. J., Carr, M. H., and Schmitt, R. J. (1990a). Effect of giant kelp on local abundance of reef fishes: The importance of ontogenetic resource requirements. *Bull. Mar. Sci.* 47, 104–114.

Holbrook, S. J., Schmitt, R. J., and Ambrose, R. F. (1990b). Biogenic habitat structure and characteristics of temperate reef fish assemblages. *Aust. J. Ecol.* 15, 489–503.

Holbrook, S. J., Swarbrick, S. L., Schmitt, R. J., and Ambrose, R. F. (1993). Reef architecture and reef fish: Correlates of population densities with attributes of subtidal rocky environments. *In* "Proceedings of the Second International Temperate Reef Symposium" (C. N. Battershill *et al.*, eds.), pp. 99–106. NIWA Marine, Wellington, Auckland, NZ.

Holbrook, S. J., Kingsford, M. J., Schmitt, R. J., and Stephens, J. S., Jr. (1994). Spatial and temporal patterns in assemblages of temperate reef fish. *Am. Zool.* 34, 463–475.

Hsieh, W. W., Lee, W. G., and Mysak, L. A. (1991). Using a numerical model of the Northeast Pacific Ocean to study the interannual variability of the Fraser River Sockeye Salmon (*Oncorhynchus nerka*). *Can. J. Fish. Aquat. Sci.* 48, 623–630.

Hughes, T. P. (1990). Recruitment limitation, mortality and population regulation in open systems: A case study. *Ecology* 71, 12–20.

Jones, G. P. (1988). Ecology of rocky reef fish of north-eastern New Zealand. *N. Z. J. Mar. Freshwater Res.* 22, 445–462.

Jones, G. P. (1991). Postrecruitment processes in the ecology of coral reef fish populations: A multifactorial perspective. *In* "The Ecology of Fishes on Coral Reefs" (P. F. Sale, ed.), pp. 294–330. Academic Press, San Diego, CA.

Keough, M. J. (1988). Benthic populations: Is recruitment limiting or just fashionable? *Proc. Int. Coral Reef Symp., 6th,* Vol. 1, pp. 141–148.

Kingsford, M. J. (1988). The early life history of fish in coastal waters of northern New Zealand: A review. *N. Z. J. Mar. Freshwater Res.* 22, 463–479.

Kingsford, M. J., Schiel, D. R., and Battershill, C. N. (1989). Distribution and abundance of fish in a rocky reef environment at the subantarctic Auckland Islands, New Zealand. *Polar Biol.* 9, 179–186.

Lewin, R. (1986). Supply-side ecology. *Science* 234, 25–27.

Mapstone, B. D., and Fowler, A. J. (1988). Recruitment and the structure of assemblages of fish on coral reefs. *TREE* 3, 72–77.

McGowan, J. A. (1984). The California El Niño. *Oceanus* 27, 48–51.

McGowan, J. A. (1985). El Niño 1983 in the Southern California Bight. *In* "El Niño North. Niño Effects in the Eastern Subarctic Pacific Ocean" (W. S. Wooster and D. L. Fluharty, eds.), pp. 166–184. Washington Sea Grant Program, University of Washington, Seattle.

Murray, S. N., and Horn, M. H. (1989). Variations in standing stocks of central California macrophytes from a rocky intertidal habitat before and during the 1982–1983 El Niño. *Mar. Ecol.: Prog. Ser.* 58, 113–122.

Mysak, L. (1986). El Niño, interannual variability and fisheries in the northeast Pacific Ocean. *Can. J. Fish. Aquat. Sci.* 43, 464–497.

Norton, J. G., and Crooke, S. J. (1994). Occasional availability of dolphin, *Coryphaena hippurus,* to Southern California commercial passenger fishing vessel anglers: Observations and hypotheses. *CalCOFI Rep.* 35, 230–239.

Quinn, W. H., Zopf, D. O., Short, K. S., and Yang, R. T. W. K. (1978). Historical trends and statistics of the Southern Oscillation, El Niño, and Indonesian droughts. *Fish. Bull.* 76, 663–678.

Raimondi, P. T. (1990). Patterns, mechanisms, consequences of variability in settlement and recruitment of an intertidal barnacle. *Ecol. Monogr.* 60, 283–309.

Roughgarden, J., Gaines, S., and Pacala, S. (1987). Supply side ecology: The role of physical transport processes. *In* "Organization of Communities Past and Present" (J. H. R. Gee and P. S. Giller, eds.), pp. 491–518. Blackwell, Oxford.

Roughgarden, J., Gaines, S., and Possingham, H. (1988). Recruitment dynamics in complex life cycles. *Science* 241, 1460–1466.

Sale, P. F. (1977). Maintenance of high diversity in coral reef fish communities. *Am. Nat.* 111, 337–359.

Sale, P. F. (1980). The ecology of fishes on coral reefs. *Oceanogr. Mar. Biol.* 18, 367–421.

Sale, P. F., ed. (1991a). "The Ecology of Fishes on Coral Reefs". Academic Press, San Diego, CA.

Sale, P. F. (1991b). Reef fish communities: Open nonequilibrial systems. *In* "The Ecology of Fishes on Coral Reefs" (P. F. Sale, ed.), pp. 564–598. Academic Press, San Diego, CA.

Schmitt, R. J., and Holbrook, S. J. (1986). Seasonally fluctuating resources and temporal variability of interspecific competition. *Oecologia* **69**, 1–11.

Schmitt, R. J., and Holbrook, S. J. (1990a). Contrasting effects of giant kelp on dynamics of surfperch populations. *Oecologia* **84**, 419–429.

Schmitt, R. J., and Holbrook, S. J. (1990b). Population responses of surfperch released from competition. *Ecology* **71**(5), 1653–1665.

Shulman, M. J. (1985). Recruitment of coral reef fishes: Effects of distribution of predators and shelter. *Ecology* **66**, 1056–1066.

Sissenwine, M. P. (1984). Why do fish populations vary? In "Exploitation of Marine Communities" (R. M. May, ed.), pp. 59–94. Springer-Verlag, Berlin.

Smith, P. E. (1985). A case history of an anti-El Niño El Niño transition on plankton and nekton distribution and abundances. In "El Niño North. Niño Effects in the Eastern Subarctic Pacific Ocean" (W. S. Wooster and D. L. Fluharty, eds.), pp. 121–142. Washington Sea Grant Program, University of Washington, Seattle.

Stephens, J. S., Jr. (1983). "The Fishes of King Harbor: A Nine Year Study of Fishes Occupying the Receiving Waters of a Coastal Steam Electric Generating Station," *Res. Dev. Ser.* 83-RD-1. Occidental College, Los Angeles, CA.

Stephens, J. S., Jr., and Zerba, K. E. (1981). Factors affecting fish diversity on a temperate reef. *Environ. Biol. Fish.* **6**, 111–121.

Stephens, J. S., Jr., Morris, P. A., Zerba, K., and Love, M. (1984). Factors affecting fish diversity on a temperate reef: The fish assemblage of Palos Verdes Point, 1974–1981. *Environ. Biol. Fish.* **11**, 259–275.

Stephens, J. S., Jr., Jordan, G. A., Morris, P. A., Singer, M. M., and McGowen, G. E. (1986). Can we relate larval fish abundance to recruitment or population stability? A preliminary analysis of recruitment to a temperate rocky reef. *CalCOFI Rep.* **27**, 65–83.

Stephens, J. S., Jr., Morris, P. A., Pondella, D. J., Koonce, T. A., and Jordan, G. A. (1994). Overview of the dynamics of an urban artificial reef fish assemblage at King Harbor, California, USA, 1974–1991: A recruitment-driven system. *Bull. Mar. Sci.* **55**, 1224–1239.

Stoner, D. S. (1990). Recruitment of a tropical colonial ascidian: Relative importance of pre-settlement vs. post-settlement processes. *Ecology* **71**, 1682–1690.

Sutherland, J. P. (1990). Recruitment regulates demographic variation in a tropical intertidal barnacle. *Ecology* **71**, 955–972.

Talbot, F. H., Russell, B. C., and Anderson, G. R. V. (1978). Coral reef fish communities: Unstable high-diversity systems? *Ecol. Monogr.* **48**, 425–440.

Tegner, M., and Dayton, P. K. (1987). El Niño effects on southern California kelp forest communities. *Adv. Ecol. Res.* **47**, 243–279.

Terry, C. B., and Stephens, J. S. J. (1976). A study of the orientation of selected embiotocid fishes to depth and shifting seasonal vertical temperature gradients. *Bull. South. Calif. Acad. Sci.* **75**, 170–183.

Underwood, A. J., and Fairweather, P. G. (1989). Supply-side ecology and benthic marine assemblages. *TREE* **4**, 16–19.

Victor, B. C. (1983). Recruitment and population dynamics of a coral reef fish. *Science* **219**, 419–420.

Victor, B. C. (1986). Larval settlement and juvenile mortality in a recruitment-limited coral reef fish population. *Ecol. Monogr.* **56**, 145–160.

Ware, D. N., and Thomson, R. E. (1991). Link between long-term variability in upwelling and fish production in the Northeast Pacific Ocean. *Can. J. Fish. Aquat. Sci.* **48**, 2296–2306.

Warner, R. R., and Chesson, P. L. (1985). Coexistence mediated by recruitment fluctuations: A field guide to the storage effect. *Am. Nat.* **125**, 769–787.

Warner, R. R., and Hughes, T. P. (1988). The population dynamics of reef fishes. *Proc. Int. Coral Reef Symp.*, *6th*, Vol. 1, pp. 149–155.

Wellington, G. M., and Victor, B. C. (1985). El Niño mass coral mortality: A test of resource limitation in a coral reef damselfish population. *Oecologia* **68**, 15–19.

Wellington, G. M., and Victor, B. C. (1988). Variation in components of reproductive success in an undersaturated population of coral-reef damselfish: A field perspective. *Am. Nat.* **131**, 588–601.

Young, C. M. (1987). Novelty of "Supply-side ecology." *Science* **235**, 415–416.

CHAPTER 3

Long-Term Studies of Northern Temperate Freshwater Fish Communities

ALLEN KEAST

Biology Department, Queen's University, Kingston, Ontario K7L 3N6, Canada

I. INTRODUCTION

Both the Nearctic and Palearctic regions are rich in aquatic environments with abundant lakes, rivers, and streams, and the northern temperate freshwater fish faunas are diverse throughout the area, occupying a broad belt across the entire Holarctic. Although sharing a common origin early in the Tertiary, the faunas of North America and Europe are now largely distinct taxonomically. On both continents the freshwater fish faunas are dominated by primary division forms, i.e., groups that are exclusively freshwater and/or have had their origins in freshwater.

In this chapter I explore some of the major features of northern temperate freshwater fish assemblages, based on 20 years of work by myself and students on eastern Ontario lakes and streams. Among the subjects I cover are seasonal adaptations (annual cycle) and the seasonal nature of temperate freshwater fish communities; community structure; habitat use patterns; and diets and trophic relations of component species and their specific age classes. Special attention is given to how the partitioning of available food resources occurred among species and their various age classes, and how this is in turn related to the seasonally fluctuating prey resource base. These subjects remain inadequately studied. Here, examples are drawn mainly from two systems that have been under long-term study: the fish communities of Jones Creek (44.5° N, 76.0° W) and Lake Opinicon (44.8° N, 76.6° W), Ontario. The latter system forms the central component of a book currently in preparation.

The fishes of lakes, ponds, and streams, partially isolated as they are, provide unusual opportunities for exploring the preceding aspects of community structure. These faunas and their component species can be quantitatively sampled for composition, abundances, diets, and habitat use by a variety of techniques including selective netting, electro-shocking, and underwater transect counts using divers (e.g., Werner *et al.*, 1977; Keast *et al.*, 1978). Zoobenthic and zooplanktonic prey organisms can also readily be quantified.

A. Historical Aspects and Faunal Origins

The fossil record and phylogenetic history of fish lineages assumes special importance relative to contemporary communities with the realization that the structuring of communities has a historical component (Ricklefs, 1987; Brooks and McLennan, 1992, 1994). Origins and fossil histories of the main northern freshwater fish faunas have been reviewed by the various authors in Hocutt and Wiley (1986), Holčik (1989), Bănărescu (1991), and Wilson and Williams (1992). The primitive paddlefishes, sturgeons, gars, and bowfins, along with eels and perches, go back to the Cretaceous. Herrings, suckers, pikes, and ancestral North American ictalurid catfishes are known from the Palaeocene, salmonids and endemic North American centrarchids from the Eocene, and cyprinids from the Oligocene. The perfection of a highly specialized predator morphology of the pikes by the Cretaceous (Wilson and Williams, 1992) indicates that they had already assumed the role of top piscivores by that time. The ictalurid catfishes, chemosensory feeders morphologically adapted for feeding at night and in muddy waters (Keast, 1985a), also have occupied their adaptive zone over an extensive geological period. Amongst the dominant North American Centrarchidae, the age of major ecomorphological types, as confirmed by phylogenetic analysis (Wainwright and Lauder, 1992), is as follows: *Micropterus* (piscivore) and *Pomoxis* (planktivore), Miocene; *Ambloplites* (specialized odonate nymph and crayfish eater), Pliocene; and *Lepomis* (small invertebrate predator and molluscivore), Pleistocene. In turn, the radiation of the highly ecologically diversified Cyprinidae, the "rodents" of the aquatic system, extends back to the Miocene.

Clearly there was dynamic evolution of freshwater fish faunas throughout the Tertiary, relative to the major ecological and trophic opportunities available, along with phylogenetic radiations. There were also massive extinction events (Wilson and Williams, 1992), resulting in a secondary confinement of amiids and gar-pikes to North America. The long geological history of many contemporary ecomorphological types of fishes proves that the utilization of their contemporary adaptive zones is old. Interspecific competition in contemporary systems, hence, can at best have had only a partial or secondary role in fashioning their niches. The point is self-evident but warrants stressing.

In both North America and Europe two semi-distinct latitudinally distributed fish faunas occur. There is a far northern one dominated by salmonids and coregonids inhabiting Arctic and subarctic regions with, in Europe, isolated outliers in the alpine lakes to the south. A more southerly fauna, taxonomically diversified and very rich in species, extends over the remainder of the continent. A few of these southern elements penetrate the subarctic to a marginal extent. North American ichthyologists have long recognized the same two contrasting categories of fish, as physiological cold-water versus warm-water forms (Hubbs and Lagler, 1958). In the southern Ontario sites discussed here, cold-adapted forms like the yellow perch (*Perca flavescens*) and northern pike (*Esox lucius*) are at the southern edge of their range where they occur with a majority of warm-water species of Mississippi and Missouri basins origin that are near their northern range limits (Keast, 1978a). The cold-water species feed in winter, but warm-water species either do not feed then or crop only minor amounts of prey at rare intervals (Keast, 1968). Within these communities of mixed origins there are, hence, some seasonal differences in feeding and growth times and opportunities.

Fish distributions in both the Old and New Worlds were severely disrupted by glaciation during the Pleistocene, which led to wholesale geographical displacements. In North America, each time the ice sheet advanced the fishes were displaced southwards, and with each northwards retreat of the ice the fish recolonized northwards. Since the major river system of the continent, the Mississippi-Missouri, drains southwards, the dispersive pathway was oriented north-south. In Europe, by contrast, ice cover was restricted to the north and northwest, and to the Alps and other high mountains oriented generally east-west. There was an east-west component to dispersal, but peri-glacial conditions selecting for cold-temperature adaptedness must have covered a wide area.

B. Community Composition and Species Richness

In contemporary North America, the Mississippi-Missouri region is faunistically the richest in species, with many endemics amongst the 31 families, 69 genera, and 375 species. High levels of species richness and endemism occur also in the eastern and southeastern Atlantic rivers, in the Rio Grande Basin, and in the West (i.e., California and Nevada; see chapters in Mayden, 1992). The Great Lakes basin, within which the fish associations discussed in this chapter lie, has 27 families, 67 genera and 168 species. Since the area was ice-covered in times of glacial maxima, the fauna is more recent and has been secondarily derived; there apparently has not been time for endemics to arise. In Europe, the Danube basin, which remained largely intact as a fish habitat during the Pleistocene, today has the richest aquatic fauna on the continent (Bănărescu, 1991). Other regions to the east, which also endured minor disruption during the glacials, are likewise major centers of currently high diversity, and they include the Balkans, the eastern Black Sea and Caspian Basins, Anatolia, and the Tigris-Euphrates region.

The contemporary fish fauna of North America is dominated by cyprinids, salmonids, catastomids, centrarchids, etheostomine darters, and ictalurid catfishes; in the Palearctic the first two taxa, along with gobiids, cobitids, and herrings, are dominant. Cyprinids are small-bodied species, mostly planktivores and small invertebrate feeders, that commonly show marked habitat specialization; they are dominant on both continents, though proportionately more important in Europe. Small lakes on both continents characteristically contain 10–20 fish species; in European lakes, characteristically, are found the dominant cyprinids, two to three percids and salmonids, and an eel, pike, gadidid, and gobiid (e.g., Herzig-Straschil, 1989; Ritterbusch-Nauwerck, 1991). Small North American temperate water bodies usually are more diversified taxonomically, with centrarchids, cyprinids, percids, and ictalurid catfishes, as well as representatives of many additional small families.

II. THE SEASONAL NATURE OF NORTHERN FRESHWATER FISH COMMUNITIES

Three special features characterize the northern fish communities studied here: a) their recency, with a maximum post-glacial age of 10,000–8,000 years, b) their highly seasonal nature, and, as a result of the latter, c) the co-occurrence of distinct size cohorts in most species because growth is limited to 2–3 months a year. All three factors have major implications for species and community ecology.

A. Climatic Limitations

In Ontario ice cover on the lakes extends for 5–7 months; in the lakes of interest here, from mid- to late December until mid- to late April. Bodily processes in invertebrates and fish alike are greatly slowed in winter. For example, the common amphipod *Hyallela azteca*, a prominent fish food item in summer, takes 40 days to complete an instar at 10°C and 4 days at 25°C (Cooper, 1965). Fish feeding in general is minimal to absent at temperatures near zero (Keast, 1968, 1985b). Thus, there is characteristically a window of only 4–6 months of warm temperatures, within which all major annual activities must be completed. The northern lake systems, however, have an extraordinarily high spring and summer productivity, and a great diversity of prey organisms proliferate as temperatures warm. Thereafter, high invertebrate population turnover rates are maintained all summer.

B. The Annual Cycle in Lake Fishes

Figure 1 documents the annual activity cycle (feeding, reproduction, growth) in the common bluegill (*Lepomis macrochirus*) in Lake Opinicon (surface area, 890 ha; maximum depth, 11 m). Its cycle is representative of that of most other species in the lake. Individuals move into the littoral shallows in mid-May as the waters reach 15–18°C. Their bodies are initially somewhat emaciated, with body fat and nitrogen levels low (A. Keast, unpublished data). Accelerated feeding begins on the proliferating biomasses of chironomid larvae, amphipods, *Bosmina longirostris*, and other organisms reaching their maximum size then, and by late May the pre-winter body "condition factor" (the ratio of weight/length × 100) has been restored (Booth and Keast, 1986). Growth (in terms of body length increases) begins in early June, and maximum body fat levels are restored by late June. Spawning, of which there are three to four bouts at 7- to 10-day intervals, extends from early or mid-June to mid-July and produces huge batches of larval progeny. Growth ceases in both old and young-of-year individuals in late August (Keast and Eadie, 1984), and the fish return to the deeper water in September and October as inshore temperatures fall.

The annual cycle of the winter-feeding yellow perch contrasts to this scenario

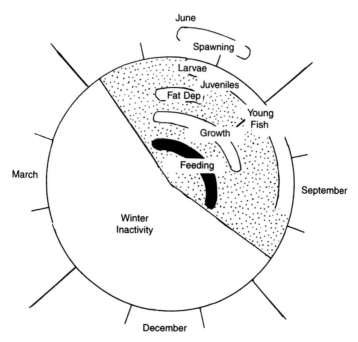

FIGURE 1. Annual cycle of the bluegill, *Lepomis macrochirus*, a common centrarchid and post-glacial colonizer of Ontario from the Mississippi refuge. As in comparable species all major phases are compressed into about five months.

in that gonad production is largely completed during the previous summer and autumn (Keast, 1978a), not in spring. This allows for a massive production of eggs over a brief 7–10 period, beginning several days after ice melt.

C. Timing of Reproduction

One feature mediating the co-occurrence of many fish species is that spawning is, at least in part, seasonally sequential for species [Fig. 2 for Jones Creek and Lake Opinicon data in Keast (1978a)]. In Jones Creek the piscivorous grass pickerel (*Esox vermiculatus*), produces larvae in April, followed by the mud minnow (*Umbra limi*) and a sequence of other taxa. This sequencing is important in that the newly hatched young of all species are limited in prey size by their small mouths to specific prey categories, such as the 0.10- to 0.15-mm-long copepod nauplii, copepods, and *Bosmina longirostris* (Keast, 1980). The sequencing of hatching spreads harvesting intensity; by an age of 8–10 days the

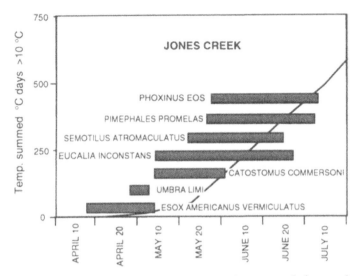

FIGURE 2.   Sequential spawning in the fishes of Jones Creek, as is typical of many other northern systems, related to summed temperatures above 10°C. This temperature is used as a baseline because most fish species are physiologically inactive below this temperature.

young of most species have shifted to larger prey items, and no longer compete with the later hatchlings. Piscivores are early spawners and, by an age of 3–4 weeks, are large enough to feed on the newly hatched larvae of their major prey species (Forney, 1976; Keast, 1985c). The staggering of spawning times in different fish presumably reflects physiological adjustment to a variety of optimum temperatures (see temperature summation line in Fig. 2.)

III. THE FOOD NICHE AND RESOURCE DIVISION IN NORTHERN FISH COMMUNITIES

A. Trophic Opportunities for Stream Fishes

Some typical patterns representative of resource use over the seasons by different fish species may be illustrated by the community in Jones Creek. This stream enters the St. Lawrence River near Brockville, and consists of series of weeded ponds separated by narrow sections and, 0.5 km downstream, expansive open-water pools. Twelve fish species in 11 genera and seven families make up the fauna. Small-bodied species predominate, with the long-lived white sucker (*Catostomus commersonnii*), pumpkinseed sunfish (*Lepomis gibbosus*), brown bullhead (*Ictalurus nebulosus*), and creek chub (*Semotilus atromaculatus*) being represented by multiple size cohorts. Population sizes and community structure were determined monthly, from May to October, by the "removal-count-restore" method during two consecutive summers, by repeated netting in six individual ponds of each type, after blocking off the ponds.

Species occurrences and abundances are plotted in Figure 3, from which it is clear that the weeded and open-water ponds supported different species associations. Creek chub, dace, fathead minnow (*Pimephales promelas*), and brook stickleback (*Eucalia inconstans*) were largely confined to the former; golden shiner (*Notemigonus crysoleucas*), grass pickerel, and pumpkinseed, to the latter

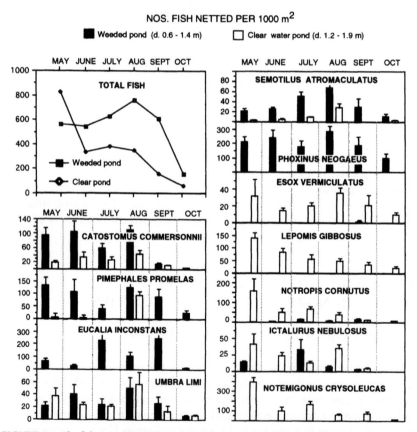

FIGURE 3. The fish assemblages for weedy and open-water ponds are compared for Jones Creek as derived by the removal-count-restore method and averaged as numbers per 1000 m². Note seasonal shifts in abundances and withdrawal of most fish before winter.

(Fig. 3). Total fish numbers were highest in May when populations of the golden shiner, common shiner (*Notropis cornutus*), and pumpkinseed sunfish peaked. The summer associations were relatively stable, with all species well-represented. A sharp drop in abundance occurred in autumn when some species moved downstream before the upper reaches of the creek froze.

B. Diet Differences among Species

Diets are expressed as pie diagrams in Figures 4 and 5, with pie "slices" representing percentage total wet weight and volume of each prey type. An analysis of the two communities, in clear-water and weeded ponds, showed that:

(1) Each species had an identifiable "food niche" (Elton, 1958).

(2) There were two predominantly algal feeding species (redbelly and finescale dace, *Phoxinus eos* and *P. neogaeus*); two detritus feeders (fathead minnow and white sucker); a chironomid larva feeder (brook stickleback); a part-time planktivore (golden shiner); four generalized invertebrate predators (mud minnow, common shiner, pumpkinseed sunfish, brown bullhead) that took isopods, amphipods, molluscs, Coleoptera, small odonates; and molluscs; and two piscivores (grass pickerel and creek chub).

(3) Diet specialists were represented in the communities (e.g., herbivorous dace and fathead minnow; piscivorous grass pickerel), as were opportunistic feeders (i.e., most other species).

(4) Diet generalists like the brown bullhead, mud minnow, and smaller creek chub consumed up to 10–12 taxonomic categories of prey (≥ 3% by weight) per month, but the predominant prey types and their proportions in the diet varied by species. They also varied somewhat seasonally. For example, the diet of 51- to 70-mm common shiner in May was dominated by chironomid pupae and adults, and ostracods; in June, by algae and beetles; in July, by algae and mayflies; in August (when young-of-the-year fish became dominant), by Cladocera; in September, by adult diptera, beetles, and algae; and in October, by beetles.

(5) Age-class diets differed only slightly in the brook stickleback, which grew to a mere 50 mm; moderately in the longer-lived mud minnow; and strikingly in the creek chub. This species is a small invertebrate predator when young, but subsequently it becomes a piscivore.

(6) The largely herbivorous species (white sucker and finescale dace) took some animal food in May, probably to satisfy a post-winter protein need.

(7) Some prey organisms (chironomids and amphipods) were cropped seasonally by several species, resulting in increased levels of diet overlaps between species at such times.

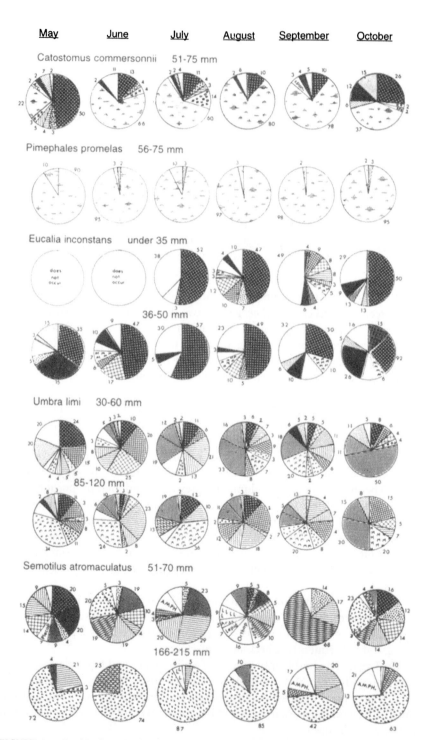

FIGURE 4. Monthly diets, based on samples of 60–90 individuals for the Jones Creek assemblage, May to October (derived, with modification, from figures in Keast, 1966). Sizes of the pie slices represent the percentage wet weight of the particular prey type consumed. For identifications of invertebrate prey types see key on page 60.

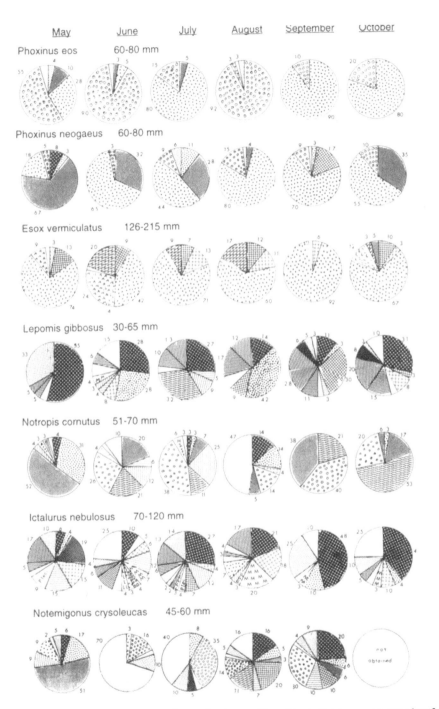

FIGURE 5. Monthly diets of Jones Creek assemblage, May to October, part 2. For identification of invertebrate prey types see key on page 60.

59

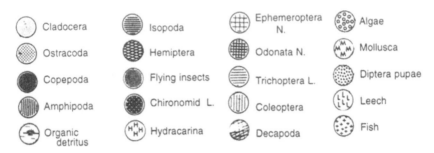

KEY: Invertebrate prey types (see Figures 4 and 5).

With respect to overlap in diet among species, it is of interest to determine whether increased diet overlap occurred because food shortages forced several species to one or two food types, or because the consumers were simultaneously attracted to newly abundant resources. Figure 6 compares the frequencies of six prey categories in the benthos with those in total fish stomachs on a month-by-month basis (with some size cohorts not illustrated). Prey taken is estimated by determining prey numbers per stomach for each category of fish, multiplying by numbers of fish, and adding consumption by all species to arrive at the totals shown.

Ostracod numbers peaked in August and September; the highest consumption by fish was in May and August. Pumpkinseed sunfish took them mainly in June and August, common shiner in May and August, and golden shiner in May. Amphipod numbers peaked in the environment in July and October, and fish consumption, mainly due to increasing predation by mud minnow, pumpkinseed, and brown bullhead, also was highest in July and August.

Chironomid larvae (see Diptera L. & P.) show a low pre-emergence peak in May, and a marked peak from August through October reflecting the massive hatches of the late summer generation. Total consumption by fish was highest in May and again in August; a significant part of the diets of white sucker, mud minnow, and creek chub in May, and of common and golden shiners, brown bullhead, and pumpkinseed sunfish in late summer, was chironomids.

Zygoptera nymphs were abundant in June and again from August through October with hatching of the young-of-the-year. Their consumption by fish was highest in May, August, and September. Pumpkinseed sunfish harvested them in August and September; golden shiner and stickleback, in August; and mud minnow, in June, August, and October.

Trichoptera larvae numbers peaked in May, July, and October. Harvesting by fish was greatest in May and July. Bullhead took them mainly in May and June; mud minnow, in July, August, and September; and creek chub, in May, July, and September. The numbers of Mollusca both in the benthos and in fish stomachs

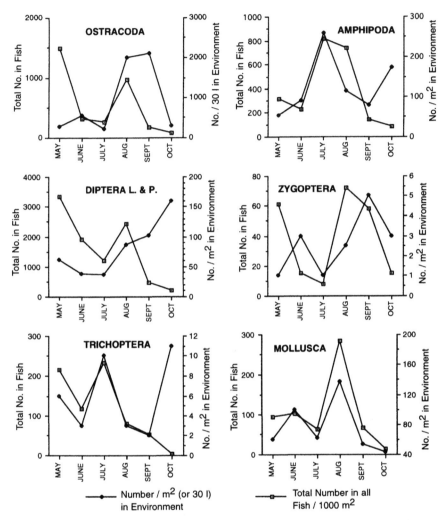

FIGURE 6.   Invertebrate abundances in the environment from May through October for six major fish prey types, relative to total cropping by all fish (see text). High numbers of a resource are commonly correlated with increased feeding.

peaked in June and August; however, the mud minnow, a mollusc specialist, took them all summer.

From Figure 6, it is apparent that increased overlap in diet occurred when fish were drawn simultaneously to an abundant resource rather than being forced in unison to share a declining prey base. The kinds of feeding specializations and resource utilization patterns documented here have been found to apply generally in northern systems, for example, in lakes of contrasting trophic status (Complak, 1982), northern Ontario lakes (Eadie, 1982), beaver ponds (Keast and Fox, 1990; Snetsinger, 1992), and contrasting stream types (Bankey, 1979; Payson, 1982).

C. Implications of Diet Differences among Size Cohorts to the Species Food Niche Concept

In lake communities, fish biomass is dominated by longer-lived species, of which a high proportion have multiple size cohorts. These size cohorts invariably differ in diet, if only by prey proportions. Diet changes with age can be illustrated in two co-occurring centrarchids, the bluegill and rock bass (*Ambloplites rupestris*) (Keast, 1977a, 1978b). Mean year-to-year body sizes for the two species are: (1) Year 0 (end of first summer—age 2 and 3 months)—average lengths 42 and 48 mm, respectively; (2) Year I (age 14 and 15 months)—59 and 66 mm; (3) Year II, 80 and 85; and (4) Year VI, 132 and 138 mm. There is a corresponding increase in mouth width, especially in the large-mouthed rock bass.

In the bluegill, diet changes progressively with age, entailing shifts in the proportions taken from the basic prey set of mainly Cladocera, chironomids, amphipods, and Trichoptera larvae. In the rock bass, by contrast, three complete diet changes occur. Year 0–1 fish take mainly chironomids and amphipods, fish in Years II–III eat mainly Anisoptera nymphs, and in Years VI–VIII, crayfish and fish. Three distinct "Eltonian" food niches are thus utilized.

The implications of such multiple diets to fish communities are discussed in Keast (1977a). One consequence is that the number of "tropic species" exceeds the number of taxonomic species; in Lake Opinicon, for example, 19 fish species corresponded to about 30 "trophic species." Polis (1984) has also discussed multiple food niches in fish.

Interspecific diet relationships are thus much more complicated in fish than in homeotherms (or in any taxon, such as birds or mammals, in which individuals spend most of their lives at a constant size). In this event, interspecific food competition can occur between adults, between equivalent age classes of different species, or between an older age class of one species and a younger age group of another. Age-equivalent and cross-age diet overlaps between species have been determined as commonly reaching 0.4–0.6 (scale of 0–1; Levins, 1968; see also Keast, 1977a,b, 1978a). They also fluctuate seasonally. It is important to provide a visual example of this relative to trophic relationships.

The members of a typical littoral zone fish community, in terms of coexisting species and size ranges, are illustrated in Figure 7 for Birch Bay, Lake Opinicon,

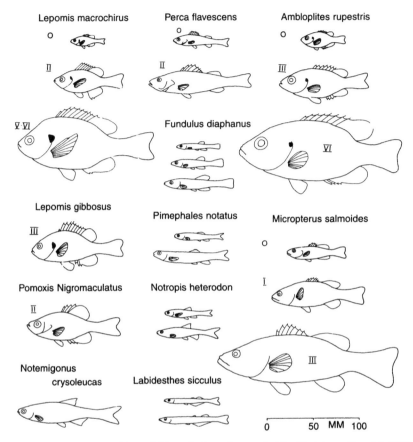

Lepomis macrochirus

Perca flavescens

Ambloplites rupestris

Fundulus diaphanus

Lepomis gibbosus

Pimephales notatus

Micropterus salmoides

Pomoxis Nigromaculatus

Notropis heterodon

Notemigonus crysoleucas

Labidesthes sicculus

0   50   MM 100

FIGURE 7. A typical assemblage of fish in the littoral zone of a lake (Opinicon, Ontario), 1965–1990. Co-occurring are small-bodied species and series of age (size) cohorts of large-bodied species. Year 0 (first summer) bluegill, yellow perch, and other species only became members of the assemblage about 1992, following siltation and marked macrophyte development.

adjacent to the Queen's University Biological Station. Small-bodied species are present, along with various series of age (size) cohorts of larger-bodied species. Year 0 (first summer) bluegill, yellow perch, and other species became more prominent members of this assemblage for the first time about 1992, following siltation as a result of increased development of macrophytes in the littoral. Dietary similarity relationships in the community, by percent composition of food items, are expressed by a cluster diagram in Figure 8. The diet data were previously published in Keast (1965, 1978a); here the spring (May) and summer (July) data are tallied separately. For methodological information on cluster diagrams, see Pielou (1984), Jain et al. (1986), and Jackson et al. (1989). Similar

Allen Keast

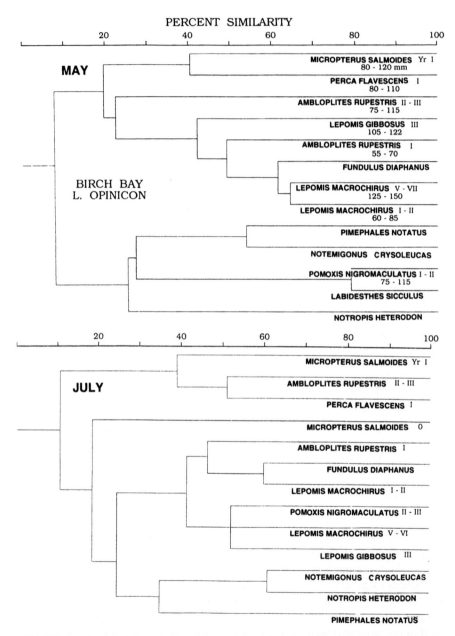

FIGURE 8. Dietary relationships (percentage overlaps, based on wet weight/volume) of species and size cohorts in Birch Bay, Lake Opinicon (Fig. 7), expressed as cluster diagrams. Spring and summer are compared. In longer-lived species the dietary relationships commonly change with age and season.

basic groupings apply in both months; there are primary separations of the semi-piscivores, largemouth bass and yellow perch year I, at the top of the figures, and of the planktivores (blunt-nose minnow (*Pimephales notatus*), golden shiner, small black crappie (*Pomoxis nigromaculatus*), brook silverside (*Labidesthes sicculus*), and blackchin shiner (*Notropis heterodon*) at the bottom.

A particular interest, however, lies in the various ways in which the different age classes of species cluster relative to each other and how this changes seasonally. In May, large and small bluegill and the small-bodied killifish shared relatively similar diets (ca. 65%), but in July the diet of the larger bluegills was closest to those of black crappie and pumpkinseed sunfish. In May, the black crappie clustered with the small-bodied planktivores, especially the silverside (ca. 80%), and pumpkinseeds were considerably removed from the bluegills earlier in the season. The diet of yellow perch was most similar (at 40%) to that of the largemouth bass in May but to Years II–III rock bass in July (ca. 50%). Even amongst planktivores, such as blunt-nose minnow and blackchin shiner, diet similarity levels may vary seasonally; between May and July, this variation was due to different levels of predation on two common zooplankters, *Bosmina longirostris* and *Chydorus sphaericus,* and to different degrees of insectivory in these species.

IV. HABITAT SEPARATION: ASSEMBLAGES AND SPECIES

As in terrestrial vertebrate systems (e.g., Cody, 1985), fish communities classify and species group according to habitat. The basic features of aquatic habitats are, of course, quite different from those of the land and include factors such as water flow, depth, substrate, and emergent vegetation. Streams, in the course of their length, provide a series of distinct habitats: entrant headwaters, ponds, gravel riffles, and embayments near the mouth, each with a different fish assemblage (e.g., Payson, 1982). Within lakes, the littoral and limnetic zones support distinct communities, and between lakes fish assemblages may differ according to limnological characteristics and the "trophic age" of the system (Colby *et al.,* 1972; Tonn and Magnuson, 1982).

In lakes such as Opinicon, with extensive, well-oxygenated and highly productive littoral zones, 80% of species present are shallow-water dwellers, and fish diversity drops rapidly beyond depths of 3.0 m (Keast and Harker, 1977). Within the littoral zone there is much substrate differentiation. Some species, like bluegill, pumpkinseed, and largemouth bass, are habitat generalists, and their distributions do not correspond to substrate differences (Keast *et al.,* 1978). On the other hand, other species are habitat specialists and segregate relative to substrate type. The blackchin shiner is restricted to inshore weedbeds; the log perch (*Percina caprodes*), to gravel shoals; and mud minnow, to areas with cattail stands. However, the young age-classes of most species are weedbed dwellers.

Limnetic species are either large-bodied, or they form schools and hence are relatively predator-free. They mainly feed from the water column. In Lake Opinicon the alewife (*Alosa pseudoharengus*), black crappie, golden shiner, and brook silverside exhibit both diel and seasonal movement patterns throughout the lake, and daily cycles may include nocturnal movements inshore (Keast and Fox, 1992).

Age classes within species commonly occupy different habitats (see Keast, 1978a, detailing this in Lake Opinicon fishes). Larval bluegill, for example, occur as loose aggregations in the inshore open waters; juveniles and smaller fish are confined to weedbeds, and larger fish occur throughout the lake, including the limnetic zone.

With habitat generalists found in a variety of sites within the lake, and variation in habitat preferences within species based on age differences, it is clearly difficult to regard fish assemblages as distinct communities. With movements of fish among habitats with season and age, the definition of the habitat and ecological niche is further complicated. Are habitat utilization patterns still in post-glacial flux, or are they relatively stabilized? The habitat descriptions for widespread fishes given in standard reference works (e.g., Scott and Crossman, 1973) indicate that in many species there is commonly a constant link with a specific habitat type throughout the species range. Such species, then, must have brought their habitat preferences north with their northerly post-glacial expansion. Are interspecific interactions a factor in habitat choice, influencing habitat utilization patterns? Such effects have been documented, as interference competition, by several workers (e.g., Werner, 1977; Werner and Hall, 1979, 1988; Persson, 1983, 1986; Persson *et al.*, 1992). Predators commonly affect the distributional occurrence of their prey, at least in the short term. There are various examples, especially from Europe, of the introduction of a second species to a single-species system leading subsequently to partial habitat segregation. Studies of the factors governing distributions over habitat should take into account the full suite of biological interactions among species, although this is rarely done.

V. HISTORICAL CHANGES IN THE LAKE ECOSYSTEM

In the last hundred years, human activity has drastically changed aquatic habitats in both North America and Europe, forcing levels of adjustment in their resident communities that probably sometimes exceed ice age effects. Eutrophication of lake systems is perhaps among the milder of these changes.

In Lake Opinicon, three major changes have occurred since my long-term studies began in 1960. About 1964 the alewife established itself, colonizing and capitalizing upon the under-utilized deep-water sections of the lake. Its effect on other lake fishes, hence, was minimal.

In 1975 an exotic macrophyte, the Eurasian milfoil *Myriophyllum spicatum,* colonized the lake. Within two years it formed dense walls of growth at 2.0–3.5 m depths along the edge of the littoral zone. A detailed study of its effects on fish has shown that the invasion disadvantaged larger-bodied fish of open water but increased available habitat for their cover-dependent juveniles (Keast, 1984). However, the *M. spicatum* beds had lower invertebrate productivity than the native macrophytes. In addition, the plant displaced bluegill from the zone of their preferred spawning depths. Notwithstanding these changes, no major deleterious effects on the fish of the lake have been recognized.

There was, however, an important indirect effect of this invasion in the slowing down of water movements along the shoreline. Increased silt deposition and decaying *Myriophyllum* changed some inshore areas, such as Birch Bay, from an expanse of sandy shallows to weedbed. The fish community that had been stable through successive surveys since 1965 suddenly changed. This is reflected in Fig. 9 for two species, bluegill and yellow perch. Through 1978, the smallest age-classes of these fish were absent or hardly present in the bay,

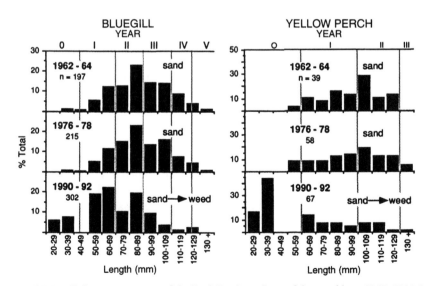

FIGURE 9. Shifts in composition of the Birch Bay littoral zone fish assemblage, 1965–1994. In the middle 1970s, invasion of the lake by the Eurasian milfoil, *Myriophyllum spicatum,* with subsequent impediment to waterflow and increased siltation, changed this sandy bay to a weeded one. It then became a major habitat of cover-dependent Year 0 fish.

but from 1990 onward Year 0 fish of both these species, as well as of others, became prominent. Correspondingly, there was a proportionate withdrawal of the largest size-classes (Fig. 8) and of larger-bodied species with open-water preferences in general.

VI. DISCUSSION

The northern temperate freshwater fish faunas of the Nearctic and Palearctic include the descendants of lineages that emerged early in the Tertiary and have been able to utilize freshwater habitats on the two continents ever since. Both continental faunas had to withstand major distributional challenges from the Pleistocene glaciations. Today they occupy a highly seasonal environment that, for a major segment of species, restricts active life to 4–6 months a year. There has been a high priority on evolving adaptations to maximize on the opportunities and ameliorate the constraints of this time window.

The co-occurrence of multiple age (size) cohorts within species is a prominent aspect of these fish associations, and it contrasts them with tropical communities. In the tropics, growth is possible for much of the year, and hence there are not persistent or pulsed assemblages of young, small-bodied individuals; their ecological role at low latitudes is performed by additional, small-bodied species. Dietary differences as a function of age- or size-difference potentially reduce intraspecific diet overlaps in northern species. Ultimately, of course, the basis of size-selective predation is physiological; it is energetically advantageous for fish to constantly switch to prey of increasing size as they grow (Paloheimo and Dickey, 1968).

The seemingly complicated diet and habitat utilization patterns of these northern aquatic systems could not be sustained if they were not anchored in an environment in which the major biological events are repeated annually with considerable predictability. This applies particularly to the life cycles of the prey invertebrates, tied to increasing water temperatures following ice melt in the spring. Fish tend to consume predictably the same diet items at equivalent times from year to year (e.g. diet data on yellow perch, in Keast, 1977b).

The long-term data base on fish communities presented here has wide implications for such fundamental ecological concepts as the species niche (Elton, 1958); the controversy as to whether or not communities are structured (MacArthur, 1972; Cody, 1974; Wiens, 1988; Strong et al., 1984, and many others); the guild concept (Simberloff and Dayan, 1991); food webs (Pimm, 1982; Abrams, 1993; Polis, 1994); and trophic cascade models (e.g., Carpenter and Kitchell, 1993). These subjects cannot be explored here. Wiens (1977, 1988) has pointed out that interspecific competition could more effectively structure communities in closed systems that do not vary seasonally.

Neither of these preconditions is met here. Other workers, of course, have thought of community structure in a quite different context, by the linkage of species through common environmental needs (Jackson et al., 1989, 1992).

How are new species incorporated into pre-existent systems? For island birds, Diamond (1975) argued that newcomers can only establish if they can "fit into" the system, i.e., if ecological opportunities are under-exploited. The lake community discussed previously presumably retains the capacity to absorb newcomers such as alewife and *Myriophyllum*. On the other hand, there is also surely an upper limit to, for example, the number of specialized piscivore species that a system can absorb.

Are these northern aquatic systems immature and unstable? Cody (1976) attributed his inability to find clearcut habitat separations in European sylviid warblers to there having been no time, post-glacially, for these to evolve. Obviously evolving systems can, at least temporarily, absorb additional species. Presumably the fishes studied here colonized north as loose species clusters that together already filled all the major adaptive zones. To this extent the systems could well be "mature."

VII. SUMMARY

The northern temperate freshwater fish associations have progressively evolved through the later Tertiary, and some of the ecomorphological types dominating present-day faunas are verifiably old. The fish today occupy a highly seasonal environment necessitating, for many species, compression of the annual cycle into a spring–summer window some 4–6 months long. Summer environments are, however, very productive, and the species represent a wide range of trophic types, with each utilizing a certain, preferred range of prey.

Some of the major features of northern temperate freshwater fish assemblages are presented in this chapter based on two decades of studies on the fish communities of lakes and streams in eastern Ontario; the seasonal nature of these habitats, and adaptations to the seasonal environment, receives particular attention. Associated with the brief growing season, many species have multiple size cohorts that commonly differ in resource partitioning, in terms of diets and the use of space, so that some species occupy distinct niches at different ages. Special attention is given to the partitioning of available food resources among species and their various age classes. The patterns of resource utilization between species and between age classes of species are complex, but they occur in environments that are both predictable and annually repetitive. Interspecies trophic interactions cannot be the major molding force in the structuring of these fish communities.

Acknowledgments

This research was carried out while the writer held grants from the Canadian National Science and Engineering Research Council, and I should like to thank this body very much for its support. The work would not have been possible if not for the facilities provided by the Queen's University Biological Station, Chaffey's Locks, Ontario. I would like to thank the Director, Prof. Raleigh Robertson, and Manager, Mr. Frank Phelan, for the use of these facilities. Many students participated in the longer term program. Laura Pearce constructed Fig. 3, 6, and 9. Finally I would like to thank Denise Cameron for typing the manuscript.

References

Abrams, P. A. (1993). Effects of increased productivity on the abundance of trophic levels. *Am. Nat.* **141**, 351–371.

Bănărescu, P. (1991). "Zoogeography of Fresh Waters. 2. Distribution and Dispersal of Freshwater Animals in North America and Europe." Aula-Verlag, Wiesbaden.

Bankey, T. A. (1979). The feeding relationships of cohabiting fish in a Lake Ontario stream. Hons. B.Sc. Thesis, Queen's University, Kingston, Ontario.

Booth, D. J., and Keast, A. (1986). Growth energy partitioning by juvenile bluegill sunfish, *Lepomis macrochirus* Rafinesque. *J. Fish Biol.* **28**, 37–45.

Brooks, D. R., and McLennan, D. A. (1992). Historical ecology as a research program. *In* "Systematics, Historical Ecology, and North American Freshwater Fishes" (R. L. Mayden, ed.), pp. 76–113. Stanford Univ. Press, Stanford, CA.

Brooks, D. R., and McLennan, D. A. (1994). Historical ecology as a research program: Scope, limitations, and future. *In* "Phylogenetics and Ecology," pp. 1–27. Linnean Society of London.

Carpenter, S. R., and Kitchell, J. F., eds. (1993). "The Trophic Cascade in Lakes." Cambridge Univ. Press, Cambridge, UK.

Cody, M. L. (1974). "Competition and the Structure of Bird Communities." Princeton Univ. Press, Princeton, NJ.

Cody, M. L. (1976). Habitat selection and interspecific interactions among Mediterranean sylviid warblers. *Oikos* **27**, 210–238.

Cody, M. L., ed. (1985). "Habitat Selection in Birds." Academic Press, San Diego, CA.

Colby, P. J., Spangler, G. R., Hurley, D. A., and McCombie, A. M. (1972). Effects of eutrophication on salmonid communities in oligotrophic lakes. *J. Fish. Res. Board Can.* **29**, 975–983.

Complak, J. A. (1982). Relationship between lake type and the performance (growth, body condition, feeding patterns) of four littoral zone fish species. M.Sc. Thesis, Queen's University, Kingston, Ontario.

Cooper, W. E. (1965). Dynamics and production of a natural population of the fresh-water amphipod, *Hyalella azteca. Ecol. Monogr.* **35**, 377–394.

Diamond, J. M. (1975). Assembly of species communities. *In* "Ecology and Evolution of Communities" (M. L. Cody and J. M. Diamond, eds.), pp. 342–444. Harvard Univ. Press, Cambridge, MA.

Eadie, J. M. (1982). Fish community structure in northern lakes: The role of competition, resource diversity, and environmental variability. M.Sc. Thesis, Queen's University, Kingston, Ontario.

Elton, C. (1958). "The Ecology of Invasions by Plants and Animals." Methuen, London.

Forney, J. L. (1976). Year-class formation in the walleye (*Stizostedion vitreum vitreum*) population of Oneida Lake, New York, 1966–73. *J. Fish. Res. Board Can.* **33**, 783–792.

Herzig-Straschil, B. (1989). Die entwicklung der fischfauna des Neusiedler Sees. *Vogelschutz Oesterr.* **3**, 19–22.

Hocutt, C., and Wiley, E. O., eds. (1986). "The Zoogeography of North American Freshwater Fishes." Wiley, New York.

Holčik, J., ed. (1989). "The Freshwater Fishes of Europe," Vol. 1, Part II. Aula-Verlag, Wiesbaden.

Hubbs, C. L., and Lagler, K. F. (1958). "Fishes of the Great Lakes Region." Univ. of Michigan Press, Ann Arbor.

Jackson, D. A., Somers, J. M., and Harvey, H. H. (1989). Similarity coefficients: Measures of co-occurrence and association or simply measures of occurrence. *Am. Nat.* **133**, 436–453.

Jackson, D. A., Somers, K. M., and Harvey, H. H. (1992). Null models and fish communities: Evidence of non-random patterns. *Am. Nat.* **139**, 930–951.

Jain, N. C., Indrayan, A., and Goel, L. R. (1986). Monte Carlo comparisons of six hierarchical clustering methods on random data. *Pattern Recognition* **19**, 95–99.

Keast, A. (1965). Resource subdivision amongst coexisting fish species in a bay, Lake Opinicon, Ontario. *Great Lakes Inst. Proc.* **13**, 106–132.

Keast, A. (1966). Trophic interrelationships in the fish fauna of a small stream. *Great Lakes Inst. Proc.* **15**, 51–79.

Keast, A. (1968). Feeding of some Great Lakes fishes at low temperatures. *J. Fish. Res. Board Can.* **25**, 1199–1218.

Keast, A. (1977a). Mechanisms expanding niche width and minimizing intraspecific competition in two centrarchid fishes. *Evol. Biol.* **10**, 333–395.

Keast, A. (1977b). Feeding and food overlaps between the year classes relative to the resource base, in the Yellow Perch, *Perca flavescens*. *Environ. Biol. Fishes* **2**, 55–70.

Keast, A. (1978a). Trophic and spatial interrelationships in the fish species of an Ontario temperate lake. *Environ. Biol. Fishes* **3**, 7–31.

Keast, A. (1978b). Feeding interrelationships between age groups of Pumpkinseed (*Lepomis gibbosus*) and comparisons with Bluegill (*L. macrochirus*). *J. Fish. Res. Board Can.* **35**, 12–27.

Keast, A. (1980). Food and feeding relationships of young fish in the first weeks after the beginning of exogenous feeding in Lake Opinicon, Ontario. *Environ. Biol. Fishes* **5**(4), 305–313.

Keast, A. (1984). The introduced aquatic macrophyte, *Myriophyllum spicatum*, as habitat for fish and their invertebrate prey. *Can. J. Zool.* **62**, 1289–1303.

Keast, A. (1985a). Implications of chemosensory feeding in catfishes: An analysis of the diets of *Ictalurus nebulosus* and *I. natalis*. *Can. J. Zool.* **63**, 590–602.

Keast, A. (1985b). Growth responses of the brown bullhead to temperature. *Can. J. Zool.* **63**, 1510–1515.

Keast, A. (1985c). The piscivore feeding guild of fishes in small freshwater ecosystems. *Environ. Biol. Fishes* **12**, 119–129.

Keast, A., and Eadie, J. (1984). Growth in the first summer of life: A comparison of nine co-occurring fish species. *Can. J. Zool.* **62**, 1242–1250.

Keast, A., and Fox, M. G. (1990). Fish community structure, spatial distribution and feeding ecology in a beaver pond. *Environ. Biol. Fishes* **27**, 201–214.

Keast, A., and Fox, M. G. (1992). Space and resource use patterns in an offshore open-water lake fish community. *Environ. Biol. Fishes* **34**, 159–170.

Keast, A., and Harker, J. (1977). Fish distribution and benthic invertebrate biomass relative to depth in an Ontario lake. *Environ. Biol. Fishes* **3**, 235–240.

Keast, A., Harker, J., and Turnbull, D. (1978). Nearshore fish habitat utilization and species association in Lake Opinicon (Ontario, Canada). *Environ. Biol. Fishes* **3**, 173–184.

Levins, R. (1968). "Evolution in Changing Environments." Princeton Univ. Press, Princeton, NJ.

MacArthur, R. J. (1972). "Geographical Ecology." Harper & Row, New York.

Mayden, R. L., ed. (1992). "Systematics, Historical Ecology, and North American Freshwater Fishes." Stanford Univ. Press, Stanford, CA.

Paloheimo, J. E., and Dickie, L. E. (1968). Food and growth of fishes. III. Relations among food, body size, and growth efficiency. *J. Fish. Res. Board Can.* 23, 1209–1248.

Payson, P. D. (1982). Fish assemblages in Medway Creek relative to prey resource background and habitat availability: Are the assemblages structured? M.Sc. Thesis, Queen's University, Kingston, Ontario.

Persson, L. (1983). Effects of intra- and inter-specific competition on dynamics and size structure of a perch (*Perca fluviatilis*) and a roach (*Rutilus rutilus*) population. *Oikos* 42, 126–132.

Persson, L. (1986). Effects of reduced interspecific competition on resource utilization in perch (*Perca fluviatilis*). *Ecology;* 67, 355–364.

Persson, L., Diehl, S., Johansson, L., Andersson, G., and Hamrin, S. (1992). Trophic interactions in temperature lake ecosystems—a test of food chain theory. *Am. Nat.* 140, 59–84.

Pielou, E. C. (1984). "The Interpretation of Ecological Data: A Primer on Classification and Ordination." Wiley, New York.

Pimm, S. L. (1982). "Food Webs." Chapman & Hall, New York.

Polis, G. A. (1984). Age structure component of niche width and intraspecific resource partitioning by predators: Can age groups function as ecological species. *Am. Nat.* 123, 541–564.

Polis, G. A. (1994). Food webs, trophic cascades and community structure. *Aust. J. Ecol.* 19, 121–136.

Ricklefs, R. E. (1987). Community diversity: Relative roles of local and regional processes. *Science* 235, 167–171.

Ritterbusch-Nauwerck, B. (1991). Die beschaffenheit des Mondseeufers und seine bedeutung fur die fishchfauna. *Oesterr. Fishcherei* 44, 100–104.

Scott, W. B., and Crossman, E. J. (1973). "Freshwater Fishes of Canada." Fisheries Research Board of Canada, Ottawa.

Simberloff, D., and Dayan, T. (1991). The guild concept and the structure of ecological communities. *Annu. Rev. Ecol. Syst.* 22, 115–143.

Snetsinger, M. A. (1992). Resource division in a pond impoundment fish community with comparisons to other small water body systems. M.Sc. Thesis, Queen's University, Kingston, Ontario.

Strong, D. R., Jr., Simberloff, D., Abele, L. G., and Thistle, A. B., eds. (1984). "Ecological Communities: Conceptual Issues and the Evidence." Princeton Univ. Press, Princeton, NJ.

Tonn, W. M., and Magnuson, J. J. (1982). Patterns in the species composition and richness of fish assemblages in northern Wisconsin lakes. *Ecology* 63, 1149–1166.

Wainwright, P. C., and Lauder, G. V. (1992). The evolution of feeding biology in Sunfishes (Centrarchidae). *In* "Systematics, Historical Ecology, and North American Freshwater Fishes" (R. L. Mayden, ed.), pp. 472–491. Stanford Univ. Press, Stanford, CA.

Werner, E. E. (1977). Species packing and niche complementry in three sunfishes. *Am. Nat.* 111, 554–578.

Werner, E. E., and Hall, D. J. (1979). Foraging efficiency and habitat switching in competing sunfishes. *Ecology* 60, 256–264.

Werner, E. E., and Hall, D. J. (1988). Ontogenetic habitat shifts in the bluegill: The foraging rate predation risk trade-off. *Ecology* 69, 1352–1366.

Werner, E. E., Hall, D. T., Laughlin, D. R., Wagner, D. J., Wilsmann, L. A., and Funk, F. C. (1977). Habitat partitioning in a freshwater fish community. *J. Fish. Res. Board Can.* 34, 360–370.

Wiens, J. A. (1977). On competition and variable environments. *Am. Sci.* 65, 590–597.

Wiens, J. A. (1988). "The Ecology of Bird Communities 1." Cambridge Univ. Press, Cambridge, UK.

Wilson, M. V. H., and Williams, R. R. G. (1992). Phylogenetic, biogeographic, and ecological significance of early fossil records of North American freshwater teleostean fishes. *In* "Systematics, Historical Ecology, and North American Freshwater Fishes" (R. L. Mayden, ed.), pp. 227–247. Stanford Univ. Press, Stanford, CA.

CHAPTER 4

Structure and Dynamics of Reef Fish Communities
A Biogeographical Comparison

PETER F. SALE

Department of Biological Sciences, University of Windsor,
Windsor, Ontario N9B 3P4, Canada

I. INTRODUCTION

Tropical coral reefs support diverse assemblages of fishes in shallow, clear, warm water, in close association with spatially complex substrata. Study of reef fishes is perhaps easier for the ecologist than is the case for any other fish assemblage. The environment lends itself to direct observation of the fishes, most of which conduct their lives on relatively small spatial scales (10s of meters) and are scarcely impeded by the presence of a diver. In addition, manipulations of the fishes or their physical habitat are readily accomplished, permitting direct experimental tests of hypotheses. In this regard, reef fishes may be superior to birds—the other group of vertebrates which has provided ecologists with such outstanding possibilities for the direct observation of ecology in progress. I sympathize with my colleagues who struggle with the secretive (or at least hidden) actions of small mammals, herpetiles, or all those fish of muddy estuaries.

The diversity of fish on coral reefs led early to an interest in the structure and dynamics of these assemblages. When assemblages routinely contain hundreds of species, certain questions quickly come to the fore: How are all these species assembled? How do they interact with each other? How, if at all, do these interactions influence assemblage structure, or the changes to that structure that occur through time? (More impoverished systems do not provoke such thoughts so frequently!)

The first efforts on this topic were launched from a framework of ideas in which current conditions were expected to be representative of long-term conditions, and explanations of species coexistence and relative abundance were couched in terms of niche partitioning and resultant steady-state equilibria. Inevitably, reef fish ecologists gradually changed their thinking in recognition of the variable dynamics which seem to characterize these systems (Sale, 1980). Elsewhere I have commented on how slowly we seemed to stumble upon what should have been obvious from the beginning, while noting that we have been no less astute than other ecologists in this regard; scientists are very good at seeing clearly whatever they expect to be present but less able to recognize the unexpected (Sale, 1988a).

Studies of assemblage structure and dynamics have used four main approaches. The first of these is the preferably quantitative description of structure at one place and time. Scientists using this approach frequently infer dynamics but cannot measure them directly. A second approach is to test specific hypotheses by means of field manipulations. These experiments can be very effective, but they can only test hypotheses that deal with small-scale processes which operate over relatively short time-scales. The third approach makes use of longer-term monitoring of conditions at a site by means of repeated field visits. This essentially descriptive approach provides direct evidence of dynamics as well as information on structure. The final approach is to compare, quantitatively and in detail, assemblage structure and/or dynamics in different locations, sometimes in biogeographically distinct regions. Often the locations compared are chosen because they differ in some way, and the study is reported as a "natural experiment." This approach can test hypotheses about larger-scale processes, or processes which act only over long periods of time, but the test is necessarily much weaker than in the case of a true manipulation.

In setting out these four approaches, I do not intend that they be seen as mutually exclusive nor the only possible approaches to use. Indeed, some of us have spent time modeling hypotheses generated from field experience as an additional approach (Sale, 1982; Doherty et al., 1985; Warner and Hughes,

1988), and many research programs have used combinations of these approaches. A "snapshot" description at one site and time frequently provides the data for a subsequent planned comparison among sites or for monitoring studies or manipulations. During a long-term monitoring study over several years, it is possible to conduct a series of field experiments testing hypotheses which the monitoring has helped to generate. One recent experiment (Doherty and Fowler, 1994) compared current age distributions for seven populations of the damselfish, *Pomacentrus moluccensis*, with predictions based on annual recruitment success at those locations. It was possible only because a precursor monitoring study had measured recruitment success anually for 10 years.

Despite this range of approaches and types of study, and the growing effort in this field, most studies of reef fish ecology remain localized to particular sites and of short duration—an inevitable consequence of the limits set by research funding policies and the logistic demands of field work in remote sites. An additional difficulty concerns the lack of standardized field methods used in this type of ecology because it is simply not true that every sampling procedure measures reef fish assemblages in the same way (Sale, 1980, 1991). Integrating data from separate studies is very difficult when each study has used different methods. My goal in this chapter is to show the value of longer-term studies and to advocate more deliberate comparisons among coral reef regions. In the following two sections I briefly review reef fish biogeography and the currently accepted interpretation of the structure and dynamics of typical reef fish assemblages. I then review what little is known about the differences between reef fish assemblages in the Caribbean and on the Great Barrier Reef. The chapter concludes with a look at an ongoing project which will permit a detailed comparison of Australian and Caribbean assemblages on a specific type of reef habitat. The conclusions emerging from this study may even turn upside down the old cliché about diversity and stability.

II. BIOGEOGRAPHIC PATTERNS

While reef fish assemblages can be characterized as diverse, there is an order of magnitude variation in total species richness from place to place (Sale, 1980). Globally, there are two regions of high species richness—the Southwest Pacific (bounded by the Philippines, Malaysia, Indonesia, and northern Australia) and the Caribbean (Briggs, 1974; Springer, 1982). As one travels away from either region in any direction, species richness falls until, in places like Easter Island, southern Japan, Lord Howe Island, or Bermuda, fish species richness is little greater than it is on temperate rocky shores, despite the existence of coral reefs and a fish fauna of reef-associated taxa. The Southwest Pacific (and indeed much of the broad Indo-West Pacific region) is notably richer than the Caribbean.

More than 2000 species of fish have been recorded in the Philippines alone compared to 600 species at best in equivalently sized regions of the Caribbean.

Regardless of their diversity, reef fish assemblages throughout the world share a common taxonomy at the levels of family and genus (Choat and Bellwood, 1991; Thresher, 1991). However, at the species level there is an almost complete separation between the Caribbean and the Pacific. While many species are more restricted in distribution, there are numerous species broadly distributed through the Indian and Pacific oceans, some ranging from East Africa and the Red Sea to Hawaii and Easter Island. Many Caribbean species similarly occur throughout that region and sometimes beyond its borders to Bermuda in the north and the eastern coast of Brazil in the south. However, only a handful of species of reef fish are circumtropical.

This near-total difference in species composition arises because the Caribbean and the Indo-West Pacific were cut off from one another from the early Oligocene when India's northward movement inexorably closed the Tethys Seaway that served to link them until then (Choat and Bellwood, 1991). Comparisons of Caribbean and Indo-West Pacific reef fish assemblages are comparisons of systems which have evolved separately for about 40 million years. They are separate solutions to the problem of how to populate a coral reef with fishes.

More than the Central American isthmus, the broad, deep, island-free East Pacific separates Caribbean and Indo-West Pacific, because it is too broad a reach of deep water for coastal species to cross easily. Recent studies suggest that periodic El Niño Southern Oscillation (ENSO) events may have been crucial to the colonization of the western coast of Central America by those Indo-West Pacific species which are found there. During ENSO events, changed meteorological conditions may reverse the usual westward flow of the Pacific Equatorial Current and transport Pacific species easterly to western Central America (Glynn, 1988). The combination of the rarely crossed East Pacific Barrier and the Central American isthmus has made the Pacific coastal fauna of Central America and the Galapagos a depauperate amalgam of Indo-West Pacific, Caribbean, and northern and southern temperate forms (Thresher, 1991).

Thresher (1991) has recently examined patterns in the biogeography of reef fishes. Not only do the Caribbean and Indo-West Pacific contain different species, but different families are species-rich in each region. In addition, differences in faunal composition within the Indo-West Pacific are sufficient to permit recognition of several sub-regions which differ in diversity, in species composition, and in the pattern of distribution of species among families. Because of their very separate histories, ecological comparisons of Caribbean and Pacific sites are particularly attractive. However, biogeographic comparisons need not be restricted to these. By choosing study sites appropriately, there is ample opportunity in coral reef systems to compare fish assemblages of differing richness but similar faunal composition, of similar richness and faunal composition but in different climatic or geographic regions, or of completely distinct species composition but equivalent richness. These are just a few of the possibilities; they have been made use of, but opportunities abound for further use of this approach.

III. THE NATURE OF REEF FISH ASSEMBLAGES

A. Open, Non-equilibrial Systems in a Patchy Environment

The early expectation was that reef fish assemblages would be stable in structure with presence, absence, and abundance of particular species determined primarily through the biotic interactions among assemblage members. This was not supported by the data that accumulated once researchers began to collect other than one-site, one-time snapshot descriptions. Instead, on the local scale of single stretches of reef habitat, these systems are now viewed as predominantly open systems, with non-constant dynamics and structure (Fig. 1). This change of perspective took place during the late 1970s and early 1980s. It is documented in Sale (1988b), and data supporting the change are discussed in a number of papers (Sale, 1980, 1988a, 1991; Mapstone and Fowler, 1988; Doher-

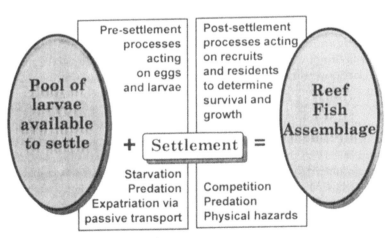

FIGURE 1. The structure and dynamics of a reef fish assemblage are driven by processes acting before and after settlement from the pelagic phase. Local assemblages are open systems subject, via the settlement process, to factors acting outside their boundaries. As such, they can be expected to be non-equilibrial and variable in structure.

ty, 1991; Doherty and Williams, 1988; Fowler, 1990; Hixon and Beets, 1993; Doherty and Fowler, 1994). I will spare readers another swim through this story.

The open, non-equilibrial nature of reef fish assemblages arises because of one feature of their habitat and three life history features. The habitat is patchy, and the fish are fecund but produce dispersive larvae which mostly die young. Their habitat is quintessentially patchy, whether viewed on scales of centimeters, meters, 10s or 100s of meters, kilometers, or 10s, 100s, or 1000s of kilometers. It is a habitat which is patchy to its inhabitants regardless of the spatial scales over which they move in the course of their daily lives. Its inhabitants respond appropriately by patterning their movements with reference to patch borders. They live within the larger, and sometimes even the smaller, patches. The fish themselves are typically fecund, producing numerous clutches, each of numerous eggs. These hatch to produce larvae which are pelagic for periods of days to months depending on the species, but nearly all the larvae die before completing larval life. It is the survivors among the larvae which serve to connect adult breeding populations isolated from one another by the patchy environment. With adults living within habitat patches and spawning numerous larvae which all disperse and mostly die, it is not too surprising to find that dynamics of reef fish populations are not closely related to local reproductive success (Sale, 1991).

Structure and dynamics of assemblages are necessarily the sum of structure and dynamics of component populations. Numerous, perhaps important data have been obtained to confirm that recruitment of reef fishes does vary markedly in space and time (Williams and Sale, 1981; Victor, 1986; Sale et al., 1984; Doherty and Williams, 1988; Fowler et al., 1992; Milicich et al., 1992; Planes et al., 1993; Robertson et al., 1993, among many) irrespective of the species, time, or place scrutinized. In recent years, the concern among many reef fish ecologists has been not whether spatially and temporally variable patterns of settlement of larvae were important in determining assemblage structure and dynamics, but whether anything other than settlement variation played a role. Some (Doherty, 1983, 1991; Victor, 1986; Meekan et al., 1993) have advocated the concept of the "recruitment-limited" system in which patterns of larval addition alone define subsequent assemblage structure and dynamics, while others (Jones, 1987, 1990, 1991; Hixon, 1991; Hixon and Beets, 1993) have pleaded for recognition of processes which act to modify patterns of survivorship after settlement has taken place. Doherty and Fowler (1994) have directly linked recruitment variation to subsequent age structure of reef fish populations. But Jones (1987, 1991), Forrester (1990), and several others (e. g., Eckert, 1987; Shulman and Ogden, 1987; Robertson, 1988; Caley, 1993) have documented equally clearly that mortality factors operating after settlement can substantially alter spatial and temporal patterns of abundance set initially at settlement. Indeed, there is evidence (Aldenhoven, 1986; Fowler, 1990) that demographic parameters can vary substantially between local populations of a species separated from each other by no more than a kilometer.

What reef fish ecologists have come to recognize (Warner and Hughes, 1988; Jones, 1991; Sale, 1991) is that a diversity of factors may operate to modify abundances of cohorts of fish and that which factors are important changes among times, places, and species. Worth emphasizing in this context are the order of magnitude differences in a wide range of life history features among reef fish species (Table I). Each species has evolved its own set of traits for surviving in this risky environment. Undoubtedly, these different traits can be read as

TABLE I Variation among Reef Fishes in Life History Features

Life history feature	Range of values among reef fish species
Spawning mode	Mid-water with no parental care
	Demersal, scattered adhesive eggs, no parental care
	Demersal, with parental care until hatching
Clutch size	200–1,000,000 eggs per spawning
Spawning frequency	Daily, every few days, fortnightly, monthly, annually
Spawning season	1–12 months
Duration of larval stage	9–100+ days
Settlement pattern	Continuous, episodic, semilunar cycle, lunar cycle
Size at settlement	8–200 mm SL
	2–50% of maximum adult size
Postsettlement mortality	High in first few days, then lower
	4.9–69.5% per annum for adult Labridae
Longevity	1–20+ years after settlement

	Examples of differences within and among species	
Fish species	Duration of larval stage	Size at settlement
Pomacentridae		
Amphiprion: 4 Pacific species	10.1–11 d	7.0–7.8 mm SL
Pomacentrus: 6 Pacific species	19.1–26.0 d	11.9–15.0 mm SL
Dascyllus aruanus	20–26 d	7.0 mm SL
Chrysiptera: 2 Pacific species	17.4 & 18.5 d	10.0 & 10.9 mm SL
Labridae		
Haemulon flavolineatum	14–18 d	6.9 mm SL
Halichoeres poeyi	21–28 d	12 mm SL
Thalassoma bifasciatum	38–78 d	12 mm SL
Balistidae		
Balistes vetula	75 d	60 mm SL
Diodontidae		
Diodon hystrix	70+ d	180–191 mm SL

Note. Table was compiled from Thresher and Brothers, 1989; Jones, 1991; Leis, 1991; Victor, 1991; and unpublished data.

solutions to different sets of mortality factors operating on populations of the different species. They may also be alternative solutions to the same factors.

B. Structure and Dynamics of Australian Patch Reef Assemblages

One long-term study which has helped to document the structure and dynamics of reef fish assemblages was done over 10 years using assemblages formed on small lagoonal patch reefs at One Tree Reef, southern Great Barrier Reef, from 1977 to 1987 (Sale and Douglas, 1984; Sale et al., 1994). For this monitoring study we selected 20 patch reefs on a shallow (3–5 m in depth) sandy lagoon floor that were isolated from one another and from larger areas of contiguous reef habitat by 5 m or so of open sand. None of the 20 reefs was emergent at low tide. All were comprised of dead, rocky surfaces with some areas of live coral. They averaged 2–3 m^2 in basal area. Given their isolation and the life cycles of reef fish species, the patch reefs were occupied by fish which arrived almost entirely from the plankton and which remained resident on single patch reefs throughout life. (A few species remained on single patch reefs until they grew so big that they abandoned patch reefs in favor of larger areas of reef substratum.) In this way these patch reefs functioned much as microcosms of the coral reef environment.

Fish occupying the patch reefs were censused a total of 20 times over the period, although not on a regular 6-month basis. At each census, a diver enumerated all fish present and visible on a patch reef on three separate occasions, taking 15–45 min to do so each time and not bringing data from previous enumerations for comparison during the second and third visits. At most census times two divers split the effort so that each reef was enumerated at least once by each diver. Precision of the census procedure was tested in an early experiment using a small group of similar patch reefs (Sale and Douglas, 1981). It was found to be comparable to a careful rotenone collection from a patch reef, after the reef had been encircled by a barrier net, although it documented a slightly different set of species. Both techniques identified about 82% of the species present on the patch reef. No attempts were made to mark fish. Care was taken to avoid disturbing the reefs other than to fasten marker buoys. That One Tree Reef was a closed scientific reserve within the Great Barrier Reef Marine Park for much of this time, and was rarely visited by people other than scientists prior to this, helped ensure that these were undisturbed sites.

The data set provided a matrix of 20 reefs by 20 censuses by 141 taxa of fish. These taxa were predominantly single species, but 6 were groups pooled because of difficulty discriminating among them in the field. Altogether some 165 species of fish were seen during the study, although the reefs undoubtedly also supported a small number of cryptic and strictly nocturnal species never seen in our censuses. At a single census, the average reef supported 125 fish of 20 taxa.

These data have permitted us to explore the nature of assemblage structure and dynamics with reference to the species present, species' absolute and relative abundances, species richness, and change in these from census to census (Sale and Douglas, 1984; Sale and Steel, 1986, 1989; Sale and Guy, 1992; Sale et al., 1994). We have examined structure with respect to reef size and structure (Sale and Douglas, 1984), and to season of year (Sale et al., 1994). The message to emerge has been one of change and of different patterns of change among neighboring patch reefs. The patch reefs have not even held a consistent number of fish from census to census.

The most surprising result emerging from these data was that the composition of an assemblage varied substantially from one census to the next. This was so, despite the fact that fish were not dispersing among patch reefs and in many cases were living long enough to show up in several censuses. Much of our analytical effort has centered on understanding the nature and extent of this temporal variation in assemblage structure. Very briefly, structure (as measured by the relative abundances of component species) varied independently among patch reefs (Fig. 2). It did not vary on a detectable seasonal cycle, and it varied to differing degrees among patch reefs (Sale and Douglas, 1984; Sale et al., 1994). Abundant species varied in the degree to which their overall numbers varied among censuses and in the degree to which their distribution across the patch reefs varied (Fig. 3). Some were consistently abundant and consistently distributed among patch reefs, while others were much more chaotically present (Sale et al., 1994). Among the 141 taxa we were not able to locate small groups which consistently occurred together or which consistently remained apart—as they probably would have done if they out-competed each other in certain habitats (Sale and Steel, 1986, 1989). While it is true that there were large numbers of relatively rare taxa present on these patch reefs, the overall structural variation was not due to the comings and goings of these. It was due to the variation through time in the distribution and abundance of the more common species (Sale et al., 1994).

In one evaluation of the data, we asked whether it was possible that our focus on the species was inappropriate in these diverse systems, and we searched for structure at a "higher" ecological level. We reclassified the fishes present on the patch reefs into a small number (13) of trophic guilds rather than the 141 taxa to see if a greater degree of temporal predictability might exist (Sale and Guy, 1992). The 10-fold reduction in the number of categories of fish automatically resulted in average between-census similarity values being higher, but variation among censuses in assemblage structure was no different when fish were assigned to real trophic categories than when they

FIGURE 2. Graphs depicting the average degree of similarity (Czekanowski index, Feinsinger *et al.*, 1981) of each census to all other censuses of that patch reef, for 20 patch reefs censused 20 times over 10 years at One Tree Reef. The patch reefs were in two groups of 10, located about 2 km apart on opposite sides of the lagoon of One Tree Reef. The lagoon is sand-floored, about 3–5 m deep, and contains numerous large and small patch reefs. It is surrounded by a tidally emergent reef flat. Average degree of similarity varies among patch reefs but seldom ranges above 0.6 or below 0.4. Degree of fluctuation varies among patch reefs and is nonsynchronous (from Sale *et al.*, 1994).

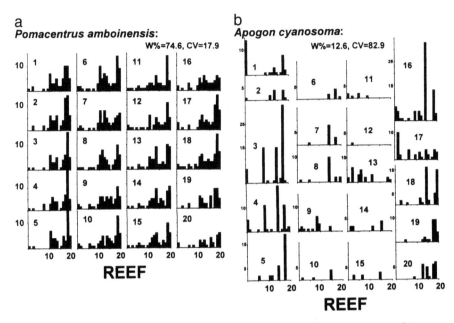

FIGURE 3. Graphs reporting the distribution and abundance of (a) *Pomacentrus amboinensis* and (b) *Apogon cyanosoma* over 20 censuses among 20 patch reefs at One Tree Reef. For each histogram, number of fish is on the ordinate, and the 20 reefs are on the absissa. Plots are numbered to correspond to the 20 censuses. *P. amboinensis* was one of the most consistently distributed, while *A. cyanosoma* was one of the least consistently distributed of the species commonly present. These differences are quantified by values for Kendall's concordance coefficient (W%), which compares the distributions across reefs at each census, and the coefficient of variation (CV) which compares the total number of fish sighted (all reefs pooled) among censuses (from Sale *et al.*, 1994).

were assigned to randomly generated amalgams in which the same numbers of species were lumped to form 13 ecologically meaningless groupings (null guilds; Fig. 4). These assemblages do not exhibit more intrinsic structure when viewed at the "higher" guild level than at the "lower" species level.

Our principal conclusions from this long-term monitoring study were that fish species settling to patch reefs behaved largely independently of one another, their habitat preferences were broad enough that most of our patch reefs could be settled by most of our taxa, and larger patch reefs tended to receive larger numbers of fish. Any sense that interactions among the members were important in structuring an assemblage vanished, as did any sense of an assemblage structure which could usefully be predicted from information about the habitat or about the assemblage as seen there on a prior occasion.

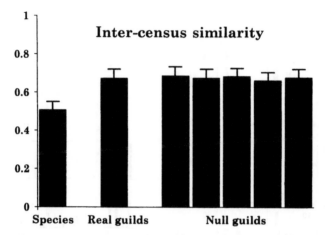

FIGURE 4. The mean similarity (Czekanowski's index, Feinsinger *et al.*, 1981) among censuses of single patch reefs at One Tree Reef is significantly increased if fish are classified into 13 trophic guilds rather than 141 species. However, this increase is due solely to the reduction in number of categories of fish because five different sets of 13 "null" guilds—random groupings of species with no trophic basis—yield similar values to those obtained using "real" trophic guilds. These assemblages are not structured more deterministically when viewed at the coarser ecological scale of "guild." (Data from Sale and Guy, 1992).

C. Patch Reef Assemblages as a Model System

Are patch reef assemblages a good model for reef fish in general? It is probably true that the strongly marked edges of patch reefs make dispersal of fish among neighboring patch reefs less likely than might be the case on equivalent areas of contiguous reef. [The difficulty of censusing marked areas within contiguous reef (how do you deal with the fish swimming across borders?) has postponed comparative work.] If the chance of dispersal among neighboring sites were higher than on our patch reefs, local differences set up by differing patterns of settlement might be broken down over time, and the very different time courses we observed among neighboring patch reefs might not occur. This might be the case over areas of contiguous reef. However, there already exist several studies documenting dramatically different demographies for particular reef fish species at nearby sites in more contiguous reef habitat (Aldenhoven, 1986; Eckert, 1987; Fowler, 1990; Jones, 1991), and differing sets of demographies must yield differing assemblage dynamics. I anticipate that patch reef assemblages will prove to be a good model of the much larger assemblages which occur on larger-scale areas of reef habitat; as on patch reefs, assemblage dynamics on these larger-scale habitats will be variable in space and time and not closely regulated by biotic processes among residents.

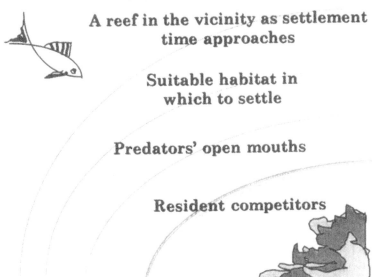

A reef in the vicinity as settlement
time approaches

Suitable habitat in
which to settle

Predators' open mouths

Resident competitors

FIGURE 5.   The settling reef fish larva faces a series of obstacles, any one of which may prevent its survival, growth, or reproduction. Reef-based predators and competitors, in particular, might be expected to act selectively on particular species of recruit. They provide a mechanism with the potential to make reef fish assemblages strongly deterministic in structure, yet a mechanism that appears not to play an important role in shaping patch reef assemblages.

I reach this view primarily because of my continuing surprise at the lack of importance of biotic interactions among resident fish in regulating assemblage structure on the patch reefs. Reef fishes settle to patch reefs when they are small and vulnerable, and they are met there by residents of their own and other species which are larger, more robust, and at home (Fig. 5). If we were to try and design a system in which interspecific interactions could play a major structuring role, this would have to be it. Yet that role is not played except in occasional instances (Shulman *et al.*, 1983; Caley, 1993).

IV. THE GREAT BARRIER REEF
AND THE CARIBBEAN

There are signs that reef fish assemblages in the Indo-West Pacific and the Caribbean are organized in essentially similar ways. Consider the relation of species richness to number of individuals present. Preliminary data suggested that patch reef assemblages at least show similar patterns despite the substantially greater overall richness typical of Pacific sites. This feature may first have been reported by Smith (1978) for patch reefs of similar size in the Bahamas and the Society Islands. Thresher (1991) lists several subsequent comparisons of this type which led to the same conclusion. Sale (1980, 1991) claimed that published data on patch reef assemblages from a wide range of geographic locations showed a consistent relationship between the number of fish occupying a patch reef and the number of species they comprised. Data published during the 1970s and 1980s also showed that Caribbean patch reef assemblages exhibited comparable levels of temporal fluctuation in structure as did assemblages at One Tree Reef, although the authors did not always interpret their data this way [see discussion in Sale (1980) and Sale and Douglas (1984)].

On the other hand, there are a number of reasons why dynamics, in particular, may be different. The lower overall diversity alone makes it unlikely that there can be as much change through time on patch reefs in the Caribbean as on the Great Barrier Reef. The Caribbean palette simply has fewer colors. Secondly, as Thresher (1982, 1991) and coworkers (Thresher and Brothers, 1989) have documented, the species of at least some families in the Caribbean show consistent differences from equivalent Indo-West Pacific species in several life history traits (Table II). Differences among demersally spawning forms are coherent (with the exception of the pomacentrid genus *Stegastes*, which does the reverse!) and suggest that the Caribbean forms produce more and smaller larvae than their Pacific counterparts. Such larvae, with smaller reserves of nutrients and energy than carried by the larger larvae typical of Pacific species, would be at greater risk of early death. Thirdly, the fauna present in the Caribbean exhibits a quite different distribution of species among families from that on the Great Barrier Reef (Thresher, 1991). We might not expect that assemblages composed of different mixes of taxa will be organized in similar ways. Finally, most Caribbean sites are physically and climatically more benign than are Great Barrier Reef sites, with lesser tidal fluctuations, calmer average wind conditions, and possibly less annual temperature variation and fewer severe storms. In a less changeable environment, assemblage dynamics might be more predictable.

This mixture of observations, giving evidence on opposite sides of an argument, is not helpful. What is needed is carefully planned comparisons of assemblage structure and dynamics in the two locations. The results will be interesting and informative whatever the outcome. If these two evolutionarily distinct lineages have resulted in assemblages showing essentially the same patterns of structure and dynamics, we may infer that structure and dynamics are defined by the general nature of the tropical reef environment and by the requirements it imposes on the life histories of the reef fish species present. This conclusion would have implications for ecological theory, and would

TABLE II   Interoceanic Differences in Life History Traits

	Egg size (mm³, with range)				
	Caribbean		Western Pacific		P
Demersal spawners					
Apogonidae	0.014	(0.007–0.028)	0.18	(0.034–1.15)	<0.01
Blenniidae	0.076	(0.050–0.139)	0.243	(0.102–0.832)	<0.01
Gobiidae	0.718	(0.311–1.76)	0.912	(0.086–4.52)	NS
Pomacentridae	0.156	(0.093–0.261)	0.524	(0.091–1.527)	<0.01
Pelagic spawners					
Pomacanthidae	0.312	(0.165–0.382)	0.210	(0.113–0.322)	NS
Serranidae	0.338	(0.150–0.524)	0.445	(0.408–0.451)	NS

	Duration of larval stage (days, with range or ± 2 SE)				
	Caribbean		Western Pacific		
Demersal spawners					
Apogonidae	23.0	(±0.87)	19.2	(±1.87)	<0.05
Apogon	23.7	(20.0–26.2)	21.6	(13.0–28.4)	<0.05
Pomacentridae	29.0	(±1.58)	20.9	(±1.6)	<0.001
Abudefduf	27.4	(27.1–27.7)	22.0	(18.3–24.2)	<0.05
Chromis	32.6	(31.0–35.1)	27.1	(20.3–36.1)	<0.05
Stegastes	27.6	(26.6–29.0)	32.1	(29.7–34.7)	<0.05
Pelagic spawners					
Pomacanthidae	25.1	(±5.65)	27.0	(±2.12)	NS
Centropyge	36.8		32.3	(25.3–38.3)	NS
Holacanthus	24.9	(22.4–29.6)	25.0		NS
Pomacanthus	19.5	(17.7–21.3)	20.6	(18.0–22.8)	NS
Labridae	27.5	(±3.13)	32.9	(±4.17)	NS
Halichoeres	21.9	(20.0–24.3)	28.6	(21.0–37.0)	NS
Thalassoma	39.8		53.1	(45.2–68.5)	NS

Note. Demersal spawners in the Caribbean tend to lay smaller eggs and produce larvae with longer pelagic stages than is the case in the Pacific. These differences are not apparent in pelagic spawners. (Data were summarized from Thresher and Brothers, 1989.)

probably heighten ecologists' faith in the power of natural selection and in the adaptiveness of community-level features. This conclusion would also be of considerable significance for management. It would mean that management principles developed for one coral reef region could likely be transported to others around the world without retesting.

Alternatively, if it turned out that the evolutionarily distinct Caribbean and Great Barrier Reef systems exhibited different patterns of structure and dynamics, it would free up our notions of what is possible when approximately similar species are assembled in approximately similar environments and might further erode our faith in determinism. It might lead to a heightened awareness that reefs can be different from one another and that too free a generalization of results obtained at one site can be dangerous. It would also caution managers to plan some on-site research before applying imported techniques or concepts because what is effective in one coral reef system need not be effective in another. In truth, of course, we are unlikely to find completely identical, or totally different, structure and dynamics. We will be able to revel in the full spectrum of degrees of similarity rather than confront the digital precision of yes/no, black/white dichotomous data.

V. A COMPARISON OF PATCH REEF ASSEMBLAGES

A. Comparing Assemblages in the Virgin Islands and on the Great Barrier Reef

In 1991, I commenced a study intended to compare precisely the structure and dynamics of patch reef assemblages in the Virgin Islands and the Great Barrier Reef. I chose Teague Bay, St. Croix, in the U.S. Virgin Islands as my study site and selected 20 small patch reefs comparable in size and structure to the ones I had earlier monitored in the lagoon of One Tree Reef. Like those at One Tree Reef, the Virgin Islands patch reefs were nonemergent and on a sandy floor at a depth of 3–5 m at high tide. They were in a protected back-reef location, physically not unlike the lagoonal location of the One Tree reefs. At the start of the study, I measured a number of attributes of each patch reef using the same procedures I had used at One Tree Reef (Sale and Douglas, 1984), although the species of corals and other substrata that I recorded were naturally totally different. Table III shows that the sets of patch reefs in the two locations were physically quite similar.

TABLE III   Characteristics (Mean ± SE) of the 20 Patch Reefs Monitored at Each of One Tree Reef, Australia, and Teague Bay, St. Croix, U.S. Virgin Islands

Characteristic	One Tree Reef		Teague Bay	
Diameter	2.34 m	±0.17	2.12 m	±0.16
Surface area	8.70 m²	±1.75	10.50 m²	±1.28
Topographic complexity	1.60	±0.06	1.36	±0.03
Substratum diversity	2.85	±0.14	3.19	±0.27
Percentage dead rock	48.15	±2.85	60.18	±5.36
Percentage living corals.	17.56	±3.05	23.04	±3.77

Note. Methods for measuring these features are given in Sale and Douglas (1984).

Beginning in July 1991, I have censused fish on each of the patch reefs twice a year, in July and January. The census procedure used has also been the one used in the One Tree Reef study, although the species lists are totally non-overlapping. Monitoring will continue for 5 years (totaling 10 censuses). Although I am not manipulating the patch reef assemblages in any way, I cannot be quite as certain as I was at One Tree Reef that they are not being disturbed by others. There is a small trap fishery in the area, but the fish associated with the patch reefs will, for the most part be too small to be caught; nor are they likely to be taken by anglers or spear fishermen.

Preliminary results are interesting. After the first census, I was aware of the lower species richness of this system. Although numbers of species per patch reef were not much lower than on the Australian reefs, there was a sameness among the patch reefs—each had much the same species. This shows clearly when species abundance curves are plotted based on pooled censuses of all 20 reefs in each system (Fig. 6). After the first 4 censuses, and 9074 fish recorded, I had a species list of 67 species in Teague Bay. This compared to a list of 141 species after counting 50,124 fish in 20 censuses at One Tree Reef. The One Tree curve is also composed of more equitably abundant species, although there remains a long tail of quite rare forms. If we reduce sample size in the One Tree data by using the first 4 censuses, we still obtain more species (103

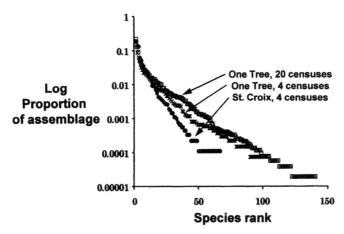

FIGURE 6. Species abundance curves based on the censuses of patch reef assemblages at One Tree Reef and at Teague Bay, St. Croix. Two curves are shown for the former location—one based on all 20 censuses (50,124 fish of 141 species sighted) and one based on the first 4 censuses (10,181 fish of 103 species sighted). The St. Croix curve is for 4 censuses (9074 fish of 67 species sighted). The greater richness and evenness evident in the One Tree Reef data are clearly not simply a function of the larger sample size.

species for 10,182 fish counted) and a species abundance curve with obviously greater equitability than that at Teague Bay.

This lower species richness has revealed an unexpected difference between One Tree Reef and Teague Bay in the relationship between species richness and number of individuals in assemblages. The larger data set now available (20 censuses of One Tree patch reefs and 6 censuses of Teague Bay patch reefs) discloses a slight but clear difference in the number of species per individual (Fig. 7). On a log–log plot, the regression for One Tree Reef data lies parallel to and above that for Teague Bay data. The difference is equivalent to the presence of about 5 "extra" species on an Australian patch reef supporting 100 fish compared to a St. Croix patch reef supporting the same number of fish. When the difference is examined for data pooled over 20 such patch reefs, the "enrichment" is more like 17 species. My earlier claim that species-per-individual relationships appeared to be uniform for patch reef assemblages in many different geographic locations is clearly not correct.

Of most interest is the question of whether intercensus similarity (across-censuses, within-patch reefs) is comparable at One Tree Reef and at Teague Bay. Table IV suggests that similarity is greater at Teague Bay, although the difference is slight.

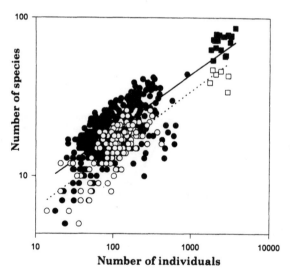

FIGURE 7. Comparison of the relation between number of species and number of individual fish for patch reef assemblages at One Tree Reef, and at Teague Bay, St. Croix. Data plotted are for (●) single patch reef and (■) all patch reefs pooled at one census for One Tree Reef and for (○) single patch reef and (□) all patch reefs pooled at one census at St. Croix. Regression lines are descriptive only and based on all One Tree or all St. Croix data despite the lack of independence of single patch reef and single census date. Assemblages at One Tree Reef are proportionately richer than those at St. Croix over a broad range of assemblage sizes. Note that both axes are logarithmic.

TABLE IV   Mean Similarity among Censuses of Single Patch Reefs

Comparison	Mean similarity[a]
One Tree Reef, fish classified as 141 species	0.508 ± 0.043
One Tree Reef, fish classified as 13 trophic guilds	0.673 ± 0.047
Teague Bay, fish classified as 67 species	0.554 ± 0.037

Note. Data shown are the average values over 20 patch reefs for mean similarity (± SE) in each case. Means calculated using similarity values for 10 randomly selected but independent pairs of censuses of each patch reef at One Tree Reef, and using similarity values for all 15 comparisons possible among the 6 censuses of each patch reef at Teague Bay, St. Croix. Data summarized from Sale and Guy (1992), and unpublished St. Croix data.
[a]Czekanowski's index (Feinsinger *et al.*, 1981) was used to calculate mean similarity.

B. Similarity, Constancy, Predictability, and Stability

There are problems in interpreting this apparently straightforward comparison which arise because of our failure to discriminate clearly between predictability in a purely numerical/statistical sense and predictability in an ecological sense subsuming regulation and determinism. The greater between-census similarity of the Teague Bay reefs could arise because structure is more strongly regulated by biotic processes determining the relative abundances of different species—that is, because the structure of patch reef assemblages at the Caribbean site is regulated more precisely by various deterministic processes, such as density-dependent competition or predation. In this case, the Caribbean assemblages could be termed more predictable in an "ecological" sense as well as in the strictly probabilistic sense. Alternatively, the greater between-census similarity may be due simply to the fact that there are fewer alternative states in a system with fewer species available. In this case, the greater similarity between censuses is due simply to the fact that there is less opportunity for change, even though the assemblage is behaving in exactly the same way ecologically with the same degree of regulation by biotic factors as is the case at One Tree Reef. Similarity indices are always sensitive to the number of categories into which the individuals are classified, and these data sets are quite different in that respect (Teague Bay: 9074 fish among 67 species; One Tree Reef: 50,124 fish among 141 species).

In an earlier study (Sale and Guy, 1992) we had drastically reduced the number of taxa in the One Tree data by classifying fish into 13 guilds instead of into 141 species. Between-census, within-reef similarities with guild data yielded a mean value of 0.673 compared to 0.508 for data classified as species. These two mean values for One Tree Reef data bracket the mean value obtained from the Teague Bay data (Table IV). We are at present manipulating these data sets to explore the responsiveness of the Czekanowski similarity index to species richness. Will the Teague Bay data exhibit greater inter-census similarity when fish are classified into trophic groups of comparable type and number as the ones used for the One Tree Reef study? Only if they do will we be able to conclude that assemblages on the Teague Bay reefs behave in a way which is ecologically more predictable as fish colonize and die out from census to census.

There is a simple, but perhaps useful, general message here. Comparisons of data collected at a site at two or more different times can show patterns of change. Similarity indices (or simple variance measures) can quantify the extent of change which occurs. Higher levels of between-census similarity mean that we have a greater ability to predict assemblage structure at one census, based on knowledge of its structure at earlier censuses. Conversely, lower levels of inter-census similarity mean that we are less able to predict future conditions. Ecologists tend to equate the degree of "statistical predictability" with degree of "ecological stability" in structure and to assume that greater stability in structure arises for "biological" reasons. In fact the term "stability" usually implies the existence of mechanisms (such as density-dependent mortality) which act to regulate or control structure. We need to remember that measures of similarity, or of temporal constancy, are just that. They do not measure stability, and temporal constancy can arise for any of several reasons: (1) for purely statistical reasons as discussed, (2) because forces which might cause change are not currently acting on the assemblage, (3) because there are ecological processes acting to regulate (stabilize) assemblage structure, or (4) because the time span over which the observations have been made is inadequate given the rates at which biotic processes take place in that system (Connell and Sousa, 1983).

If the variation in Table IV turns out to be explained by the differing species richness alone, and if assemblages on Teague Bay patch reefs are typical of Caribbean reef fish assemblages (I know this is rampant generalization), we will be forced to conclude that the Caribbean is more predictable than the Great Barrier Reef precisely because it is less diverse. Far from diversity promoting stability in ecological systems as was once believed, we may come to find that diversity promotes unpredictability!

VI. THE FUTURE

Lest you conclude from the past section that there is nothing more to be learned from biogeographic comparisons, let me remind you I have speculated grandly. I have data from a few censuses of fish on patch reefs at one site in each of two different oceans. Even if dynamics are strongly stochastic in both places, I doubt they are identical. We can probably get beyond the simple question of whether reef fish assemblages are stable or not fairly quickly. But a lot of intriguing questions remain. How do the dynamics of species with differing fecundity, larval duration, and post-settlement survivorship vary; and what is the consequence of this on dynamics of the assemblages to which they belong? Are Caribbean demersal spawners, with their more numerous, smaller eggs and longer larval durations, a different kind of fish to the demersal spawners of the Great Barrier Reef? Are Caribbean species, on average, more dispersive—either because of life history traits or of physical and climatic conditions—and does this result in assemblages with different structure or dynamics? Do assemblages, such as those in the Caribbean, that are subject to strong fishing pressure exhibit dynamic behavior which differs from that on the Great Barrier Reef, where human impacts still remain generally low? Without detracting in any way from the value of other kinds of studies, there are a number of questions concerning the ecology of reef fishes which can profitably be studied by means of planned biogeographic comparisons. We have been given many different coral reef systems in the world. We should celebrate their differences by using them to advance our overall understanding.

VII. SUMMARY

The rich assemblages of fish on coral reefs provide an ideal system in which to explore experimentally the structure and dynamics of diverse communities. Here, I make a case for the value of long-term monitoring studies to disclose patterns in assemblage structure and dynamics, and for the deliberate use of biogeographical patterns to set up large-scale, long-term comparisons of ecologically and evolutionarily separate assemblages.

The fishes of the vast Indo-West Pacific and the Caribbean (and western Atlantic) comprise two groups, taxonomically quite similar at the levels of genus and family, but virtually separate at the level of species. This separation of the world's reef fish fauna is a direct consequence of its Eocene evolution and the subsequent closure of the Tethyan seaway, but it provides the possibility of ecological comparisons between evolutionarily separate yet ecologically similar systems. Biogeographic variations within the Indo-West Pacific and, to a lesser extent, within the Caribbean offer other useful contrasts awaiting planned evaluation.

Reef fish assemblages are open and non-equilibrial because these demersal, site-attached species live in a quintessentially patchy environment with which they cope by producing numerous, dispersive larvae. In recent years, ecologists have documented the important role played by variability in recruitment success in determining species distribution and abundance. (Recruitment is usually measured at or shortly after the time of settlement from plankton to reef habitat.) Some have claimed, with supporting evidence, that reef fish assemblages are "recruitment limited" in that recruitment variation is the primary, or even the only, factor playing a role in determining assemblage structure. Others recognize that important post-settlement processes may also act to influence species abundances on reef habitat, and a consensus is growing that a multiplicity of factors may operate differently in different times and places to determine this. A long-term monitoring study of fish assemblages on small Australian patch reefs is used to illustrate the changeable nature of assemblage structure in this system.

I am currently making a comparable study of patch reef assemblages in the U.S. Virgin Islands with the goal of carefully contrasting structure and dynamics in these two locations. Preliminary results are discussed from two perspectives: How similar are structure and dynamics of patch reef assemblages of fishes in these biogeographically separate locations? What are the relationships among ecological constancy, statistical predictability, and biological determinism in studies of community organization? It may be that Caribbean patch reef assemblages are more predictable in structure than are Australian ones. If this is the case, it may be only because the Caribbean palette of species offers fewer options for change.

Acknowledgments

My Australian research reported here was done with the generous support of the Australian Research Grants Scheme, and the Marine Science and Technologies Grants Scheme, with the facilities of Sydney University's One Tree Island Research Station, and with the help of numerous people including Rand Dybdahl, Bill Douglas, Tim Jones, Doug Ferrell, Jeff Guy, and Warren Steel. The research at St. Croix has been generously supported by NSF grant OCE-9018724, by NOAA-NURP funds through the Caribbean Marine Research Center (to Sale and T. D. Kocher), and by grant OGP0154284 from the National Science and Engineering Research Council of Canada. In it, I have been ably assisted by Jeff Guy and Kelly Gestring, and by my students, Phil Levin, Mike Ganger, Nick Tolimieri, Rick Nemeth, Dan Ha, and Andrea Risk. This manuscript has been helped by the critical comments of Bret Danilowitz and the editors. To the agencies and to these people, my thanks. The distorted views and misinterpretations remain my own.

References

Aldenhoven, J. M. (1986). Local variation in mortality rates and life-expectancy estimates of the coral-reef fish *Centropyge bicolor* (Pisces: Pomacanthidae). *Mar. Biol.* 92, 237–244.

Briggs, J. C. (1974). "Marine Zoogeography." McGraw-Hill, New York.

Caley, M. J. (1993). Predation, recruitment and the dynamics of communities of coral reef fishes. *Mar. Biol.* 117, 33–43.

Choat, J. H., and Bellwood, D. R. (1991). Reef fishes: Their history and evolution. In "The Ecology of Fishes on Coral Reefs" (P. F. Sale, ed.), pp. 39–66. Academic Press, San Diego, CA.

Connell, J. H., and Sousa, W. P. (1983). On the evidence needed to judge ecological stability or persistence. Am. Nat. 121, 789–824.

Doherty, P. J. (1983). Tropical territorial damselfishes: Is density limited by aggression or recruitment? Ecology 64, 176–190.

Doherty, P. J. (1991). Spatial and temporal patterns in recruitment. In "The Ecology of Fishes on Coral Reefs" (P. F. Sale, ed.), pp. 261–293. Academic Press, San Diego, CA.

Doherty, P. J., and Fowler, A. J. (1994). An empirical test of recruitment limitation in a coral reef fish. Science 263, 935–939.

Doherty, P. J., and Williams, D. McB. (1988). The replenishment of coral reef fish populations. Oceanogr. Mar. Biol. 26, 487–551.

Doherty, P. J., Williams, D. McB., and Sale, P. F. (1985). The adaptive significance of larval dispersal in coral reef fishes. Environ. Biol. Fishes 12, 81–90.

Eckert, G. J. (1987). Estimates of adult and juvenile mortality for labrid fishes at One Tree Reef, Great Barrier Reef. Mar. Biol. 95, 167–171.

Feinsinger, P., Spears, E. E., and Poole, R. W. (1981). A simple measure of niche breadth. Ecology 62, 27–32.

Forrester, G. E. (1990). Factors influencing the juvenile demography of a coral reef fish. Ecology 71, 1666–1681.

Fowler, A. J. (1990). Spatial and temporal patterns of distribution and abundance of chaetodontid fishes at One Tree Reef, southern GBR. Mar. Ecol.: Prog. Ser. 64, 39–53.

Fowler, A. J., Doherty, P. J., and Williams, D. McB. (1992). Multi-scale analysis of recruitment of a coral reef fish on the Great Barrier Reef. Mar. Ecol.: Prog. Ser. 82, 131–141.

Glynn, P. W. (1988). El Nino-Southern Oscillation 1982-1983: Nearshore population, community, and ecosystem responses. Annu. Rev. Ecol. Syst. 19, 309–345.

Hixon, M. A. (1991). Predation as a process structuring coral reef fish communities. In "The Ecology of Fishes on Coral Reefs" (P. F. Sale, ed.), pp. 475–508, Academic Press, San Diego, CA.

Hixon, M. A., and Beets, J. P. (1993). Predation, prey-refuges, and the structure of coral-reef fish assemblages. Ecol. Monogr. 63, 77–101.

Jones, G. P. (1987). Competitive interactions among adults and juveniles in a coral reef fish. Ecology 68, 1534–1547.

Jones, G. P. (1990). The importance of recruitment to the dynamics of a coral reef fish population. Ecology 71, 1691–1698.

Jones, G. P. (1991). Postrecruitment processes in the ecology of coral reef fish populations: A multifactorial perspective. In "The Ecology of Fishes on Coral Reefs" (P. F. Sale, ed.), pp. 294–328. Academic Press, San Diego, CA.

Leis, J. M. (1991). The pelagic stage of reef fishes: The larval biology of coral reef fishes. In "The Ecology of Fishes on Coral Reefs" (P. F. Sale, ed.), pp. 183–230. Academic Press, San Diego, CA.

Mapstone, B. D., and Fowler, A. J. (1988). Recruitment and the structure of assemblages of fish on coral reefs. Trends Ecol. Evol. 3, 72–77.

Meekan, M. G., Milicich, M. J., and Doherty, P. J. (1993). Larval production drives temporal patterns of larval supply and recruitment of a coral reef damselfish. Mar. Ecol.: Prog. Ser. 93, 217–225.

Milicich, M. J., Meekan, M. G., and Doherty, P. J. (1992). Larval supply—a good predictor of recruitment of 3 species of reef fish (Pomacentridae). Mar. Ecol.: Prog. Ser. 86, 153–166.

Planes, S., Levefre, A., Legendre, P., and Galzin, R. (1993). Spatio-temporal variability in fish recruitment to a coral reef (Moorea, French Polynesia). Coral Reefs 12, 105–113.

Robertson, D. R. (1988). Abundances of surgeonfishes on patch-reefs in Caribbean Panama: Due to settlement or post-settlement events? *Mar. Biol.* **97**, 495–501.

Robertson, D. R., Schober, U. M., and Brawn, J. D. (1993). Comparative variation in spawning output and juvenile recruitment of some Caribbean reef fishes. *Mar. Ecol.: Prog. Ser.* **94**, 105–113.

Sale, P. F. (1980). The ecology of fishes on coral reefs. *Oceanogr. Mar. Biol.* **18**, 367–421.

Sale, P. F. (1982). Stock-recruit relationships and regional coexistence in a lottery competitive system. *Am. Nat.* **120**, 139–159.

Sale, P. F. (1988a). Perception, pattern, chance and the structure of reef fish communities. *Environ. Biol. Fishes* **21**, 3–15.

Sale, P. F. (1988b). What coral reefs can teach us about ecology. *Proc. Int. Coral Reef Symp., 6th*, Vol. 1, pp. 19–27.

Sale, P. F. (1991). Reef fish communities: Open nonequilibrial systems. *In* "The Ecology of Fishes on Coral Reefs" (P. F. Sale, ed.), pp. 564–598. Academic Press, San Diego, CA.

Sale, P. F., and Douglas, W. A. (1981). Precision and accuracy of visual census technique for fish assemblages on coral patch reefs. *Environ. Biol. Fishes* **6**, 333–339.

Sale, P. F., and Douglas, W. A. (1984). Temporal variability in the community structure of fish on coral patch reefs and the relation of community structure to reef structure. *Ecology* **65**, 409–422.

Sale, P. F., and Guy, J. A. (1992). Persistence of community structure: What happens when you change taxonomic scale? *Coral Reefs* **11**, 147–154.

Sale, P. F., and Steel, W. J. (1986). Random placement and the structure of reef fish communities. *Mar. Ecol.: Prog. Ser.* **28**, 165–174.

Sale, P. F., and Steel, W. J. (1989). Temporal variability in patterns of association among fish species on coral patch reefs. *Mar. Ecol.: Prog. Ser.* **51**, 35–47.

Sale, P. F., Doherty, P. J., Eckert, G. J., Douglas, W. A., and Ferrell, D. J. (1984). Large scale spatial and temporal variation in recruitment to fish populations on coral reefs. *Oecologia* **64**, 191–198.

Sale, P. F., Guy, J. A., and Steel, W. J. (1994). Ecological structure of assemblages of coral reef fishes on isolated patch reefs. *Oecologia* **98**, 83–99.

Shulman, M. J., and Ogden, J. C. (1987). What controls tropical reef fish populations: Recruitment or benthic mortality? An example in the Caribbean reef fish *Haemulon flavolineatum. Mar. Ecol.: Prog. Ser.* **39**, 233–242.

Shulman, M. J., Ogden, J. C., Ebersole, J. P., McFarland, W. N., Miller, S. L., and Wolf, N. G. (1983). Priority effects in the recruitment of juvenile coral reef fishes. *Ecology* **64**, 1508–1513.

Smith, C. L. (1978). Coral reef fish communities: A compromise view. *Environ. Biol. Fishes* **3**, 109–128.

Springer, V. G. (1982). Pacific plate biogeography, with special reference to shore-fishes. *Smithson. Contrib. Zool.* **367**, 1–182.

Thresher, R. E. (1982). Interoceanic differences in the reproduction of coral-reef fishes. *Science* **218**, 70–72.

Thresher, R. E. (1991). Geographic variability in the ecology of coral reef fishes: Evidence, evolution, and possible implications. *In* "The Ecology of Fishes on Coral Reefs" (P. F. Sale, ed.), pp. 401–436. Academic Press, San Diego, CA.

Thresher, R. E., and Brothers, E. B. (1989). Evidence of intra- and inter-oceanic differences in the early life history of reef-associated fishes. *Mar. Ecol.: Prog. Ser.* **57**, 187–205.

Victor, B. C. (1986). Larval settlement and juvenile mortality in a recruitment-limited coral reef fish population. *Ecol. Monogr.* **56**, 145–160.

Victor, B. C. (1991). Settlement strategies and biogeography of reef fishes. *In* "The Ecology of Fishes on Coral Reefs" (P. F. Sale, ed.), pp. 231–260. Academic Press, San Diego, CA.

Warner, R. R., and Hughes, T. P. (1988). The population dynamics of reef fishes. *Proc. Int. Coral Reef Symp., 6th,* Vol. 1, pp. 149–155.

Williams, D. McB., and Sale, P. F. (1981). Spatial and temporal patterns of recruitment of juvenile coral reef fishes to coral habitats within "One Tree Lagoon," Great Barrier Reef. *Mar. Biol.* 65, 245–253.

CHAPTER 5

Dynamic Diversity in Fish Assemblages of Tropical Rivers

Kirk O. Winemiller

*Department of Wildlife and Fisheries Sciences, Texas A&M University,
College Station, Texas 77843*

I. INTRODUCTION

This chapter examines temporal stability of fish assemblage structure in two tropical rivers and briefly reviews similar information from other tropical rivers. The factors associated with population and assemblage change plus those associated with population persistence and long-term stability of assemblage structure are discussed.

We know that many, and probably most, biotic communities change over time. Changes in community composition and structure are the hallmarks of processes like vegetative succession (Clements, 1949) and ephemeral habitats such as residual pools in the Florida everglades (Kushlan, 1976). Yet much of the basic and applied literature either assumes or strongly implies the existence of stable systems and chronic density-dependence. Traditional fisheries management, in particular, is rife with assumptions of density-dependence and assemblage stability as, for example, in stock indices for small impoundments

(Anderson and Gutreuter, 1983) and the application of stock-recruit models for estimation of optimal yields (critique in Fletcher and Deriso, 1988). The present-day search for indicators of ecosystem health and integrity has resulted in the acceptance of methods that contain strong assumptions about the structure and stability of natural systems. Karr et al.'s (1986) index of biotic integrity (IBI) for North American streams implies that these fish assemblages have regular, stable, and, hence, predictable structures that can be assessed against standards across different settings.

At the same time, however, some ecologists are evaluating the degree of temporal consistency observed in biotic communities. For example, the stability assumption has been questioned for North American stream fish assemblages (Grossman et al., 1982, 1990). When temporal change is observed, it begs the following questions: To what extent are stochastic environmental and demographic factors indicated, and to what extent are deterministic biotic factors indicated? Virtually all of the chapters contained in this volume offer insights into the question of whether communities represent "chance aggregations" or "stable systems" (following Keast's terminology in Chapter 3 of this volume).

Climatic regime and the abiotic environment can directly influence mortality and reproduction. In harsh or unpredictable climates, environmental factors can cause apparent stochastic mortality. This idea dates back at least to Darwin's time and was well articulated by Dobzhansky (1950). Consequently, in more moderate or stable climates, biotic factors that are largely deterministic may have dominant effects on populations and communities. Availability of space and other resources, competition, predation, and other biotic interactions can influence mortality, growth, reproduction, resource use, and niche segregation. However, the contrast between stochastic and deterministic factors can be blurred, as largely stochastic abiotic factors may interact with density-dependent biotic factors. Appleblossom thrips show density-dependent mortality in response to cold temperatures because at higher densities more individuals are excluded from microhabitats that serve as safe havens from freezing (Davidson and Andewartha, 1948).

A. Temperate Fish Assemblages

The assumption of temporal stability of temperate fish assemblages was put to a test by Grossman (1982; Grossman et al., 1982). He found high temporal stability and resiliency in intertidal fish assemblages but not in a Midwestern stream assemblage. Because his interpretations challenged traditional thinking, they promptly received strong challenges (Herbold, 1984; Rahel et al., 1984; Yant et al., 1984). Yet clearly there is much evidence for the disruptive influence of floods and droughts on stream biotas (Harrell, 1978; Matthews, 1986; Resh et al., 1988; Yount and Niemi, 1990; Reice, 1994).

Part of the controversy is due to variation in the scale of analysis (assemblage, space, time) and its influences, but part is derived also from differences in biological interpretation of the patterns. Subsequent research on the temporal stability of North American stream fish assemblages has led to the general conclusion that both stochastic and deterministic factors influence all of these systems, but the degree and scale of their effects differ from place to place and from time to time. For example, Matthews (1986) showed how a stream fish assemblage was changed by flooding yet returned to a more or less consistent structure during a period of relative habitat stability. Several studies highlight how stochastic disturbances alter stream fish assemblages in the short term, though in the longer term assemblages are persistent and relatively stable (Moyle and Vondracek, 1985; Ross et al., 1985; Freeman et al., 1988; Matthews et al., 1988; Yount and Niemi, 1990). Here persistence and stability follow the operational definitions of Connell and Sousa (1983): persistence is indicated by consistent species presence, and stability by consistent patterns of species relative abundance.

B. Tropical Fish Assemblages

Long-term studies of river fish assemblages in tropical rivers are relatively scarce. This is not surprising given the fact that many fish biologists equate the state of ichthyological knowledge of the Neotropics with that of North America during the early nineteenth century. Rapidly changing taxonomy, incomplete biogeographical knowledge, and the paucity of even the most basic faunal surveys over vast regions of the tropics have all contributed to a heavy emphasis on sampling large areas at the expense of repeated sampling within sites. Yet a few exceptional studies can be cited; Lowe-McConnell (1964) was among the first to discuss the temporal dynamics of fish ecology in a tropical region, namely the Rupununi savanna district of Guyana. Although her study did not track the dynamics of a designated study site, she discussed numerous ecological factors associated with wet/dry seasonality, including fish migrations, respiratory adaptations, population dynamics, diet shifts, and species diversity (see also Lowe-McConnell, 1979).

Bonetto et al. (1969) were perhaps the first to examine the temporal stability of fish assemblages at fixed locations within a tropical river system. They surveyed assemblage composition of floodplain pools on islands of the Río Paraná in northern Argentina, a region with strong wet/dry seasonality. Their study highlighted the random nature of species strandings in isolated pools but at the same time noted how deterministic processes, especially predation, interact with pool conditions to influence development of fish assemblages subsequent to pool isolation. Welcomme (1969) examined the temporal dynamics of fish populations in a stream flowing into Lake Victoria, Africa. As in Low-McConnell's regional study, Welcomme's findings highlighted strong seasonal dynamics and the importance of longitudinal migrations, especially those associated with wet-season spawning and return to dry-season refugia.

In a similar study, Rodríguez and Lewis (1990) surveyed fish assemblages during two consecutive dry seasons in 20 oxbow lagoons in the floodplain of the lower Río Orinoco, Venezuela. They found much greater variation between lakes within years compared to variation within lakes between years. Species richness and species composition within lakes between years was characterized as strikingly similar, and fish-habitat associations were hypothesized to account for this pattern. Similarly, Chapman and Chapman (1993) analyzed the stability of fish assemblages in residual floodplain pools of the River Sokoto, Nigeria, based on data from a fisheries survey conducted by Holden (1963). They identified some of the same random elements discussed by Bonetto et al. but also found a high degree of interannual consistency in the species present in pools; like Rodriguez and Lewis, they postulated habitat selection as a mechanism.

Zaret and Rand's (1971) study of a fish assemblage in a small Panamanian stream was perhaps the first to examine the effect of seasonality on patterns of niche segregation. They observed that most fishes expanded resource use during the wet season when most resources are more abundant and showed a trend to niche compression and segregation during the dry season. Because of its pointed claim to support the "competitive exclusion principle," this study has been widely cited by both fish and non-fish ecologists. Despite the fact that similar patterns have been demonstrated in other tropical fish assemblages (e.g., McKaye and Marsh, 1983; Winemiller, 1989a), several authors have expressed skepticism over Zaret and Rand's study due to their small sample sizes and use of inappropriate statistics for comparing niche overlaps. With a focus on the use of terrestrial resources rather than niche segregation, Goulding (1980) examined seasonal diet shifts among fishes of the Río Madeira in Brazil's Amazon basin. Goulding's study clearly revealed a great reliance by many fishes on food resources of the flooded forest and identified the potential of fishes as seed dispersers for riparian plants.

Several studies associated with international development projects have examined seasonal ecological dynamics (e.g., Lagler et al., 1971) or faunal changes in association with river impoundments (e.g., Balon and Coche, 1974). Zaret and Paine (1973) discussed the changes that occurred in the food-web structure of Lake Gatún, Panama (an impoundment of the Río Chagras in the Canal Zone), following the introduction of exotic Cichla ocellaris, a voracious South American piscivore. They noted how the lake's fish assemblage changed over a period of just four years with major effects on non-fish taxa, both aquatic and terrestrial. Recently, Agostinho et al. (1992) examined the influence of the Itaipu Reservoir on faunal changes in the Upper Río Paraná and associated wetlands in Brazil. They presented evidence that exotic piscivores, brought in by the submergence of a natural barrier to upstream dispersal, have contributed to changes in the fish assemblage of the upper Paraná.

Over the past 10 years, I have studied the fish assemblage of Caño Maraca, a swamp creek in the western llanos of Venezuela. Based primarily on data obtained over 12 months of 1984, various aspects of the seasonal dynamics have been published, including resource utilization (Winemiller, 1989a), life-history strategies (Winemiller, 1989b), the influence of respiratory adaptations on population dynamics (Winemiller, 1989c), and food-web structure (Winemiller, 1990, 1996). This chapter examines assemblage stability based on survey data acquired over the full 10-year interval 1984–1994. In addition, I will compare data from fish surveys conducted in the Upper Zambezi River and associated floodplain in Zambia, Africa during 1966 (Kelley, 1968) with those from a very similar survey that I conducted in the same area in 1989 (Winemiller, 1991a; Winemiller and Kelso-Winemiller, 1994).

II. METHODS

A. Study Regions

1. South American Llanos

During January 1984–June 1994, surveys were conducted on the fish assemblage of Caño Maraca, a lowland stream (caño) in the Venezuelan llanos (Fig. 1). The study site is located in the upper reaches where this low-gradient stream has a large, seasonally-inundated floodplain (additional descriptions of the habitat, community, and food-web appear in Winemiller, 1989b,c, 1990). Highly seasonal rainfall produces large within-year variation in the physical and biological attributes of the site. During the harsh dry season (January through early May), many aquatic plants and invertebrates (e.g., diatoms, rooted macrophytes, molluscs, entocostrachans, conchostracans, annual killifishes) survive in sediments in a state of quiescence or arrested early development. During the early stages of wet-season flooding, these populations have very rapid production.

Sources of basal ecosystem production differ markedly between seasons (Winemiller, 1990, 1996). During the wet season (late May–August), the vast flooded plain contains very high densities of aquatic and emergent macrophytes. Relatively few invertebrates or fishes feed directly on these plants, but numerous herbivores consume periphyton (diatoms and filamentous algae) attached to macrophytes. Wet-season primary production is transferred to the upper food web via abundant invertebrates and juvenile fishes. During the transition to dry season (September–December), increasing desiccation causes dieback in the accumulated macrophyte biomass. Macrophyte decomposition

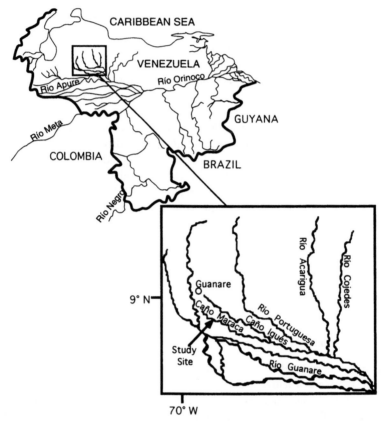

FIGURE 1. Map showing the location of Caño Maraca within the Río Portuguesa basin of the western llanos in Venezuela.

and the confinement of water in isolated pools and small channels results in reduced dissolved oxygen (DO) and higher concentrations of hydrogen sulfide. Some of the resident fishes survive hypoxia by using special respiratory adaptations and are able to feed and, in some cases, reproduce (Winemiller, 1989b). Detritus from aquatic macrophytes is the principal source of basal production during the dry season. Some fishes have seasonal migrations and leave the site during the onset of harsh dry-season conditions, while other species remain throughout the year; some of these suffer high mortality during the dry season when aquatic macrophyte cover is reduced and fishes are at their highest densities.

Diverse life-history strategies have been documented for fishes at the site (Winemiller, 1989b). Some species remain throughout the dry season and, by virtue of their special respiratory adaptations, survive harsh conditions. Other populations remain during the dry season, suffer high mortality from hypoxic conditions and predation, but later rebound with a burst of reproduction following the first floods. Other year-round residents suffer high dry-season mortality and rebound during the wet season either by virtue of an opportunistic strategy of rapid maturation and multiple spawnings of small clutches or by way of an equilibrium strategy of less frequent spawning, production of larger eggs, extended parental care, and enhanced juvenile survivorship. Finally, the fishes that leave the site during the dry season return during the floods and spawn large numbers of eggs, usually by scattering them among submerged plants.

2. Barotse Floodplain of South-Central Africa

During April–December 1989, fishes and their habitats were surveyed in the Upper Zambezi River and associated Barotse floodplain in Zambia's western province (Fig. 2). These surveys were comparable to very similar studies performed in the same region during 1966 (Kelley, 1968). Like the llanos, the Upper Zambezi River has a very strong seasonal cycle of flooding and desiccation (additional descriptions appear in Winemiller, 1991a). As in the llanos, aquatic macrophytes, periphyton, and much of the detritus from aquatic plants originate on the floodplain during high water (January–April). Basal production is transferred to the upper food web primarily in floodplain habitats (i.e., lagoons, sloughs, canals), and to a lesser extent within the primary river channel (Winemiller, 1996). During the period of falling water (May–August), there is a massive movement of organisms from the floodplain to the river channel. There, as also in the llanos, detritus assumes greater importance as a basal input, while macrophyte standing stocks decline with falling water levels.

B. Field Surveys

Fish and habitat surveys were conducted at the Caño Maraca site during each month of 1984 and at several irregular intervals thereafter. Methods for habitat parameters are described in detail elsewhere (Winemiller, 1989c). Each survey required one day, used the same sampling methods, and was performed with the same criterion, namely to sample until one hour of sampling yielded no additional (rare) species. In some instances, a representative sample that reflected species' relative abundance rankings was retained and preserved (in place of sacrificing all captured specimens). During each survey, fishes were captured from a "large open pool" area with a 20-m seine (12.7-mm mesh), from "open pool" areas and "vegetation mats" with a 2.5-m seine (3.2-mm mesh), and from "shallow vegetated edge" habitats with a large dipnet (3.2-mm mesh). Total sampling effort varied between dates depending on habitat conditions, but the methods permitted good assessment of species richness without sacrificing information on relative abundances. Fishes were ultimately fixed in formalin and preserved in ethanol; most specimens were ultimately deposited in natural history collections. During four surveys (in December 1984, January 1988, October 1993, and June 1994), quantitative seine hauls also were made in open water and densely vegetated habitats. These samples permitted comparisons of population densities on a per-unit-area basis between habitats on the same dates, and between dates within these habitats. Except for several large specimens that were identified, measured, and weighed in the field prior to release, fishes from the quantitative samples were retained, and species numbers and biomasses were determined in the laboratory.

During April–December 1989, fishes were surveyed at locations scattered throughout the Upper Zambezi River and Barotse floodplain from Sioma in the South to the juncture of the Kabompo and Zambezi Rivers in the North (Fig. 2). All habitats were targeted, including the main river channel, lagoons, canals, creeks, and flooded vegetation. Fishes in deepwater habitats were captured with monofilament experimental gillnets (50 × 2 m, containing segments of 5.1-, 10.2-, and 15.2-cm mesh) operated overnight. Angling supplemented some of the river channel samples. Fishes were also captured using seines (30.5 × 2 m, 25-mm mesh; 6 × 1.5 m, 6-mm mesh), castnets (25-mm mesh), and dipnets (3.2-mm mesh). Samples of fishes leaving the flooded plain during the early falling-water period (May–July) were obtained by purchasing maalelo catches from local fishermen; maalelo are dams constructed of grass mounds or reed fences that direct fishes into a trap as they attempt to move from flooded grasslands to the permanent water bodies.

Very similar methods were employed by Kelley (1968) to survey the same areas of the Upper Zambezi/Barotse plain during 1966. The goals of the two surveys were to obtain samples representative of fish densities, population size structures, and habitat associations during different seasons. Kelley used a 25-m (12.7-mm mesh) and a 3.3-m (12.7-mm mesh) seine and several 30-m gillnets with meshes ranging between 3.8 and 15.2 cm. Fish abundances were totalled from data reported by Kelley (1968) based on 119 overnight sets of 100 m of gillnet, 43 seine hauls, and nine maalelo samples. His flood-season (March–April) gillnet sets ($N = 40$) caught relatively few fishes. The 1989 abundances were derived from 40 sets of 100 m of gillnet (earliest set was June 28), 40 seine hauls, and two maalelo samples, plus specimens taken by castnet, dipnet, and hook-and-line. Kelley's 1966 survey produced a total of 25,508 specimens, and my 1989 Barotse survey produced 14,504 specimens.

C. Data Analyses

Spearman's rank correlations and Pearson's product-moment correlations were used as estimates of sample similarity based on species abundances. Rank cor-

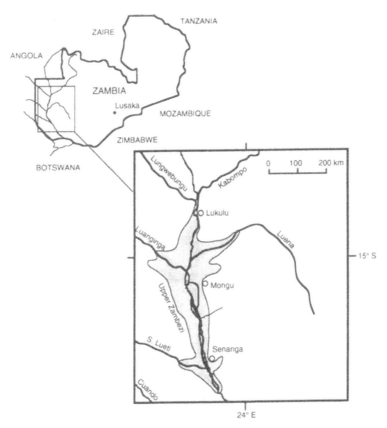

FIGURE 2. Map showing the location of the Upper Zambezi River and Barotse floodplain in western Zambia.

relation has been used extensively in studies of temporal stability of vertebrate assemblages (Grossman *et al.*, 1990) because it reduces the effects of subtle changes in species abundances that may be influenced by sampling bias. I chose to use rank correlations as a conservative test and also parametric correlation to highlight the effect of the magnitude of differences in species' relative abundances on assemblage similarity. The latter was used primarily in cluster analyses ("average linkage" method) to show hierarchies of assemblage similarities, following the approach of Ross *et al.* (1985).

   The criterion (or relative abundance scale) used to exclude rare species influences assemblage patterns and their interpretation (Rahel *et al.*, 1984). No consensus exists for this criterion; most investigators choose either to exclude those species comprising <1% of the total fishes surveyed (e.g., Ross *et al.*, 1985; Bass, 1990), or to examine only the dominant 10 species (Matthews *et al.*, 1988; Chapman and Chapman, 1993). Inclusion of many species that are always rare tends to inflate similarity and estimates of assemblage stability based on correlative methods (Rahel *et al.*, 1984; Grossman *et al.*, 1990). Conversely, to eliminate all but a handful of dominant species imposes a rigid criterion that severely reduces the chance of identifying high similarity based on abundance estimates from field samples. Rather than select a single arbitrary criterion, I examined the influence of scale on assemblage patterns by making comparisons and performing analyses based on several different relative abundance criteria for species inclusion. All available species-abundance data were compiled, and average monthly abundances of species were plotted by rank (Figs. 3 and 4). I looked for inflections in these frequency distributions and analyzed several data sets based on such "natural" breaks. Most of the patterns and interpretations were the same, and therefore results from only two of these data sets are presented here to contrast low and high inclusion criteria: the first inflection that grouped at least 10 dominant species ($N = 14$ for both Venezuela and Zambia studies), and a later inflection that grouped many species, but still eliminated the long tail of the distribution representing very rare taxa ($N = 58$ for Venezuela, $N = 31$ for Zambia). Kelley (1968) failed to report certain species either because they were undescribed at the time and confused with known taxa (e.g., *Serranochromis altus*, Winemiller and Kelso-Winemiller, 1991) or because they were difficult to key out (e.g., *Synodontis* spp.). Thus certain 1989 data had to be combined in order to match the taxonomic units reported by Kelley (e.g., *Synodontis* spp., *Barbus* spp., *Aplocheilichthys* spp.), thereby reducing the number of taxa from 68 to 38 (Fig. 4).

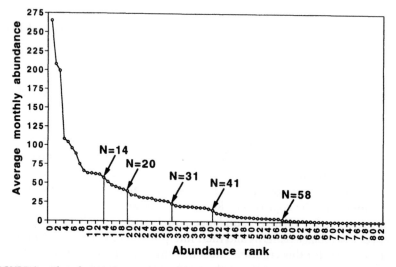

FIGURE 3. Plot of species' average monthly abundances by abundance rank for Caño Maraca fishes based on 1984 surveys.

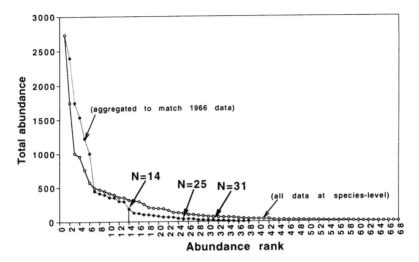

FIGURE 4. Plots of species' average monthly abundances by abundance rank for Upper Zambezi River/Barotse floodplain fishes based on 1989 surveys. One plot contains species aggregations that allow comparisons between the 1989 survey and Kelley's 1966 survey.

Following Winemiller (1989b) and Grossman *et al.* (1990), the coefficient of variation (CV) of monthly population abundance estimates was also used to examine the temporal stability of the Venezuelan fish assemblage. Population CVs (100 · SD/mean abundance) were calculated based on the 1984 monthly samples (to estimate within-year seasonal variation during a typical year) and also based on the full 10-year data set (to estimate overall within- and between-year variation). To examine the relationship between population variability, abundance, and persistence, the CV was regressed against mean abundance and the percentage of samples in which a species was present. In addition, correlations between population abundance estimates and several key environmental variables were performed to examine the influence of habitat conditions on population dynamics.

III. RESULTS

A. Venezuela

Of the 87 species documented at Caño Maraca, characiforms were dominant both in terms of species richness and numbers of individuals, a reflection of their prominent representation in the South American ichthyofauna (Lowe-McConnell, 1975). The nine top-ranked species in overall abundance were all characiforms (Table I, Fig. 3). Based on the 18 samples collected during the

TABLE I Species Relative Abundance Ranks and Percentage Relative Abundances across Sampling Dates at Caño Maraca

Species	Jan84 Rank	Jan84 %	Feb84 Rank	Feb84 %	Mar84 Rank	Mar84 %	Apr84 Rank	Apr84 %	May84 Rank	May84 %	June84 Rank	June84 %	Jul84 Rank	Jul84 %	Aug84 Rank	Aug84 %	Sep84 Rank	Sep84 %
Steindachnerina argentea	2	9.23	2	6.94	21	1.43	7	6.71	15	1.06	1	13.60	6	6.79	2	8.47	2	9.47
Odontostilbe pulcher	1	23.70	8	5.88	1	12.41	5	7.84	5	6.45	4	10.65	1	11.07	9	3.74	7	4.88
Ctenobrycon spilurus	3	9.13	3	6.67	1	12.41	2	9.66	23	0.44	2	10.96	4	7.86	3	5.97	1	10.97
Roeboides dayi	8	2.91	26	0.79	32	0.43	32	0.35	15	1.06	10	3.72	7	5.69	1	8.88	8	4.53
Gephyrocharax valenciae	8	2.91	16	1.93	25	1.12	47	0.00	37	0.00	7	5.02	5	7.54	10	3.69	6	5.07
Astyanax bimaculatus	7	3.16	38	0.35	37	0.25	32	0.35	31	0.09	3	10.70	2	9.62	8	3.79	3	8.05
Aphyocharax alburnus	5	4.91	20	1.67	23	1.18	29	0.49	27	0.27	20	0.57	16	2.14	12	3.38	4	6.54
Markiana geayi	11	2.72	43	0.09	33	0.37	36	0.20	37	0.00	5	8.79	8	5.60	14	2.70	5	6.38
Thoracocharax stellatus	15	2.04	5	6.50	5	5.09	34	0.30	37	0.00	50	0.00	47	0.06	29	0.99	29	0.72
Caquetaia kraussii	18	1.70	14	2.02	28	0.87	16	1.87	26	0.35	22	0.52	3	7.98	11	3.58	12	2.41
Triportheus sp.	24	0.78	38	0.35	30	0.56	41	0.10	37	0.00	14	1.29	9	4.43	19	1.92	26	0.99
Bunocephalus amaurus	16	1.94	6	6.23	5	5.09	3	9.62	3	11.40	29	0.26	39	0.22	34	0.73	13	2.39
Aequidens pulcher	4	5.29	14	2.02	10	3.60	25	0.79	19	0.88	22	0.52	10	2.86	5	4.62	10	3.14
Corydoras habrosus	23	0.83	1	7.11	3	6.33	4	8.09	2	16.96	50	0.00	57	0.00	62	0.00	40	0.29
Prochilodus mariae	10	2.87	11	3.42	14	2.23	18	1.78	9	3.71	13	2.12	11	2.73	24	1.45	19	1.39
Otocinclus sp.	6	4.47	9	5.27	7	4.97	6	7.10	1	17.05	36	0.10	43	0.09	41	0.26	40	0.29
Rineloricaria caracasensis	35	0.24	7	5.97	9	3.91	11	2.37	4	6.98	50	0.00	15	2.26	7	3.90	16	1.50
Ochmacanthus alternus	34	0.29	4	6.58	8	4.47	27	0.69	22	0.71	40	0.05	21	0.94	25	1.35	11	3.08
Hoplias malabaricus	17	1.80	32	0.53	22	1.24	22	1.43	37	0.00	8	4.91	14	2.29	6	3.95	22	1.23
Cichlasoma orinocense	22	1.07	20	1.67	18	1.86	17	1.82	17	0.97	9	4.50	26	0.63	13	3.12	36	0.51
Loricariichthys typus	55	0.00	28	0.61	14	2.23	12	2.17	10	3.53	11	3.15	17	1.82	19	1.92	14	2.25
Pyrrhulina lugubris	21	1.31	19	1.76	26	1.06	41	0.10	37	0.00	36	0.10	52	0.03	34	0.73	55	0.03
Pterygoplichthys multiradiatus	49	0.05	12	3.34	14	2.23	15	2.02	7	4.15	6	7.14	18	1.63	21	1.71	50	0.11
Gymnotus carapo	55	0.00	17	1.84	33	0.37	20	1.73	14	2.03	28	0.31	22	0.88	22	1.56	21	1.37
Astronotus ocellatus	29	0.49	24	0.88	37	0.25	25	0.79	31	0.09	36	0.10	43	0.09	4	4.68	55	0.03
Corydoras aeneus	11	2.72	32	0.53	19	1.49	1	10.11	12	2.12	36	0.10	52	0.03	62	0.00	63	0.00
Hypostomus argus	39	0.15	12	3.34	12	2.48	13	2.07	8	3.98	12	2.43	33	0.38	18	1.97	44	0.24

Charax gibbosus	28	0.53	48	0.00	51	0.00	47	0.00	37	0.00	25	0.36	19	1.23	23	1.51	18	1.45
Apistogramma hoignei	19	1.51	10	5.09	13	2.36	24	0.84	20	0.80	50	0.00	52	0.03	41	0.26	47	0.16
Pimelodella sp.2	36	0.19	40	0.18	30	0.56	30	0.44	31	0.09	50	0.00	35	0.28	56	0.05	9	3.70
Microglanis iheringi	20	1.36	17	1.84	26	1.06	8	4.04	12	2.12	33	0.16	27	0.57	15	2.60	24	1.15
Poecilia reticulata	49	0.05	22	1.32	4	5.46	44	0.05	37	0.00	18	0.83	12	2.36	28	1.04	30	0.70
Hemigrammus sp.	44	0.10	43	0.09	48	0.06	47	0.00	37	0.00	30	0.21	47	0.06	16	2.23	15	1.88
Corydoras septemtrionalis	13	2.19	28	0.61	17	2.11	9	3.06	17	0.97	50	0.00	43	0.09	47	0.16	40	0.29
Characidium sp.1	26	0.68	32	0.53	40	0.19	34	0.30	23	0.44	40	0.05	57	0.00	36	0.62	16	1.50
Eigenmannia virescens	30	0.44	40	0.18	40	0.19	31	0.39	20	0.80	40	0.05	47	0.06	36	0.62	28	0.75
Hoplosternum littorale	31	0.39	23	1.14	11	2.92	10	2.47	6	5.65	25	0.36	40	0.19	47	0.16	63	0.00
Pimelodella sp.3	26	0.68	24	0.88	23	1.18	23	1.04	10	3.53	50	0.00	23	0.72	41	0.26	39	0.35
Pygocentrus cariba	55	0.00	48	0.00	51	0.00	47	0.00	37	0.00	19	0.72	20	1.07	30	0.88	22	1.23
Parauchenipterus galeatus	24	0.78	32	0.53	35	0.31	18	1.78	23	0.44	24	0.41	31	0.41	40	0.47	25	1.05
Tetragonopterus argenteus	55	0.00	48	0.00	51	0.00	47	0.00	37	0.00	50	0.00	35	0.28	62	0.00	19	1.39
Cheirodontops geayi	13	2.19	37	0.44	19	1.49	28	0.64	37	0.00	50	0.00	57	0.00	62	0.00	55	0.03
Rhamdia sp.	44	0.10	28	0.61	44	0.12	21	1.48	29	0.18	16	1.03	31	0.41	46	0.21	37	0.48
Rachovia maculipinnus	55	0.00	48	0.00	51	0.00	47	0.00	37	0.00	15	1.24	13	2.33	27	1.30	45	0.19
Synbranchus marmoratus	44	0.10	48	0.00	44	0.12	47	0.00	27	0.27	33	0.16	29	0.47	31	0.83	30	0.70
Hypopomus sp.1	36	0.19	32	0.53	35	0.31	36	0.20	31	0.09	40	0.05	57	0.00	53	0.10	45	0.19
Schizodon isognathus	55	0.00	48	0.00	51	0.00	47	0.00	37	0.00	33	0.16	37	0.25	47	0.16	27	0.88
Serrasalmus irritans	55	0.00	48	0.00	51	0.00	47	0.00	37	0.00	30	0.21	24	0.66	25	1.35	34	0.64
Leporinus friderici	39	0.15	48	0.00	51	0.00	47	0.00	37	0.00	40	0.05	57	0.00	47	0.16	34	0.64
Serrasalmus medinai	49	0.05	48	0.00	51	0.00	47	0.00	37	0.00	50	0.00	29	0.47	17	2.03	51	0.08
Ancistrus sp.	39	0.15	27	0.70	37	0.25	13	2.07	29	0.18	50	0.00	57	0.00	62	0.00	63	0.00
Hypoptopoma sp.	36	0.19	28	0.61	40	0.19	47	0.00	37	0.00	50	0.00	37	0.25	33	0.78	47	0.16
Brycomamericus beta	55	0.00	48	0.00	51	0.00	47	0.00	37	0.00	50	0.00	24	0.66	31	0.83	47	0.16
Adontosternarchus devananzii	49	0.05	48	0.00	51	0.00	47	0.00	37	0.00	40	0.05	57	0.00	38	0.52	32	0.67
Entomocorus gameroi	49	0.05	48	0.00	51	0.00	47	0.00	37	0.00	50	0.00	47	0.06	41	0.26	32	0.67
Triportheus angulatus	32	0.34	43	0.09	29	0.62	36	0.20	37	0.00	17	0.93	28	0.50	62	0.00	63	0.00
Hopleryhrinus unitaeniatus	55	0.00	48	0.00	40	0.19	39	0.15	37	0.00	25	0.36	57	0.00	56	0.05	63	0.00
Serrasalmus rhombeus	55	0.00	48	0.00	51	0.00	47	0.00	37	0.00	50	0.00	33	0.38	38	0.52	40	0.29
Total sample abundance	2059		1139		1611		2028		1132		1934		3181		1925		3729	

(continues)

111

Table I *continued*

Species	Oct84 Rank	Oct84 %	Nov84 Rank	Nov84 %	Dec84 Rank	Dec84 %	Jan88 Rank	Jan88 %	Jan89 Rank	Jan89 %	Jan92 Rank	Jan92 %	Oct93 Rank	Oct93 %	Feb94 Rank	Feb94 %	Jun94 Rank	Jun94 %
Steindachnerina argentea	3	6.87	4	5.71	1	18.06	7	6.54	11	3.06	2	15.61	1	29.64	4	0.00	17	1.15
Odontostilbe pulcher	1	23.68	3	5.81	2	8.89	1	8.03	3	8.89	3	9.70	2	13.81	4	0.00	4	8.08
Ctenobrycon spilurus	2	10.18	5	5.61	4	5.30	2	7.73	24	0.19	15	1.27	3	13.57	4	0.00	6	5.00
Roeboides dayi	5	6.60	9	3.80	8	2.92	15	2.38	5	7.74	9	3.80	5	9.31	4	0.00	2	15.00
Gephyrocharax valenciae	4	6.79	6	5.34	7	3.87	3	7.44	12	2.77	3	9.70	7	4.17	4	0.00	27	0.38
Astyanax bimaculatus	11	2.37	11	3.19	5	4.21	11	3.51	20	0.38	6	4.64	9	2.12	4	0.00	5	6.54
Aphyocharax alburnus	7	2.84	18	2.12	24	1.32	4	7.14	17	0.86	15	1.27	6	6.41	4	0.00	3	8.85
Markiana geayi	25	0.92	1	6.42	6	3.93	13	2.80	8	3.73	22	0.84	17	0.35	4	0.00	27	0.38
Thoracocharax stellatus	30	0.47	21	1.88	11	2.46	14	2.50	31	0.00	22	0.84	4	11.25	4	0.00	1	16.54
Caquetaia kraussii	10	2.50	10	3.36	17	1.81	8	5.71	6	11.76	5	5.49	10	0.90	4	0.00	11	2.69
Triportheus sp.	19	1.58	2	6.05	3	6.68	44	0.00	6	6.12	30	0.00	8	2.16	4	0.00	9	3.08
Bunocephalus amaurus	6	3.45	13	3.02	20	1.40	25	0.77	20	0.38	25	0.42	30	0.07	4	0.00	36	0.00
Aequidens pulcher	9	2.60	12	3.16	25	1.12	12	3.09	7	5.07	11	2.11	19	0.28	1	83.33	8	4.62
Corydoras habrosus	17	1.74	36	0.57	10	2.49	5	6.84	31	0.00	10	3.38	48	0.00	4	0.00	36	0.00
Prochilodus mariae	13	2.00	25	1.41	12	2.44	5	6.84	28	0.10	30	0.00	38	0.04	4	0.00	14	1.54
Otocinclus sp.	36	0.32	38	0.50	13	1.98	31	0.30	31	0.00	30	0.00	48	0.00	4	0.00	36	0.00
Rineloricaria caracasensis	22	1.24	46	0.24	23	1.38	37	0.12	31	0.00	30	0.00	32	0.05	4	0.00	9	3.08
Ochmacanthus alternus	8	2.82	31	0.94	17	1.81	20	1.61	31	0.00	30	0.00	48	0.00	4	0.00	36	0.00
Hoplias malabaricus	27	0.66	14	2.92	31	0.77	23	1.07	15	1.34	15	1.27	25	0.11	2	8.33	12	1.92
Cichlasoma orinocense	32	0.42	32	0.81	35	0.57	18	2.08	4	8.70	7	4.22	20	0.26	4	0.00	36	0.00
Loricariichthys typus	13	2.00	26	1.31	36	0.54	37	0.12	31	0.00	25	0.42	32	0.05	4	0.00	17	1.15
Pyrrhulina lugubris	20	1.53	8	4.13	9	2.81	19	1.78	14	1.43	1	17.72	11	0.79	4	0.00	36	0.00
Pterygoplichthys multiradiatus	23	1.21	27	1.24	51	0.17	33	0.18	22	0.29	30	0.00	25	0.11	4	0.00	22	0.77
Gymnotus carapo	18	1.66	20	2.02	20	1.40	17	2.14	13	2.10	7	4.22	22	0.16	4	0.00	36	0.00
Astronotus ocellatus	46	0.08	35	0.67	40	0.46	30	0.36	1	24.47	15	1.27	42	0.02	4	0.00	36	0.00
Corydoras aeneus	26	0.76	19	2.08	56	0.09	27	0.59	31	0.00	30	0.00	21	0.25	4	0.00	36	0.00
Hypostomus argus	24	1.05	24	1.44	34	0.63	44	0.00	31	0.00	30	0.00	30	0.07	4	0.00	17	1.15

Species																		
Charax gibbosus	12	2.08	23	1.68	16	1.86	33	0.18	31	0.00	30	0.00	15	0.49	4	0.00	36	0.00
Apistogramma hoignei	38	0.18	28	1.18	45	0.34	9	5.53	24	0.19	14	1.69	12	0.71	4	0.00	36	0.00
Pimelodella sp.2	35	0.37	16	2.65	15	1.89	44	0.00	31	0.00	30	0.00	48	0.00	4	0.00	22	0.77
Microglanis iheringi	32	0.42	40	0.37	46	0.32	33	0.18	24	0.19	30	0.00	48	0.00	4	0.00	27	0.38
Poecilia reticulata	21	1.45	50	0.07	43	0.40	40	0.06	19	0.57	11	2.11	32	0.05	2	8.33	27	0.38
Hemigrammus sp.	15	1.87	44	0.27	29	0.83	10	3.87	30	0.76	30	0.00	48	0.00	4	0.00	27	0.38
Corydoras septemtrionalis	37	0.26	39	0.44	20	1.40	21	1.37	18	0.00	30	0.00	16	0.37	4	0.00	36	0.00
Characidium sp.1	16	1.82	30	1.01	26	1.00	24	0.89	31	0.10	25	0.42	13	0.62	4	0.00	36	0.00
Eigenmannia virescens	42	0.13	7	4.37	14	1.92	40	0.06	28	0.00	30	0.00	25	0.11	4	0.00	36	0.00
Hoplosternum littorale	48	0.05	42	0.30	32	0.66	26	0.65	31	0.00	15	1.27	42	0.02	4	0.00	36	0.00
Pimelodella sp.3	34	0.39	34	0.74	26	1.00	15	2.38	10	3.44	30	0.00	38	0.04	4	0.00	36	0.00
Pygocentrus cariba	50	0.00	17	2.28	19	1.69	31	0.30	31	0.00	30	0.00	48	0.00	4	0.00	12	1.92
Parauchenipterus galeatus	44	0.11	22	1.78	29	0.83	44	0.00	16	1.15	30	0.00	42	0.02	4	0.00	22	0.77
Tetragonopterus argenteus	27	0.66	15	2.79	28	0.89	44	0.00	31	0.00	30	0.00	28	0.09	4	0.00	36	0.00
Cheirodontops geayi	38	0.18	50	0.07	41	0.43	44	0.00	31	0.00	30	0.00	14	0.51	4	0.00	36	0.00
Rhamdia sp.	38	0.18	41	0.34	49	0.26	29	0.48	31	0.00	22	0.84	48	0.00	4	0.00	36	0.00
Rachovia maculipinnus	50	0.00	54	0.03	68	0.00	44	0.00	31	0.00	30	0.00	48	0.00	4	0.00	27	0.38
Synbranchus marmoratus	31	0.45	44	0.27	56	0.09	27	0.59	31	0.00	30	0.00	22	0.16	4	0.00	36	0.00
Hypopomus sp.1	42	0.13	37	0.54	37	0.49	21	1.37	22	0.29	15	1.27	38	0.04	4	0.00	36	0.00
Schizodon isognathus	49	0.03	29	1.11	46	0.32	40	0.06	31	0.00	30	0.00	48	0.00	4	0.00	36	0.00
Serrasalmus irritans	50	0.00	60	0.00	61	0.03	44	0.00	31	0.00	30	0.00	48	0.00	4	0.00	14	1.54
Leporinus friderici	50	0.00	32	0.81	37	0.49	37	0.12	31	0.00	30	0.00	48	0.00	4	0.00	27	0.38
Serrasalmus medinai	50	0.00	48	0.20	53	0.11	40	0.06	31	0.00	30	0.00	48	0.00	4	0.00	17	1.15
Ancistrus sp.	50	0.00	60	0.00	56	0.09	44	0.00	31	0.00	11	2.11	38	0.04	4	0.00	36	0.00
Hypoptopoma sp.	44	0.11	54	0.03	53	0.11	44	0.00	31	0.00	30	0.00	32	0.05	4	0.00	6	5.00
Bryconamericus beta	46	0.08	54	0.03	56	0.09	33	0.18	31	0.00	30	0.00	24	0.14	4	0.00	22	0.77
Adontosternarchus devananzii	50	0.00	60	0.00	32	0.66	44	0.00	24	0.19	30	0.00	48	0.00	4	0.00	36	0.00
Entomocorus gameroi	38	0.18	54	0.03	41	0.43	44	0.00	31	0.00	30	0.00	42	0.02	4	0.00	27	0.38
Triportheus angulatus	50	0.00	60	0.00	68	0.00	44	0.00	31	0.00	30	0.00	48	0.00	4	0.00	36	0.00
Hoplerythrinus unitaeniatus	50	0.00	60	0.00	68	0.00	44	0.00	9	3.63	15	1.27	48	0.00	4	0.00	36	0.00
Serrasalmus rhombeus	50	0.00	46	0.24	48	0.29	44	0.00	28	0.10	30	0.00	48	0.00	4	0.00	27	0.38
Total sample abundance	3801		2976		3489		1681		1046		237		5661		12		260	

Note. Only the 58 most abundant species are listed (total species = 87).

10-year study, both persistent and nonpersistent species were common. The frequency distribution of the number of samples in which a species was present was strongly bimodal with approximately half of the species present in 50% or more of the collections (Fig. 5). That so many fishes were represented in so many samples is rather remarkable given that the habitat becomes very harsh during the dry season. During years of unusually low rainfall, the site can dry up completely, as was nearly the condition during the February 1994 visit when only three fish species (*Hoplias malabaricus, Poecilia reticulata, Aequidens pulcher*) were encountered in a single residual pool just 30 cm deep. Recent deforestation in the watershed may be contributing to increasingly harsh conditions during the driest period (March–April), yet subsequent immigration and spawning reestablish a diverse fish assemblage each year.

Within an annual cycle, a direct relationship exists between environmental parameters that indicate the quality of the aquatic habitat and fish diversity and abundance (Fig. 6). Due to the very low relief and gradient of the llanos landscape, the area of aquatic habitat greatly expands with increasing water depth. Depth and pH had the highest correlations with species richness (Table II). Depth and pH covary strongly at this site (Table II), so that high species richness during the wet season probably results from increased water depth and volume rather than low pH. Temperatures can exceed 35°C in shallow dry-season pools, yet many fishes appear to thrive. Dissolved oxygen (DO) is frequently cited as one of the most critical environmental parameters influencing short-term survivorship of aquatic organisms. Yet owing to special respiratory adaptations, diverse llanos fishes can persist in nearly anoxic conditions,

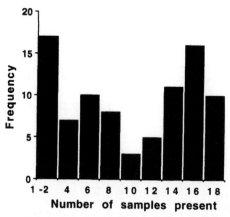

FIGURE 5. Frequency distribution of the number of samples in which individual fish species were present (maximum possible = 18).

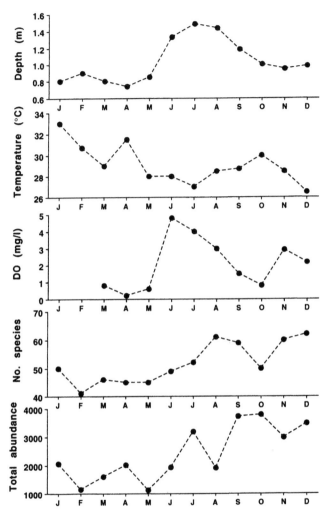

FIGURE 6.   1984 plots of monthly variation in environmental parameters and species and individual fish abundances at Caño Maraca.

and in some cases they are even able to feed and reproduce. For example, the October 1993 sample (surface DO = 0.2 mg/l) contained 47 species and 5661 individuals. Most of the characid fishes sampled had dermal lip protuberances, a morphological response to hypoxia that presumably enhances aquatic surface respiration (Winemiller, 1989c). The low correlations between DO and species richness and abundance (Table II) were influenced by the presence of special respiratory adaptations in a large number of llanos fish species.

TABLE II   Correlations between Environmental Parameters, Fish Species Richness, and Total Fish Abundance Based on 16 Sampling Dates at Caño Maraca

	Depth	Temperature	DO	pH	Velocity	Richness
Temperature	−0.363					
DO	0.614*a*	−0.508*a*				
pH	−0.764*	0.582*	−0.397*a*			
Velocity	0.806*b***	−0.383	0.821*b***	−0.534*b*		
Richness	0.621*	−0.165	0.035*a*	−0.641*	0.178*b*	
Abundance	0.221	0.090	−0.318*a*	−0.188	−0.043*b*	0.025

*a*df = 14 due to lack of field measurement for parameter.
*b*df = 13.
*P <0.025,   **P <0.0001.

Clearly this fish assemblage varies over the short term in response to dynamic changes in habitat driven by markedly seasonal precipitation. The more vexing questions are the following: How do highly persistent species differ from nonpersistent species, and to what extent does the assemblage return to a predictable composition during the late wet season and transition to dry season? The frequency distribution for sample abundance CVs approximates a log-normal distribution, which suggests that very few populations are highly stable, many populations are moderately variable, and many others are highly variable (Fig. 7). The difference between the distribution based on abundant species and that based on all 87 species suggests that extremely large variability is associated with rarity (Fig. 7), but to some extent this is an artifact of

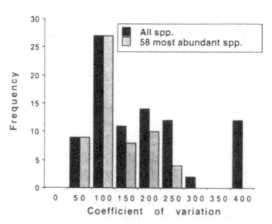

FIGURE 7.   Frequency distribution of the coefficient of variation for species at Caño Maraca based on all species and the dominant 58 species over the 10-year period.

sample size. The relationship between average monthly abundance and the CV was inverse and rather weak for both the data set based on the 12 monthly surveys of 1984 and that based on the full 18 samples from the 10-year survey (Fig. 8). The slope was steeper and the correlation higher for the 1984 data series, an indication that species that tended to remain resident year-round also had lower variability in local density; this pattern is confirmed by the plots in Fig. 9. The most abundant species were the most persistent and also tended to be the least variable over the 10-year study.

Only one species, the cichlid *Aequidens pulcher,* was present in all 18 samples. This fish is a trophic generalist with an "equilibrium-type" life-history characterized by relatively small broods, large eggs, brood defense, and asea-

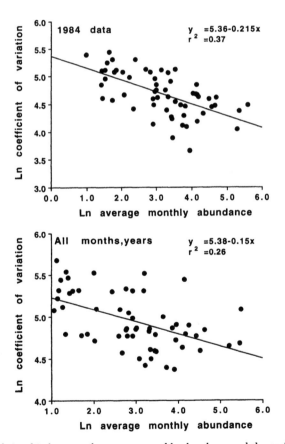

FIGURE 8.   Relationship between the average monthly abundance and the coefficient of variation based on the 1984 surveys and on the 10-year survey period.

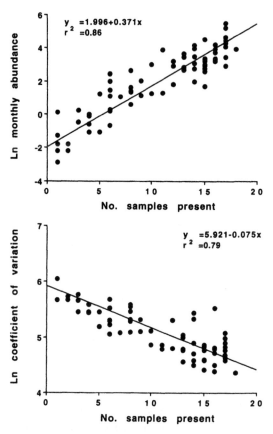

FIGURE 9. The average monthly abundance and the coefficient of variation of sample abundance plotted against the number of samples in which a species was present.

sonal reproduction (Winemiller, 1989b). Because at least some individuals are reproducing during any given period, the population has a strongly persistent size-structure. Of the eight species that were present in 17 of the samples, one characiform has a relatively equilibrium-type strategy (*Hoplias malabaricus*), two characiforms have relatively "periodic-type" strategies (*Steindachnerina argentea, Astyanax bimaculatus*), and five species have relatively "opportunistic-type" strategies (four characids, one poeciliid). Periodic strategists (the "seasonal strategy" of Winemiller, 1989b) spawn only once or twice during the early wet season, shed large clutches of small eggs, and have little or no parental care (Winemiller, 1992; Winemiller and Rose, 1992). Opportunistic fishes mature at small sizes and produce small broods at frequent intervals over extended breeding seasons.

To shed further light on the attributes of stable versus variable fish populations at this site, the effects of one phylogenetic and three ecological variables
on average sample abundance and its CV were examined. Assignment of life-
history strategies was based on data and patterns reported in Winemiller
(1989b) and conform to the following criteria: equilibrium-type (maximum SL
>50 mm, mature egg diameter >1.50 mm, with brood guarding), opportunis-
tic-type (maximum SL <50 mm, number of spawning bouts per year ≥2), and
periodic-type (all others, with few exceptions species with maximum SL >50
mm, 1 spawning bout per year, egg diameter <1.50 mm, no brood guarding).
Species were assigned to trophic guilds (detritivore-algivore, omnivore, inver-
tebrate-feeder, piscivore) based on volumetic dietary data (Winemiller and
Pianka, 1990). Fishes were further categorized into two groups: those possessing special accessory respiratory adaptations and those not currently known to
possess such attributes. Finally, fishes were segregated based on classification
(or not) in the order Characiformes. The results are shown in Table III; neither
monthly sample abundance nor its CV were significantly affected by life history, trophic guild, respiratory adaptations, or taxonomic order. The effect of
trophic guild on abundance was nearly significant at $P = 0.051$, and a Tukey-
HSD test showed that omnivores were more abundant than invertebrate feeders ($P = 0.04$).

The results of Spearman's rank correlations, r_s based on fish abundances
among all 18 monthly samples, revealed no general relationship between similarity and duration of the time interval between samples (Table IV). Similarity
was nearly as high between-years (especially among samples taken during the
same season) as within-years (among samples taken at different seasons) (Table IV). Similar qualitative results were obtained for the 14 and 58 species data

TABLE III   Results of ANOVA and t-Tests for the Effects of Life-History Strategy,
Trophic Guild, Accessory Respiration, and Taxonomic Order on the Average Abundance
and Coefficient of Variation Based on 58 Species at Caño Maraca during 1984

	Abundance			CV		
	F	df	P	F	df	P
Life-history strategy	0.258	2, 55	0.77	0.942	2, 55	0.40
Trophic guild	2.76	3, 54	0.051	2.30	3, 54	0.09
	t	df	P	t	df	P
Accessory respiration	1.340	28	0.186	0.723	28	0.47
Taxonomic order[a]	1.122	28	0.27	1.440	28	0.155

Note. Average monthly abundances and CVs were ln-transformed.
[a]Characiforms vs all other orders.

TABLE IV   Spearman's Rank Correlations for 10 Selected Sampling Dates at Caño Maraca, Based on Numeric Percentages of the 14 and 58 Dominant Species

	Based on 14 dominant species								
	Jan84	Feb84	Jun84	Oct84	Jan88	Jan89	Jan92	Oct93	Feb94
Feb84	0.157								
Jun84	0.673*	−0.185							
Oct84	0.713*	0.322	0.548*						
Jan88	0.598*	0.372	0.399	0.596*					
Jan89	0.101	−0.443	0.214	0.220	−0.084				
Jan92	0.509	0.212	0.448	0.560*	0.613*	0.308			
Oct93	0.631*	0.176	0.593*	0.534*	0.380	0.117	0.370		
Feb94	0.241	0.000	−0.207	−0.034	−0.103	0.172	−0.035	−0.310	
Jun94	0.380	−0.183	0.100	0.051	0.026	0.099	0.007	0.595*	0.034

	Based on 58 dominant species								
	Jan84	Feb84	Jun84	Oct84	Jan88	Jan89	Jan92	Oct93	Feb94
Feb84	0.598*								
Jun84	0.303*	0.203							
Oct84	0.603*	0.621*	0.469*						
Jan88	0.661*	0.601*	0.356*	0.685*					
Jan89	0.404*	0.295*	0.538*	0.404*	0.452*				
Jan92	0.431*	0.491*	0.400*	0.463*	0.618*	0.670*			
Oct93	0.631*	0.383*	0.344*	0.654*	0.533*	0.421*	0.520*		
Feb94	0.105	0.087	0.191	0.147	0.060	0.236	0.236	0.068	
Jun94	0.242	0.161	0.393*	0.330*	0.182	0.205	0.107	0.322*	0.197

*$P < 0.01$.

sets; the correlation between two r_s matrices containing all possible pairwise combinations among 18 samples was 0.69 ($df = 152$, $P < 0.0001$). The first column of Table IV is selected to provide one example (lower half, results based on 58 species): sample Jan84 was most similar to Jan88 followed by Oct93 and Oct84; aside from Feb94 (the sample from the nearly dry pool), the two early wet-season samples (Jun84, Jun94) were the most dissimilar to Jan84.

The same qualitative pattern resulted from Pearson correlations based on sample abundances; it is illustrated in Fig. 10 by dendrograms using the average clustering algorithm. Based on the dominant 14 species, the four 1984 dry-season samples form a tight cluster on branch D; Jan84, Jan88, and Oct84 cluster on branch H; a combination of wet-1984, transition-, and dry-season samples cluster on branch G; and Jun94 and Feb94 are outliers (see Fig. 10). Based on 58 dominant species, Feb94, Jan89, Jun94, and Jan92 are outliers;

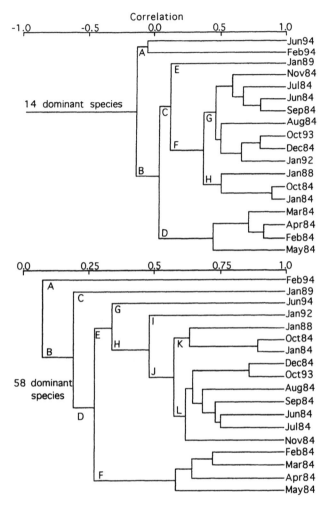

FIGURE 10.   Dendrograms showing similarity of fish assemblage structure among surveys conducted over the 10-year period. Top dendrogram is based on the 14 dominant species, and the bottom dendrogram is based on the 58 dominant species. In each case, Feb94, Jun94, and Jan89 are outliers; dry season 1984 samples are grouped; and wet/transition season samples are grouped.

with the four dry-season 1984 samples clustering on branch F; Jan84, Jan88, and Oct84 clustering on K; and all wet season 1984 and the remaining transition-season samples clustering on branch L (Fig. 10).

Absolute fish densities (numbers or biomass/m²) were estimated for four samples and are plotted in Fig. 11. By either measure, total fish densities were greatest during the transition season (samples Dec84 and Oct93), lower dur-

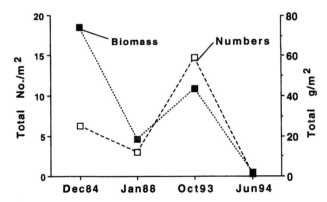

FIGURE 11.   The total abundance and biomass of fishes per unit area from four quantitative density estimates performed at Caño Maraca.

ing the dry season (Jan88), and extremely low during the wet season (Jun94). Biomass was larger relative to numbers in the Dec84 sample compared with the Oct93 sample, and this may reflect the influence of piscivory during the transition season. Late transition-season samples, Nov84 and Dec84, contained many large fishes (Table I), some that were piscivores (e.g., *Pygocentrus cariba, Hoplias malabaricus, Caquetaia kraussii*) and others that were nonpiscivores but possessing armor or spines (*Hoplosternum littorale, Hypostomus argus*). The abundance of the two dominant fish orders varied greatly in these four density samples (Fig. 12). Characiforms (tetras and related fishes), followed by siluriforms (catfishes), dominated all four periods in terms of numbers, biomass, and species richness. Characiform numbers were greater in sample Oct93 than Dec84, but the reverse trend occurred for biomass density. This trend supports the transition season piscivory hypothesis. Siluriform numbers show a small graded decline from sample Dec84 to sample Jun94, but biomass follows the characiform pattern. Except for the wet-season sample (Jun94), perciform (cichlid) density varied little.

In three samples the fish density data were segregated by two major pool habitats, open water ("open") and dense vegetation ("veg"). Clusters based on correlation between the numerical versus biomass densities of the dominant 58 species were different (Fig. 13). Open water samples from the three periods formed a tight cluster (branch C) when the analysis was based on biomass data. Clustering based on numerical densities had the veg-Dec84 sample most similar to open-Jan88, and the two Jun94 samples clustered together (Fig. 13). The differences between the two dendrograms is due to the influence of the Jun84 sample. The early wet-season sample (Jun94) contained mostly small juvenile fishes, most of which used dense vegetation both as a foraging sub-

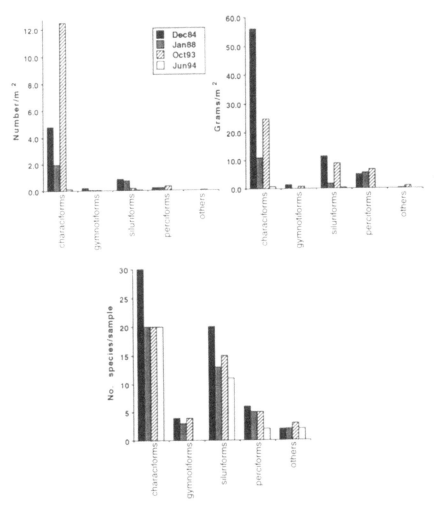

FIGURE 12.   The numerical, biomass, and species densities of fish orders from four quantitative density estimates performed at Caño Maraca.

strate containing microcrustacea and other juvenile food resources and as a refuge from predators. Growth is rapid, and many species undergo ontogenetic niche shifts involving less frequent occupation of vegetation and greater feeding on terrestrial arthropods, detritus, seeds, or fishes (Winemiller, 1989a). The habitat affinities of adult size classes were very predictable for many species. For example, most of the knifefishes (weakly electric gymnotiforms) reside almost exclusively in vegetation, whereas redbelly piranhas, *Pygocentrus*

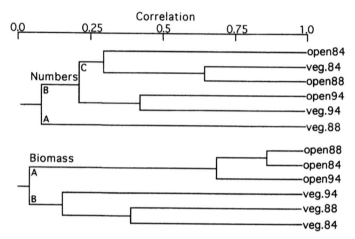

FIGURE 13. Dendrograms showing the similarity of three open water and three vegetation samples from Caño Maraca calculated from quantitative density estimates of fish numbers and biomass on samples Dec84, Jan88, and Jun94 with 58 dominant species.

cariba (Characidae), are schooling midwater fishes. Several habitat generalists also occur at this site, including the very common species *Aequidens pulcher* and *Odontostilbe pulcher.*

B. Zambia

The 68 species documented from the Upper Zambezi River/Barostse floodplain during 1989 were aggregated into 39 taxonomic units for comparison with Kelley's 1966 data. In addition to undescribed taxa, five of the 39 aggregate taxa were not mentioned by Kelley, either because they were not collected (*Afromastacembelus frenatus, Hippopotamyrus ansorgii*) or they were confused with other species (*Tilapia ruweti* similar to juvenile *T. sparrmanii*). The distribution of species' relative abundances was approximately log-normal as illustrated by the strong negative exponential function for the ranked abundance distribution (Fig. 4). Although characiforms are an important component of the Upper Zambezi fish fauna, they do not dominate as they do in South America. The most abundant species in both Zambezi samples was the small cichlid *Pseudocrenilabrus philander,* and the next five highest ranking species were the same for both years, with taxa ranking second and fourth switched between years (Table V).

Spearman's rank correlation between the two annual samples was 0.87 using 14 dominant species, 0.87 with 25 species, and 0.86 with 31 species (P

TABLE V   Comparison of Species Abundance Ranks and Percentage Relative Abundances between Year-Long Surveys during 1966 and 1989 on the Upper Zambezi/Barotse Floodplain

Species	1966		1989	
	Rank	% abundance	Rank	% abundance
Pseudocrenilabrus philander	1	21.707	1	18.829
All Aplocheilichthys spp.	2	11.185	4	10.597
Schilbe mystus	3	8.633	3	12.038
All Barbus spp	4	8.527	2	16.526
All Synodontis spp.	5	8.515	5	8.432
Tilapia sparrmanii	6	6.143	6	6.922
All Clarias spp.	7	4.046	12	2.055
Marcusenius macrolepidotus	8	3.438	7	3.075
Hydrocynus forksahli	9	3.356	10	2.441
Serranochromis macrocephalus	10	3.289	13	2.020
Tilapia rendalli	11	3.266	9	2.716
Micralestes acutidens + Rhabdalestes maunensis	12	2.799	24	0.407
Pollimyrus castelnaui	13	2.070	8	2.847
Oreochromis andersonii	14	1.937	27	0.255
Serranochromis codringtoni	15	1.529	11	2.427
Petrocephalus catastomus	16	1.501	16	0.820
Serranochromis altus + S. angusticeps	17	1.494	20	0.531
Ctenopoma multispinis	18	1.407	21	0.455
Serranochromis giardi	19	1.353	19	0.607
Hemichromis elongatus	20	1.015	17	0.696
Hepsetus odoe	21	0.533	14	1.276
Pharyngochromis darlingi	22	0.423	28	0.248
Serranochromis robustus	23	0.388	18	0.683
Auchenoglanis ngamensis	24	0.369	33	0.083
Labeo lunatus	25	0.243	30	0.145
Oreochromis macrochir	26	0.216	22	0.427
Serranochromis carlottae	27	0.204	26	0.262
Mormyrus lacerda	28	0.141	15	0.869
Hemigammocharax multifasciatus	29	0.125	32	0.124
Hemigrammocharax machadoi	30	0.074	38	0.007
Chiloglanis sp.	31	0.043	37	0.014
Brycinus (=Alestes) lateralis	32	0.016	34	0.076
Hippopotamyrus discorhynchus	33	0.012	25	0.303
Ctenopoma intermedium	34	0.004	23	0.414
Tilapia ruweti		0.000	29	0.159
Afromastacembelus frenatus		0.000	31	0.138
Amphilius uranoscopus		0.000	35	0.041
Serranochromis longimanus		0.000	36	0.028
Hippopotamyrus ansorgii		0.000	39	0.007
Total sample abundance		25,508		14,504

<0.0001 in each case). The correlations were even greater for Pearson correlations using ln-transformed sample abundances (0.90 with 14 species, 0.91 with 25 species, and 0.92 with 31 species, where P <0.0001 in each case). Whereas some shuffling of species' abundance ranks has occurred, the high correspondance between these large-scale surveys conducted nearly a quarter of a century apart is rather remarkable, especially in the presence of artisanal and commercial fishing. Abundance does not seem to be related to phylogeny, life history, trophic niche, or body size; representatives of all these categories are well dispersed throughout the rankings in Table V. The greatest drops in abundance rank were *Micralestes acutidens* (from 12th in 1966 to 24th in 1989) and *Oreochromis andersonii* (from 14th to 27th). *Micralestes acutidens* might have been undersampled in 1989 since this small schooling characid appeared to favor shallow water areas between sand ridges along the edge of the river channel. Most of Kelley's *M. acutidens* came from river channel samples, and this habitat was not extensively seined during 1989. *Oreochromis andersonii* should have been captured just as effectively as other large tilapine cichlids within lagoons, river edge habitats, and flooded terrain. Large tilapine cichlids are among the most valued fishes, but why this species alone might have suffered reductions from overharvest is unclear.

IV. DISCUSSION

A. The Influence of Seasonality on Assemblage Patterns

The annual changes that occur in the abiotic environment of floodplain rivers are very predictable in tropical regions (Lowe-McConnell, 1975, 1979; Welcomme, 1979; Junk et al., 1989). Strongly unimodal patterns of annual precipitation (or bimodal patterns in some equatorial regions) result in sheet flooding followed by gradual recession of waters until the only aquatic habitats that remain are permanent stream channels, lagoons, and isolated floodplain pools. Primary production is strongly correlated with initial flooding and the area of aquatic habitat. Secondary production, both in the form of invertebrates and algivorous fishes, rapidly follows suit; degradation of aquatic habitats ensues with the onset of the dry season, with profound effects on fish assemblages.

In the Venezuelan llanos, high temperatures, aquatic hypoxia, habitat reduction, and isolation contribute to a general reduction in local fish populations in the dry season, with changes in assemblage structure and, in many cases, niche shifts (Winemiller, 1989a,b, 1990). Despite the fact that many species possess adaptations for survival under extreme hypoxia, fish mortality is very high during the llanos dry season. Occasionally all aquatic habitat may be eliminated locally, as was nearly the case in sample Feb94; if this happens, the subsequent wet-season assemblage is reconstructed entirely from immigration. At least two Caño Maraco fishes, an annual killifish (*Rachovia maculipinnis*) and a swamp eel (*Synbranchus marmoratus*), can survive for several months in the absence of surface water.

The Caño Maraca fish assemblage showed its greatest changes in association with seasons. In terms of species' relative abundances, the assemblage was more similar within seasons over periods separated by as much as 10 years than it was between seasons during the same year. Species biomass distributions between open water and dense vegetation habitats were conserved between seasons and years. Species' relative abundance patterns at Caño Maraca indicated that the assemblage was persistent from year to year, especially during the late wet season and transition (falling water) season. Despite recent deforestation in this watershed, it appears that the species composition of this assemblage can be predicted with a high degree of accuracy each year, especially during the transition period (August through December). But since the timing of hydrological events in the llanos cannot be predicted with great accuracy, it would be naïve to assume that the assemblage will return to the same composition and structure each year on the same month. For example, December and January seem to be pivotal months in the transition to dry-season conditions, during which the habitat is shrinking but pools are not yet isolated nor severely degraded.

Within the seasonal progression, late September through October is an unusual period each year in that hypoxic conditions appear for a few weeks then disappear until the peak dry season (February through April). This circumstance occurred during both 1984 (Fig. 6; Winemiller, 1989c) and 1993, and it seems to result from a pulse of microbial respiration associated with the decomposition of aquatic macrophytes that die en masse as large tracts of higher terrain are drained. Because of the brief occurrence of dry-season features within the middle of the transition season, October samples tended to cluster with January (the early dry season) in Fig. 10.

Reproduction by most fishes is strongly tied to precipitation and hydrology, and even those species that spawn year-round have bursts of spawning activity during the early wet period. This pattern of seasonal reproduction has been described for tropical floodplain regions throughout the world (Lowe-McConnell, 1964; Welcomme, 1969; Winemiller, 1989b, 1991a). It is actually rather surprising that any fishes would attempt to reproduce under the crowded conditions of the transition season (see following discussion) or the harsh conditions of the dry season. Nonetheless, small opportunistic fishes, such as the guppy and annual killifishes, maximize their lifetime fitnesses and population growth rates by adopting a strategy of rapid maturation and continual high reproductive effort. Some species with brood guarding also spawn during the dry season (e.g., the cichlids *Aequidens pulcher* and *Caquetaia krausii*). In the Upper Zambezi/Barotse system, many of the mouth-brooding cichlids spawn once just before the beginning of the flood period (Winemiller, 1991a). Presumably, mate selection is facilitated by high fish densities during the low water period, and developing eggs and larvae can be transported onto the flooded plains by the brooding parent. The large eggs of these species allow the larvae to develop from endogenous reserves for many days while protected inside the parent's mouth, so that exogenous feeding can be delayed whenever floods come later than usual.

B. The Influence of Biotic Interactions on Assemblage Patterns

Biotic interactions influence the structure of tropical floodplain fish assemblages throughout the year, but they seemingly play the largest role during the transition season. During the wet season, large fish populations and biomass build up as a result of immigration, reproduction, and rapid growth in the vast and productive aquatic environment. As floodwaters recede, this fish biomass and diversity are concentrated in ever increasing densities; at the same time, many food resources become more limited. For example, there are fewer submerged surfaces for the attachment and growth of algae and associated benthic fauna, and a reduced water surface area results in fewer terrestrial insects and seeds reaching the aquatic system. For most fishes, resource competition becomes more intense with the progression of the transition season. Most benthic algivores shift to detritivory (Winemiller, 1990, 1996), and numerous invertebrate feeders and omnivores show increased food and habitat segregation (Winemiller and Pianka, 1990). Even piscivores increase interspecific segregation by diet during the transition season, and this despite the fact that fish prey should be more available in the shrinking habitat (Winemiller, 1989a). Ontogenetic diet shifts are largely responsible for this pattern, with wet season diets strongly influenced by juveniles feeding on invertebrates, and transition season diets dominated by specialized piscivory by juveniles and adults both (Winemiller, 1989a).

Theoretically, the fish assemblage should reveal the greatest between-year similarity during the transition season because this is the period when biotic interactions should eliminate vagrant or fugitive species and adjust the relative abundances of the resident species that are well suited to the local conditions. Conversely, dry-season assemblages should be determined to a greater extent by stochastic abiotic factors like pool strandings. Determination of wet-season assemblages, especially during early flooding, should contain a large random element associated with the timing of immigration and spawning. Transition-season samples were similar within and between years (Fig. 10, Table III). Dry-season samples formed a tight cluster within the year 1984; however, there was relatively low between-year clustering of dry-season samples (Fig. 10, Table III). With the exception of the June 1994 sample (very early wet season), wet-season samples tended to be similar to transition-season samples, particularly early transition season, and to those from the same year.

It would be misleading to suggest that the fish assemblage of Caño Maraca is completely determined by ecological factors such as habitat selection, competition, and predation. Several surprises were encountered during the later surveys. For example, the large cichlid, *Astronotus ocellatus,* was a common but not dominant species in the 1984 transition-season samples; in the Jan89 sample, *Astronotus* was the most abundant fish. No explanation for this apparent explosion of the local *Astonotus* population can be provided. Likewise, only a few *Pseudoplatystoma fasciatum* (Pimelodidae), a large piscivorous catfish, were captured at the site during 1984 surveys. This species did not reproduce at the site and occurred in only one sample; it can be classified as a vagrant species. *Pseudoplatystoma* was never again encountered at Caño Maraca (although it occurs in the channel several kilometers downstream), but several individuals of another large piscivorous catfish, *Pseudopimelodus apurensis* (Pimelodidae), were captured at the site in October 1993. Many of these *Pseudopimelodus* were dead or in distress from hypoxia. Presumably, both of these large catfishes migrated from downstream into the estero region of the creek where they fed on the abundant fishes. In the case of *Pseudopimelodus,* many individuals failed to exit prior to the hypoxia that occurs each October. Quite likely, even a relatively small number of these large predatory catfishes could have a significant impact on the local fish assemblage. It is very doubtful that a few vagrant omnivores or algivores have any influence at all on the ecosystem. Another somewhat unpredictable source of mortality is piscivorous wading birds such as ibises, egrets, herons, and storks. These birds congregate each year in greatest numbers during the late transition and dry seasons, and they certainly have an impact on llanos fish populations.

C. An Integrated View of Tropical Fish Assemblage Dynamics

Tropical freshwater fish assemblages are, taxonomically and ecologically, among the most diverse vertebrate assemblages anywhere in the world (Roberts, 1972; Lowe-McConnell, 1975; Goulding *et al.,* 1988). The underlying basis for phylogenesis is geographical and historical, but the generation and maintenance of physiological, morphological, and behavioral diversity associated with this phylogenetic diversity is largely a function of ecology and history (Roberts, 1972; Lowe-McConnell, 1975; Winemiller, 1991b; Lundberg, 1993). Perhaps we can never fully understand the biogeographic/ecological histories that produced extant regional fish assemblages. However, we can begin to develop a model of the factors that interact to produce local assemblage patterns on much shorter ecological time scales. The combination of predictable, seasonal abiotic factors; short-term, unpredictable abiotic factors; and responses by a diverse biota make tropical floodplain fish assemblages rich systems for study.

Lowe-McConnell (1975) provided one of the first conceptual models of fish responses to tropical floodplain ecosystem dynamics. This model has been elaborated upon by Welcomme (1979) and Junk *et al.* (1989), among others. The model assumes a strong unimodal pattern of annual precipitation, as would be encountered in the llanos or Barotse floodplain. During the annual floods, nutrients enter the aquatic ecosystem from the adjacent and newly submerged terrestrial landscape, and this drives a sustained pulse of aquatic primary production in the form of aquatic vegetation and microbes. In some areas, water quality may be poor, even hypoxic, during the initial stages of inundation when microbes process newly flooded detritus (Lagler *et al.*, 1971). Many fishes migrate from rivers and lagoons onto floodplains where a burst of spawning takes place. Others spawn in the channels of larger rivers, and eggs and larvae drift into floodplain habitats, some perhaps located many kilometers downstream from the spawning site. Many fishes, larger species in particular, migrate long distances for spawning.

Each fish species seeks a particular kind of local habitat within the floodplain drainage basin, and species vary greatly in their degree of habitat selectivity. Some species move farther up tributaries to small floodplain habitats (like Caño Maraca); others seek particular environmental conditions in flooded areas and lagoons along major river channels (e.g., Upper Zambezi; see also Rodríguez and Lewis, 1990; Chapman and Chapman, 1993). Most locations also contain a population of year-round residents, mostly smaller species, many possessing accessory respiratory adaptations.

Microcrustacean and aquatic insect production is very high on the flooded plains, and this serves as a food resource for juvenile fishes. Terrestrial arthropods, terrestrial vegetation, and seeds are important food resources for many tropical river fishes, and these are more available to fishes during the annual floods. During the initial stages of the flood season, fish survivorship is high because encounter rates with predators are low within the vast habitat with its complex matrix of submerged macrophytes.

As flood waters recede, fishes are concentrated at higher densities, and many migrate back into permanent river channels. During this time, predatory fishes and birds often congregate at major points of entry to larger rivers. The migratory fishes of the larger rivers may swim hundreds of kilometers to find optimal habitats. During this transition season, young-of-the-year continue to grow, encounter rates with preferred food resources decline, and encounter rates with competitors and predators increase. Mortality gradually increases, especially for small fishes. Local floodplain habitats may experience acute hypoxia as water recession results in the death of aquatic macrophytes (as described for Caño Maraca). Biotic interactions during this period impose selection for efficient use of space and food resources, as well as effective means for predator avoidance and defense. Members of diverse piscivore assemblages probably gain a very significant fraction of their annual food intake during this period, and this places a premium on efficient foraging, even when fish prey are relatively abundant. In many regions (like Caño Maraca and the Upper Zambezi), densities of small fishes are greatly reduced by piscivores within a period of just 3–4 months [see Jackson (1961) for a discussion of the influence of piscivores on African river fish ecology].

The major changes that unfold in floodplain ecosystems during the falling water period do not result in random fish assemblages; rather, selection for ecological performance results in local assemblages with similar structures year after year. The intensity of this selection and associations between optimal habitat conditions and fixed geographical locations probably vary from year to year depending on the timing and height of the floods. To some extent, fishes must respond to a moving target on these highly heterogeneous floodplain landscapes. Ultimately, many aquatic organisms become trapped in isolated pools or channel segments, where peak dry-season conditions cause deterioration of water quality and the death of large numbers of fishes. Within these isolated dry-season habitats, physiological stress interacts with predation as a major mortality agent. Whereas strandings contain a large random element, interspecific variation in resistance to degraded conditions and predation is nonrandom, and this creates a degree of stability in the composition and structure of dry-season fish assemblages.

V. SUMMARY

Fishes in tropical floodplain habitats respond to changes in their abiotic and biotic environments, many of which are predictable on a seasonal or annual basis but unpredictable in the short term. The fish assemblage and habitat characteristics of a seasonal creek and its floodplain in the Venezuelan llanos were examined over 10 years. Each year, aquatic populations undergo massive die-offs and emigrations in response to the drying of the floodplain. Each year, the rainy season causes extensive flooding and a renewal of aquatic habitats and populations. Levels of variability in the structure of the fish assemblage were similar within and between years, and this was true when comparisons were based on either the dominant 14 or 58 species. Hierarchical clustering tended to group assemblages based on season rather than year. Species richness was positively associated with water depth, and the persistence of individual species varied widely. The most persistent species tended to be the most abundant with the least variability. Persistence and abundance were not significantly associated with life history, trophic niche, accessory respiratory adaptations, or taxonomic order. Species biomass distributions between open water and dense vegetation habitats were conserved between seasons and years. Fish assemblages of the Upper Zambezi River and floodplain in Zambia were compared between surveys conducted in 1966 and 1989. Assemblage structure was highly correlated, with only two species showing large shifts in relative abundance ranks. The patterns from both areas indicate that tropical floodplain fish assemblages are structured by a blend of deterministic and stochastic processes operating across a broad range of temporal and spatial scales. Fish assemblage variability is increased by the short-term unpredictability of precipitation and stochastic mortality associated with dry season conditions. Interannual variation is reduced by habitat selection during wet-season colonization, population resiliency, and species interactions during the late wet season.

References

Agostinho, A. A., Julio, H. F., Jr., and Borghetti, J. R. (1992). Considerações sobre os impactos dos represamentos na ictiofauna e medidas para sua atenuação. Um estudo de caso: Reservatório de Itaipu. *Rev. UNIMAR; BB14,* Suppl., 89–107.

Anderson, R. O., and Gutreuter, S. J. (1983). Length, weight, and associated structural indices. *In* "Fisheries Techniques" (L. A. Nielsen and D. L. Johnson, eds.), pp. 283–300. Am. Fish. Soc., Bethesda, MD.

Balon, E. K., and Coche, A. G. (1974). "Lake Kariba, a Man-made Tropical Ecosystem in Central Africa," Monogr. Biol., No. 24. Dr. W. Junk, The Hague, The Netherlands.

Bass, G. D., Jr. (1990). Stability and persistence of fish assemblages in the Escambia River, Florida. *Rivers* 1, 296–306.

Bonetto, A., Cordiviola de Yuan, E., Opignalberi, C., and Olivieros, O. (1969). Ciclos hidrológicos del rio Paraná y las poblaciones de peces contenidas en las cuencas temporarias de su valle de inundación. *Physis (Buenos Aires)* 29, 213–223.

Chapman, L. J., and Chapman, C. A. (1993). Fish populations in tropical floodplain pools: A re-evaluation of Holden's data on the River Sokoto. *Ecol. Freshwater Fish* 2, 23–30.

Clements, F. E. (1949). "Dynamics of Vegetation." Hafner, New York.

Connell, J. H., and Sousa, W. P. (1983). On the evidence needed to judge ecological stability or persistence. *Am. Nat.* 121, 789–824.

Davidson, J., and Andewartha, H. G. (1948). The influence of rainfall, evaporation and atmospheric temperature on fluctuations in the size of a natural population of *Thrips imaginis* (Thysanoptera). *J. Anim. Ecol.* 17, 200–222.

Dobzhansky, T. (1950). Evolution in the tropics. *Am. Sci.* 267, 76–82.

Fletcher, R. I., and Deriso, R. B. (1988). Fishing in dangerous waters: Remarks on a controversial appeal to spawner-recruit theory for long-term impact assessment. *Am. Fish. Soc. Monogr.* **4**, 232–244.

Freeman, M. C., Crawford, M. K., Barrett, J. C., Facey, D. E., Flood, M. G., Hill, J., Stouder, D. J., and Grossman, G. D. (1988). Fish assemblage stability in a southern Appalachian stream. *Can. J. Fish. Aquat. Sci.* **45**, 1949–1958.

Goulding, M. (1980). "The Fishes and the Forest." Univ. of California Press, Berkeley.

Goulding, M., Carvalho, M. L., and Ferreira, E. G. (1988). "Río Negro: Rich Life in Poor Water." SPB Academic, The Hague, The Netherlands.

Grossman, G. D. (1982). Dynamics and organization of a rocky intertidal fish assemblage: The persistence and resilience of taxocene structure. *Am. Nat.* **119**, 611–637.

Grossman, G. D., Moyle, P. B., and Whitaker, J. O., Jr. (1982). Stochasticity in structural and functional characteristics of an Indiana stream fish assemblage: A test of community theory. *Am. Nat.* **120**, 423–453.

Grossman, G. D., Dowd, J. F., and Crawford, M. (1990). Assemblage stability in stream fishes: A review. *Environ. Manage. (N.Y.)* **14**, 661–671.

Harrell, H. L. (1978). Responses of the Devil's River (Texas) fish community to flooding. *Copeia*, pp. 60–68.

Herbold, B. (1984). Structure of an Indiana stream fish association: Choosing an appropriate model. *Am. Nat.* **124**, 561–572.

Holden, M. J. (1963). The populations of fish in dry season pools of the R. Sokoto. *Fish. Publ. Colonial Office, London* **19**, 1–58.

Jackson, P. B. N. (1961). The impact of predation, especially by the tigerfish (*Hydrocynus vittatus* Cast.) on African freshwater fishes. *Proc. Zool. Soc. London* **132**, 1–30.

Junk, W. J., Bayley, P. B., and Sparks, R. E. (1989). The flood pulse concept in river-floodplain ecosystems. *Can. Spec. Pub. Fish. Aquat. Sci.* **106**, 110–117.

Karr, J. A., Fausch, K. D., and Yant, P. R. (1986). Assessing biological integrity in running waters: A method and its rationale. *Spec. Publ.—Ill. Nat. Hist. Surv.* **5**, 1–29.

Kelley, D. W. (1968). "Fishery Development in the Central Barotse Flood Plain." Food Agric. Organ. Fish. U. N. Dev. Programme (Tech. Assist.) Rep., Vol. FRi/UNDP(TA) 2554, pp. 1–151. FAO/UN, Rome.

Kushlan, J. A. (1976). Environmental stability and fish community diversity. *Ecology* **57**, 821–825.

Lagler, K. F., Kapetsky, J. M., and Stewart, D. J. (1971). The fisheries of the Kafue River flats, Zambia, in relation to the Kafue gorge dam. *FAO [Tech. Rep.]* SF/Zam **11–2**, 1–161.

Lowe-McConnell, R. H. (1964). The fishes of the Rupununi savanna district of British Guiana. Pt. 1. Grouping of fish species and the seasonal cycles on the fishes. *J. Linn. Soc. London, Zool.* **45**, 103–144.

Lowe-McConnell, R. H. (1975). "Fish Communities in Tropical Freshwaters." Longman, London.

Lowe-McConnell, R. H. (1979). Ecological aspects of seasonality in fishes of tropical waters. *Symp. Zool. Soc. London* **44**, 219–241.

Lundberg, J. G. (1993). African-American freshwater fish clades and continental drift: Problems with a paradigm. *In* "The Biotic Relationships between Africa and South America" (P. Goldblatt, ed.), pp. 157–199. Yale Univ. Press, New Haven, CT.

Matthews, W. J. (1986). Fish faunal structure in an Ozark stream: Stability, persistence and a catastrophic flood. *Copeia*, pp. 388–397.

Matthews, W. J., Cashner, R. C., and Gelwick, F. P. (1988). Stability and persistence of fish faunas and assemblages in three mid-western streams. *Copeia*, pp. 947–957.

McKaye, K. R., and Marsh, A. (1983). Food switching by two specialized algae-scraping cichlid fishes in Lake Malawi, Africa. *Oecologia* **56**, 245–248.

Moyle, P. B., and Vondracek, B. (1985). Persistence and structure of the fish assemblage of a small California stream. *Ecology* **66**, 1–13.

Rahel, F. J., Lyons, J. D., and Cochran, P. A. (1984). Stochastic or deterministic regulation of assemblage structure? It may depend on how the assemblage is defined. *Am. Nat.* 124, 583–589.

Reice, S. R. (1994). Nonequilibrium determinants of biological community structure. *Am. Sci.* 82, 424–435.

Resh, V. H., Brown, A. V., Covich, A. P., Gurtz, M. E., Li, H. W., Minshall, G. W., Reice, S. R., Sheldon, A. L., Wallace, J. B., and Wissmar, R. (1988). The role of disturbance in stream ecology. *J. North Am. Benthol. Soc.* 7, 433–445.

Roberts, T. R. (1972). Ecology of fishes in the Amazon and Congo basins. *Bull. Mus. Comp. Zool.* 143, 117–147.

Rodríguez, M. A., and Lewis, W. M., Jr. (1990). Diversity and species composition of fish communities of Orinoco floodplain lakes. *Nat. Geogr. Res.* 6, 319–328.

Ross, S. T., Matthews, W. J., and Echelle, A. A. (1985). Persistence of stream fish assemblages: Effects of environmental change. *Am. Nat.* 126, 24–40.

Welcomme, R. L. (1969). The biology and ecology of fishes of a small tropical stream. *J. Zool, London* 158, 485–529.

Welcomme, R. L. (1979). "Fisheries Ecology of Floodplain Rivers." Longman, London.

Winemiller, K. O. (1989a). Ontogenetic diet shifts and resource partitioning among piscivorous fishes in the Venezuelan llanos. *Environ. Biol. Fishes* 26, 177–199.

Winemiller, K. O. (1989b). Patterns of variation in life history among South American fishes in seasonal environments. *Oecologia* 81, 225–241.

Winemiller, K. O. (1989c). Development of dermal lip protuberances for aquatic surface respiration in South American characid fishes. *Copeia*, pp. 382–390.

Winemiller, K. O. (1990). Spatial and temporal variation in tropical fish trophic networks. *Ecol. Monogr.* 60, 331–367.

Winemiller, K. O. (1991a). Comparative ecology of *Serranochromis* species (Teleostei: Cichlidae) in the Upper Zambezi River. *J. Fish Biol.* 39, 617–639.

Winemiller, K. O. (1991b). Ecomorphological diversification of freshwater fish assemblages from five biotic regions. *Ecol. Monogr.* 61, 343–365.

Winemiller, K. O. (1992). Life history strategies and the effectiveness of sexual selection. *Oikos* 62, 318–327.

Winemiller, K. O. (1996). Factors driving spatial and temporal variation in aquatic floodplain food webs. *In* "Food Webs: Integration of Patterns and Dynamics" (G. A. Polis and K. O. Winemiller, eds.). pp. 298–312. Chapman & Hall, New York.

Winemiller, K. O., and Kelso-Winemiller, L. C. (1991). *Serranochromis altus*, a new piscivorous cichlid species from the Upper Zambezi River. *Copeia*, pp. 673–684.

Winemiller, K. O., and Kelso-Winemiller, L. C. (1994). Comparative ecology of the African pike, *Hepsetus odoe*, and tigerfish, *Hydrocynus forskahlii*, in the Zambezi River floodplain. *J. Fish Biol.* 45, 211–225.

Winemiller, K. O., and Pianka, E. R. (1990). Organization in natural assemblages of desert lizards and tropical fishes. *Ecol. Monogr.* 60(1), 27–55.

Winemiller, K. O., and Rose, K. A. (1992). Patterns of life-history diversification in North American fishes: Implications for population regulation. *Can. J. Fish. Aquat. Sci.* 49, 2196–2218.

Yant, P. R., Karr, J. A., and Angermeier, P. L. (1984). Stochasticity in stream fish communities: An alternative interpretation. *Am. Nat.* 124, 573–582.

Yount, J. D., and Niemi, G. J. (1990). Recovery of lotic communities and ecosystems from disturbance—a narrative review of case studies. *Environ. Manage. (N.Y.)* 14, 547–570.

Zaret, T. M., and Paine, R. T. (1973). Species introduction in a tropical lake. *Science* 182, 449–455.

Zaret, T. M., and Rand, A. S. (1971). Competition in stream fishes: Support for the competitive exclusion principle. *Ecology* 52, 336–342.

Reptile and Amphibian Communities

CHAPTER 6

Structure and Dynamics of a Turtle Community over Two Decades

JUSTIN D. CONGDON AND J. WHITFIELD GIBBONS

University of Georgia's Savannah River Ecology Laboratory, Aiken, South Carolina 29802

I. INTRODUCTION

The space that organisms occupy throughout their lives is a critical resource. Because individuals of the same species and other taxa often co-occupy space, variation in biotic or abiotic environments and their interactions are believed to be important in determining the stability of species populations and of the communities that they form (Schluter and Ricklefs, 1993). Population and community levels of organization or association both depend on the success of the individuals that comprise coexisting populations that must persist if community stability is to be maintained. However, some conditions might act selectively and more severely on certain component populations than on entire communities. Documenting the short-term stability and dynamics of community components is a difficult task that requires a period of study that is by definition truly long-term and beyond the scope of a single investigator's lifetime. The description of a study as "long-term" is usually obtained from the investigator's perspective of the length of investigation, often with reference to the duration of similar studies. The extended longevity of individuals in some species and populations would seem to render the description long-term in relation to many community studies almost always inappropriate.

The University of Michigan's E. S. George Reserve (ESGR) provides a well-protected research area, a critical factor for long-term research. It was 42 years ago (1953) that the first turtle was marked on the ESGR. Since that time, three mark–recapture studies of turtles have been conducted that total 30 years of study. Turtles are among the longest-lived vertebrates with cohort generation times sometimes exceeding 30 years, ages at maturity in excess of 15 years (Congdon *et al.*, 1993a), and maximum longevities of greater than 75 years (Brecke and Moriarty, 1989). In long-lived organisms, adequate documentation of sources of variation in life-history trait values among individuals and within individuals as they age can require long-term studies. For example, marked individuals of two of the three primary species on the ESGR were captured as adults during all three studies (Congdon *et al.*, 1993a; Tinkle *et al.*, 1981), and there is no reason to believe that individuals of the third species were not also present during all three studies (Congdon *et al.*, 1994).

Although all three turtle studies on the ESGR were designed to investigate life histories and ecology rather than community dynamics, we used data from the present study to examine the magnitude of changes in the absolute and relative number of individuals of species populations that comprise the community and examined data from all three studies to attempt to infer whether interspecific interactions may have influenced the structure of the turtle community.

II. MATERIALS AND METHODS

A. Study Site

The ESGR is located in southeastern Michigan about 6 km west of the town of Pinckney, Livingston County, Michigan (approximately 42° 28′ N, 84° 00′ W). The 615-ha area was purchased as a game preserve by Colonel Edwin S. George in 1927–1928 and given to the University of Michigan in 1930. It is administered by the Museum of Zoology as a research site with access limited by a 12-ft-high chain link fence and locked gates.

Since 1930 terrestrial and aquatic habitats that were typical of southeastern Michigan at that time have been protected. At present the ESGR is characterized by relatively mature forests, old fields that are gradually filling in with trees, and in some areas closed-canopy stands of the introduced shrub Autumn Olive (*Elaeagnus umbellata*). Succession has eliminated some of the nesting areas previously used by the turtles. Primary aquatic areas on the ESGR (Fig. 1) are a 7.3-ha complex consisting of Southwest Swamp, Fishhook Marsh, and Crane

Pond (southwest population) and a 5-ha complex consisting of East Marsh and Cattail Marsh (southeast population). Other aquatic areas include George and Burt Ponds (0.6 ha), Hidden Lake (0.4 ha), Southeast Marsh (0.4 ha), the Canal, Big Swamp, and some smaller pot holes. When combined, these smaller wetlands represent a substantial area, but they are not considered permanent residences for many turtles (Fig. 1). Detailed descriptions of the ESGR during the 1940s can be found in Cantrall (1943); more recent descriptions can be found in Sexton (1959) and Wilbur (1975).

B. History of Turtle Studies on the E. S. George Reserve

Three species of turtles are common on the ESGR: midland painted turtles, hereafter called painted turtles (*Chrysemys picta marginata*); Blanding's turtles (*Emydoidea blandingii*); and common snapping turtles, hereafter called snapping turtles (*Chelydra serpentina*). Between 1953–1994, approximately 35,000

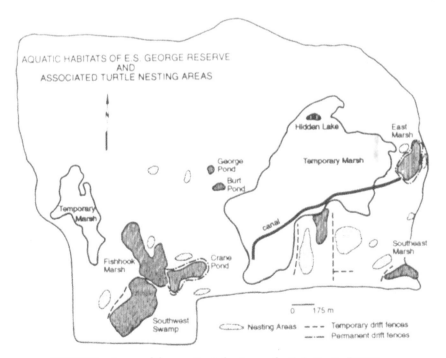

FIGURE 1. A map of the aquatic study sites on the E. S. George Reserve.

captures and recaptures were made of 8105 marked individuals (including hatchlings) on the ESGR, and an additional 1105 individuals were marked in marshes immediately adjacent to the ESGR (Table 1). Owen Sexton marked 913 painted turtles and 92 Blanding's turtles in five primary marshes (Southwest Swamp, Fishhook Marsh, Crane Pond, Cattail Marsh, and the Canal—also called the Ditch) on the ESGR between 1953–1957 (Sexton, 1959). From 1968–1973 Henry Wilbur, working primarily in the same wetlands studied by Owen Sexton, marked approximately 600 painted turtles, 60 Blanding's turtles, and 12 snapping turtles on the ESGR. Of the painted turtles captured by Henry Wilbur, 46 were recaptures of those marked during the 1950s (Wilbur, 1975). From 1975–1979, Donald Tinkle and Justin Congdon marked 1216 painted turtles (Tinkle et al., 1981; Congdon and Tinkle, 1982; Congdon et al., 1982; Breitenbach et al., 1984), 281 Blanding's turtles (Congdon et al., 1983), and 356 snapping turtles. During the period from 1980 to 1994, the study was continued by Justin Congdon and an additional 2687 painted turtles (Congdon and Gatten, 1989; Congdon and Gibbons, 1987, 1989; Congdon et al., 1993b; Scribner et al., 1993), 469 Blanding's turtles (Congdon and van Loben Sels, 1991; Congdon et al., 1983, 1993a), and 1291 snapping turtles were marked on the ESGR (Congdon et al., 1986, 1987, 1993b, 1994; Burke et al., 1993). Between 1975–1994, 13 and 95 painted turtles and 21 and 37 Blanding's turtles were recaptured that had been marked by Sexton and Wilbur, respectively. Nine snapping turtles marked by Wilbur were also recaptured during the present study.

C. Capture Methods

Between 1953–1957 and 1968–1973, turtle research was concentrated on the southwestern area of the ESGR (Fig. 1). Turtles were collected using baited hoopnet traps, basking traps, fyke nets (funnel traps), dip nets, and seines and by collecting from a boat, on land, and at drift fences.

Methods used to capture turtles during the previous studies were all used in the present study, with the major difference primarily related to the intensity of effort and inclusion of East Marsh as a primary study site and Southeast Marsh, West Marsh, and Hidden Lake as minor sites. The present study relied heavily on a 0.9-km drift fence that completely enclosed Crane Pond (1975–1982), a 1.3-km fence that enclosed East Marsh (1983–1994), and shorter drift fences installed between a marsh and adjacent nesting areas. In addition, each year from 1975 to 1986 and from 1991–1994, intensive aquatic trapping (from 75 to 140 traps) was carried out from early May through early September. Aquatic traps consisted of baited hoop traps, fykes (netting forms a funnel that is attached to the hoop trap), and drift traps (netting stretched between two posts with a hoop trap put at each end). Drift fences were usually monitored from

TABLE I Summary of Individuals (Including All Hatchlings and Juveniles)
of the Three Common Species of Turtles Marked between 1953 and 1994 on the ESGR

Years	C. picta	% of all turtles marked	C. serpentina	% of all turtles marked	E. blandingii	% of all turtles marked
1953–1957	913	—	0	—	92	—
1968–1973	600	—	12	—	60	—
1975–1979	1282	66.2	371	19.1	284	14.7
1980–1984	1135	53.1	741	34.7	261	12.2
1985–1989	635	64.7	262	26.7	84	8.6
1990–1994	950	69.2	293	21.3	130	9.5
Totals	5515		1679		911	
Mean, 1975–1994		62.2		25.9		11.8

Note. In addition, a total of 1105 individuals of all three species were marked in aquatic habitats immediately adjacent to the ESGR to determine rates of migration. Total individuals marked of all three species is 9210.

April through June and during September and October. For four years (1987–1990), field work was conducted during the nesting seasons from early May to early July. The reduced capture effort is evident in almost all population estimates made for all three species during these four years. In addition, during all nesting seasons (mid May to the beginning of July) from 1976–1994, four to seven people walked fences and searched nesting areas between 0600h and 2300h. As a result of nesting season efforts, observations were made on over 2500 nests of all three species. For more details on research methods see Congdon et al. (1983, 1987) and Congdon and van Loben Sels (1991). All juvenile and adult turtles were individually marked by notching or drilling the margins of the carapace. Prior to 1983, hatchlings were also individually marked, and from 1983 to 1992 all hatchling snapping turtles from each nest were given identical marks. Hatchling nest-cohort marks were subsequently changed to unique individual marks when a turtle was recaptured. The straight line lengths of both the plastron and carapace were measured at each capture. All turtles were then released at the point of capture.

D. Natural History of the Turtles

The three common species on the ESGR have some features in common. Sex is determined by the incubation temperature of the embryos rather than genetically (Pieau, 1972); therefore, nest site selection and annual variation in weather patterns can alter sex ratios of hatchlings. Females of all three species leave marshes and move into nesting areas to lay eggs from mid May to early July; however, the nesting season for snapping turtles is substantially shorter than that for painted and Blanding's turtles (Congdon *et al.*, 1983, 1987; Tinkle *et al.*, 1981). Across all years of the study on the ESGR, nest survivorship is highest in painted turtles, intermediate in Blanding's turtles, and lowest in snapping turtles; however, among-year variation in nest survival ranges from 0 to 67% for all species. Eggs incubate throughout the summer and hatch during late August and September and then hatchling Blanding's and snapping turtles emerge in autumn and move to aquatic areas to overwinter (Butler and Graham, 1995; Congdon *et al.*, 1983, 1987). In contrast, the majority of painted turtle hatchlings remain in the nest for the winter, emerging in the following April (Breitenbach *et al.*, 1984). The pattern of spring emergence from the nest is the most prevalent pattern in some populations (Newman, 1906; Williams, 1957; Woolverton, 1963; Christens and Bider, 1987; Breckenridge, 1970) versus fall emergence in other populations (Ernst, 1971).

Painted turtles are divided into four subspecies (eastern, southern, western and midland painted turtles). The distribution of midland painted turtles extends from southern Canada through southern New England and New York and as far south as central Mississippi. Painted turtles are one of the best known emydid turtles in North America (Ernst, 1971; Gibbons, 1968; Gibbons and Tinkle, 1969; MacCulloch and Secoy, 1983; Mitchell, 1988; Wilbur, 1975; Tinkle *et al.*, 1981; Zweifel, 1989). Female painted turtles have smaller body sizes than the other two species on the ESGR (Table II), where they reach sexual maturity at 7–12 years of age and produce one or two clutches of about seven eggs annually (Table II). Both clutch size and egg size increase significantly with body size of the females (Congdon and Gibbons, 1987). Males mature earlier and at smaller body size than do females.

Research on Blanding's turtles has increased in recent years because of concern for their conservation. Their geographic range is restricted compared to painted and snapping turtles; it extends from southeastern Ontario and adjacent Quebec, with some disjunct populations in Nova Scotia. In the eastern United States disjunct populations occur in southeastern New York, Massachusetts, New Hampshire, and Maine. The largest populations occur in Michigan and Minnesota, but the species is also found westward to western Nebraska and Iowa, and south to extreme northeastern Missouri. Blanding's turtles are gentle and seldom attempt to bite while being handled by researchers. They are intermediate in size to the other species on the ESGR (Table II), they reach maturity between 11 and 20 years (Congdon and van Loben Sels, 1993), and males and females are similar in body size (Congdon and van Loben Sels, 1991, 1993). They are extremely long-lived (Brecke and Moriarty, 1989), with cohort generation times of approximately 37 years (Congdon *et al.*, 1993a). Blanding's turtles are known to make extended overland forays in association with nesting and during other times of the year (Congdon *et al.*, 1983, 1996; Linck *et al.*, 1989;

TABLE II   Body Size, Clutch Size, and Egg Mass of Three Species of Turtles
Occupying the E. S. George Reserve

Species	Carapace length (mm)	Body mass (g)	Clutch size	Egg mass (g)
Chrysemys picta				
Mean	136.6	371.8	6.96	5.6
Standard error, N	0.28, 1015	2.24, 1015	0.05, 1015	0.08, 92
Minimum–maximum	114–164	205–615	2–13	3.0–7.5
Emydoidea blandingii				
Mean	195.2	1171.9	10.0	12.0
Standard error, N	0.63, 314	10.58, 314	0.13	0.17, 20
Minimum–maximum	166–223	730–1750	2–19	5.4–14.9
Chelydra serpentina				
Mean	253.1	4085.4	27.8	11.0
Standard error, N	1.56, 190	79.33, 190	0.46, 190	0.52, 12
Minimum–maximum	201–301	2000–8400	12–44	9.0–14.6

Ross, 1989; Ross and Anderson, 1990). Clutch size averages 10 eggs (Table II) and one or fewer clutches are produced per year (Congdon and van Loben Sels, 1993). Clutch size but not egg size increases with body size of females (Graham and Doyle, 1979; Congdon and van Loben Sels, 1993; Petokas, 1986; Mac-Culloch and Weller, 1987). Blanding's turtles are primarily carnivorous with diets composed of snails, crayfish, earthworms, insects, and vertebrates (Lagler, 1943; Kofron and Schreiber, 1985; Rowe, 1987).

Snapping turtles are the most widely distributed of the three species that occupy the ESGR. They range from southern Canada south throughout the central and eastern United States to the Gulf of Mexico and west to Iowa and New Mexico, and thence south into Central and South America. Because of their large body size snapping turtles have been historically, and presently are, harvested for meat throughout their range (Clark and Southall, 1920; Brooks *et al.*, 1988; Congdon *et al.*, 1994). In addition to the ESGR study, intensive population studies have been carried out in Ontario, Canada (Brooks *et al.*, 1991; Galbraith and Brooks, 1987; Galbraith *et al.*, 1988, 1989; Obbard, 1983). Snapping turtles are aggressive and substantially larger than either of the other species on the ESGR (Table II), with the largest males twice the body mass of large females. Females mature between 11 and 16 years of age on the ESGR, clutch size averages 28 eggs (Table II), and clutch frequency is less than annual (Brooks *et al.*, 1988; Congdon *et al.*, 1987, 1994). Snapping turtles are primarily carnivorous in some habitats (Alexander, 1943); however, a large portion of their diet can consist almost entirely of plant material in some habitats and during some seasons (Alexander, 1943; Hammer, 1969; R. Brooks, personal communication).

E. Population and Community Analyses

Substantial numbers of hatchling turtles were marked at nests or at drift fences on the ESGR. Because the catchabilities of the youngest juveniles compared to adults varied among species, we restricted the data used to estimate species population sizes to those individuals ($n = 5447$) that were greater than 60% of minimum adult size.

Population sizes were estimated using the total number of individuals captured each year adjusted by catchability indices calculated over roving 3-year periods encompassing all years of the study. Catchability estimates are the ratio of the total number of individuals captured in each of 3 sequential years divided by the total number of individuals captured during years one and three, but not in year two. Catchability indices are based on the assumption that an animal caught in year one and year three was available to be captured in year two, and estimates were used to adjust the totals of animals captured each year. For example, if a catchability estimate over a 3-year period was 0.85, the total number of individuals in the population was estimated as the total number of individuals captured in year two divided by the catchability estimate. In addition, population estimates were also calculated using repeated Lincoln indices calculated from the numbers of individuals captured and recaptured during sequential pairs of years over all years of the study.

The total number of older juvenile and adult turtles on the ESGR was the mean of the annual captures of individuals adjusted by catchability indices and the rolling Lincoln indexes for each year. Estimates of changes in the number of individuals in each population were not used in regression calculations if they were obviously influenced by the years when captures of turtles were made only during the nesting season, or if the estimates were unrealistically high. Data from eliminated points are presented in all graphs. To be conservative, changes in population sizes were considered to be significant only if they exceeded two standard errors of the mean of the 3-year catch estimates and the repeated Lincoln indices over all years of the study. The proportions of individuals of each species present each year were calculated from population estimates made during all years.

III. RESULTS

A. Mark-Recapture Studies

Six species of turtles were captured on the ESGR during the past 42 years, of which three species are common: painted turtle, Blanding's turtle, and snapping turtle. The stinkpot turtle (*Sternotherus odoratus*) is present in low numbers but is seen only during series of wet years. Over the past 20 years, 46 stinkpot turtles were marked, and approximately 32 were present from 1975–1980, which were wet years. They were not captured from 1980–1989, which for the most part were hot and dry years, but about a dozen individuals were captured between the years of 1990–1994, which were relatively wet and cool years. Two species, the spiny softshell (*Trionyx spiniferus*) and the common map turtle (*Graptemys geographica*), were rarely captured, and no individuals were recaptured in different years; therefore, individuals of these species are not considered to be residents of the ESGR wetlands.

From 1953 through 1994 approximately 8100 individuals of the three common turtle species were marked on the ESGR (Table I). Approximately two-thirds of the turtles marked and approximately three-quarters of the adults marked during the first 10 years of the study were painted turtles (Tables I and III). The relatively large differences between the percentage of all turtles marked (Table I) and the percentage of older juveniles and adults marked (Table III) are primarily due to the differences in the number of hatchlings and young juveniles of each species marked at nests and at drift fences.

Sex ratio in the painted turtle population has apparently shifted from primarily females (58.8%) in the 1950s to primarily males (67.8%) in the 1980s and 1990s (Table III). The Blanding's turtle population has maintained its female-bias (77.4% females) throughout all three studies, and the snapping turtle population has maintained an approximately equal sex ratio over the past 20 years (Table III). The highest portion of males among turtles marked for all three species occurred during the 1985 to 1989 sample (Table III).

B. Dynamics of Species Populations

Between 1975–1994, the population size of Blanding's turtles averaged 176 and 186 individuals based on individuals/year and Lincoln estimates, respectively (Table IV), and the estimated increase of 10 individuals between 1975–1994 (Fig. 2) was not significant (2 SE = 16.4, Table IV). The snapping turtle population averaged 198 and 220 individuals based on individuals/year and Lincoln estimates, respectively (Table IV), and the estimated increase of 40 individuals between 1975–1994 (Fig. 3) was significant (2 SE = 20.2, Table IV). The painted turtle population averaged 959 and 986 individuals based on individuals/year and Lincoln estimates, respectively (Table IV), and the estimated increase of 300 individuals between 1975–1994 (Fig. 4) was significant (2 SE = 16.4, Table IV).

Between 1957–1972 the estimated size of the southwest painted turtle population had apparently fallen from approximately 1000 to fewer than 200 individuals (Table V). However, our estimate of population size for the same area from

TABLE III Summary of Individuals of Three Species of Turtles Marked between 1953 and 1994 That Were Greater Than 59% of Adult Size at Initial Capture, Percentage of All Turtles Marked, and the Percentage of Turtles of Each Species That Were Males in the Population

Years	C. picta	% of all turtles marked	% males	C. serpentina	% of all turtles marked	% males	E. blandingii	% of all turtles marked	% males
1953–1957[a]	780	—	33.0	0	—	—	74	—	18.9
1968–1973[b]	540	—	54.4	12	—	—	36	—	11.0
1975–1979	931	76.8	65.9	152	12.5	50.0	130	10.7	26.2
1980–1984	534	80.7	60.3	56	8.5	41.1	72	10.9	15.3
1985–1989	403	83.4	75.4	30	6.2	70.0	50	10.3	38.0
1990–1994	618	91.3	69.7	32	4.7	68.7	27	4.0	11.1
Totals	3806			282			389		
Mean, 1975–1994		83.0	67.8		8.0	57.4		9.0	22.6

[a]Data from Sexton (1959) included.
[b]Data from Wilbur (1975) included.

TABLE IV   Summary of Population Estimates from Totals of Individuals Captured during Sequential 3-Year Periods, and 19 Simple Lincoln Indices for Three Species of Turtles on the E. S. George Reserve over the Past 20 Years (1975 to 1994)

Species	Estimate	Mean	Minimum	Maximum	Standard error
C. picta	Individuals/year	959	541	1273	12.9
	Lincoln	986	500	1535	15.3
E. blandingii	Individuals/year	176	142	207	7.6
	Lincoln	186	113	257	8.8
C. serpentina	Individuals/year	198	136	258	7.5
	Lincoln	220	133	339	12.7

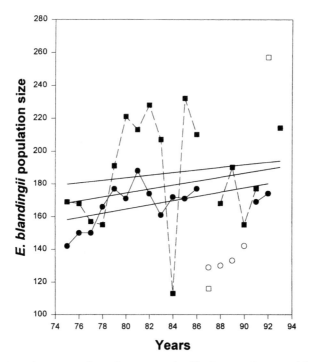

FIGURE 2.   Annual estimates of population sizes of *E. blandingii* on the ESGR (closed circles = total individuals caught over 3 years adjusted by catchability estimates; closed squares = repeated Lincoln estimates over all sequential years; and open circles and open squares = estimates from years turtles were captured only during the nesting season or were outliers, and these data were not used in the regression). Regression lines: upper = Lincoln index; lower = captures over 3 years; middle = means of Lincoln index and captures over 3 years.

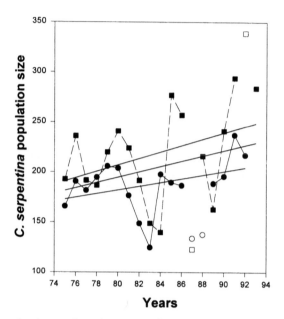

FIGURE 3.   Annual estimates of population sizes of *C. serpentina* on the ESGR (closed circles = total individuals caught over 3 years adjusted by catchability estimates; closed squares = repeated Lincoln estimates over all sequential years; and open circles and open squares = estimates from years turtles were captured only during the nesting season or were outliers, and these data were not used in the regression). Regression lines as in Fig. 2.

1975–1979 was about 460 turtles. If both estimates are accurate, the painted turtle population more than doubled in approximately 8 years (1970–1978). Over the past 20 years population estimates indicate an increase of about 100 individuals in both the southwest (Fig. 5) and southeast populations (Fig. 6), or a growth rate of less than 1% annually. Between 1981–1985 the estimated population size in East Marsh increased between 90 and 220 individuals above the mean estimate from the regression (Fig. 6) based on individuals/year (1981–1984) and Lincoln estimates (1982–1985). The increase was not evident by the end of the decade or through the early 1990s (Fig. 6). The peak estimates of population size at East Marsh occurred during hot and dry years and coincided with the absence of *S. odoratus* in East Marsh and Crane Pond.

C. Relative Species Composition of the Community

With the exception of the stinkpot turtle, which was not present on the ESGR in all years between 1975–1994, the other three species were present in all years

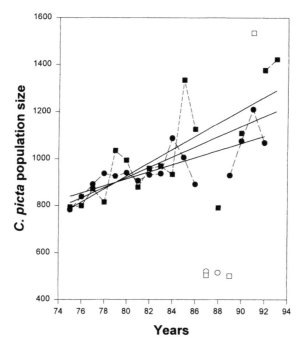

FIGURE 4. Annual estimates of population sizes of *C. picta* on the ESGR (closed circles = total individuals caught over 3 years adjusted by catchability estimates; closed squares = repeated Lincoln estimates over all sequential years; and open circles and open squares = estimates from years turtles were captured only during the nesting season or were outliers, and these data were not used in the regression). Regression lines as in Fig. 2.

studied between 1953–1994. Whereas the total number of turtles of the three common species has grown from about 1200 to 1600 individuals from 1975–1994 (Fig. 7a), the relative numbers of turtles of each species have remained fairly constant (Fig. 7b). Most of the increase seems to be caused by painted

TABLE V Estimates of Population Sizes of *C. picta* on the Southwestern Area of the ESGR during 1953–1957 and Late 1968–1972 and Mean Estimates Made from 1975–1979

Years	Estimator	Mean	Minimum	Maximum	Standard error
1953–1957[a]	Baily triple catch	981	—	—	—
1968–1972[a]	Baily triple catch	186	—	—	—
1968–1972[a]	Jolly	107	—	—	—
1975–1980	Individuals/year	463	425	503	13.9
	Lincoln	452	389	579	31.0

[a]Calculated by Wilbur, 1975.

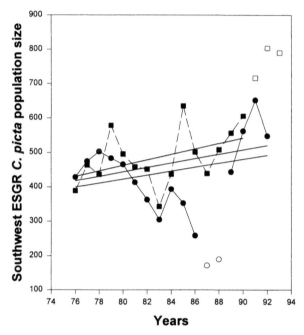

FIGURE 5. Annual estimates of population sizes of *C. picta* in the southwest population on the ESGR (closed circles = total individuals caught over 3 years adjusted by catchability estimates; closed squares = repeated Lincoln estimates over all sequential years; and open circles and open squares = estimates from years turtles were captured only during the nesting season or were outliers, and these data were not used in the regression). Regression lines as in Fig. 2.

turtles, which have increased steadily from approximately 850 to 1150 individuals (Fig. 4).

IV. DISCUSSION AND CONCLUSIONS

Among the four species of turtles considered residents of the ESGR, only the stinkpot turtle population exhibited substantial instability. The stinkpot population has ranged from total extirpation during dry years to establishing small populations that constituted less than 2% of the turtle community during wet years. Fluctuations in their presence or population sizes do not appear to be correlated with the numbers of individuals of the other species but may be related to the presence of open water and water of some critical minimum depth.

Populations of three common species make up over 98% of all resident turtles on the ESGR. The Blanding's turtle population has remained stable, and the population of common snapping turtles has exhibited slow growth rates. The

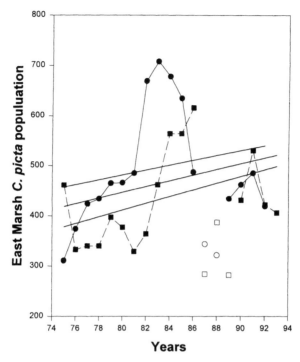

FIGURE 6. Annual estimates of population sizes of *C. picta* in the southeast population on the ESGR (closed circles = total individuals caught over 3 years adjusted by catchability estimates; closed squares = repeated Lincoln estimates over all sequential years; and open circles and open squares = estimates from years turtles were captured only during the nesting season or were outliers, and these data were not used in the regression). Regression lines as in Fig. 2.

painted turtle population, which decreased dramatically between 1957–1973, increased substantially between 1975–1994. Why populations of two species should remain at almost constant levels while one species population fluctuated is not known. Snapping turtle males are very aggressive, and male-to-male combat during the breeding season has been observed. Thus, the aggressive nature of this species may regulate densities of these turtles. However, although little is documented about the social biology of Blanding's turtles, it is well known that they are not aggressive, and so social structure is unlikely to be a major determinant of their density. A feature shared by both Blanding's and snapping turtles is that they are primarily carnivores, whereas painted turtles are omnivores.

A substantial part of the perceived mean increase in the population estimates of snapping turtles during the early years of the present study is almost certainly attributable to enhanced experience and technique: the result of our increasing

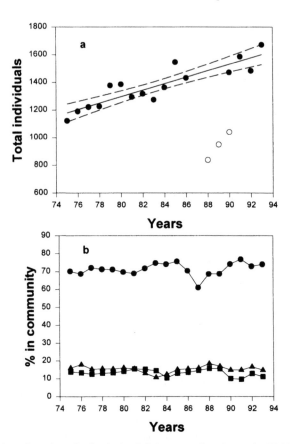

FIGURE 7.   (a) Total number of individuals of all 3 species of turtles on the ESGR (closed circles = values used in the regression; open circles = estimates from years turtles were captured only during the nesting season. (b) Proportions of turtles of each species: circles = painted turtles (*Chrysemys picta*); squares = Blanding's turtles (*Emydoidea blandingii*); and triangles = snapping turtles (*Chelydra serpentina*).

knowledge of the ESGR wetland system, the use of temporary drift fences, and the increased trapping effort and success rate between 1975–1980. In addition, between 1976–1980 the total numbers of individuals captured on land during the nesting seasons increased. Although increased capture results for painted turtles are also evident in the early years of the study, they do not account for a substantial amount of the observed increase in population estimates for this species.

Between the studies of Owen Sexton (1953–1957) and Henry Wilbur (1968–1973), the southwest population of painted turtles appeared to fall from 1000 to approximately 200 individuals. In 1957 the sex ratio reported for the southwest population was female-biased (67% females, $n = 604$ juveniles and adults; Sexton, 1959). To make comparisons among similar groups of turtles, we attempted to use the same classes of individuals that were used by Sexton (1959), although we recognize that these are not the functional adult sex ratios (Gibbons, 1990). Sex ratios among the 1880 new painted turtles first marked at greater than 90 mm in plastron length were 1.94, 1.41, 3.38, and 2.48 males to 1 female for the years 1975–1979, 1980–1984, 1985–1989, and 1990–1994, respectively. The same sex ratio pattern is evident when the southwest painted turtle population is examined separately. A shift in sex ratios in a population can result from differential mortality, emigration, immigration of the sexes, or a change in the primary sex ratio (Gibbons, 1990). Although the sex ratios of individuals captured from 1975–1994 are always male-biased, they are variable enough to suspect that environmental sex determination may be partially responsible. However, determining the sex of hatchlings of any of the three species on the ESGR is not possible in the field. Whether the sex ratio shift from female- to male-biased was related to the population reduction that took place from the 1950s to the 1970s is not known, but sex ratios have not moved back toward a female bias over the past 20 years.

Data on painted turtles from both 1968–1973 and 1975–1979 (Table IV) support the conclusion that a substantial population reduction occurred. However, the magnitude of the reductions based on the 1968–1973 and 1975–1979 data are 80 and 55–60%, respectively. If both estimates are correct, the population would have to have doubled in size over the approximately 8 years between the mean of the sampling periods. Such a high rate of increase seems unlikely. In addition, if 550 turtles were marked and the population size averaged only 200 turtles during 1968–1973, then the death and emigration rates would have had to exceed 50% during that 5-year period (a rate higher than indicated in Wilbur, 1975). Therefore, we conclude that the painted turtle population was reduced by less than 60% between the 1950s and 1968–1973. The levels of change estimated by Wilbur (1975) and by this study both represent a substantial reduction in the population size of painted turtles that was apparently catastrophic rather than gradual (Wilbur, 1975). The reason for the decline is not known, but potential explanations include the dredging of Crane Pond margins in 1966; the drying of Southwest Swamp, Crane Pond, and Fishhook Marsh in 1971 (Wilbur, 1975); or both. If we assume that drought caused the wetlands to dry, which in turn resulted in substantial numbers of turtles moving to more permanent water, it is unlikely that many individuals returned to the southwest population in subsequent years. Overland movements are associated with an increased risk of mortality, as female painted turtles are killed by predators during terrestrial movements associated with nesting (Wilbur, 1975). The data collected over the past 20 years do not indicate that the southwest painted turtle population is rapidly recovering to the level of the 1950s. Indeed, the data can be interpreted to suggest that the population might not be recovering from the population decline at all, since the apparently undisturbed East Marsh population was also growing at the same rate as the southwest population.

The peak in the East Marsh population during the dry years in the 1980s may have been due to the drying of more temporary, nearby habitats which resulted in turtles moving into marshes that still contained water. Short-distance movements among wetland sites are apparently common on the ESGR (Sexton, 1959; Wilbur, 1975; Scribner et al., 1993); however, movements between the southwest population and the southeast population occurred in less than 5% of the individuals studied over a 15-year period (Scribner et al., 1993). A pattern of immigration similar to that seen in East Marsh during dry years occurred in Sherriff's Marsh in southwestern Michigan during the 1980s (Frazer et al., 1991, 1993). In contrast to the southwest population where there is no evidence that turtles returned after the 1971 drought, annual population estimates suggest that an almost equal number of residents or immigrants apparently left East Marsh after the quality of nearby marshes improved. Whatever the reasons, the East Marsh population returned to near pre-peak levels by the late 1980s (Fig. 6).

Regardless of the cause of the decline in the southwest painted turtle population, the relative numbers of individuals of each species in the ESGR community may have been dramatically, if temporarily, altered if the impact of the event was not equal across all three common species. Observations of aquatic turtles responding to drying of aquatic habitats indicate that responses are different among species and not always consistent within species and thus could influence community structure. For example, emigrations in response to drying of a lake were observed for painted turtles and the yellow-bellied sliders (*Trachemys scripta*) but not for snapping turtles (Cagle, 1944), although the fate of the turtles was not determined. In contrast, Cahn (1937) thought that a substantial number of painted turtles died because they did not leave a drying lake in Illinois. When a bay in South Carolina dried, the primarily aquatic sliders and Florida cooters (*Pseudemys floridana*) left, but mud turtles (*Kinosternon subrubrum*) and chicken turtles (*Deirochelys reticularia*), which overwinter on land, did not (Gibbons et al., 1983). Therefore, the literature provides some evidence that painted turtles on the ESGR would have been impacted by a severe drought, but unfortunately the relative composition of the turtle community was not known during the previous studies on the ESGR.

Among the four species of turtles considered to be residents of the ESGR, only the population of stinkpot turtles which was always less than 2% of the community, exhibited substantial instability ranging from total extirpation during the dry years, with small populations becoming established during wet years. The presence or absence of stinkpot turtles on the ESGR appeared to be related to marsh conditions rather than to interactions with the three primary species, which maintained relatively constant numbers during periods of presence and absence of stinkpot turtles. In addition, deeper marshes and backwater areas of lakes near the ESGR support substantial populations of stinkpot turtles that coexist with the three dominant species of turtles on the ESGR and with softshell, common map and spotted turtles (*Clemmys guttata*). A general impression from the pattern of occurrence of turtles in Michigan is that more turtle species co-exist where the habitat types include open water such as lakes and deep marshes and flowing water such as large streams and rivers. The composition of communities of turtles that exist in areas that include a number of habitat types may be influenced by interactions among species.

Blanding's and snapping turtles have characteristics that make them of concern in relation to management and conservation. Nowhere do populations of Blanding's turtles appear to reach the high densities of many painted turtle populations. Populations of Blanding's turtles are threatened throughout their range by collecting, habitat destruction, and possibly by an increase in nest predators (Congdon *et al.*, 1993a, 1995). Due to their large body size, snapping turtles have historically been harvested for meat (Clark and Southhall, 1920) and currently still are. In some instances snapping turtles are also being killed on bird sanctuaries because they prey on ducks (Hammer, 1969; Hawthorne, 1980). Both Blanding's and snapping turtles are long-lived and have minimum ages at sexual maturity of 14 and 11 years, respectively (Congdon and van Loben Sels, 1993; Congdon *et al.*, 1993a, 1994; Brooks *et al.*, 1988; Obbard, 1983). The demographic traits of long-lived vertebrates make recovery from a major population reduction a slow process for both species, and under some conditions compensatory mechanisms do not occur in snapping turtle populations that have been reduced in size (Brooks *et al.*, 1991).

The long-term turtle studies on the ESGR provide encouraging news with respect to conservation and management programs, since the turtle populations and community composition have been essentially stable over at least 2 decades. Stability has been maintained with no provisions other than the following: (1) undisturbed and unpolluted wetland habitats and associated terrestrial areas that are necessary for nesting (Congdon *et al.*, 1983, 1987; Burke and Gibbons, 1995), (2) corridors for movement among wetlands both on and off of the ESGR, and (3) protection from direct exploitation by man.

V. SUMMARY

Three sequential studies of life histories and ecology of long-lived organisms have added to our understanding of age-specific biology and population ecology of freshwater turtles. Our attempt to use the long-term data from these studies to examine community levels of organization met with mixed results. Populations of Blanding's and snapping turtles were essentially stable, while the painted turtle population increased substantially over 20 years. A fourth species (stinkpot turtle) was present in low numbers but not in all years. Population dynamics of each species appeared to be independent of the other two species (i.e., no substantial levels of competitive or predator-prey relationships were indicated among species). Whether interspecific interaction would become prevalent and influential in structuring turtle communities in habitats having more turtle species, including congenerics, is yet to be determined. However, findings from the current study and from virtually all other long-term studies on turtles suggest that environmental factors or intraspecific interactions are more likely to influence turtle community characteristics than are interspecific interactions. In any case, documentation of community features among late-maturing, long-lived species is likely to become increasingly difficult due to the limited opportunities for the study of natural communities unaffected by human modification.

Acknowledgments

We thank the late Donald W. Tinkle for his participation in the early years of the ESGR study and Joyce Klevering for assistance with all aspects of X-radiography. We thank the University of Michigan Museum of Zoology and the E. S. George Reserve Committee for maintaining the E. S. George Reserve as a world-class research area. The following people made notable contributions to the ESGR study: Harold and Sue Avery, Vincent Burke, Margaret Burkman, Carl and Melvin Congdon, Nancy Dickson, Ruth Estes, Rose Fama, Robert Fischer, Matthew Hinz, Mark Hutcheson, Owen Kinney, David Kling, Shawn Meager, Roy Nagle, Tal Novak, Patricia Orleans, Willem Roosenburg, Tod Sajwaj, John Stegmier, Julie Wallin, and Bradley, John, and Dick Wiltse. The manuscript was improved by comments from Mike Dorcas and David Scott. Funding was provided by NSF Grants DEB-74-070631, DEB-79-06301, BSR-84-00861, and BSR-90-19771. Editing of computer files, data analysis, and manuscript preparation were aided by contract DE-AC09-76SROO-819 between the University of Georgia and the United States Department of Energy.

References

Alexander, M. M. (1943). Food habits of the snapping turtles in Connecticut. *J. Wild. Manage.* 7, 278–282.
Brecke, B., and Moriarty, J. J. (1989). *Emydoidea blandingii* (Blanding's Turtle) Longevity. *Herpetol. Rev.* 20(2), 53.
Breckenridge, W. J. (1970). "Reptiles and Amphibians of Minnesota." Univ. of Minnesota Press, Minneapolis.

Breitenbach, G. L., Congdon, J. D., and van Loben Sels, R. C. (1984). Winter temperatures of *Chrysemys picta* nests in Michigan: Effects on hatchling survival. *Herpetologica* 40, 76–81.

Brooks, R. J., Galbraith, D. A., Nancekivell, E. G., and Bishop, C. A. (1988). Developing management guidelines for snapping turtles. *In* "Management of Amphibians, Reptiles and Small Mammals in North America" (R. C. Szaro, K. E. Severson, and D. R. Patton, eds.), Gen. Tech. Rep. RM-166, pp. 174–179. U.S. Department of Agriculture, Forest Service, Washington, D.C.

Brooks, R. J., Brown, G. P., and Galbraith, D. A. (1991). Effects of sudden increase in natural mortality of adults on a population of the common snapping turtle (*Chelydra serpentina*). *Can. J. Zool.* 69, 1214–1320.

Burke, V. J., and Gibbons, J. W. (1995). Terrestrial buffer zones and wetlands conservation: A case study of freshwater turtles in a Carolina bay. *Conserv. Biol.* 9, 1365–1369.

Burke, V. J., Nagle, R. D., Osentoski, M., and Congdon, J. D. (1993). Common snapping turtles associated with ant mounds. *J. Herpetol.* 27, 114–115.

Butler, B. O., and Graham, T. E. (1995). Early post-emergent behavior and habitat selection in hatchling *Emydoidea blandingii* in Massachusetts. *Chelonian Conserv. Biol.* 3, 187–196.

Cagle, F. R. (1944). "Home Range, Homing Behavior, and Migration in Turtles," Misc. Publ. Mus. Zool., Univ. Mich., No. 61. University of Michigan, Ann Arbor.

Cahn, A. R. (1937). The turtles of Illinois. *Ill. Biol. Monogr.* 35.

Cantrall, I. J. (1943). "The Ecology of the Orthoptera and Dermaptera of the George Reserve, Michigan," Misc. Publ. Mus. Zool., Univ. Mich., No. 54. University of Michigan, Ann Arbor.

Christens, E., and Bider, R. (1987). Nesting activity and hatching success of the painted turtle (*Chrysemys picta marginata*) in southwestern Quebec. *Herpetologica* 43, 55–65.

Clark, H. W., and Southall, J. B. (1920). Freshwater turtles: A source of meat supply. *U.S. Bur. Fish. Doc.* 889, 3–20.

Congdon, J. D., and Gatten, R. E., Jr. (1989). Movement and energetics of nesting *Chrysemys picta*. *Herpetologica* 45, 94–100.

Congdon, J. D., and Gibbons, J. W. (1987). Morphological constraint on egg size: A challenge to optimal egg size theory? *Proc. Natl. Acad. Sci. U.S.A.* 84, 4145–4147.

Congdon, J. D., and Gibbons, J. W. (1989). Biomass productivity of turtles in freshwater wetlands: A geographic comparison. *In* "Freshwater Wetlands and Wildlife" (R. R. Sharitz and J. W. Gibbons, eds.), CONF-8603101, pp. 583–592. USDOE Office of Science and Technology, Oak Ridge, TN.

Congdon, J. D., and Tinkle, D. W. (1982). Reproductive energetics of the painted turtle (*Chrysemys picta*). *Herpetologica* 38, 228–237.

Congdon, J. D., and van Loben Sels, R. C. (1991). Growth and body size variation in Blanding's turtles (*Emydoidea blandingi*): Relationships to reproduction. *Can. J. Zool.* 69, 239–245.

Congdon, J. D., and van Loben Sels, R. C. (1993). Relationships of reproductive traits and body size with attainment of sexual maturity and age in Blanding's turtles (*Emydoidea blandingi*). *J. Evol. Biol.* 6, 547–557.

Congdon, J. D., Dunham, A. E., and Tinkle, D. W. (1982). Energy budgets and life histories of reptiles. *In* "Biology of the Reptilia", (G. Gans, ed.), Vol. 13, pp. 233–271. Academic Press, New York.

Congdon, J. D., Tinkle, D. W., Breitenbach, G. L., and van Loben Sels, R. C. (1983). Nesting ecology and hatching success in the turtle *Emydoidea blandingi*. *Herpetologica* 39, 417–429.

Congdon, J. D., Greene, J. L., and Gibbons, J. W. (1986). Biomass of freshwater turtles: A geographic comparison. *Am. Midl. Nat.* 115, 165–173.

Congdon, J. D., Breitenbach, G. L., van Loben Sels, R. C., and Tinkle, D. W. (1987). Reproduction and nesting ecology of snapping turtles (*Chelydra serpentina*) in southeastern Michigan. *Herpetologica* 43, 39–54.

Congdon, J. D., Dunham, A. E., and van Loben Sels, R. C. (1993a). Delayed sexual maturity and demographics of Blanding's turtles (*Emydoidea blandingii*): Implications for conservation and management of long-lived organisms. *Conserv. Biol.* 7, 826–833.

Congdon, J. D., Gotte, S. W., and McDiarmid, R. W. (1993b). Ontogenetic changes in habitat use by juvenile turtles, *Chelydra serpentina* and *Chrysemys picta*. *Can. Field Nat.* 106, 241–248.

Congdon, J. D., Dunham, A. E., and van Loben Sels, R. C. (1994). Demographics of common snapping turtles (*Chelydra serpentina*): Implications for conservation and management of long-lived organisms. *Am. Zool.* 34, 397–408.

Congdon, J. D., Graham, T. E., and Brecke, B. (1996). *Emydoidea blandingii* (Holbrook, 1842), Blanding's turtle. *In* "The Conservation Biology of Freshwater Turtles: Volume 2—New World Turtles (Neoarctic and the Neotropical)" A. Rhodin and P. Pritchard, eds.), IUCN Publ.

Ernst, C. H. (1971). Population dynamics and activity cycles of *Chrysemys picta* in southeastern Pennsylvania. *J. Herpetol.* 5, 151–160.

Frazer, N. B., Gibbons, J. W., and Greene, J. L. (1991). Growth, survivorship, and longevity of painted turtles *Chrysemys picta* in a southwestern Michigan marsh. *Am. Midl. Nat.* 125, 245–258.

Frazer, N. B., Greene, J. L., and Gibbons, J. W. (1993). Temporal variation in growth rate and age at maturity of male painted turtles, *Chrysemys picta*. *Am. Midl. Nat.* 130, 314–324.

Galbraith, D. A., and Brooks, R. J. (1987). Survivorship of adult females in a northern population of common snapping turtles, *Chelydra serpentina*. *Can. J. Zool.* 65, 1581–1586.

Galbraith, D. A., Bishop, C. A., Brooks, R. J., Simser, W. L., and Lampman, K. P. (1988). Factors affecting the density of populations of common snapping turtles (*Chelydra serpentina*). *Can. J. Zool.* 66, 1233–1240.

Galbraith, D. A., Brooks, R. J., and Obbard, M. E. (1989). The influence of growth rate on age and body size at maturity in female snapping turtles (*Chelydra serpentina*). *Copeia*, pp. 896–904.

Gibbons, J. W. (1968). Population structure and survivorship in the painted turtle, *Chrysemys picta*. *Copeia*, pp. 260–268.

Gibbons, J. W. (1990). Sex ratios and their significance among turtle populations. *In* "The Life History and Ecology of the Slider Turtle" (J. W. Gibbons, ed.), pp. 171–182. Smithsonian Institution Press, Washington, DC.

Gibbons, J. W., and Tinkle, D. W. (1969). Reproductive variation between turtle populations in a single geographic area. *Ecology* 50, 340–341.

Gibbons, J. W., Greene, J. L., and Congdon, J. D. (1983). Drought-related responses of aquatic turtle populations. *J. Herpetol.* 17, 242–246.

Graham, T. E., and Doyle, T. S. (1979). Growth and population characteristics of Blanding's turtle, *Emydoidea blandingi*, in Massachusetts. *Herpetologica* 33, 410–414.

Hammer, D. A. (1969). Parameters of a marsh snapping turtle population, Lacreek Refuge, South Dakota. *J. Wildl. Manage.* 33, 995–1005.

Hawthorne, D. W. (1980). Wildlife damage and control techniques. *In* "Wildlife Management Techniques Manual" (S. D. Schemnitz, ed.), pp. 411–439. Wildlife Society, Washington, DC.

Kofron, C. P., and Schreiber, A. A. (1985). Ecology of two endangered aquatic turtles in Missouri: *Kinosternon flavescens* and *Emydoidea blandingii*. *J. Herpetol.* 19, 27–40.

Lagler, K. F. (1943). Food habits and economic relations of the turtles of Michigan with special reference to fish managements. *Am. Midl. Nat.* 29, 257–312.

Linck, M. H., Depari, J. A., Butler, B. O., and Graham, T. E. (1989). Nesting behavior of the turtle *Emydoidea blandingii*, in Massachusetts. *J. Herpetol.* 23, 442–444.

MacCulloch, R. D., and Secoy, D. M. (1983). Demography, growth, and food of western painted turtles, *Chrysemys picta bellii* (Gray), from southern Saskatchewan. *Can. J. Zool.* 61, 1499–1509.

MacCulloch, R. D., and Weller, W. F. (1987). Some aspects of reproduction in a Lake Erie population of Blanding's turtle, *Emydoidea blandingii*. *Can. J. Zool.* 66, 2317–2319.

Mitchell, J. C. (1988). Population ecology and life histories of the freshwater turtles, *Chrysemys picta* and *Sternotherus odoratus* in an urban lake. *Herpetol. Monogr.* 2, 40–61.

Newman, H. H. (1906). The habits of certain tortoises. *J. Comp. Neurol. Physiol.* 16, 126–152.

Obbard, M. E. (1983). Population ecology of the common snapping turtle *Chelydra serpentina* in north-central Ontario. Ph.D. Dissertation, University of Guelph, Ontario, Canada.

Petokas, P. J. (1986). Patterns of reproduction and growth in the freshwater turtle *Emydoidea blandingii*. Ph.D. Dissertation, State University of New York, Binghamton.

Pieau, C. (1972). Effets de la température sur le développement des glands génitales ches les embryuonsde deux chelonians *Emys obicularis* (L.) et *Testudo graeca* (L.). *C. R. Hebd. Seances Acad. Sci.* 274, 719–722.

Ross, D. A. (1989). Population ecology of painted and Blanding's turtles (*Chrysemys picta* and *Emydoidea blandingii*) in central Wisconsin. *Trans. Wis. Acad. Sci.* 77, 77–84.

Ross, D. A., and Anderson, R. K. (1990). Habitat use, movements and nesting of *Emydoidea blandingi* in central Wisconsin. *J. Herpetol.* 24, 6–12.

Rowe, J. W. (1987). Seasonal and daily activity of Blanding's turtles (*Emydoidea blandingi*) in northern Illinois. M.S. Thesis, Eastern Illinois University, Charleston.

Schluter, D., and Ricklefs, R. E. (1993). Species diversity: An introduction to the problem. *In* "Species Diversity in Ecological Communities" (R. E. Ricklefs and D. Schluter, eds.), pp. 1–12. Univ. of Chicago Press, Chicago.

Scribner, K. T., Congdon, J. D., Chesser, R. K., and Smith, M. H. (1993). Annual differences in female reproductive success affect spatial and cohort-specific genotypic heterogeneity in painted turtles. *Evolution (Lawrence, Kans.)* 47, 1360–1373.

Sexton, O. J. (1959). Spatial and temporal movements of a population of the painted turtle *Chrysemys picta marginata* (Agassiz). *Ecol. Monogr.* 29, 113–140.

Tinkle, D. W., Congdon, J. D., and Rosen, P. C. (1981). Nesting frequency and success: Implications for the demography of painted turtles. *Ecology* 62, 1426–1432.

Wilbur, H. M. (1975). The evolutionary and mathematical demography of the turtle *Chrysemys picta*. *Ecology* 56, 64–77.

Williams, G. C. (1957). Pleiotropy, natural selection, and evolution of senescence. *Evolution (Lawrence, Kans.)* 11, 398–411.

Woolverton, E. (1963). Winter survival of hatchling painted turtles in northern Minnesota. *Copeia*, pp. 569–570.

Zweifel, R. G. (1989). Long-term ecological studies on a population of painted turtles (*Chrysemys picta*), on Long Island, New York. *Am. Mus. Novit.* 2952, 1–55.

CHAPTER 7

Predation and Competition in Salamander Communities

NELSON G. HAIRSTON, SR.

Department of Biology, University of North Carolina, Chapel Hill, North Carolina 27599

I. INTRODUCTION

A brief description of the natural history of salamanders will serve to dispel some common misconceptions about them. Only 81 species out of 355 always follow the "typical" amphibian life history of largely terrestrial adults returning to water to mate and deposit eggs which hatch into gilled larvae for part of their lives. Another 23 species include some members that follow the pattern described, while others have paedomorphic adults that retain the larval morphology and habitat throughout their lives. Thus, fewer than one-third of the species conform to the commonly accepted view of salamander life history. Of the remaining species, 48 are permanently aquatic, 23 of these retaining larval traits throughout life, and 25 remaining aquatic as adults. Most surprising are those that make up the majority of all salamander species. They have no larval stage, and they lay eggs in damp places underground or in rotten logs or other permanently moist locations. Metamorphosis occurs before hatching, with a few species retaining the eggs in the oviducts until they have hatched.

Two-thirds of all salamander species are members of one of the nine recognized families—the Plethodontidae. All members of that family lack lungs; they respire through the skin and the lining of the mouth cavity. Surprisingly, 85% of the species of Plethodontidae are exclusively terrestrial. One might suppose that the requirement of keeping the skin moist for respiration would make life difficult in a terrestrial environment, yet the family is the most widespread among the salamanders, and the terrestrial members are the only ones to have spread beyond eastern North America—the agreed location of origin of the family.

All salamanders are predacious, whether as adults or as larvae. Like most predators, they are unselective with regard to their prey, except for size. Analysis of their stomach contents and general observations show that salamanders feed on any animal that moves, provided that it is of a size that can be swallowed.

In the southern Appalachians, some species are very abundant. Jaeger (1979) estimated the density of the small *Plethodon cinereus* as 2.1–2.3 m^{-2} in the mountains of Virginia. In the mountains of North Carolina, the abundances of the large species of the same genus, *P. jordani* and *P. oconaluftee,* reach at least 0.7 m^{-2} (Hairston, 1987). These are all purely terrestrial. *Desmognathus ochrophaeus,* a streambank form, is estimated at 16 per linear meter of stream, not including the aquatic larvae. Combining all 10 species in the Coweeta Hydrologic Laboratory, a research forest in the Nantahala Mountains, the total salamander biomass is 0.6–1.0 $g \cdot m^{-2}$, more than mammals and birds combined. They are thus not mere oddities, but play a unique role in the distribution and flow of energy through the ecosystem. Their metabolic rate is very low compared to the endothermic birds and mammals. Burton and Likens (1975) calculated that at the experimental Hubbard Brook Forest in New Hampshire, the salamanders had 5 times the biomass of birds and mammals but were only responsible for 20% as much energy flow as the birds. At Coweeta, with 3 times the salamander biomass per hectare, they would be expected to contain an even larger fraction of the total animal tissue and be responsible for an appreciable part of the energy flow through the carnivore trophic level. As will be shown in the following discussion, salamander populations are remarkably stable over many years, a finding that is important to their potential for damping fluctuations in nutrient cycling and energy flow that would otherwise be caused by variations due to major changes in abundances in the remainder of the fauna.

The populations of salamanders inhabiting the forest floor or streambank habitats are more stable than those of other small vertebrates. Observations carried out over 23 years show no trends for any of the six species whose numbers were recorded (Fig. 1). The observations were made by university undergraduates in classes consisting of the same number of students, working for the same amount of time at the same locations. There have been a total of 43 independent studies since 1976 (one class study was made in 1972). The collections of *Plethodon jordani* and *P. oconaluftee* (formerly *P. glutinosus,* Hairston,

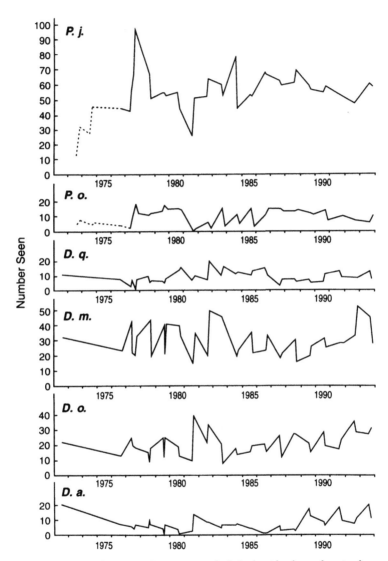

FIGURE 1. Population histories of six species of plethodontid salamanders in the southern Appalachians from 1972 to 1993. The number seen on each date represents the same effort by successive classes at the University of North Carolina. Abbreviations as follows: *Plethodon jordan*, *P. j.*; *P. oconaluftee*, *P. o*; *Desmognathus quadramaculatus*, *D. q.*; *D. monticola*, *D. m.*; *D. ochrophaeus*, *D. o.*; *D. aeneus*, *D. a.*

1993) were made in a mature forest in the Great Smoky Mountains National Park. The students were instructed to collect 10 specimens at each of six predetermined places. As those tasks were completed in approximately the same amount of time for each class, the points in the figure represent the catch per student effort. The species of *Desmognathus* were collected in the mature forest and small streams in the Coweeta Hydrologic Laboratory mentioned previously. The students were instructed to collect specimens, record the distance of the collection site from surface water, and determine the species with the help of the instructor. Again, the exercise was completed in approximately the same amount of time for each class. At both locations, collected specimens were returned to their habitats after examination (Hairston and Wiley, 1993). The fluctuations from one count to another are the result of different weather conditions (e.g., cold causing *Plethodon* to remain underground in September, 1981) or of exceptionally enthusiastic students (September 1977). The mean generation time (age of the average female when it has laid half of its total eggs) of the different species ranges from 4 to 10 years, and the mean annual survival of adults ranges from 0.5 to 0.81 (Organ, 1961; Hairston, 1983a). Thus, the fluctuations can scarcely reflect real changes in numbers actually present. The populations were followed longer than the maximum longevity of individuals and covered 1.5 to 5 population turnovers, more than adequate for establishing stability (Connell and Sousa, 1983; Schoener, 1985).

It is more difficult to estimate population trends of those amphibious species breeding in ponds. The number of adults migrating to the ponds each year is strongly dependent on the immediate weather and thus fluctuates greatly from year to year, although the adults may be long-lived (Pechmann *et al.*, 1991). Studies on the communities of these species have nearly always been conducted on the larvae; there are a few exceptions: the work of Bell (1974, 1977) and Bell and Lawton (1975) on *Triturus vulgaris* in England, Gill's (1978, 1979) work on populations of newts (*Notophthalmus viridescens*) in Virginia, and work cited by Semlitsch *et al.* (this volume).

The abundant and stable populations have made salamanders attractive to students of community ecology, especially in the southern Appalachians, where 3 to 10 species coexist in nearly all habitats. Under these conditions, it is possible to make statistically satisfactory observations that lead to hypotheses of competition and predation and to conduct equally satisfactory experiments testing for these ecological interactions between species.

It is most logical to describe separately the work on salamander communities in forest floor, streambank, and pond habitats because the interspecific interactions have been found to be quite different in the three situations. Most of this chapter is a description of the outcomes of the experiments, but there have been many observational studies, the results of which should be compared with the experimental ones, as agreement has not been universal, and reasons for disagreements are revealing about the preconceptions that led to many observational studies.

Throughout this chapter, the term "community" will be used in a loose manner to mean a group of coexisting species. The chapter divisions of this book require that the species be related taxonomically, at farthest belonging to the Order Caudata. Such an implied definition has been in common usage for 35 years, although it is not ordinarily defended logically. Several careful experiments on salamanders have shown that neither competition nor predation can be shown to take place among some of the coexisting, related species, despite detailed ecological and anatomical observations that led to the conclusion that they are in competition with each other, as will be covered in the following discussion. I do not use the term guild because predation between the member species does not appear to be part of the implied interspecific interactions.

In this chapter, I describe the ecological interactions that have been revealed between salamander species, and the word community is used in the manner stated above. Most of the work has been conducted in the eastern United States, but three descriptive studies have been made in the western United States, two in Great Britain, and one in Poland.

II. TERRESTRIAL SALAMANDERS

The first indications that salamanders compete came from studies of their ecological distributions in restricted localities. For terrestrial habitats, the pioneering work of Schmidt (1936) must be cited. His altitudinal transect of Volcan Tajumulco in Guatemala showed that each of the nine species of bolitoglossine salamanders occurs at a restricted altitude, nearly always within a specific vegetational zone. He did not consider competition to be a factor, but, in light of subsequent studies, it seems probable that it was for some combinations of species. Several observational studies have considered the possibility that competition had brought about current ecological relationships. Hairston (1949, 1964, 1973) interpreted the altitudinal distribution of two *Plethodon* species in North Carolina as being mutually restricted by competition. Dumas (1956) found that *P. vehiculum* and *P. dunni* in western Washington differed slightly in their distributions based on the moisture of the substrates on which they were found, and they also differed in their tolerances to high temperatures and low humidities in the laboratory. There was, however, wide overlap in the prey items that they had taken. He concluded that competition between them was too slight to influence their abundance or distribution. Fraser (1976b) investigated the relationship between *P. cinereus* and *P. hoffmani* in western Virginia. The two species occupied parapatric distributions with no obvious environmental factor separating them. They consumed the same kinds of food overall, but the relative contribution of different taxonomic categories differed significantly between the two salamanders, as did their relative consumption of different sizes of prey. Laboratory experiments did not reveal evidence of competition, and Fraser

suggested that there was competition for space but did not find direct evidence
for it. He also investigated the relationship between *P. hoffmani* and *P. punctatus*,
the adults of which are much larger than those of *P. hoffmani*, and again found no
evidence of competition, although the immature individuals of *P. punctatus*
overlapped in size with all *P. hoffmani* (Fraser, 1976a). Maiorana (1978) and
Lynch (1985) made detailed analyses of the food of coexisting salamanders in
California. The former concluded that food was not limiting the populations of
Aneides lugubris and *Batrachoseps attenuatus*, but on an island where food items
were small and the larger *A. lugubris* could not avoid overlap, the amount of food
limited growth and fat storage. She argued that there is competition for space.
Lynch worked with four species, *A. lugubris*, *A. flavipunctatus*, *B. attenuatus*, and
Ensatina eschscholtzii, and found differences in sizes of food items but with
much overlap. He concluded that competition for food was a reasonable possibility. Although *A. lugubris* spends much time in tree cavities, all four species are
inhabitants of the forest floor. Neither Maiorana nor Lynch reported any ecological differences other than food. Both authors related the sizes of different
salamanders to the kinds and sizes of food items eaten. From these studies, it can
be concluded that observational work has been based on the prior assumption of
competition, usually for food, but the resulting conclusions are weak.

The first experimental evidence of competition among terrestrial salamanders came from Jaeger's study in the mountains of Virginia (Jaeger, 1970,
1971a,b). He found that *Plethodon cinereus* and *P. shenandoah* occupied exclusive
distributions where a talus slope adjoined normal forest soil. In competition
experiments in cages with each type of substrate, *P. shenandoah* survived better
on talus, and *P. cinereus* better on soil, corresponding to the natural situation.
Alone, both species survived better on soil than on talus. Under direct observation, *P. cinereus* was the more aggressive and was apparently able to exclude *P.
shenandoah* from the soil (Wrobel *et al.*, 1980).

My studies in the southern Appalachians began as an investigation of the
altitudinal distribution of closely related species (Hairston, 1949). Schmidt
(1936) studied an altitudinal range of 4200 m, whereas the southern Appalachians provided a maximum range of 1500 m. I was advised by several authorities, including Schmidt, that a range of 1500 m was insufficient to be reflected in
altitudinal distributions, and general surveys had shown widely overlapping
elevations at which different species had been collected.

In contrast to haphazard collecting, my systematic survey revealed that in
single transects, with collections at each 30.5 m (100 ft) of elevation, two species
of *Plethodon* overlapped by at most 61 m on any transect in the Black Mountains
of North Carolina. For the combination of five transects, the overlap was 533.4
m, nearly 9 times as great. Moreover, a comparison of the five transects showed
that the location of overlap or near overlap was related to the direction in which
the slope faced (Fig. 2). Gaps between the distributions of the two species on

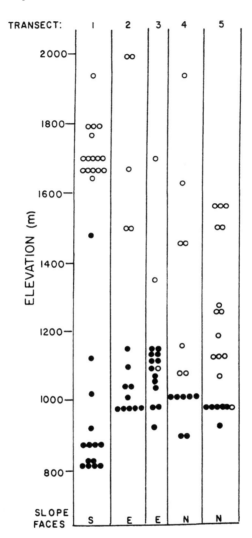

FIGURE 2. The altitudinal distribution of *Plethodon jordani* and *P. cylindraceus* in the Black Mountains of North Carolina. Five different transects are shown. Open circles, *P. jordani;* filled circles, *P. cylindraceus.* Each symbol represents one specimen.

three transects indicate my failure to find any specimens of any species. Doubtless, more thorough or repeated searches would have filled the gaps. That the overlap on any transect would have been extended by more work is doubtful, as shown by a multi-year study of one transect in the Great Smoky Mountains National Park, where the altitudinal distributions of the two species is like those

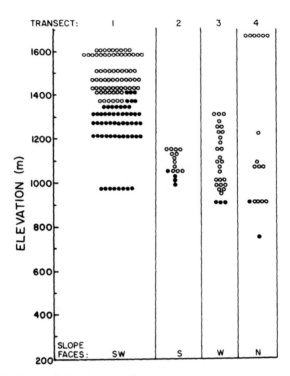

FIGURE 3. The altitudinal distributions of *Plethodon jordani* and *P. oconaluftee* in the Great Smoky Mountains National Park, Swain County, North Carolina. Open circles, *P. jordani*; filled circles, *P. oconaluftee*. Each symbol represents one specimen, except for Transect 1, where the number of symbols represents the average number seen.

in the Black Mountains (Fig. 3). The study was conducted by 43 successive classes of undergraduates from 1976 to 1994. *Plethodon jordani* and *P. oconaluftee* (formerly *P. glutinosus*) occurred together at 1406 m, 1372 m, and 1348 m; a total of 267 individuals of the former and 340 of the latter were observed at these three elevations (they were not collected but were returned to the forest). Outside of the overlap zone, up to 1622 m and down to 960 m, 1551 specimens of the two species were observed. Two of these were *P. jordani* below 1348 m, and one was *P. oconaluftee* above 1406 m. These three were not included in Fig. 3 because the symbols on that transect represent mean values. Thus, despite the prolonged intensive effort over 18 years, the zone of overlap was not extended to any appreciable degree. The evidence is very strong that the two species overlapped much more narrowly than any vegetational change would be expected to cause; interspecific competition was postulated to be the cause of the narrowness of altitudinal overlap (Hairston, 1951).

A taxonomic note is necessary here. Highton (1989) used biochemical traits to divide *Plethodon glutinosus* into 16 "species." I consider most of them to be of questionable validity, but following proper taxonomic procedure, I refer to two of the forms by the new names because no one has published reasons to invalidate them. The form east of the French Broad River in North Carolina, including the Black Mountains, is *P. cylindraceus;* the one west of that river is now *P. oconaluftee* (Hairston, 1993).

Despite the similarity in altitudinal distributions of the two species in the Black and Great Smoky Mountains, in the Balsam Mountains, which lie geographically between the other two, *P. jordani* and *P. oconaluftee* occurred together over an altitudinal range of 1365 m, including all but the highest 61 m available, where *jordani* alone was present (Fig. 4). Clearly, if interspecific competition is preventing altitudinal spread of the two species in the Blacks and Smokies, it must be much weaker, if existent, in the Balsams (Hairston, 1951).

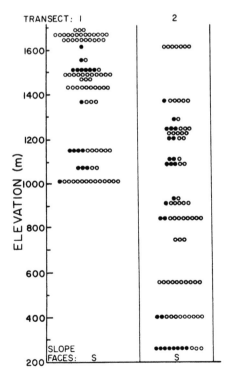

FIGURE 4. The altitudinal distribution of *Plethodon jordani* and *P. oconaluftee* in the Balsam Mountains, North Carolina. Open circles, *P. jordani;* filled circles, *P. oconaluftee.*

The hypothesis of the relative strength of competition in the three mountain ranges seems plausible but might be unconvincing to the skeptic. The following experiment was conducted as a test of the hypothesis. Experimental octagonal plots, 24.38 m (80 ft) in diameter, were established at the same elevation on south-facing slopes of the Balsam and Great Smoky Mountains. At both locations, replicate plots were selected randomly for the removal of *P. jordani* and for the removal of *P. oconaluftee*. Replicate control plots were also selected randomly. On six dates in each of the 5 years of the experiment, all plots were searched at night, and all salamanders active above ground were counted. It was necessary to continue removing the target species because the plots were unfenced and because not all individuals present on a plot were active on the surface on any given night. Five years was chosen as the duration of the experiment because that is the age at which female *P. oconaluftee* begin egg-laying, and 5 years is thus the minimum generation time for this species, the longer-lived of the two.

The results of the experiments in the two mountain ranges were as expected from the hypothesis of differing strengths of competition. In the Great Smoky Mountains, both species responded positively to the removal of the other. Where *P. jordani* was removed, *P. oconaluftee* increased in abundance by a statistically significant number compared with the control plots; where *P. oconaluftee* was removed, the proportion of hatchling and 1-year-old *P. jordani* was significantly greater than on the control plots, showing that reproduction had increased. In the Balsam Mountains, where the altitudinal overlap between the two species is nearly complete, the responses of the two species were much weaker than in the Smokies; *P. oconaluftee* increased where *P. jordani* was removed only after the end of the 4th year, and the proportion of young *P. jordani* was barely significantly greater where *P. oconaluftee* was removed than on the control plots (Hairston, 1980a).

The data from these experiments provide a challenge to the widely held assumption that coexisting related species must be in competition or must have been in competition recently enough to select for any observed differences. The two species that were removed were the most abundant ones present, accounting for 87.2% of the total on the plots in the Smokies, and 93.9% of those in the Balsams. None of the other nine species present responded to the removal of either of these two abundant species, not even the congeneric *Plethodon serratus* (Hairston, 1981).

Given that the experiments demonstrated that competition is stronger where the altitudinal overlap is narrow, an explanation of the two kinds of distributions is needed. The most obvious explanation, and one derivable from theories of community organization, is that within the zones of overlap, resource partitioning has evolved in the Balsams, with the respective requirements of the two species for one or more critical resources having become dissimilar enough to allow their extensive coexistence, whereas the situation in the Black and Great Smoky Mountains represents the original condition of intense competition. There is, however, a difficulty with the hypothesis of resource partitioning. The hypothesis requires the evolution of mutually exclusive specializations for different parts of the supposed critical resource. If the resource is consumable, as would be the case with food, it would be supposed that the two species would have become specialized to feed on different species of food animals. Such specialization seems unlikely if the food species are in limiting abundance. As all species of potential food animals fluctuate in abundance seasonally and annually, for a species to restrict its diet would most likely mean being faced with a more severe shortage when its favored food decreased in abundance. Such a tendency would surely be selected against. If the two species were to specialize on different parts of a nonconsumable resource, which would be related to their microdistributions, one would expect to be able to detect the difference by careful study. No such partitioning has been found in the Balsam Mountains, where theory leads us to expect it, as is documented in the following discussion.

It was necessary, then, to produce an alternative hypothesis to resource partitioning. I proposed that the original situation was as found in the Balsam Mountains and that one or both of the species had evolved to become more effective competitors in the Black Mountains and in the Great Smoky Mountains, thus being able to exclude each other from the elevations where each was best adapted (high altitude for *P. jordani* and low altitude for *P. oconaluftee*). With two hypotheses, it was possible to propose an experimental test. I proposed that such a test could be made by transferring *P. jordani* between additional plots in the Smokies and the Balsams (Hairston, 1973).

The Smokies were selected over the Blacks because in the Smokies, *P. jordani* has distinctive red cheeks, in contrast to the gray cheeks of specimens in the Balsams and Blacks. Thus, specimens transferred in either direction could be easily identified, as could their descendants. Transferring *P. oconaluftee* would not be feasible because there is no such convenient color or other marker differing between the two areas. After I removed the local *P. jordani* from plots in each location, I introduced specimens from the other area. The response of the local *P. oconaluftee* would permit a choice between the two hypotheses. In the Smokies, local *P. oconaluftee* should respond favorably to the introduction of *P. jordani* from the Balsams under either hypothesis because under either, the introduced *P. jordani* would be weaker competitors than the local form in the Smokies. The anticipated result was obtained. Over the 5 years of the experiment, *P. oconaluftee* increased in abundance in the presence of *P. jordani* introduced from the Balsams almost exactly as much as it did when the local *P. jordani* were removed without replacement. The critical test was the response of *P. oconaluftee* in the Balsams when faced with *P. jordani* introduced from the Smokies. If that form of *P. oconaluftee* had evolved a different requirement in the face of competition, the introduced *P. jordani* should have had little or no effect; if *P. jordani* in the Smokies had evolved a superior competitive ability, it should have had a deleterious effect on the Balsams *P. oconaluftee* which had not experienced such competition. The result of the experiment was clear. The Balsams *P. oconaluftee* decreased in abundance to a significant number relative to the control situation (Hairston, 1983b). These experiments were continued for 8 years because of difficulty in establishing enough introduced *P. jordani* for the first 3 years. Thus, in the Smokies, and presumably in the Blacks, there has been selection for increased competitive ability resulting in restricted altitudinal distributions of both species, especially of *P. jordani.*

These results were reinforced in two ways. First, careful observations at night showed that each species foraged in its characteristic way at both locations; *P. oconaluftee,* being larger, climbed more on woody plants, and *P. jordani* more on herbaceous vegetation. Second, an examination of the stomach contents of both species in both areas showed that the food was more alike in the Balsams than in the Smokies, contrary to the hypothesis of niche partitioning (Hairston *et al.,* 1987). Furthermore, laboratory experiments on their behavior showed that interspecific aggression by both species from the Smokies was as great as intraspecific aggression, whereas Balsams *P. oconaluftee,* especially, was less aggressive interspecifically than intraspecifically (Nishikawa, 1985).

Plethodon cinereus has been shown to be territorial (Jaeger, 1971a, 1974, 1979, 1981; Jaeger and Gergits, 1979; Jaeger *et al.,* 1983); our results indicate that *P. oconaluftee* and *P. jordani* are also territorial. Hence, the limiting resource is space.

III. STREAMBANK SALAMANDERS

Nearly all early observational studies of streambank salamanders concentrated on possible competitive interactions among the species. Hairston (1949, 1973), Organ (1961), Means (1975), and Krzysik (1979) compared sizes, ecological distributions, and food of coexisting species of *Desmognathus* and concluded that the differences in these attributes were the results of niche diversification under the selective pressure of competition. The different species were found at characteristic, but overlapping, distances from surface water (Table I). Corresponding to these distributions, the lengths of the heads of adults decreased in ratios close to 1.35. The largest and most aquatic species was *D. quadramaculatus,* followed by *D. monticola, D. ochrophaeus,* and the two smallest, *D. aeneus* and *D. wrighti.* The last two have no larval stage, being completely terrestrial. Their geographic distributions are almost completely

TABLE I Proportional Distribution of Four Species of *Desmognathus* with Respect to Distance Found from Water

Species	Head length (mm)	Number	0 (m)	<0.3 (m)	0.3–1.0 (m)	1.0–3.0 (m)	3.0–6.0 (m)	>6.0 (m)
D. quadramaculatus	16.8	330	0.791	0.122	0.064	0.009	0.015	0.009
D. monticola	12.3	1302	0.385	0.270	0.232	0.068	0.035	0.010
D. ochrophaeus	8.6	876	0.121	0.201	0.312	0.204	0.109	0.052
D. aeneus	5.5	307	0.0	0.013	0.085	0.182	0.261	0.459

Note. No larvae are included. Elevation 686 m in Coweeta Hydrologic Laboratory, Nantahala Mountains, North Carolina. Numbers include all class data, 1972–1994.

separate. The larval period of D. *quadramaculatus* is at least 3 years; that of *D. monticola* is 11 months; and the larval period of *D. ochrophaeus* varies from a few weeks to 11 months.

The habitats from which the respective prey of the different species had been taken corresponded fairly well with the locations of their daytime retreats, from which their distributions were originally determined; *D. quadramaculatus* had taken a disproportionate number of terrestrial prey species, but it was found later that this species forages extensively away from water at night.

Eventually, it was pointed out that niche diversification as the result of interspecific competition was not an acceptable explanation for the ecology of the members of the genus *Desmognathus* (Tilley, 1968; Hairston, 1980b). The reasoning was as follows: It is generally agreed that the ancestor was most like the two largest, and most aquatic, species. All members of the Plethodontidae are lungless, and respiration must take place through the skin or partly through the buccopharyngeal cavity. As all *Desmognathus* are approximately the same shape, small species would be at a disadvantage away from water, as water loss would be relatively greater for them than for larger species because of the larger surface-to-volume ratio of a small animal. Yet it is the smaller species that are the most terrestrial, and therefore selection for the kind of niche partitioning that had been envisioned would be opposed by selection against adopting the terrestrial habitat. An alternative hypothesis was proposed: At each speciation, the smaller of the two new species would be under predation pressure to avoid the habitat of the larger. Since the larger species was aquatic or semiaquatic, the smaller species became more terrestrial. I proposed two experiments to permit a choice between competition and predation as the driving force in the evolution of the genus (Hairston, 1980b, 1984). *Desmognathus monticola*, the second in the series from aquatic to terrestrial, was to be removed from one set of replicate plots. *Desmognathus ochrophaeus*, the next in the series, should increase in abundance under either hypothesis because it should be either a competitor or a prey item. Removal of *D. ochrophaeus* from another set of plots would provide the critical test. If *D. ochrophaeus* were a competitor, *D. monticola* should increase in abundance with the reduced competition; if *D. ochrophaeus* were a prey of *D. monticola*, the latter should not benefit, and might decrease in abundance, if the smaller *D. ochrophaeus* were an important item of food. The experiments were conducted in a block design with each of three small streams in the Coweeta Hydrologic Laboratory being a block. The experimental plots extended 7 m beyond the stream on each side and 10 m along the stream. The plots were separated by at least 30 m. One plot was designated as a control where no manipulations were done; one was for the removal of *D. monticola*, and one was for the removal of *D. ochrophaeus*.

The results of the experiments confirmed the predation hypothesis; *D. ochrophaeus* increased significantly in abundance on the plots from which *D. monticola* was removed; *D. monticola* declined significantly on the plots from which *D. ochrophaeus* was removed, as did *D. quadramaculatus,* leading to the conclusion that the smaller species is an important prey item for the two larger ones. Removal of *D. monticola* also led to a slight and barely significant increase in the abundance of *D. quadramaculatus* relative to the control plots, suggesting competition between them (Hairston, 1986). The importance to the community of this interaction is problematical, as Southerland (1985) has observed *D. quadramaculatus* preying on *D. monticola* in the field, and cannibalism is common among captive specimens of all three species.

I now describe three studies that showed direct and indirect effects of aggression among coexisting species of *Desmognathus.* Keen and Sharp (1984) tested individuals of *D. monticola* and *D. fuscus* in small laboratory arenas that had been occupied for several days by individuals of either one or the other species. Intruding individuals of the other species were then introduced. They were attacked in 101 of 204 trials. Southerland (1986) found that the smaller *D. fuscus* chose cover objects different from those it preferred when it was in the presence of the larger *D. quadramaculatus* and *D. monticola.* In a different experiment, he found that in confinement with a number of congeners and numerous cover objects, two individuals, regardless of species, were found together under one cover object significantly less frequently than expected by chance. He also confined *D. monticola* separately with *D. ochrophaeus, D. fuscus* and *D. quadramaculatus* in matched cages at locations appropriate to the natural distributions of the several species. Survival of *D. monticola* was worst in combination with the larger *D. quadramaculatus* (Southerland, 1985).

An unexpected result of my removal experiments was the failure to detect any effect on the population of the smallest species, *D. aeneus.* In the Nantahala Mountains of North Carolina, where the experiments were conducted, *D. aeneus* is more secretive at night than any other salamander species with which I have worked. In contrast with all other species, it is harder to find *D. aeneus* active above ground at night than it is to find it under cover during the day. The same species has been found very active at night near Tuscaloosa, Alabama, where it occurs in an isolated population (Wilson, 1984). As none of the larger species of *Desmognathus* occur with it there, I concluded that predation has made *D. aeneus* secretive in North Carolina (Hairston, 1987), and I postulated that we are seeing "the ghost of predation past" in the failure of *D. aeneus* to respond to the removal of either *D. monticola* or *D. ochrophaeus.*

Beachy (1993) has demonstrated that there is significant predation by the larvae of *D. quadramaculatus* on those of another plethodontid, *Eurycea wilderae,* in the same area as my experiments were conducted. In addition, although there is no local evidence, the large plethodontid *Gyrinophilus por-*

phyriticus is present and has frequently been reported to prey on other salamanders, especially *D. ochrophaeus,* most recently in the laboratory experiment of Formanowicz and Brodie (1993).

Combining the known interactions among the members of the streambank community (Fig. 5), we find a picture that would have been difficult to suggest had we been guided by the earlier interpretations of several careful ecological studies, my own included. The coexisting species of *Desmognathus* provide an excellent example of the kinds of data that during the 1960s and 1970s had been claimed to be the result of niche partitioning under the selection provided by competitive interactions. Experimental tests with this genus have shown that with respect to the two competing hypotheses, predation is an overwhelmingly important factor and competition is a relatively minor one, found only between two of the six species involved. Is *D. aeneus* a member of this community? Is there any basis on which it could have been excluded before the experiment? I think the answer is no to both questions. It could be included only through the argument that all organisms in the area must interact through the flow of energy or the cycling of materials. If accepted, such an argument would invalidate the title of this book, as "vertebrate communities" would then be only parts of communities.

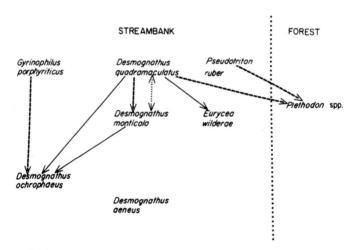

FIGURE 5. The known interactions among the species of plethodontid salamanders inhabiting the streambank habitat, including two instances of predation on species from the forest floor. Solid lines represent predation quantified by field experiments; broken lines represent unquantified observations of predation; dotted line represents experimentally confirmed competition. Arrows point in the direction of the prey or the competitor. Approximate mean lengths of adults: *G. porphyriticus,* 16 cm; *D. quadramaculatus,* 14 cm; *D. monticola,* 10 cm; *D. ochrophaeus,* 7.4 cm; *D. aeneus,* 5.3 cm; *P. ruber,* 9 cm; *E. wilderae,* 7 cm; *Plethodon* spp., 11–14 cm.

IV. POND SALAMANDERS

A long-term study of salamander populations in a pond in the Savannah River Site was reported by Pechmann *et al.* (1991). The data were obtained by trapping migrating specimens against a drift fence as they came to the pond to breed or as they left as newly metamorphosed individuals. The number migrating and the reproductive success rate were strongly influenced by weather conditions at the time of migration and during the larval period. The actual size of the adult population could not be found by the methods used, and the striking fluctuations in the numbers obtained were not representative of the stability or lack of stability of the populations. The only species showing a trend in numbers over the 12 years, *Ambystoma opacum*, increased in the number trapped. The other two species, *A. talpoideum* and *A. tigrinum*, showed no consistent trends in abundance. Persistence of the populations is a more accurate expression than stability. Such observations are typical for many communities of pond-dwelling salamanders.

Like observational studies in terrestrial and streambank salamander communities, many such studies in ponds represented searches for ecological differences that could account for coexistence of the species involved. The interpretations were not always in agreement even when the compositions of the studied communities were much alike. Three studies of European members of the newt genus *Triturus* serve as an example. Szymura (1974) sampled populations of *T. cristatus, T. alpestris, T. vulgaris,* and *T. montandoni* in three ponds in Poland. The differences in size distribution and in ecological distribution were thought to be sufficient to reduce competition for food, assumed to be the limiting resource. The food items taken by two of these species, *T. cristatus* and *T. vulgaris,* plus *T. helveticus* were studied by Avery (1968) in two ponds in England. He recognized that there was too much similarity among the diets for niche partitioning to be invoked, and he hypothesized that temperature relations provided a better separation. Griffiths (1986) also found the food of *T. vulgaris* and *T. helveticus* to be very similar and could obtain no evidence of any niche partitioning.

In the eastern United States, there have been at least five studies of different combinations of species of *Ambystoma*, sometimes accompanied by *Notophthalmus viridescens*. In all of these studies, evidence of niche partitioning was sought, and it was claimed in several. Spatial separation of species, especially where they fed in the water column and when they fed at night, has been studied in New Jersey by Anderson and Graham (1967) and Hassinger *et al.* (1970). The larvae of most species spent the day under cover at the bottom of the pond and fed in the water column at night. *Ambystoma opacum, A. tigrinum,* and *A. jeffersonianum* migrated to the upper layer of the water column in a regular sequence with only one species being there at a time. A fourth species, *A. maculatum*, did not migrate; nor did the aquatic adults of *Notophthalmus viridescens*. The authors concluded that the sequence might reduce competition and increase utilization of the pond ecosystem. In Mississippi, Branch and Altig (1981) obtained contradictory evidence. The larvae of *A. maculatum* did migrate into the water column, and their distribution overlapped widely with those of *A. opacum* and *A. talpoideum*. There were differences, but they were too slight to obviate competition. In Maryland, Worthington (1968, 1969) found sequential hatching of *A. opacum*, *A. maculatum* and *N. viridescens* with the result that although their larvae achieved the same sizes, that never happened at the same time; *A. opacum* was 26 mm in snout-vent length when *A. maculatum* hatched at 9 mm; and *A. maculatum* was 23 mm in snout-vent length, when *N. viridescens* was 10 mm. He concluded that the sequence had evolved to reduce competition. Heyer (1976) observed microhabitat differences between the two *Ambystoma* species and concluded that effectively they did not inhabit the same ponds. In the only observational study of pond salamanders where predation was examined, Stewart (1956) showed that *A. opacum* larvae preyed on *A. maculatum* larvae in ponds in North Carolina.

In experiments on different combinations of species of *Ambystoma*, both predation and interspecific competition have been demonstrated repeatedly. Wilbur (1972) experimented with the members of a four-species community in ponds in southern Michigan. He used replicated pens constructed of window screening in a natural pond. *Ambystoma maculatum*, *A. laterale*, and *A. tremblayi* are closely related taxonomically. *Ambystoma tigrinum* is quite distinct, and it was shown experimentally to act as a predator on the others. Experiments on two different densities of each pair of the first three species revealed complex competitive interactions among them. Survival, growth, and the duration of the larval period were observed. An increase in the number of any of the three species in two-way competition prolonged the larval period of the others; in the three-way experiments, an increase in the number of *A. maculatum* had that effect on the other two, but other possible effects were not observed. An increase in the number of *A. maculatum* depressed the growth of both of the other two, and an increase in the number of *A. tremblayi* depressed the growth of *A. laterale*. In the three-way experiment, the overall results were affected by the densities of both *A. laterale* and *A. tremblayi*; *A. maculatum* was important through its interaction with the other two species. Differences between replicates made survival difficult to analyze, and only an effect of *A. laterale* on *A. tremblayi* could be documented. There were several non-linear results of adding further densities of each species. These revealed "higher-order interactions." They can be visualized as "coalitions" that either increased the effect of two species on the third beyond the effect of the sum of their separate effects or reduced the effect of one species when both of the others were present.

Three other combinations of species have been the subjects of experiments testing for competition. In North Carolina, Stenhouse et al. (1983) found mutual negative effects on the growth of A. opacum and A. maculatum both in field enclosures and in laboratory containers. Intraspecific density effects in A. opacum were more prominent, however, and predation by A. opacum on A. maculatum was the most important interaction, as described in the following discussion. Some of the various experiments have been interpreted as indicating that exploitative competition was involved. Smith (1990), however, found that whereas larger larvae of A. opacum reduced the growth of smaller ones in 21-cm plastic bowls, that did not happen when they were separated by a partition with mesh large enough to permit passage of the prey (Daphnia pulex). Thus, behavioral interference was the competitive mechanism. Interactions between A. maculatum and A. talpoideum have been investigated in Louisiana, with interspecific aggression by A. talpoideum being the most prominent feature of the competition, although A. maculatum inhibited the growth of A. talpoideum (Walls and Jaeger, 1987). A comparison between the competitive abilities of the respective populations from Mississippi and South Carolina (Semlitsch and Walls, 1993) revealed no geographic variation; A. talpoideum was the superior competitor based on the effect of each on the other's growth and size at metamorphosis. There was no significant effect of either on the survival of the other.

Most studies have concentrated on larvae as being more likely to show ecological interactions, but four studies have attempted to discover if metamorphosed individuals interfered with one another. Martin et al. (1986) matched adult A. texanum in laboratory arenas and found no evidence of any behavioral interference. Ducey and Ritsema (1988), using similar methods, observed intraspecific aggression among adult males of A. maculatum, and Ducey (1989) observed at least some intraspecific aggression in A. opacum, A. macrodactylum, A. tremblayi, and A. talpoideum. Walls (1990) investigated the possibility of interspecific interference between juveniles of A. maculatum and A. talpoideum. She found both aggressive and submissive behavior. Response to conspecifics was greater than to congeners. Unlike the larvae, juveniles of A. maculatum were more aggressive toward juveniles of A. talpoideum than the reverse.

Mutual competitive effects between the larvae of A. tigrinum and adults and immature N. viridescens have been found experimentally (Morin, 1983), but the effects are best described as part of predation. Morin used large metal tanks, each holding 1 m³ of water, as replicates. Each tank was provided with natural assemblages of aquatic insects and crustaceans plus 1200 frog tadpoles of six species as food for the salamanders. Four tanks had four adult newts each; three had four larvae of A. tigrinum; and three had four newts and four larvae of A. tigrinum. The newts reduced the size of A. tigrinum at metamorphosis but did not affect survival. The larvae of A. tigrinum reduced the production of juvenile newts to little more than one-fourth that of the control tanks (27 vs 101). The reduced densities of newt larvae allowed the production of much larger young, showing that there was intraspecific competition in the controls. In a natural pond where both species were present, Morin was able to show that adult newts preyed so extensively on the eggs of A. tigrinum that few larvae reached metamorphosis. By protecting the gelatinous egg masses in mesh cages, he increased the survival of A. tigrinum.

In ponds where more than one species of Ambystoma breeds, there is normally a seasonal progression of egg-laying dates. As most of the ponds are not permanent, filling with autumn or winter rains and drying up by midsummer, there is an advantage to early breeding. The disadvantage lies in the possibility that the shallow pond will freeze to the bottom after the early-breeding species have laid eggs. Where A. tigrinum occurs, it is frequently the earliest to breed. The eggs hatch early, and the larvae have frequently grown to a size large enough to prey on the later-breeding species. Such was the case in Michigan, where Wilbur (1972) showed that numbers of all three congeners were greatly reduced by predation by A. tigrinum.

Ambystoma opacum has pushed early breeding as far as it is possible. The adults mate in early fall, and the females lay their eggs near the future water line before the pond fills with autumn rains. As water covers the eggs, they hatch; they have grown to a size at which they can prey on the larvae of A. maculatum when the latter hatch in early spring. The quantitative aspects of this relationship were explored observationally and experimentally by Stenhouse et al. (1983) and Stenhouse (1984, 1985). The use of a drift fence around a pond permitted the calculation that the survival to metamorphosis of A. maculatum was 0.0011 (4 juveniles out of 3672 eggs laid); survival of A. opacum was 0.0882 (230 juveniles out of 2607 eggs laid). In laboratory experiments, 1, 2, or 4 newly hatched A. maculatum larvae were kept in plastic dishes 21 cm in diameter with 1, 2, or 4 A. opacum in a completely crossed experiment. The dishes were supplied with enough tubificid worms to satiate two A. opacum. Survival of the A. maculatum larvae was inversely proportional to the number of A. opacum present, all deaths being from predation. The daily mortality was 0.059 with one A. opacum, 0.134 with two, and 0.170 with four. The density of A. opacum in the pond was estimated to be 33 m^{-2}, or slightly higher than the lowest density in the plastic dishes. If predation by A. opacum were the only cause of mortality, the expected survival of A. maculatum over the 60 days when both species were present would be 0.029; with the intermediate density of A. opacum, it would be 0.0003. Thus, the laboratory conditions appeared to be reasonably realistic. A further experiment was conducted in the field using circular cages 0.5 m² in area. The bottom of the cages had natural cover of dead leaves. Forty A. maculatum were introduced with either 10 or 20 A. opacum (0.61 or 1.22 times the natural density). In control pens, the daily mortality of A. maculatum was 0.074; with 10 A. opacum, it was 0.162; and with 20 A. opacum, it was 0.195. These were about 6 times as great as at comparable densities in the laboratory, suggesting that the food provided in the laboratory was unnaturally generous, and that the A. opacum searched more diligently in the field. The experiments demonstrated that A. opacum was an important natural predator on A. maculatum. There are several studies based on the examination of stomach contents, but the populational effect cannot be estimated from such data.

Table II shows interspecific competition and predation among the species of salamanders dwelling in ponds. It should be noted that four of the references in the table are not mentioned in the text because they record anecdotal information only. When we consider that the "observed" examples of competition are based on the assumption that ecological differences evolved to reduce competition, predation and competition are about equal in frequency. The "observed" examples of predation are based on examination of stomach contents of predators or on direct observation of the act of predation, and they are more reliable than are claims of competition not determined by experiments.

V. DISCUSSION

This review has shown that competition and predation influence the composition of different salamander communities to differing degrees, depending largely on the kind of habitat in which experiments testing for the interactions have been conducted. Among exclusively terrestrial salamanders, competition has been demonstrated experimentally in two combinations of species: Plethodon cinereus and P. shenandoah in Virginia, and P. jordani and P. oconaluftee in North Carolina. In five other studies, evidence of niche partitioning was sought but was either not found or the evidence showed much overlap in resource use. Where competition was demonstrated experimentally, the species coexisting with the competitors could not be shown to be affected by either of them. Within this community, no evidence of predation has been found. Streambank salamanders, in contrast with forest forms, have been found to be involved in a number of examples of interspecific predation, but competition has been demonstrated between only two of the seven species investigated. It is an interesting anomaly that the smallest species present could not be shown to be affected by any of the others, either as prey or as a competitor. This example makes it especially clear that generalizations about

TABLE II Predation and Competition among Salamander Species in Ponds

Prey or competitor	Predator						
	Ambystoma tigrinum	A. opacum	A. tremblayi	A. laterale	A. texanum	A. talpoideum	Notophthalmus viridescens
A. tigrinum							pre (2) e
A. opacum	com (11) o						com (12) o
A. tremblayi	pre (1) e						
A. latrerale	pre (1) e		com (1) e	can (10) o			
A. jeffersonianum	com (11) o	pre (7) o com (11) o					
A. texanum		pre (7) o	com (1) e	com (1) e	can (9) o		
A. maculatum	pre (1) e	pre (3, 8) e com (3) e	com (1) e	com (1) e	com (1) e	com (4, 5, 6) e	com (12) o
Notophthalmus viridescens	pre (2) e com (2) e						

Note. Symbols: pre, predation; com, competition; can, cannibalism; e, experimentally determined; o, based on observation. Sources: (1) Wilbur, 1972; (2) Morin, 1983; (3) Stenhouse et al., 1983; (4) Walls, 1990; (5) Walls and Semlitsch, 1991; (6) Semlitsch and Walls, 1993; (7) Walters, 1975; (8) Stewart, 1956; (9) Hay, 1889; (10) Bishop, 1941; (11) Anderson and Graham, 1967; (12) Worthington, 1968.

interactions that determine community composition are likely to be invalidated in unexpected ways, in this case by selection for a highly secretive behavior, and differ from those observed in the same species at a distant location where it coexists with a different combination of species.

In ponds, salamanders interact both as competitors and as predators and prey. Like most species that have been investigated, pond salamanders are aggressive, both intraspecifically and interspecifically. When two individuals are near the same size, the effect of the aggression is one of competition; where there is a large enough size difference, the aggression becomes predation. Indeed, in all examples of salamander competition in whatever habitat, aggressive interference has been found to be the important feature of the interaction.

Most ecological studies gain in value the longer they are continued. The obvious reasons are that annual and seasonal changes are revealed, as are the responses of the organisms to such changes, in reproduction and survival as well as in physiological processes. If observations are not prolonged over a number of years, there is a strong temptation to assume that the period during which the study was made was representative of all such periods and that conclusions reached by induction from the study are generally applicable. If the populations were temporarily increasing, the conclusions about the importance of competition or predation might be very different from a study of declining populations.

The only way to avoid this pitfall is to continue estimating populations long enough to establish any trends or to show a repeatable pattern in changes in numbers or age structure. The requirement is that the study continue until all of the organisms present at the beginning of the study have died (Connell and Sousa, 1983). The reason for the requirement is that repeatedly counting the same individuals in the original population gives a false impression of stability. A more realistic requirement is that more than 95% of the original individuals are no longer present (Schoener, 1985).

That requirement is difficult to meet for salamanders because of their longevity, which is greater than that of most other small vertebrates. For example, half of the adult *Plethodon jordani* are more than 9 years old, and adult *P. glutinosus* or *P. oconaluftee* are older. That means that observations should continue for 11–13 years before population stability could be claimed with assurance or that trends or regular changes could be claimed. To judge by survival estimated from Twitty's (1966) recovery of marked individuals of the red-bellied newt, *Taricha rivularis*, those populations should be followed even longer than the species of *Plethodon*; adult survival of *T. rivularis* is 0.91 per year; that of *P. jordani* is 0.81 per year. For *Desmognathus* species, which do not live as long, observations over 4–6 years would still be required, and that is longer than most ecological studies.

Salamander populations have been the subject of more long-term studies than most other vertebrates, but some notable examples should be mentioned: the turtles of the University of Michigan's E. S. George Reserve, observed since 1953 (Sexton, 1959; Wilbur, 1975; Congdon and Gibbons, 1990, Chap. 6 this volume), and the experiments conducted on granivorous rodents in the Chihuahuan Desert of Arizona since 1977 (Heske *et al.*, 1994; Valone and Brown, Chap. 17 this volume).

The value of such studies has been stressed repeatedly, but financial and academic support for them has been meager.

VI. SUMMARY

Despite their classification as Amphibia, most salamander species are not amphibious. The majority (191 of 355) are exclusively terrestrial, and a substantial number (48) never leave the water (Hairston, 1987). Two-thirds of the species belong to the lungless Plethodontidae which do not appear to be handicapped by the loss as they are the most widespread and abundant members of the Order Caudata. In the Appalachian Mountains, some species reach densities of $0.7-2.3$ m^{-2}, more than birds and mammals combined. Populations of terrestrial and streambank forms are unusually stable. There have been no trends in abundance over 23 years for six species that have been counted 2–4 times annually since 1976. It is more difficult to establish the population sizes of pond-dwelling species because of the wide fluctuations in the number coming to breed. The fluctuations are caused by variation in rainfall just before and during the short breeding season.

Different kinds of interspecific interactions have been found to be characteristic of the different habitats.

Although five comparative studies of the forest floor have revealed nothing conclusive (morphological and ecological differences between species can always be found), two sets of experiments have produced convincing evidence of interspecific competition. Both were stimulated by local ecological distributions: *Plethodon cinereus* and *P. shenandoah* are sharply divided at the border between deep soil and talus. Each was found to be the superior competitor in experimental cages placed in their respective kinds of soil. *Plethodon jordani* and *P. oconaluftee* are separated by altitude in the Great Smoky Mountains, occurring together only over about 75 m vertically; in the adjacent Balsam Mountains, the same two species overlap by 1365 m. Removal experiments over 5 years showed strong competition in the Smokies and much weaker competition in the Balsams. The basis for the difference was established by experiments that involved substituting *P. jordani* populations between plots in each of the two mountain ranges. There was no evidence that resource partitioning in the Balsams was involved. Instead, the two species had evolved to become more effective interspecific competitors in the Smokies. That had been achieved by their becoming more aggressive against the other species. The other nine species on the removal plots were unaffected by the removal of either *P. jordani* or *P. oconaluftee,* which are the two most abundant species present.

Preliminary streambank observations seemed to show that four coexisting species of *Desmognathus* represented a classical example of niche partitioning. The species have different but overlapping distributions on the aquatic-terrestrial streambank gradient, and their sizes decline along this gradient in nearly constant ratios of 1.35. Niche partitioning was challenged by the hypothesis that predation by the larger, more aquatic species forced the smaller ones to become adapted to the less favorable terrestrial habitats. Experiments conducted to choose between the hypotheses resulted in showing that the predation hypothesis is correct. When a small, abundant species was removed from replicated plots, the next larger ones both decreased in abundance, whereas if the competition theory were true, they should have responded favorably. The small species was evidently an important prey item. Another experiment showed that larvae of the largest *Desmognathus* prey on those of a different genus, *Eurycea.*

Like comparative studies of salamanders in other habitats, the focus of pond studies has been on searching for ecological differences in the expectation that they could account for the coexistence of the different species. Differences were found, but the conclusions were reached by weak inference and were not easily defendable. Two studies in England were exceptions in that evidence of niche partitioning was lacking. It remained for experimentation to reveal the true ecological interactions, and such interactions between larvae are the universal results of all experiments. Aggression is always present, and it has been observed intraspecifically as well as interspecifically. As the different coexisting species frequently oviposit sequentially, the extra time for growth makes the early-laying species large enough to act as a predator. The predators are *Ambystoma opacum,* which lays eggs in autumn, and *A. tigrinum,* which lays in late winter. The other species lay eggs in early or mid spring. Experiments have shown that interspecific competition occurs in nine combinations of species, and interspecific predation in six combinations.

Salamanders are longer-lived than most other small vertebrates. For example, half of the adults of *Plethodon jordani* are more than 9 years old, and *P. glutinosus* and *P. oconaluftee* are at least a year older at adulthood. This means that population changes will be detected slowly, and, to be convincing of stability, population studies should continue until the great majority of individuals alive at the start of the study have died. For the two species mentioned, population counts less than 11–13 years apart would not establish stability. For most other salamander species, the necessary span between counts would be less, but it would still be formidable: 4–6 years for different species of

Desmognathus and less for some *Ambystoma,* with the caution that the counts for these species are unreliable as population estimates. Studies lasting longer than one might expect are therefore highly necessary to establish population stability. Without such a demonstration, we could not be truly confident of our conclusions about the importance of any interspecific interactions to the community under study.

References

Anderson, J. D., and Graham, R. E. (1967). Vertical migration and stratification of larval *Ambystoma. Copeia,* **1967,** 371–374.

Avery, R. A. (1968). Food and feeding relations of three species of *Triturus* (Amphibia Urodela) during the aquatic phases. *Oikos* **19,** 408–424.

Beachy, C. K. (1993). Guild structure in streamside salamander communities: A test for interactions among larval plethodontid salamanders. *J. Herpetol.* **27,** 465–468.

Bell, G. (1974). The reduction of morphological variation in natural populations of smooth newt larvae. *J. Anim. Ecol.* **43,** 115–128.

Bell, G. (1977). The life of the smooth newt (*Triturus vulgaris*) after metamorphosis. *Ecol. Monogr.* **47,** 279–299.

Bell, G., and Lawton, J. H. (1975). The ecology of the eggs and larvae of the smooth newt [*Triturus vulgaris* (Linn.)]. *J. Anim. Ecol.* **44,** 393–423.

Bishop, S. C. (1941). "The Salamanders of New York." New York State Museum, Albany.

Branch, L. C., and Altig, R. (1981). Nocturnal stratification of three species of *Ambystoma* larvae. *Copeia,* **1981,** 870–873.

Burton, T. M., and Likens, G. E. (1975). Energy flow and nutrient cycling in salamander populations in the Hubbard Brook Experimental Forest, New Hampshire. *Ecology* **56,** 1068–1080.

Congdon, J. D., and Gibbons, J. W. (1990). The evolution of turtle life histories. *In* "The Life History and Ecology of the Slider Turtle" (J. W. Gibbons, ed.). Smithsonian Institution Press, Washington, DC.

Connell, J. H., and Sousa, W. P. (1983). On the evidence needed to judge ecological stability or persistence. *Am. Nat.* **121,** 789–824.

Ducey, P. K. (1989). Agonistic behavior and biting during intraspecific encounters in *Ambystoma* salamanders. *Herpetologica* **45,** 155–160.

Ducey, P. K., and Ritsema, P. (1988). Intraspecific aggression and responses to marked substrates in *Ambystoma maculatum* (Caudata: Ambystomatidae). *Copeia,* **1988,** 1008–1013.

Dumas, P. C. (1956). The ecological relations of sympatry in *Plethodon dunni* and *Plethodon vehiculum. Ecology* **37,** 485–495.

Formanowicz, D. R., Jr., and Brodie, E. D., Jr. (1993). Size-mediated predation pressure in a salamander community. *Herpetologica* **49,** 265–270.

Fraser, D. (1976a). Coexistence of salamanders of the genus *Plethodon:* A variation of the Santa Rosalia theme. *Ecology* **57,** 238–251.

Fraser, D. (1976b). Empirical evaluation of the hypothesis of food competition in salamanders of the genus *Plethodon. Ecology* **57,** 459–471.

Gill, D. E. (1978). The metapopulation ecology of the red-spotted newt, *Notophthalmus viridescens* (Rafinesque). *Ecol. Monogr.* **48,** 145–166.

Gill, D. E. (1979). Density dependence and homing behavior in adult red-spotted newts *Notophthalmus viridescens* (Rafinesque). *Ecology* **60,** 800–813.

Griffiths, R. A. (1986). Feeding niche overlap and food selection in smooth and palmate newts, *Triturus vulgaris* and *T. helveticus*, at a pond in Mid-Wales. *J. Anim. Ecol.* **55**, 201–214.

Hairston, N. G. (1949). The local distribution and ecology of the plethodontid salamanders of the southern Appalachians. *Ecol. Monogr.* **19**, 47–73.

Hairston, N. G. (1951). Interspecies competition and its probable influence upon the vertical distribution of Appalachian salamanders of the genus Plethodon. *Ecology* **32**, 266–274.

Hairston, N. G. (1964). Studies on the organization of animal communities. *J. Anim. Ecol., Suppl.* **33**, 227–239.

Hairston, N. G. (1973). Ecology, selection and systematics. *Breviora* **414**, 1–21.

Hairston, N. G. (1980a). The experimental test of an analysis of field distributions: Competition in terrestrial salamanders. *Ecology* **61**, 817–826.

Hairston, N. G. (1980b). Species packing in the salamander genus *Desmognathus*: What are the interspecific interactions involved? *Am. Nat.* **115**, 354–366.

Hairston, N. G. (1981). An experimental test of a guild: Salamander competition. *Ecology* **62**, 65–72.

Hairston, N. G. (1983a). Growth, survival and reproduction of *Plethodon jordani*: Trade-offs between selective pressures. *Copeia*, **1982**, 1024–1035.

Hairston, N. G. (1983b). Alpha selection in competing salamanders: Experimental verification of an a priori hypothesis. *Am. Nat.* **122**, 105–113.

Hairston, N. G. (1984). Inferences and experimental results in guild structure. *In* "Ecological Communities: Conceptual Issues and the Evidence" (D. R. Strong, Jr., D. Simberloff, L. G. Abele, and A. B. Thistle, eds.), pp. 19–27. Princeton Univ. Press, Princeton, NJ.

Hairston, N. G. (1986). Species packing in *Desmognathus* salamanders: Experimental demonstration of predation and competition. *Am. Nat.* **127**, 266–291.

Hairston, N. G. (1987). "Community Ecology and Salamander Guilds." Cambridge Univ. Press, New York.

Hairston, N. G. (1993). On the validity of the name *teyahalee* as applied to a member of the *Plethodon glutinosus* complex (Caudata: Plethodontidae): A new name. *Brimleyana* **18**, 65–69.

Hairston, N. G., and Wiley, R. H. (1993). No decline in salamander (Amphibia: Caudata) populations: A twenty-year study in the southern Appalachians. *Brimleyana* **18**, 59–64.

Hairston, N. G., Nishikawa, K. C., and Stenhouse, S. L. (1987). Evolution of competing species of terrestrial salamanders: Niche partitioning or interference? *Evol. Ecol.* **1**, 247–262.

Hassinger, D D., Anderson, J. D., and Dolrymple. 1970. The early life history and ecology of *Ambystoma tigrinum* and *Ambystoma opacum* in New Jersey. *Am. Midl. Nat.* **84**, 474–495.

Hay, O. P. (1889). Notes on the habits of some ambystomas. *Am. Nat.* **23**, 602–612.

Heske, E. J., Brown, J. H., and Mistry, S. (1994). Long-term experimental study of a Chihuahuan Desert rodent community: 13 years of competition. *Ecology* **75**, 438–445.

Heyer, W. R. (1976). Studies in larval habitat partitioning. *Smithson. Contrib. Zool.* **242**, 1–27.

Highton, R. (1989). Biochemical evolution in the slimy salamanders of the *Plethodon glutinosus* complex in the eastern United States. I. Geographic protein variation. *Ill. Biol. Monogr.* **57**, 1–78.

Jaeger, R. G. (1970). Potential extinction through competition between two species of terrestrial salamanders. *Evolution* **24**, 632–642.

Jaeger, R. G. (1971a). Competitive exclusion as a factor influencing the distribution of two species of terrestrial salamanders. *Ecology* **52**, 632–637.

Jaeger, R. G. (1971b). Moisture as a factor influencing the distributions of two species of terrestrial salamanders. *Oecologia* **6**, 191–207.

Jaeger, R. G. (1974). Interference or exploitation? A second look at competition between salamanders. *J. Herpetol.* **8**, 191–194.

Jaeger, R. G. (1979). Seasonal spatial distributions of the terrestrial salamander *Plethodon cinereus*. *Herpetologica* **35**, 90–93.

Jaeger, R. G. (1981). Dear enemy recognition and the costs of aggression between salamanders. *Am. Nat.* 117, 962–974.

Jaeger, R. G., and Gergits, W. F. (1979). Intra- and interspecific communication through chemical signals on the substrate. *Anim. Behav.* 27, 150–156.

Jaeger, R. G., Nishikawa, K. C., and Barnard, D. E. (1983). Foraging tactics of a terrestrial salamander: Costs of territorial defense. *Anim. Behav.* 31, 191–198.

Keen, W. H., and Sharp, S. (1984). Responses of a plethodontid salamander to conspecific and congeneric intruders. *Anim. Behav.* 32, 58–65.

Krzysik, A. J. (1979). Resource allocation, coexistence, and the niche structure of a streamside salamander community. *Ecol. Monogr.* 49, 173–194.

Lynch, J. F. (1985). The feeding ecology of *Aneides flavipunctatus* and sympatric plethodontid salamanders in northwestern California. *J. Herpetol.* 19, 328–352.

Maiorana, V. C. (1978). Difference in diet as an epiphenomenon: Space regulates salamanders. *Can. J. Zool.* 56, 1017–1025.

Martin, D. L., Jaeger, R. G., and Labat, C. P. (1986). Territoriality in an *Ambystoma* salamander? Support for the null hypothesis. *Copeia*, 1986, 725–730.

Means, D. B. (1975). Competitive exclusion along a habitat gradient between two species of salamanders (*Desmognathus*) in western Florida. *J. Biogeogr.* 2, 253–263.

Morin, P. J. (1983). Competitive and predatory interactions in natural and experimental populations of *Notophthalmus viridescens dorsalis* and *Ambystoma tigrinum*. *Copeia*, 1983, 628–639.

Nishikawa, K. C. (1985). Competition and the evolution of aggressive behavior in two species of terrestrial salamanders. *Evolution (Lawrence, Kans.)* 39, 1282–1294.

Organ, J. A. (1961). Studies of the local distribution, life history, and population dynamics of the salamander genus *Desmognathus* in Virginia. *Ecol. Monogr.* 31, 189–220.

Pechmann, J. H. K., Scott, D. E., Semlitsch, R. D., Caldwell, J. P., Vitt, L. J., and Gibbons, J. W. (1991). Declining amphibian populations: The problem of separating human impacts from natural fluctuations. *Science* 253, 892–895.

Schmidt, K. P. (1936). Guatemalan salamanders of the genus *Oedipus*. *Field Mus. Nat. Hist. Publ., Zool. Ser.* 20, 135–166.

Schoener, T. W. (1985). Are lizard population sizes unusually constant through time? *Am. Nat.* 126, 633–641.

Semlitsch, R. D., and Walls, S. C. (1993). Competition in two species of larval salamanders: A test of geographic variation in competitive ability. *Copeia*, 1993, 587–595.

Sexton, O. J. (1959). A method of estimating the age of painted turtles for use in demographic studies. *Ecology* 40, 716–718.

Smith, C. K. (1990). Effects of variation in body size on intraspecific competition among larval salamanders. *Ecology* 71, 1777–1788.

Southerland, M. T. (1985). Organization in desmognathine salamander communities: The roles of habitat and biotic interactions. Ph.D. Thesis, University of North Carolina, Chapel Hill.

Southerland, M. T. (1986). Coexistence of three congeneric salamanders: The importance of habitat and body size. *Ecology* 67, 721–728.

Stenhouse, S. L. (1984). Coexistence of the salamanders *Ambystoma maculatum* and *Ambystoma opacum*: Predation and competition. Ph.D. Thesis, University of North Carolina, Chapel Hill.

Stenhouse, S. L. (1985). Interdemic variation in predation on salamander larvae. *Ecology* 66, 1706–1717.

Stenhouse, S. L., Hairston, N. G., and Cobey, A. E. (1983). Predation and competition in *Ambystoma* larvae: Field and laboratory experiments. *J. Herpetol.* 17, 210–220.

Stewart, M. M. (1956). The separate effects of food and temperature differences on the development of marbled salamander larvae. *J. Elisha Mitchell Sci. Soc.* 72, 47–56.

Szymura, J. M. (1974). A competitive situation in the larvae of four sympatric species of newts (*Triturus cristatus*, *T. alpestris*, *T. montandoni*, and *T. vulgaris*) living in Poland. *Acta Biol. Cracov. Ser. Zool.* 17, 235–252.

Tilley, S. G. (1968). Size-fecundity relationships and their evolutionary implications in five desmognathine salamanders. *Evolution* 22, 806–816.

Twitty, V. (1966). "Of Scientists and Salamanders." Freeman, San Francisco.

Walls, S. C. (1990). Interference competition in postmetamorphic salamanders: Interspecific differences in aggression by coexisting species. *Ecology* 71, 307–314.

Walls, S. C., and Jaeger, R. G. (1987). Aggression and exploitation as mechanisms of competition in larval salamanders. *Can. J. Zool.* 65, 2938–2944.

Walls, S. C., and Semlitsch, R. D. (1991). Visual and movement displays function as agnostic behavior in larval salamanders. *Copeia,* 1991, 936–942.

Walters, B. (1975). Studies of interspecific predation within an amphibian community. *J. Herpetol.* 9, 267–279.

Wilbur, H. M. (1972). Competition, predation, and the structure of the *Ambystoma-Rana sylvatica* community. *Ecology* 53, 3–21.

Wilbur, H. M. (1975). The evolutionary and mathematical demography of the turtle *Chrysemys picta. Ecology* 56, 64–77.

Wilson, E. O. (1984). "Biophilia." Harvard Univ. Press, Cambridge, MA.

Worthington, R. D. (1968). Observations on the relative sizes of three species of salamander larvae in a Maryland pond. *Herpetologica* 24, 242–246.

Worthington, R. D. (1969). Additional observations on sympatric species of salamander larvae in a Maryland pond. *Herpetologica* 25, 227–229.

Wrobel, D. I., Gergits, W. F., and Jaeger, R. G. (1980). An experimental study of interference competition among terrestrial salamanders. *Ecology* 61, 1034–1039.

CHAPTER 8

Long-Term Changes in Lizard Assemblages in the Great Victoria Desert
Dynamic Habitat Mosaics in Response to Wildfires

ERIC R. PIANKA

Department of Zoology, University of Texas, Austin, Texas 78712

I. BACKGROUND

The Australian deserts house the most diverse lizard assemblages on planet Earth. Nowhere else can so many kinds of lizards be found together—at least 47 species (now 53 species; see Appendix) coexist at a sandridge site in the Great Victoria Desert. These include agamids such as the ant-eating "thorny devil," *Moloch horridus*; 6 species of predatory monitor lizards such as the gigantic perentie, *Varanus giganteus*; a dozen species of exquisitely beautiful nocturnal geckos; more than a dozen wary and secretive skinks; and many other species including several snake-like, flap-footed, legless pygopodid lizards.

At least a dozen factors contribute to the very high lizard diversity in the

Australian deserts (Pianka, 1989). One of the most important is frequent natural wildfires, which generate a patchwork of habitats in different states of recovery, each of which favors a different subset of lizard species (Pianka, 1992). Inland Australia is one of the last remaining areas where natural wildfires remain a regular and dominant feature of an extensive semi-pristine natural landscape largely undisturbed by humans. In this region, an important fire succession cycle generates spatial and temporal heterogeneity in microhabitats and habitats. Habitat-specialized species can go locally extinct within a given habitat patch (fire scar) but persist in the overall system by periodic reinvasions from adjacent or nearby patches of suitable habitat of different age. Such spatial-temporal regional processes facilitate local diversity. This system is currently being studied at the local level in the field in Western Australia and at the regional level at the University of Texas at Austin using aerial photography and Landsat multispectral satellite (MSS) imagery. Satellite imagery offers a powerful way, heretofore underutilized by biologists, to acquire regional-level data on the frequency and phenomenology of wild fires and thus on the system-wide spatial–temporal dynamics of disturbance.

Compared to Australia, North American deserts are impoverished, with a mere dozen species of lizards (Pianka, 1986). On 12 North American study sites, lizard species diversity[1] ranges from 1.4 to 4.9 with an average of 3.0. In the Kalahari semi-desert of southern Africa, only 20 species occur (at 10 Kalahari sites, lizard diversity varies from 2.5 to 8.7; mean = 6.3). Ten study sites in Australia's Great Victoria Desert support from 15 to 47 species of lizards (now 53 species; see Appendix)—species diversity there ranges from 6.2 to 14.4 (mean = 8.6). Many species of Australian lizards are quite uncommon. There are two components to species diversity: (1) number of species, or species richness, and (2) equitability of relative abundances, which can be measured by the ratio of observed diversity divided by maximum possible diversity (i.e., species richness). Averages of such ratios for all study sites in North America, the Kalahari, and Australia, are, respectively, 0.44 (SD = 0.19), 0.42 (SD = 0.10), and 0.31 (SD = 0.11). Equitability is lower in Australia than in either of the other continental desert-lizard systems, an indication that there are more rare species in Australia than in North America or in the Kalahari.

Why are the Australian deserts so rich in lizard species? Explaining this high diversity, and understanding what goes on between and among component species, is exceedingly difficult. How do so many species of lizards avoid competition and manage to coexist in the Australian desert? How do they partition resources such as food and microhabitats? What historical and ecological fac-

---

[1]All diversity and niche breadth measurements reported here were calculated from relative abundances or resource utilization spectra using the reciprocal of the diversity index of Simpson (1949).

tors have led to the evolution and maintenance of such high biodiversity? Ecologists still know surprisingly little about exactly how diverse natural ecological systems function—such ecological understanding is much needed and will obviously be crucial to our own survival as well as that of other species of plants and animals. In fact, the Australian deserts may well offer the last opportunity to study the regional effects of disturbance on local diversity.

In the Great Victoria Desert (Fig. 1), the rusty-red sands are rich in iron and laced with long, undulating, stable sandridges parallel to prevailing winds. Evergreen marble gum trees (*Eucalyptus gongylocarpa*) with smooth white bark adorn this landscape. Various species of *Acacia* and *Eremophila* shrubs, plus shrub-like large mallee, as well as an occasional mulga tree, are also scattered

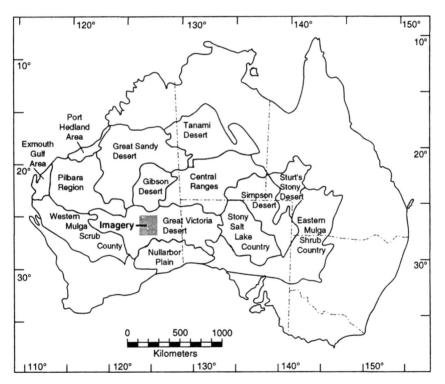

FIGURE 1. Map showing locations of various arid zone regions within Australia. The Great Victoria Desert stretches from western South Australia over 1000 kilometers into south-central Western Australia. Other areas referred to in the text include the Central Ranges and Tanami Desert in the Northern Territory and the Simpson Desert (southeastern Northern Territory, north-western South Australia, and southwestern Queensland). Approximate position of imagery is shown. See color plate (Fig. 2) facing page 196.

about. The dominant ground cover is porcupine grass or spinifex (*Triodia base-dowi*), a fire-prone perennial grass that grows in hummocks or tussocks. Rain is infrequent and clear, blue skies prevail most of the time.

II. FIRE ECOLOGY IN SPINIFEX GRASSLANDS

Spinifex tussocks are almost perfectly designed for combustion, consisting of hemispherical clumps of numerous blades of dry grass filled with flammable resins that are rolled into hollow match-stick sized cylinders, loosely inter-penetrating one another and laced with ample air spaces among blades. Spinifex is nearly optimal tinder (Burrows *et al.*, 1991) and has been called an "ideal pyrophyte" (Pyne, 1991). Following a fire caused by lightning, sands are fre-quently wetted by thundershowers, facilitating rapid regeneration of spinifex, both from live roots as well as from seedling establishment (Burbidge, 1943). Newly-burned areas are usually quite open with extensive bare ground and tiny, well-spaced, clumps of *Triodia*. Unburned patches, in contrast, are composed of large ancient tussocks, frequently quite close together with little open space between them. As time progresses, *Triodia* clumps grow, simultaneously increas-ing the amount of combustible material and reducing gaps between tussocks. Both increase the probability that a new fire will spread. Marble gum trees are fire-resistant and usually survive a hot but brief ground fire carried by the exceedingly combustible spinifex grasses. The above-ground parts of mallees, however, burn but the root bole survives and sends up new growth after a fire.

Deserts in other regions, such as the Sonoran desert of North America, do not usually accumulate enough combustible material to carry fires. Fires do, how-ever, occur regularly in the Kalahari semidesert of southern Africa, an open savanna woodland. Fires are a predictable event in arid Australia, particularly in spinifex grasslands (Catling and Newsome, 1981; Gill, 1975; Gill *et al.*, 1981; Pyne, 1991; Winkworth, 1967). Bush fires are usually started by lightning, raging completely out of control for weeks on end across many square kilome-ters of desert. Thunderstorms and lightning are regular during summer. Light-ning-induced fires are most frequent during the months of November and December (Ralph, 1984). Over the 35-year-period from 1950 through 1984, approximately 60% of natural wild fires in the Northern Territory were esti-mated to have been caused by lightning (Griffin *et al.*, 1983, 1988). During that same period, there was a movement of aborigines out of the bush, reducing the numbers of fires started by humans. Aborigines have increasingly been moving back to reside on tribal lands, except that nowadays aborigines seldom walk, preferring to ride in vehicles. They still set fires but seldom very far away from roads. Most early explorers and historians of Australia commented on the extent to which the aborigines exploited fire. Spinifex grasses give off dark smoke that can be seen from afar. Australian aborigines used fire to send long-distance smoke signals, to manage habitats and keep terrain open, as well as to facilitate capture of various animals for food. Some think that the extensive grasslands in Australia were formed and maintained by regular aboriginal burning over many thousands of years and that aborigines acted to select members of plant communities for resistance to fire or for an ability to come back quickly following a fire.

Some Australian land management authorities are attempting to mimic aboriginal burning patterns to create a spatial mosaic of fire-disturbed habitat patches of differing ages and sizes. By acting as fire breaks, small burns prevent extensive burns from taking place which homogenize the landscape (Minnich, 1983, 1989).

All spinifex communities are in a state of cyclic development from fire to fire (Winkworth, 1967). In the Northern Territory, only about 20% of some 150,000 km² of spinifex habitat is in a "mature" climax state, with the other 80% either in regenerative stages following fires or in a degenerative state owing to drought. A single fire in this region covered about 10,000 km² (Winkworth, 1967). An even larger fire in 1982–1983 burned 30,000 km², an area more than twice the size of the state of Massachusetts. Over the 35-year-period from 1950 through 1984, an average of 143 fires per year occurred in this region, most set by lightning (Griffin et al., 1983; Ralph, 1984). Such an exceedingly high level of disturbance requires a rapid rate of recovery. In the Northern Territory, fire-return intervals are short. Some areas will carry another burn within 4–5 years (Perry and Lazarides, 1962). Burned plots converge on their original state quickly; in 7 years, dry weight production of spinifex totalled 823 kilograms per hectare, approximately one quarter of the standing crop of "mature" stands at a nearby site (Winkworth, 1967). Fire-return intervals are longer in the Great Victoria Desert due to lower precipitation. Time required for a burned stand to reach maturity is a function of precipitation and can require 20–25 years or longer. The probability of a burn is directly proportional to accumulated precipitation since the last fire and generally increases with the time since the last burn.

Spinifex fires in the Central Ranges and the Tanami Desert of the Northern Territory were also studied by Griffin et al. (1983, 1988) and Allan and Griffin (1986). Using aerial photography, every identifiable fire in the southern half of the Northern Territory between 1950 and 1984 (more than 5000 fires in some 750,000 km²) was mapped. Certain informative relationships between fire patch dynamics and rainfall were discovered. At wetter, more productive areas, fires were not only more frequent and more numerous but also patches were more variable in size and tended to be younger on average. Following periods of low rainfall, fires were less frequent and homogeneously smaller. Fuel load recovery is a function of time and total rainfall. Cumulative millimeters of precipitation is a useful temporal productivity metric against which to calibrate fires and vulnerability to fire. Approximately 63 cm of rain are necessary for a site to accumulate sufficient fuel to burn again under extreme summer conditions (Griffin et al., 1988).

III. FIRE GEOMETRY

Bush fires generate a spatial mosaic of patches of habitat at various stages of post-fire succession. The geometry of burns is rich in detail and exceedingly varied (Fig. 2). As a regular agent of disturbance, fires contribute substantially to maintaining a spatially heterogeneous patchwork of habitats, hence fostering diversity in Australian desert lizards (Pianka, 1986, 1989). Fires vary considerably in intensity and extent.

Grassland fires burn along two "fronts" that are essentially unidimensional, each at approximately a right angle to prevailing winds burning away from one another. The "backfire" burns slowly into the wind, whereas the "headfire" burns faster, racing with the wind. Airborne flaming materials, termed "firebrands," sometimes jump over unburned areas to rekindle new fires on the downwind (leeward) side of a fire, sometimes resulting in establishment of multiple fire fronts (these can extinguish one another when one runs into another's swath).

Fires frequently reticulate, missing an occasional isolated grass tussock or even large tracts embedded within or immediately adjacent to burns (Fig. 2). Upon ignition, isolated grass tussocks generate an egg-shaped thermal field. At low wind velocities, isotherms for such fields are symmetric and close to combustibles, but as wind speed increases, thermal fields become asymmetric and extend farther out, especially downwind. If other tussocks are within such a field's threshold temperature for ignition, they too are lighted and the fire spreads. Due to the geometry of such thermal fields, fires tend to burn along broad continuous fronts at low wind velocities, whereas at higher wind velocities elongate narrow tongues of flame are produced. Fires are also more likely to reticulate at higher wind velocities due to these elongate tongues of flame.

Previously burned areas with sparser vegetation act as fire breaks. Sandridges and termitaria create smaller vegetationless areas that can also stop fires from spreading but with fundamentally different geometries and at much more local spatial scales. "Sleepers," embers created from burning hard wood of Eucalyptus, lay dormant in burned areas. Some of these hot coals at the edges re-ignite new secondary bush fires days or even weeks later when strong winds come up. Interesting laciniated patterns are generated that constitute qualitatively and spatially different sorts of refuges.

Major factors that determine the frequency, extent and geometry of grassfires include temperature, humidity, precipitation, combustibility, plant biomass and

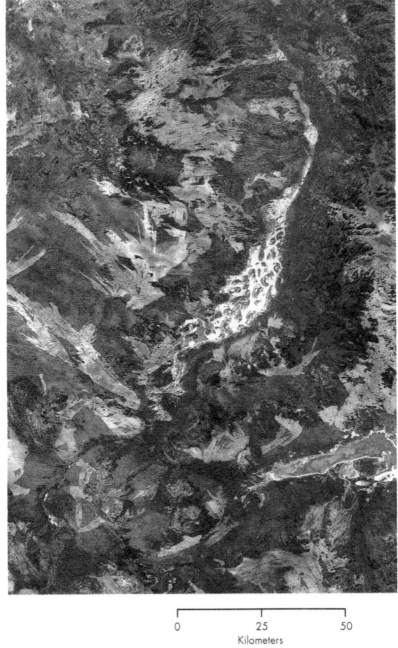

N

FIGURE 2. Landsat false color satellite image showing the habitat mosaic. Fire scars are light beige and pale green. Dry lakebeds are blue and white (Lake Throssell and Lake Yeo). Sandridges are fine striations. Nonburnable shrub-*Acacia* habitats are brown and dark spots.

spatial distribution, natural firebreaks, and, of course, winds, which as explained above, are of utmost importance. With a lot of combustible material, fire fronts burn virtually everything in their paths, leaving behind almost completely burned swaths. However, when grass is green or wet and therefore less flammable and/or if tussocks are widely spaced and/or if winds are weak, fires may falter and die out. Strong winds, by supplying oxygen, "feed" fires making them burn both faster and hotter. With sufficient wind, even a relatively uncombustible area can burn. With little or no wind, however, a fire may not "take" even in a fairly combustible situation. Fires often die out at night when lower temperatures cause relative humidity to increase and winds to die down. Headfires will sometimes burn even when backfires will not. Studies of thousands of forest fires have shown that their extent is directly proportional to wind velocity (Barrows, 1951), allowing area to be used as an after-the-fact, indirect estimate of wind speed at the time of a burn. Moreover, wind direction can usually be estimated from the long axis of a fire scar.

The extent to which grassfires reticulate appears to vary between geographic regions with differences in vegetation structure. In the savanna woodland of the Kalahari semidesert in Botswana, southern Africa, for example, fires are frequent but continuous, seldom reticulating (Short et al., 1976, Plate 342, p. 377). Because Australian spinifex habitats are relatively more open and because they contain various sorts of natural firebreaks, fires in those habitats reticulate more than they do in other more homogeneous areas. The unique growth form and extreme flammability of spinifex doubtlessly contributes to fire laciniation in Australia as well. Intercontinental differences in winds could also be crucial in the generation of different fire geometries.

Throughout this process of fire succession, cover, microhabitats, and associated food resources change gradually along with reflectance properties. The prey spectrum for a recently burned area differs from those of more mature areas with relatively fewer termites but comparatively more spiders (Pianka, 1989). Presumably, by destroying litter and spinifex, fires reduce cellulose availability, hence reducing food supplies for termites. Why spiders increase in abundance is unknown. Through succession, relative abundances of various lizard species fluctuate along with such changes in available resources, with some common species becoming quite rare and vice versa. A particular habitat-specialized species can go extinct within a given area but, by surviving in an adjacent patch of appropriate habitat, still persist in the region. Such regional processes facilitate local diversity because periodic dispersal from such "source" habitats and recolonization of "sink" habitats allows such species to continue to exist as a metapopulation in the overall landscape (Pulliam, 1988; Pulliam and Danielson, 1991). An appropriate mix of subdivision or patchiness plus disturbance and dispersal can promote coexistence in competitive systems (Kareiva, 1990).

Combined effects of fires on lizards and their microhabitats are drastic yet exceedingly heterogeneous in space (Caughley, 1985; Fyfe, 1980; Longmore and Lee, 1981; James, 1994). Certain lizard species are arboreal or associated with trees. These species are often relatively unaffected by fire. Many or even most individual lizards sometimes survive such burns, although survivorship is doubtlessly reduced for some time afterwards. Fires also attract hawks and crows which feed on fire-killed animals and take advantage of the lack of cover to catch survivors. Ectothermy may allow many lizards to become inactive and stay underground for a month or more until the vegetation and insect fauna recover.

IV. LIZARD STUDIES

Over a 16-month period from late 1966 to early 1968, I censused lizard faunas at seven study sites within the Great Victoria Desert plus one site (the A-area) slightly outside (Pianka, 1969, 1975, 1986). Study sites were chosen to represent various vegetation types and included a dry lakebed shrub-*Acacia* (mulga) site, ecotonal sites between shrubby and spinifex habitats, flat sandplain spinifex sites both with and without trees and shrubs, as well as two sandridge sites. Half a dozen undescribed species of *Ctenotus* skinks were discovered. The number of lizard species (lizard species richness) in the saurofaunas of these sites varied from 15 to 39—lizard species diversity varied between 6.3 and 11.3 (Simpson's index). Considerable habitat specificity was evident, with 8 species being restricted to shrub habitats, 10 species to sandplain spinifex habitats, and 8 species to sandridges (Pianka, 1969, 1972). Still other species are habitat generalists occurring in several habitat types.

During the past quarter of a century, my assistants and I have monitored the saurofauna at a flat sandplain study site located about 38 km east of Laverton, Western Australia (appropriately designated the L-area). The vegetation of the L-area is complex, consisting of a spinifex understory with scattered marble gums, mallee, and small *Acacia* shrubs plus an occasional mulga tree. During our first visit in 1966–1968 (hereafter referred to as 1967 for brevity), the presence of many large mature spinifex tussocks suggested it had been at least a decade since the area burned (an old reticulated fire scar is visible in a 1959 high-altitude aerial photograph). A 1969 high-altitude aerial photograph shows that a fire occurred sometime between early 1968 (when we last visited the area) and September 1969. This fire scar was laciniated, leaving behind numerous unburned refugia of varying sizes and shapes, ranging from single tussocks to areas covering many hectares. A decade later, I returned to the L-area in June 1978 and censused lizards until April 1979 (for brevity, this 1978–1979 sampling period will be referred to as 1978).

Over the decade between 1967 and 1978, lizard species richness increased from 30 to 34, but lizard species diversity decreased from 9.9 to 6.2. Another fire burned most, but again not all, of the L-area in 1983, again leaving behind many unburned refugia of varying sizes and shapes. A third, more extensive sampling effort was undertaken during late 1989 through mid 1992 when lizard species diversity was again found to be 9.9 (the L-area is now ready to burn again). Over the past 28 years, some 37 different species of lizards have been collected at the L-area (Table I), many of which are rare. One species, a large nocturnal skink (*Egernia kintorei*), had not been seen since 1967, when it was observed just once. Another large diurnal skink, *Tiliqua occipitalis*, was not seen prior to 1990 and has been observed only twice since then. Relative abundances of many lizard species have fluctuated dramatically over this period (Table I). The abundant agamid *Ctenophorus isolepis* appears to have first increased and then decreased in abundance. Five species that have apparently increased in abundance from 1966 to 1991 are *Ctenotus calurus*, *C. quattuordecimlineatus*, *C. schomburgkii*, *Lerista bipes*, and *Diplodactylus conspicillatus*. Four species that appear to have decreased in abundance over that 25-year interval include *C. grandis*, *C. helenae*, *Gehyra purpurascens*, and *Rhynchoedura ornata*.

A detailed comparative analysis of relative abundances and patterns of resource utilization was undertaken for the L-area data sets of 1966–1968 and 1978–1979 (Pianka, 1986).[2] Patterns of resource utilization in 1978 were compared with comparable information gathered in 1967. Considerable species-specificity and substantial fidelity in the use of food types and microhabitats was evident among many of these Australian desert lizards. For example, some species are termite specialists, whereas others virtually never touch termites. The ant specialist *Moloch horridus* consumed nothing but ants during both 1967 and 1978. Diets of saurophagous species such as the pygmy monitor *Varanus eremius* also varied little. Diet and microhabitat utilization patterns of most species, even those of generalists such as the gecko *Gehyra variegata*, proved to be fairly consistent in time, too. Overall estimates of resource utilization based upon all specimens of each species varied relatively little between 1967 and 1978. Also, site-to-site comparisons show that each species is typically its own closest neighbor in niche space. A few shifts in resource utilization were evident in certain species, however (see Pianka, 1986, for details).

On the L-area, 530 lizard specimens representing 29 species were collected in 1967 (Table I). Tracks of the wary and large monitors *Varanus gouldi* and *V. giganteus* were regularly noted, although no specimens of these very wary lizards were actually sighted. Specimens of the elusive spinifex gecko *Diplodactylus elderi* were not collected either, but this species was listed as "highly expected on the basis of geographic range, habitat, autecology, and micro-

---

[2]Unfortunately, due to lack of funds I have not yet been able to finish such an analysis for lizards collected during 1989–1992.

TABLE I   Relative Abundance (Percentages) of Lizard Species on the L-Area
in 1966–1968, 1978–1979, 1989–1992, and 1966–1992

	1966–1968	1978–1979	1989–1992	1966–1992
Ctenophorus inermis	1.69	0.32	0.20	0.44
Ctenophorus isolepis	10.34	33.82	12.37	20.31
Moloch horridus	1.69	0.57	0.25	0.56
Pogona minor	1.50	0.77	0.20	0.59
Varanus eremius	0.56	0.83	0.50	0.64
Varanus giganteus	tracks	tracks	tracks	tracks
Varanus gouldi	tracks	0.26	0.60	0.42
Varanus tristis	1.13	1.15	0.20	0.68
Ctenotus ariadnae	0.94	0.70	1.80	1.27
Ctenotus calurus	2.07	9.38	15.02	11.18
Ctenotus grandis	8.46	2.49	0.85	2.47
Ctenotus helenae	9.96	1.98	1.05	2.56
Ctenotus pantherinus	1.50	1.34	2.25	1.81
Ctenotus piankai	0.38	0.19	2.40	1.29
Ctenotus quattuordecimlineatus	8.65	11.68	20.78	15.72
Ctenotus schomburgkii	1.32	2.55	4.21	3.20
Cryptoblepharus plagiocephalus	5.08	3.38	3.21	3.52
Egernia inornata	0.38	0.13	0.15	0.17
Egernia kintorei	0.19	0.0	0.0	0.02
Egernia striata	6.96	4.34	4.11	4.57
Lerista bipes	1.13	3.38	8.06	5.37
Lerista muelleri	0.0	0.38	1.05	0.66
Menetia greyi	0.56	0.26	4.06	2.15
Morethia butleri	0.0	0.06	0.10	0.07
Tiliqua occipitalis	0.0	0.0	0.10	0.05
Delma butleri	0.0	0.26	0.35	0.27
Delma nasuta	0.0	0.0	0.05	0.02
Lialis burtonis	0.38	0.13	0.25	0.22
Pygopus nigriceps	1.50	0.19	0.20	0.37
Diplodactylus conspicillatus	1.13	1.72	3.11	2.32
Diplodactylus elderi	0.0	0.26	0.15	0.17
Diplodactylus stenodactylus	0.0	0.13	0.05	0.07
Gehyra purpurascens	19.43	11.89	5.86	10.57
Gehyra variegata	2.0	1.00	0.50	0.24
Heteronotia binoei	0.19	0.06	0.10	0.10
Nephrurus levis	1.13	0.26	2.90	1.66
Rhynchoedura ornata	9.40	4.08	2.85	4.18
Total number of lizards	530	1565	1997	4092
Lizard species diversity	9.9	6.2	9.9	9.8
Lizard species richness	30	34	36	37

habitat" (Pianka, 1969). During 1978, some 1565 new lizard specimens representing 34 species were captured on the L-area (Table I). Only 1 species collected in 1967 was not encountered in 1978 (the large nocturnal skink *Egernia kintorei* mentioned previously); 5 new species were recorded in 1978, including *Varanus gouldi* and *Diplodactylus elderi*, both of which were expected in 1967 as explained above. Tracks of the enormous *Varanus giganteus* were seen again, but these exceedingly intelligent lizards evaded sighting and eluded capture. During 1989–1992 (hereafter referred to as 1990), 36 species were observed (Table I).

Relative abundances of various species on the L-area, as reflected in the numbers actually collected, did not remain constant but fluctuated fairly substantially (Table I; Pianka, 1986). Abundances of some species shifted upwards or downwards by factors of 3 or more, but most species abundances varied to a lesser degree. Regardless of the direction (increase versus decrease), the average magnitude of change in relative abundance from 1967 to 1978 was 2.70 (SD = 1.55, N = 28). Average change in relative abundance from 1978 to 1990 was 3.01 (SD = 3.46, N = 31). Change in relative abundance from 1978 to 1990 is plotted against the change from 1967 to 1978 in Fig. 3. The two species that increased markedly in abundance over the second time interval are the tiny skink *Menetia greyi* and the knob-tailed gecko, *Nephrurus levis*. Apparent changes in abundances of uncommon species could easily be artifacts and should not be taken too seriously.

Some 26 species are represented in Fig. 3. Under the null hypothesis that densities are changing randomly, the expectation would be 8.5 species in each of the four quadrants. Observed numbers of 9, 6, 3, and 8 do not differ significantly from expected (χ^2 = 4.35, df = 3, P = 0.25). The presence of species in the northwest and southeast quadrants suggests density-dependent feedback that would confer community stability. However, species in the southwest and northeast quadrants pose a dilemma: why is it that species in the southwest quadrant (decreasing over both intervals) do not go extinct and those in the northeast quadrant (increasing over both intervals) do not take over? The proportion of species that increased over the 1967–1978 interval was only 9/26 = 0.35, whereas over the 1978–1990 interval 15 out of 26 species (0.58) increased in abundance. The greatest disparities from expected in changing abundances were the 3 species that were increasing over the 1967–1978 interval but decreasing over the 1978–1990 interval (southeast quadrant).

From 1967 to 1978, at least two species increased markedly in abundance (*Ctenophorus isolepis* and *Ctenotus calurus*). Four other species declined during this period (*Ctenotus grandis*, *Ctenotus helenae*, *Gehyra variegata* and *Rhynchoedura ornata*). In the light of these apparent changes in relative abundance, patterns of resource utilization in both diets and microhabitats among eight abundant species were scrutinized closely (Pianka, 1986). Diets of five food specialists changed little, but proportional representation of various prey items

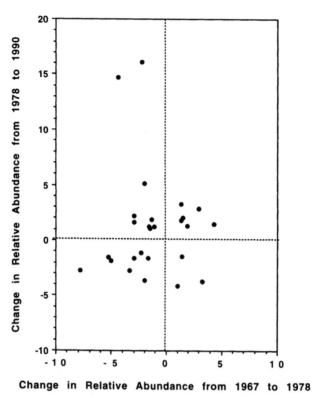

FIGURE 3.   Change in relative abundance of L-area lizards between 1978–1979 and 1989–1992 plotted against the change in relative abundance from 1966–1968 to 1978–1979.

in the diets of three more generalized feeders shifted somewhat. Similarly, microhabitat utilizations varied little in five species (two arboreal plus three terrestrial), but changed in three other terrestrial species (for further details, see Pianka, 1986, and discussion following).

Foods available to these lizards, as reflected in what the animals actually ate, appeared to have undergone some changes between 1967 and 1978. Two observations suggested that the food supply available to lizards was diminished in 1978. First, average stomach volume decreased from 0.47 cc (1968) to 0.31 cc (1978). Second, prey diversity increased: the diversity of foods eaten by all lizards increased slightly from 3.48 to 4.28, partially due to the decrease in the importance of termites. The L-area is unusual among my Australian study areas in its very high values for termite consumption (Pianka, 1989). Percentage compositions of prey categories consumed by all species changed by less than an order of magnitude, so these changes may be deemed as relatively conservative.

Nonetheless, consumption of grasshoppers and crickets and insect larvae dropped precipitously, whereas consumption of ants and vertebrates showed strong increases (the apparent change in vertebrate foods is an artifact to the extent that it reflects my own heightened effort to collect *Varanus*).

Although there is little reason to suspect that availabilities of microhabitats should have altered appreciably over the decade between my two visits, data on microhabitat utilization of the entire L-area saurofauna in 1967 versus 1978 did show some changes (Pianka, 1986). The fraction of lizards first sighted in the sunshine at the edge of porcupine grass tussocks increased 4-fold, whereas those observed in grass shade decreased. Overall diversity of microhabitats used by all individuals of all lizard species decreased from 6.36 to 5.37.

A simplistic first hypothesis might be that lizard abundances fluctuate directly with prey availability. If so, a doubling of the availability of ants would be expected to lead to a doubling of the density of myrmecophagous species. Likewise, abundances of termite specialists would be expected to "track" termite availabilities. Assuming that my estimates based on the diets of all lizards reflect real changes in prey availabilities (a sort of a "bioassay"), this hypothesis is easily tested and rejected: *Moloch horridus*, an obligate ant specialist, decreased by a factor of 3 even though ants actually increased more than 2-fold in the overall diet of all lizards. Also, although termites decreased slightly (from 50 to 42%) in the overall diet of all lizards (Pianka, 1986), the relative abundance of an obligate termite specialist, *Diplodactylus conspicillatus*, increased by 50%. Another termite specialist, *Rhynchoedura ornata*, fluctuated in the opposite direction, declining to less than half its former abundance. A related observation of interest can be made on *Ctenotus calurus*, the species that exhibited the most dramatic increase (4.5-fold, or 450%) from 1967 to 1978; this tiny blue-tailed skink almost doubled its consumption of termites (from 44.3 to 81.2%), despite the fact that termites decreased in the overall diet of all lizards. The fraction of insect larvae in its diet fell from 51.3 to only 2.0% while percentage representation of larvae in the overall diet of all lizards fell from 5.7 to 1.0%.

Yet another interesting, although unfortunately uncommon, species is the flap-footed legless lizard, *Pygopus nigriceps*, a nocturnal denizen of the open spaces with an unusually high consumption of scorpions and spiders (the diet of 16 individuals consisted of 34% scorpions and 26% spiders by volume). In the overall diet of all lizards, the importance of scorpions and spiders was trivial and did not change appreciably (only 1.5 to 1.7%, and 1.2 to 0.7%, respectively). Nonetheless, *Pygopus* relative abundance declined by almost an order of magnitude from 1.5 to a mere 0.2%.

Dietary and microhabitat niche breadths of L-area lizard species, and changes therein, were summarized by Pianka (1986, Table 10.6). The average magnitude of observed change in niche breadths (irrespective of sign) among all species was similar for foods (1.51) and microhabitats (1.46). Average food niche breadths declined slightly (2.55 to 2.13), whereas average microhabitat niche breadths remained approximately constant (2.30 versus 2.32). Variances in niche breadths among species were lower in microhabitat than in diet (Pianka, 1986). Across species, dietary niche breadth is weakly but significantly positively correlated with microhabitat niche breadth ($r = 0.268$, $P < 0.02$), an indication that food specialists tend to be restricted to fewer microhabitats than food generalists.

Optimal foraging theory predicts that diets should contract when foods are abundant but expand when foods are scarce (MacArthur and Pianka, 1966). Relative abundances of consumers might also be expected to vary directly with prey abundance. Of the 7 species that increased in abundance, food niche breadths decreased as predicted in 4, increased in 2, and stayed constant in 1 (the termite specialist *Diplodactylus conspicillatus*). Abundances of 2 species changed little and their food niche breadths remained fairly constant. Of the 15 species that declined in abundance, diets expanded (as expected) in only 2, contracted in 6, and changed little among 7 others (the latter included both food specialists and food generalists).

The L-area 1967 and 1978 data sets were also exploited to test the hypotheses that abundances of ecologically similar species fluctuate either (1) in phase with one another, or (2) out of phase with one another, more so as compared with ecologically dissimilar species (Pianka, 1986). The first hypothesis emerges from a non-competition argument asserting only that species track available resources, whereas the second suggests interactions between species of the mutually detrimental class such as might arise from interspecific competition. Both hypotheses could easily be fatal oversimplifications if the majority of species experience diffuse competition and both positive as well as negative indirect effects on one another (see also the following discussion).

To perform such a test (Pianka, 1986), I computed the direction and magnitude of changes in relative abundances among all possible pairs of species. Using lumped data from both visits, overall ecological overlap was estimated as the product of dietary overlap times microhabitat overlap. The relative change in the abundances of each pair of species was expressed as the ratio of the change in each ($\Delta N_i/\Delta N_j$). No correlation emerged from comparison of this matrix of changes in abundance versus the above-mentioned matrix of overall ecological overlap, either among all 33 species ($r = -0.011$) or using just 11 species for which sample sizes were more adequate ($r = -0.001$). Nor did the elements in the matrix of changes in abundance correlate with dietary overlap ($r = +0.006$). Finally, the correlation between the elements in the above matrix of changes in abundance versus elements in the inverse of the matrix of overall ecological overlap was also negligible ($r = +0.005$). While these results are less than satisfying, at least they seem to support neither of the above hypotheses. Rather, they suggest either that stochasticity in this system is considerable, or that abundances of each species vary more or less independently of those of other species. Alternatively, species abundances might vary in response to resources and to one another, but their interactions might be aligned along many niche dimensions with some direct competitors yielding positive density effects (indirect "competitive mutualisms"), whereas others yield net negative density effects; under such circumstances, the overall total effects might appear to be mere "noise."

Another study site, Red Sands, located about 100 km to the east of the L-area (8 km west of new Yamarna Homestead), was chosen as a study site because it contains large stable sandridges. Interdunal flats at Red Sands have a complex vegetation similar to that of the L-area consisting of scattered marble gum trees, mallees, and various shrubs with an understory of spinifix. Slopes and crests of sandridges support relatively less spinifex but more woody shrubs. The saurofauna of Red Sands (Table II) was first censused in 1978–1979 when 42 species were observed (diversity = 8.0). Much of the Red Sands area burned in 1982, but some extensive patches of old spinifex escaped the fire. A subsequent, more complete census undertaken in 1989–1992 demonstrated a sharp increase in species diversity (12.0) plus marked changes in the relative abundances of about a dozen species (Table II). Abundances changed more drastically at Red Sands than they did on the L-area, and changes were much more variable; average change in relative abundance from 1978 to 1990 at Red Sands was 8.26 (SD = 15.26, N = 39), as compared to the L-area changes of 2.70 (SD = 1.55) and 3.01 (SD = 3.46) for 1967–1978 and 1978–1990, respectively. A total of 47 species of lizards have now been collected at Red Sands (see Appendix), several of which may be accidental transients since they were collected only once or twice [these include *Ctenophorus fordi*, which appears to have undergone a local extinction on the study area, and *Nephrurus vertebralis* and *Varanus gilleni*, both of which are shrub-*Acacia* species (Pianka, 1969, 1986)]. Ten species that appear to have increased in abundance at Red Sands are *Ctenophorus clayi, Ctenotus ariadnae, Ctenotus brooksi, Ctenotus colletti, Ctenotus piankai, Egernia inornata, Lerista bipes, Lerista desertorum, Menetia greyi,* and *Diplodactylus conspicillatus.* Five species that appear to have decreased in abundance include *Ctenophorus isolepis, Moloch horridus, Diplodactylus ciliaris, Gehyra purpurascens,* and *Rhynchoedura ornata.*

Two other study sites on flat sandplains support pure stands of spinifex, with no trees and just a very few scattered shrubs. An informative comparison can be made between these two sites, one of which was newly burned and the other long unburned (Table III). The newly burned site, known as the N-area (supporting 16 species of lizards with a diversity of 8.5), is located about 100 km to the east (8 km west of Neale Junction) and was censused during 1967. A comparable, but long-unburned, flat sandplain site, the B-area, located about 4

TABLE II   Relative Abundances (Percentages) of Lizard Species at Red Sands in 1978–1979, 1989–1992, and 1978–1992 (see Appendix)

	1978–1979	1989–1992	1978–1992
Ctenophorus clayi	1.11	4.22	3.26
Ctenophorus fordi	0.07	0.00	0.02
Ctenophorus inermis	1.04	1.50	1.36
Ctenophorus isolepis	14.55	1.63	5.64
Gemmatophora longirostris	1.81	0.22	0.71
Moloch horridus	4.87	0.78	2.05
Pogona minor	0.70	1.13	0.99
Varanus brevicauda	0.00	0.13	0.09
Varanus eremius	0.49	0.25	0.32
Varanus giganteus	0.07	0.13	0.11
Varanus gilleni	0.07	0.00	0.02
Varanus gouldi	0.35	0.84	0.69
Varanus tristis	2.65	0.34	1.06
Ctenotus ariadnae	0.07	0.63	0.45
Ctenotus brooksi	0.28	3.91	2.79
Ctenotus calurus	0.97	0.91	0.93
Ctenotus colletti	0.56	3.54	2.61
Ctenotus dux	7.80	8.89	8.55
Ctenotus grandis	0.35	0.34	0.35
Ctenotus helenae	1.81	0.97	1.23
Ctenotus pantherinus	1.74	0.38	0.80
Ctenotus piankai	0.56	6.54	4.69
Ctenotus quattuordecimlineatus	6.27	9.20	8.29
Ctenotus schomburgkii	0.00	0.13	0.09
Cyclodomorphus branchialis	0.21	0.06	0.11
Egernia inornata	0.63	3.57	2.66
Egernia striata	0.91	0.66	0.73
Eremiascincus richardsoni	0.14	0.03	0.07
Lerista bipes	0.35	19.65	13.66
Lerista desertorum	0.07	1.44	1.02
Menetia greyi	0.07	5.69	3.95
Morethia butleri	0.07	0.34	0.26
Delma fraseri	0.21	0.00	0.07
Delma nasuta	0.00	0.03	0.02
Lialis burtonis	0.14	0.13	0.13
Pygopus nigriceps	0.21	0.28	0.26
Diplodactylus ciliaris	2.99	0.69	1.40
Diplodactylus conspicillatus	0.70	3.97	2.96
Diplodactylus damaeus	1.39	1.85	1.71
Diplodactylus elderi	0.91	0.16	0.39
Diplodactylus strophurus	0.84	0.06	0.30
Gehyra purpurascens	26.32	3.04	10.26
Gehyra variegata	0.00	0.16	0.11
Heteronotia binoei	0.07	0.03	0.04

(*continues*)

TABLE II *continued*

	1978–1979	1989–1992	1978–1992
Nephrurus laevissimus	13.44	10.29	11.27
Nephrurus vertebralis	0.00	0.06	0.04
Rhynchoedura ornata	2.16	1.25	1.53
Total number of lizards	1436	3196	4632
Lizard species diversity	8.0	12.0	14.4
Lizard species richness	42	44	47

km south of Red Sands (12 km west of new Yamarna Homestead), was censused during 1992–1993 and had a lizard species diversity of 7.7. This site
contained old spinifex of two different ages, the oldest of which were probably
about 40 years old (judged from counts of growth rings on an old mallee
adjacent to the site). Some 29 species of lizards, including several probable
"accidentals," were collected on the B-area (another species, *Varanus gouldi*,
was also present as judged by numerous, very characteristic diggings). Although nearly twice as many lizard species were present at the B-area than at
the N-area, species diversity was actually lower (7.7 versus 8.5). Two of the 30
species present at the B-area would appear to be accidentals, both because they
were collected only once and because of their habitat requirements (*Ctenotus
greeri* is a shrub-*Acacia* species and *Ctenotus leae* is a sandridge species). Four
species of *Ctenotus* appear to increase in abundance as spinifex matures; these
are *Ctenotus calurus*, *Ctenotus grandis*, *Ctenotus pantherinus*, and *Ctenotus piankai*. Five lizard species that appear to decrease in abundance as spinifex
matures are *Ctenophorus inermis*, *Ctenophorus isolepis*, *Egernia inornata*, *Heteronotia binoei*, and *Rhynchoedura ornata*. If the B-area does not burn naturally
on its own accord, it will be burned experimentally in a controlled burn; lizard
relative abundances will be monitored as often as possible over the next two
decades in an effort to document how post-fire abundances change in time.

Percentages of lizards of 43 species that were captured at four different
locations (flat, base, slope, and crest) on the two sandridge sites (the E-area
and Red Sands) plus a nearby sandplain site (the B-area) are summarized in
Table IV. Note that 7 species are virtually restricted to flats (*Varanus brevicauda*,
Ctenotus ariadnae, *Ctenotus calurus*, *Ctenotus grandis*, *Ctenotus pantherinus*,
Delma butleri and *Heteronotia binoei*), 4 other species occur on flats and/or
near the bases of sandridges (*Ctenophorus isolepis*, *Ctenotus helenae*, *Ctenotus
schomburgki*, *Egernia striata*), whereas 8 other species are quite strongly associated with sandridges (*Ctenophorus fordi*, *Diporiphora winneckei*, *Gemmatophora
longirostris*, *Ctenotus brooksi*, *Ctenotus leae*, *Diplodactylus damaeus*, *Diplodactylus elderi* and *Nephrurus laevissimus*). Fourteen other species have somewhat

TABLE III   Relative Abundances (Percentages)
of Lizard Species at a Recently Burned Site
(N-Area) and a Long-Unburned Site (B-Area)

	N-area	B-area
Ctenophorus clayi		0.41
Ctenophorus inermis	2.74	0.41
Ctenophorus isolepis	24.66	11.26
Pogona minor		0.69
Varanus brevicauda		2.20
Varanus eremius	1.37	2.20
Varanus gouldi	1.37	diggings
Ctenotus ariadnae		8.38
Ctenotus calurus	6.85	19.78
Ctenotus grandis	4.11	21.02
Ctenotus greeri		0.14
Ctenotus helenae	1.37	0.41
Ctenotus leae		0.14
Ctenotus pantherinus	2.74	10.17
Ctenotus piankai	6.85	10.85
Ctenotus quattuordecimlineatus	1.37	2.20
Ctenotus schomburgkii		0.14
Egernia inornata	8.22	0.14
Egernia striata	5.48	2.06
Lerista bipes		0.55
Menetia greyi	4.11	3.71
Tiliqua multifasciata		0.14
Delma butleri		0.96
Delma nasuta		0.14
Diplodactylus conspicillatus		0.14
Diplodactylus damaeus		0.28
Gehyra variegata		0.28
Heteronotia binoei	4.11	1.10
Nephrurus laevissimus		0.14
Nephrurus levis	10.96	
Rhynchoedura ornata	13.70	0.14
Total number of lizards	73	729
Lizard species diversity	8.5	7.7
Lizard species richness	16	29

broader habitat niche breadths, occurring with fairly high frequencies in all four habitats (these include *Ctenophorus clayi, Ctenophorus inermis, Moloch horridus, Pogona minor, Varanus eremius, Varanus gouldi, Varanus tristis, Ctenotus colletti, Ctenotus dux, Ctenotus quattuordecimlineatus, Egernia inornata, Lialis burtonis, Pygopus nigriceps,* and *Gehyra purpurascens*). The frequency distribution of habitat niche breadths is shown in Fig. 4. Across species,

TABLE IV Percentage of Lizards Captured at Four Different Habitat Locations, Total Number of Lizards, and Habitat Niche Breadth

	Flat	Base	Slope	Crest	Total	Niche breadth[a]
Ctenophorus clayi	26.7	34.0	15.8	23.5	143	3.74
Ctenophorus fordi	0.0	2.7	53.1	44.2	113	2.09
Ctenophorus inermis	44.9	24.7	18.0	12.4	89	3.22
Ctenophorus isolepis	61.3	19.8	15.1	3.8	450	2.28
Diporiphora winneckei	0.0	0.0	16.7	83.3	18	1.38
Gemmatophora longirostris	12.0	16.3	16.3	55.4	46	2.67
Moloch horridus	19.8	16.6	43.6	20.0	174	3.37
Pogona minor	37.5	25.0	17.9	19.6	56	3.66
Varanus brevicauda	90.0	10.0	0.0	0.0	20	1.22
Varanus eremius	53.0	14.0	17.0	16.0	50	2.68
Varanus giganteus	27.3	18.2	9.1	45.5	6	3.10
Varanus gouldi	35.0	27.0	24.0	14.0	50	3.67
Varanus tristis	41.7	28.0	16.7	13.7	60	3.34
Ctenotus ariadnae	90.8	2.6	2.6	3.9	76	1.21
Ctenotus brooksi	0.0	3.1	16.0	80.8	159	1.47
Ctenotus calurus	92.5	6.4	0.8	0.3	194	1.16
Ctenotus colletti	10.5	38.3	25.8	25.4	105	3.46
Ctenotus dux	15.9	31.7	25.5	26.9	412	3.80
Ctenotus grandis	96.6	2.3	1.1	0.0	174	1.07
Ctenotus helenae	65.6	18.0	9.4	7.0	64	2.10
Ctenotus leae	4.0	0.0	32.0	64.0	25	1.95
Ctenotus pantherinus	87.9	5.5	4.3	2.3	128	1.29
Ctenotus piankai	46.1	32.9	13.0	8.0	257	2.91
Ctenotus quattuordecimlineatus	33.0	34.3	15.9	16.7	392	3.57
Ctenotus schomburgkii	67.9	21.4	3.6	7.1	14	1.95
Egernia inornata	13.5	45.8	28.0	12.7	174	3.10
Egernia striata	72.5	25.5	0.0	2.0	51	1.69
Lerista bipes	3.7	38.8	15.0	42.5	614	2.82
Lerista desertorum	1.3	42.3	10.3	46.2	39	2.48
Menetia greyi	22.3	49.8	8.3	1.97	211	2.92
Morethia butleri	0.0	66.7	13.3	20.0	15	1.99
Delma butleri	80.0	0.0	15.0	5.0	10	1.50
Lialis burtonis	18.8	31.2	25.0	25.0	8	3.88
Pygopus nigriceps	23.1	34.6	23.1	19.2	13	3.80
Diplodactylus ciliaris	7.6	17.1	25.9	49.4	85	2.89
Diplodactylus conspicillatus	24.9	53.0	9.6	12.4	125	2.72
Diplodactylus damaeus	7.2	17.8	17.1	57.9	76	2.49
Diplodactylus elderi	11.1	5.6	22.2	61.1	18	2.28
Diplodactylus strophurus	25.0	6.2	31.2	37.5	16	3.28
Gehyra purpurascens	40.3	14.0	22.0	23.7	513	3.49
Heteronotia binoei	91.7	4.2	4.2	0.0	12	1.19
Nephrurus laevissimus	2.9	20.3	20.9	55.9	599	2.51
Rhynchoedura ornata	50.7	21.7	8.6	19.1	76	2.88
Total number of lizards	1876	1461	1045	1548	5930	

Note. Niche breadth varies from 1 to 4 (based on all lizards captured on the B-area, the E-area, and Red Sands).

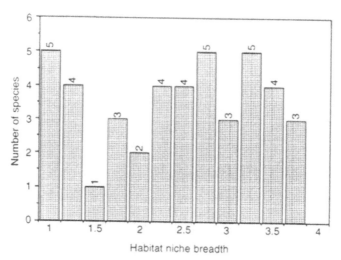

FIGURE 4. Histogram of habitat niche breadths among 43 species of Australian desert lizards (data from Table IV). Note that about 10 species are habitat specialists (niche breadths ≤1.5), while many other species are habitat generalists (niche breadths >1.5).

neither dietary niche breadth nor microhabitat niche breadth is correlated with habitat niche breadth.

Macrohabitat and microhabitat relationships among lizards at a sandridge site in the eastern Simpson Desert in southwestern Queensland were reported by Downey and Dickman (1993). Interestingly, most of the species they captured (30 out of 36) were species that were also present at my own study sites halfway across Australia in the western Great Victoria Desert of Western Australia. Macrohabitat usage by these wide-ranging lizard species is generally quite similar in both regions (*Varanus brevicauda* is an exception). Some species are habitat specialists closely associated with the crests of sandridges, whereas many other species have broader macrohabitat niche breadths and are found in swales (flat sandplains), slopes, as well as on the crests of sandridges.

Lizards were also censused 40 km south of Alice Springs in central Australia by James (1994). He recorded 39 species, 31 of which occurred at my study sites in the Great Victoria Desert, about 1000 km to the west. He found that lizard faunas varied seasonally and that sandridge saurofaunas differed from those on flat sandplains. Moreover, he interpreted the unusual lizard composition on one subsite (his site 1) to be the result of a long period of shelter from fire as compared to other subsites. James (1994) also suggested that saurofaunas change over the long term as spinifex habitats regenerate after fires. Apparently, saurofaunas of sandy spinifex/sandridge habitats are very similar across arid Australia from the eastern Simpson Desert in Queensland to some 1500 kilometers to the west in the western Great Victoria Desert.

Population densities of certain species of lizards, such as *Ctenophorus fordi* and *Ctenophorus isolepis*, are actually greater on recently burned areas (Cogger, 1969, 1972; Griffin *et al.*, 1988). Some species of burrowing lizards such as *Lerista* spp. are also more abundant on recently burned areas than they are on nearby unburned sites. Many other desert lizard species with open habitat requirements, including *Ctenophorus inermis*, presumably reinvade and/or repopulate burned areas rapidly, quickly reaching high densities in their "preferred" open habitat (such species usually persist at very low densities even in mature stands of dense, closed-in spinifex). Other lizard species, such as *Delma butleri, Delma nasuta, Cyclodomorphus branchialis,* and *Diplodactylus elderi* require large spinifex tussocks for microhabitats and nearly vanish over extensive open areas following a burn. However, such "climax" species can continue to exist as relicts in the inevitable isolated, unburned pockets and patches of habitat that escaped burning.

Species present on recently burned areas represent a subset of those present on more mature climax sites, both in lizards (Pianka, 1989) and in birds (Pianka and Pianka, 1970). As noted above, relative abundances fluctuate substantially through succession, with some common species becoming quite rare as succession progresses (compare N- and B-areas in Table III). Rare species do not always remain rare and may be vitally important to hold an ecosystem together (Main, 1982), allowing the system to respond to changing environmental conditions. In the Great Victoria Desert, there are, quite literally, waves of lizards constantly moving about the landscape.

Evidence for metapopulation structure can be gleaned from comparison of the saurofaunas of two nearby sandridge study sites separated by a region of shrub-*Acacia* habitat. Eight sandridge-specialized species were present at one of these sandridge study sites, the E-area 30 km east of Red Sands, censused during 1966–1968. However, three of these eight species were not present 30 km west at Red Sands, where ample appropriate habitat was available. During 1978, one newly hatched juvenile of one of the three missing species was present (*Ctenophorus fordi*), but none was encountered despite concerted efforts in several subsequent years of sampling thousands of lizards during 1978–1979 and 1989–1992. A solitary individual, *Ctenotus leae*, also a missing sandridge species, was pit-trapped on the B-area 4 km south of Red Sands in 1993. Such missing species can be considered to be temporarily extinct locally—given enough time, presumably such species will eventually reinvade and reestablish viable populations. Likewise, populations of such habitat-specialized species could dwindle to precariously low levels during droughts and go temporarily extinct locally.

During four years of field work in arid Australia over a quarter of a century, I have personally witnessed several instances of potential dispersal propagules—lizards wandering through inappropriate habitats where that species normally does not occur. Two were sandridge species, *Ctenophorus fordi* and

Ctenotus leae, both of which were a kilometer or more away from the nearest sandridge, well out into flat sandplain habitat. At a sandplain-sandridge site (Red Sands), I found two individuals of the shrub-*Acacia* species, *Nephrurus vertebralis,* one on the crest of a sandridge, the other at the base of a sandridge. Both individuals were in open areas with little plant cover, reminiscent of conditions within their preferred shrub-*Acacia* habitat, patches of which occurred about a kilometer away. On another occasion, at Red Sands, I collected a large male *Varanus gilleni,* which I strongly suspect is also a shrub-*Acacia* species, as I have captured 50 *Varanus tristis* at this site and never encountered another *V. gilleni.* Even though all five of these individual lizards were far away from appropriate habitat, they were in good physical condition and could easily have survived to reach another patch of appropriate habitat. Such a dispersal episode by a gravid female could start a new local population and result in windfall reproductive success (although there would be a genetic founder effect). Modern molecular techniques, such as DNA fingerprinting via restriction fragment length polymorphisms (RFLPs), offer a powerful new methodology for determination of relatedness, which might allow inference of such past genetic bottlenecks.

Empirical study of lizard metapopulations in the Great Victoria Desert is exceedingly difficult due to the long temporal and extensive spatial scales of this dynamic habitat mosaic, as well as to the infrequency of local extinction and dispersal-colonization events. Detailed analyses of satellite imagery in progress should enable us to describe the actual spatial-temporal phasing of the fire-succession cycle, which will then be used as a realistic habitat mosaic background in computer simulation studies of metapopulation dynamics. To what extent do such landscape dynamics impose particular biological attributes on component species? Given various degrees of extinction proneness, I plan to explore on computers what reproductive capacities and dispersal abilities are required in order for various habitat specialists to persist in this ever-changing landscape. For example, one would predict that habitat specialists must have higher powers of dispersal than habitat generalists. Moreover, due to the dynamic habitat mosaic and local extinctions, populations will be below their respective carrying capacities to varying degrees that will depend upon their relative abilities to track their preferred habitats in this ever-changing habitat mosaic.

What constitutes a "long-term" study? Some would say a year, others decades, but even millennia pale against the geological record. Humans live such short lives that it becomes an absurd anthropocentrism even to contemplate the long term. The desert-lizard/spatial-temporal habitat mosaic described here clearly fluctuates over decades and very probably even over centuries— although I amassed what appears to be a considerable amount of data during 28 years, a great deal more information will be required before a reasonable degree of understanding of this very complex system can be reached. I intend to devote the remainder of my life to the continuation of this challenging study.

V. SUMMARY

Australian desert lizards have evolved a high degree of habitat specificity. Some species are restricted to flat spinifex sandplains, others to shrub-*Acacia* (mulga) habitats, and still others to sandridges. Lightning sets hundreds of wildfires annually in inland arid Australia, generating an ever-changing spatial-temporal patchwork of habitats that differ in their state of post-fire recovery. Different species of desert lizards have differing post-fire habitat requirements. Species that require open areas increase in abundance immediately following fire. Other species that require long-unburned spinifex tussocks disappear after fire. However, fires frequently reticulate, leaving behind refuges that allow the latter species to persist in isolated pockets of suitable old spinifex habitat. Sandridges and mulga habitats add two more important phases to this complex mosaic of habitats. Lizard populations in this region form metapopulations, temporarily going extinct within some isolated patches of habitat, but periodically re-invading from others. Such regional processes foster local diversity. Waves of lizards are constantly moving about this landscape. Long-term studies at two sites document changing relative abundances. Comparisons between sites in different stages of post-fire recovery suggest that some lizard species decrease in abundance following fire, whereas other species increase in abundance. Further long-term studies are needed to document changing relative abundances during this dynamic fire succession cycle.

Acknowledgments

My research has been financed by grants from the National Geographic Society, the John Simon Guggenheim Memorial Foundation, a senior Fulbright Research Scholarship, the Australian-American Educational Foundation, the University Research Institute of the Graduate School at the University of Texas at Austin, the Denton A. Cooley Centennial Professorship in Zoology at the University of Texas at Austin, the U.S. National Science Foundation, the U.S. National Aeronautics and Space Administration, as well as my own pocketbook. For logistic support, I am indebted to the staffs of the Department of Zoology at the University of Western Australia and the Western Australian Museum plus the staffs of Conservation and Land Management, both at Kalgoorlie and at the Western Australian Wildlife Centre at Woodvale. Aussies who gave generously of their time and greatly assisted in this research effort include June Anderson, Ken Aplin, Jean Bannister, Andrew Burbidge, Andy Chapman, Ian Keally, Bert Main, Lee Ann Miller, David Pearson, Nancy Scade, Laurie Smith, and the late Glen Storr. For much assistance in the field, I thank Helen Dunlap, Karen Pianka, Gretchen Pianka, François Odendaal, William Giles, Rand Dybdahl, the late Wayne Sanders, Daniel Bennett, and Magnus Peterson.

Appendix

As of April 1996, six additional species have been trapped at the Red Sands site bringing the total species richness for this site to 53 species of lizards. Two of these were expected at this site (*Ctenotus leae* and *Tiliqua multifasciata*) and two are "accidentals" from nearby mulga habitats (*Ctenotus leonhardii* and *Ctenophorus scutulatus*). The remaining two species are cryptic and were previously confused with others at the site.

References

Allan, G. E., and Griffin, G. F. (1986). Fire ecology of the hummock grasslands of central Australia. *Proc. Bienn. Conf. Austr. Rangeland Soc., 4th.* Armidale, Australia, *1986*, pp. 126–129.

Barrows, J. S. (1951). Fire behavior in northern rocky mountain forests. *USDA For. Serv., Stn. Pap.* No. 29.

Burbidge, N. T. (1943). Ecological succession observed during regeneration of *Triodia pungens* after burning. *J. R. Soc. West. Aust.* 28, 149–156.

Burrows, N., Ward, B., and Robinson, A. (1991). Fire behaviour in spinifex fuels on the Gibson Desert Nature Reserve, Western Australia. *J. Arid Environ.* 20, 189–204.

Catling, P. C., and Newsome, A. E. (1981). Responses of the Australian vertebrate fauna to fire: An evolutionary approach. *In* "Fire and the Australian Biota" (A. M. Gill, R. H. Groves, and I. R. Noble, eds.), Chapter 12, pp. 273–310. Aust. Acad. Sci., Canberra.

Caughley, J. (1985). Effect of fire on the reptile fauna of Mallee. *In* "Biology of Australasian Frogs and Reptiles" (G. Grigg, R. Shine, and H. Ehmann, eds.), pp. 31–34. Surrey Beatty & Sons, Chipping Norton, New South Wales.

Cogger, H. G. (1969). A study of the ecology and biology of the mallee dragon *Amphibolurus fordi* (Lacertilia: Agamidae) and its adaptations to survival in an arid environment. Ph.D. Thesis, Macquarie University, New South Wales.

Cogger, H. G. (1972). Thermal relations of the mallee dragon *Amphibolurus fordi* (Lacertilia: Agamidae). *Aust. J. Zool.* 22, 319–339.

Downey, F. J., and Dickman, C. R. (1993). Macro- and micro-habitat relationships among lizards of sandridge desert in central Australia. *In* "Herpetology in Australia. A Diverse Discipline" (D. Lunney and D. Ayers, eds.), pp. 133–138. Trans. R. Zool. Soc., New South Wales.

Fyfe, G. (1980). The effect of fire on lizard communities in central Australia. *Herpetofauna* 12, 1–9.

Gill, A. (1975). Fire and the Australian flora: A review. *Aust. For.* 38, 4–25.

Gill, A. M., Groves, R. H., and Noble, I. R., eds. (1981). "Fire and the Australian Biota." Australian Academy of Science, Canberra.

Griffin, G. F., Price, N. F., and Portlock, H. F. (1983). Wildfires in the central Australian rangelands 1970–1980. *J. Environ. Manage.* 17, 311–323.

Griffin, G. F., Morton, S. R., and Allan, G. E. (1988). Fire-created patch-dynamics for conservation management in the hummock grasslands of central Australia. *Proc. Int. Grassl. Symp.,* Huhhot, China.

James, C. D. (1994). Spatial and temporal variation in structure of a diverse lizard assemblage in arid Australia. *In* "Lizard Ecology: Historical and Experimental Perspectives" (L. J. Vitt and E. R. Pianka, eds.), Chapter 13, pp. 287–317. Princeton Univ. Press, Princeton, NJ.

Kareiva, P. (1990). Population dynamics in spatially complex environments: Theory and data. *Philos. Trans. R. Soc. London, Ser. B* 330, 175–190.

Longmore, R., and Lee, P. (1981). Some observations on techniques for assessing the effects of fire on reptile populations in Sturt National Park. *Aust. J. Herpetol.* 1, 17–22.

MacArthur, R. H., and Pianka, E. R. (1966). On optimal use of a patchy environment. *Am. Nat.* 100, 603–609.

Main, A. R. (1982). Rare species: Precious or dross? In "Species at Risk: Research in Australia" (R. H. Graves and W. D. L. Ride, eds.), pp. 163–174. Aust. Acad. Sci., Canberra.

Minnich, R. A. (1983). Fire mosaics in southern California and northern Baja California. *Science* 219, 1287–1294.

Minnich, R. A. (1989). Chaparral fire history in San Diego County and adjacent northern Baja California: An evaluation of natural fire regimes and the effects of suppression management. In "The California Chaparral: Paradigms Reexamined" (S. C. Keeley, ed.), Sci. Ser. No. 34, pp. 37–47. Natural History Museum of Los Angeles County.

Perry, R. A., and Lazarides, M. (1962). Vegetation of the Alice Springs area. *CSIRO Aust. Land Resour. Ser.* 6, 208–236.

Pianka, E. R. (1969). Habitat specificity, speciation, and species density in Australian desert lizards. *Ecology* 50, 498–502.

Pianka, E. R. (1972). Zoogeography and speciation of Australian desert lizards: An ecological perspective. *Copeia*, pp. 127–145.

Pianka, E. R. (1975). Niche relations of desert lizards. In "Ecology and Evolution of Communities" (M. Cody and J. Diamond, eds.), pp. 292–314. Harvard University Press, Cambridge, MA.

Pianka, E. R. (1986). "Ecology and Natural History of Desert Lizards. Analyses of the Ecological Niche and Community Structure." Princeton Univ. Press, Princeton, NJ.

Pianka, E. R. (1989). Desert lizard diversity: Additional comments and some data. *Am. Nat.* 134, 344–364.

Pianka, E. R. (1992). Fire ecology. Disturbance, spatial heterogeneity, and biotic diversity: Fire succession in arid Australia. *Res. Explor.* 8, 352–371.

Pianka, H. D., and Pianka, E. R. (1970). Bird censuses from desert localities in Western Australia. *Emu* 70, 17–22.

Pulliam, H. R. (1988). Sources, sinks, and population regulation. *Am. Nat.* 132, 652–661.

Pulliam, H. R., and Danielson, B. J. (1991). Sources, sinks, and habitat selection: A landscape perspective on population dynamics. *Am. Nat.* 137, Symp. Suppl., S50–S66.

Pyne, S. J. (1991). "Burning Bush: A Fire History of Australia." Holt, New York.

Ralph, W. (1984). Fire in the centre. *Ecos: CSIRO Environ. Res.* 40, 3–10.

Short, N. M., Lowman, P. D., Freden, S. C., and Finch, W. A. (1976). "Mission to Earth: Landsat Views of the World." Natl. Aeronaut. Space Admin., Washington, DC.

Simpson, E. H. (1949). Measurement of diversity. *Nature (London)* 163, 688.

Winkworth, R. E. (1967). The composition of several arid spinifex grasslands of central Australia in relation to rainfall, soil water relations, and nutrients. *Aust. J. Bot.* 15, 107–130.

CHAPTER 9

Structure and Dynamics of an Amphibian Community

Evidence from a 16-Year Study of a Natural Pond

RAYMOND D. SEMLITSCH,* DAVID E. SCOTT,†
JOSEPH H. K. PECHMANN,†
AND J. WHITFIELD GIBBONS†

*Division of Biological Sciences, University of Missouri, Columbia, Missouri 65211;
and †University of Georgia's Savannah River Ecology Laboratory, Aiken,
South Carolina 29802

I. INTRODUCTION

We examined 16 years of census data on the amphibians in a natural pond ecosystem under the paradigm that competition, predation, and disturbance are interacting mechanisms regulating the abundance and distribution of species at the local level. Although much evidence has been accumulated over the last

25 years on the effects of each of these mechanisms and their interactions on particular species, few studies of amphibians have examined them within a whole natural community (but see Smith, 1983; Hairston, 1986). Most of our knowledge comes from detailed experimental studies of single or mixed species in cages or artificial pond communities (e.g., Brockelman, 1969; Jaeger, 1971; Wilbur, 1972, 1987; Morin, 1981, 1983a, 1987; Travis *et al.*, 1985a,b; Souther-land, 1986; Van Buskirk, 1988; Wilbur and Fauth, 1990). Such experimental studies have provided a wealth of information on processes that may structure amphibian communities, and which can be used to generate a set of predictions to be tested in natural communities.

Numerous amphibian species use temporary ponds for a portion of their complex life cycle, primarily for mating, oviposition, and larval growth. During the nonbreeding season, most pond-breeding amphibians live in the terrestrial habitat surrounding a pond. In a typical year, adults migrate to the pond during favorable weather conditions to breed. Mating and oviposition usually occur in the pond, and then adults return to their terrestrial habitat. After hatching, the aquatic larvae feed, grow, and develop until metamorphosis, after which they emigrate as juveniles to terrestrial habitats.

Because of the high diversity of species and the high density of larvae in many ponds, species interactions and density dependence are likely to be important in population and community dynamics (Wilbur, 1980). Several excellent field studies have clearly demonstrated the primacy of density-dependent growth and survival in larval amphibians (e.g., Smith, 1983; Petranka and Sih, 1986; Petranka, 1989; Scott, 1990, 1994; Van Buskirk and Smith, 1991), yet there is little information on whether population regulation occurs only in the larval stage, the adult stage, or both and in which stages community structure is determined (Istock, 1967; Wilbur, 1980, 1996).

A species' use of ponds is constrained by the relationship between charac-teristics of the pond related to hydrological dynamics and by species characteris-tics such as physiological tolerances, genetic ability for local adaptation, and life-history requirements. Heyer *et al.* (1975) and Wilbur (1980, 1984) sug-gested that ponds that are extremely ephemeral (<30 days) or permanent (>1 year) are used by fewer species of amphibians than ponds with intermediate hydroperiods. If rapid drying and pond permanence are considered as two ends of a disturbance continuum, this is tantamount to suggesting that species diver-sity is maximized at intermediate levels of disturbance (Levin and Paine, 1974; Gibbons, 1976; Connell, 1978; Odum *et al.*, 1979). Pond drying may be viewed as an extreme disturbance for larval amphibians, as larvae may be killed if the pond dries before metamorphosis can occur. Likewise, pond permanence also severely affects larval amphibians by permitting many predators, such as fishes, to persist. In addition, ponds vary in hydroperiod not only across a spatial landscape but also temporally, so that species must cope with unpredictable annual fluctuations within a pond. Limitation on the use of a pond, once colonized, is determined by synchrony between the availability of water and the season of reproduction. Following successful mating and oviposition, species' characteristics such as larval requirements for food, temperature tolerance, predator avoidance, and length of the larval period all interact to determine larval success along the gradient of pond hydroperiod.

In communities of pond-breeding amphibians, predation and competition interact within the context of a disturbance gradient related to pond hydroperiod (Morin, 1981, 1983b; Smith, 1983; Wilbur, 1987; Werner and McPeek, 1994; Skelly, 1995). Because all salamanders (larval and adult) are carnivorous and frequently occur at high densities, they can exert strong predation pressure within the amphibian community (e.g., Morin, 1981), especially on small herbivorous tadpoles. Predatory salamanders can persist only in ponds with long hydroperiods, and predatory fish only in permanent ponds (occasionally they colonize temporary ponds after flooding). These predators can reduce or completely eliminate other species of salamanders or anurans with small vulnerable larvae or those without effective skin secretions or behaviors that reduce the chance of being eaten (e.g., Morin, 1986; Kats et al., 1988; Lawler, 1989). In ponds with short hydroperiods that lack predatory fish or salamanders, explosive breeding species produce large numbers of fast developing larvae that compete for limited food through exploitative and interference competition, sometimes even switching from herbivory to carnivory and cannibalism (Collins and Cheek, 1983; Newman, 1989; Pfennig, 1990). Competition for food reduces growth and developmental rates, increasing the length of the larval period, and hence vulnerability to desiccation in a drying pond or exposure to predators in more permanent ponds (Wilbur, 1987, 1988; Wilbur and Fauth, 1990). Reduced growth and developmental rates also reduce body size at metamorphosis, which in turn may increase age at first reproduction and decrease size at first reproduction, survival to first reproduction, and fecundity (Berven and Gill, 1983; Smith, 1987; Berven, 1988, 1990; Semlitsch et al., 1988; Scott, 1994).

Interactions between predation and competition have been demonstrated experimentally. For example, Morin (1981) showed that, in the absence of salamander predators, some anuran species whose larvae forage efficiently can outcompete other species and dominate in numbers. In the presence of salamander predators, however, competitively superior species may be preferentially eaten, allowing other competitively inferior species to increase in relative abundance. Wilbur (1987) also demonstrated that predation can ameliorate the effects of competition at high densities by removing larvae from the community and lowering effective density, thus allowing survivors to grow more rapidly and metamorphose before ponds dry.

We used the general results of the previously cited experimental studies on competition, predation, pond drying, and their interactions to generate a set of predictions. We then used the data we accumulated from a natural amphibian community during the past 16 years to examine these predictions:

1. Variation in breeding population sizes is related to the effects of environmental variation both on breeding activity and on past juvenile recruitment.

2. There are long-term trends in the breeding population sizes of some species and hence in community structure.

3. Pond hydroperiod is highly variable but is a significant predictor of the total numbers of metamorphosing juvenile amphibians and their species diversity.

4. Within each group of amphibians (salamanders and anurans), the density of competitors (potentially also predators, in the case of salamanders) is a significant predictor of the number of metamorphosing juveniles per breeding female.

5. The density of salamanders (predators) is a significant predictor of the number of metamorphosing anurans (prey) per breeding female.

Each of these predictions is related to general questions and themes of this volume. Predictions 1, 2, and 3 address the question of whether species composition and diversity change over time in the community. Predictions 1, 3, 4, and 5 are related to whether changes in the community can be predicted from changes in the environment. Predictions 4 and 5 address whether the effects of density compensation and predation can be detected by a correlative analysis of changes over time. In addition to the results of these predictions, we will discuss the consequences of our results for community structure and organization.

II. STUDY SYSTEM

A. The Pond

The study was conducted for 16 years (1979–1994) at Rainbow Bay in Barnwell County, South Carolina, on the U.S. Department of Energy's Savannah River Site. Rainbow Bay is a relatively undisturbed freshwater wetland known as a Carolina bay (Sharitz and Gibbons, 1982; Ross, 1987). Carolina bays are natural elliptical depressions that vary in size (long axis extremes from 50 m to 8 km; Sharitz and Gibbons, 1982) and in the degree to which they retain water. They serve as the primary breeding sites for many species of amphibians indigenous to the southeastern Atlantic Coastal Plain. Most Carolina bays are not connected to stream systems and thus are filled by rainfall and ground water recharge.

Rainbow Bay is a temporary pond with a surface area of approximately 1 ha and a maximum water depth of 1.04 m. The low, center portions of the pond are vegetated with rush (*Juncus repens*), spike-rush (*Eleocharis* sp.), bulrush (*Scirpus cyperinus*), panic grass (*Panicum verrucosum*), and knotweed (*Polygonum* sp.). Several pond cypress (*Taxodium ascendens*) also occur near the center. The periphery and higher portions of the pond are vegetated by buttonbush (*Cephalanthus occidentalis*), sweetgum (*Liquidambar styraciflua*), black gum (*Nyssa biflora*), water oak (*Quercus nigra*), Darlington oak (*Q. laurifolia*), and red maple (*Acer rubrum*). The understory around the edge of the pond consists of wax myrtle (*Myrica cerifera*), greenbriers (*Smilax* spp.), and blackberry (*Rubus* sp.). Prior to 1951, Rainbow Bay was surrounded by agricultural fields and pastures, but the area is now dominated by slash pine (*Pinus elliottii*) and loblolly pine (*P. taeda*) plantations (35–40 years old). This study site is not unusual in any obvious manner relative to other amphibian breeding ponds of its size in this region (Sharitz and Gibbons, 1982), except for its relatively undisturbed condition and protected status during the past 40 years. The 27 species of amphibians we observed in our study are representative of the diversity found in southern regions of the U. S. For example, Dodd (1992) collected 16 species of amphibians during a 6-year study at a small pond in the north Florida sandhills, and Wiest (1982) found 15 species of anurans during just one year in a series of temporary ponds in east-central Texas.

B. Sampling Technique

We sampled the amphibians migrating to and from Rainbow Bay using a terrestrial drift fence with pitfall traps (Gibbons and Bennett, 1974; Shoop, 1974; Gill, 1978). The pond was encircled by a drift fence of aluminum flashing (440 m long, 50 cm high, buried 10–15 cm deep in the ground). Pitfall traps (40-liter buckets) were buried inside and outside the fence flush to the ground and next to the fence at 10-m intervals (Gibbons and Semlitsch, 1982). These traps were checked daily from 21 September 1978 through 1 July 1994. All amphibians were identified, toe-clipped for future identification, and immediately released on the opposite side of the fence. For many species, this sampling technique provided a nearly complete annual census of the number of breeding adults and of juvenile recruitment. Adults of species that are proficient climbers (e.g., treefrogs) or jumpers (e.g., bullfrogs) could trespass the fence without being captured (Gibbons and Semlitsch, 1982; Dodd, 1991; R.D.S., J.H.K.P. personal observation); however, few juveniles of these species were able to cross the fence undetected. Species for which nearly all adults trespassed the drift fence were excluded from data analyses. For the remaining species, we assumed that variation among years in trespass rates was random and therefore unlikely to systematically bias annual comparisons or to create biased trends over time.

C. Data and Statistical Analyses

Census data for 13 amphibian species and hydrological data for the pond (Tables I and II) were used in this study. The other 14 species collected at Rainbow Bay were excluded from the data analyses because they reproduced too infrequently (or not at all) at the study site, were represented only by a few adults, or could not be censused by the drift fence. We treated data for each year as an independent observation and used years as replicates in all analyses. Data for juvenile recruitment each year were independent because the pond dried annually, which separated larval dynamics into discrete episodes. Data for breeding population sizes were not strictly independent because some individuals were present during more than one year, and because numbers in one year may be affected (positively or negatively) by the numbers available to produce offspring in previous years. Hydrological dynamics in each year also may be influenced by hydrological dynamics in previous years. In addition, temporal autocorrelation of environmental variables such as rainfall and temperature may compromise the independence of breeding population sizes and hydrological data among years.

We defined the amphibian reproductive year as 1 September through 31

TABLE I   The 13 Species of Amphibians Breeding at Rainbow Bay Selected for the Study

Species	Breeding season	Females		Metamorphs	
Caudata					
Ambystoma opacum	Sept.–Dec	3,426	(.08)	27,908	(.13)
Ambystoma talpoideum	Sept.–Apr	10,638	(.25)	43,528	(.20)
Ambystoma tigrinum	Nov.–Mar	362	(.01)	2,536	(.01)
Eurycea quadridigitata	Aug.–Jan	7,282	(.17)	3,902	(.02)
Notophthalmus viridescens	Sept.–May	10,833	(.26)	43,208	(.20)
Anura					
Bufo terrestris	All year	816	(.02)	693	(<.01)
Gastrophryne carolinensis	May–Oct	3,072	(.07)	2,930	(.01)
Pseudacris crucifer	Nov–Apr	2,499	(.06)	9,245	(.04)
Pseudacris nigrita	Nov–Apr	318	(.01)	237	(<.01)
Pseudacris ornata	Nov–Apr	1,592	(.04)	19,182	(.09)
Rana clamitans	All year	106	(<.01)	2,214	(.01)
Rana utricularia	All year	605	(.01)	56,225	(.26)
Scaphiopus holbrooki	All year	197	(<.01)	3,483	(.02)
Totals (all species)		41,776		216,251	

Note. Breeding season of each species was defined by the earliest and latest months breeding adults were captured during the study period. Relative abundance according to the number and the proportion (in parentheses) of captures over 16 years is indicated for all 13 species. The total of all species includes six uncommon species that represent a combined total of less than 1%.

TABLE II   Dates of Filling and Drying and the Number of Days That Rainbow Bay Contained Standing Water, i.e. the Hydroperiod, for Each Year

	Year															
	1979	1980	1981	1982	1983	1984	1985	1986	1987	1988	1989	1990	1991	1992	1993	1994
Date filled	8 Feb	6 Sep	12 Feb	1 Jan	10 Dec	20 Dec	6 Feb	30 Nov	12 Dec	24 Apr	9 Apr	8 Dec	11 Oct	29 Dec	22 Nov	28 Jan
Date dried	3 Aug	16 Jun	7 May	6 Jul	11 Jul	27 Sep	4 Apr	24 Apr	3 Jun	26 Apr	28 Apr	4 Apr	5 Nov	18 May	25 Jun	17 May
First hydroperiod	177	285	85	187	214	283	58	146	174	3	20	118	391	142	216	110
Date refilled	6 Sep	18 Jun	8 Jun	10 Jul				20 Aug	19 Jun		2 May	22 Aug		9 Jun		
Date redried		27 Jun	22 Jun	14 Sep				24 Aug	25 Jun		5 May	21 Sep		8 Jul		
Second hydroperiod	30	10	15	67				5	7		4	31		30		

Note. To correspond with the amphibian reproductive year, fillings that occurred after 1 September were included with the following calendar year. The period 6 September–5 October 1979 was included in both the 1979 and the 1980 hydroperiods because the pond held water continuously from 6 September 1979 to 1980 and was used by both summer- and autumn-breeding species during this period.

August of the following calendar year based on the natural phenology of species. Rainbow Bay filled and dried more than one time during some years. To account for the effect of hydrological dynamics on all 13 species, we calculated the first, second, and total hydroperiods for each year (Table II). First hydroperiod was defined as the number of days the pond held water (i.e., visible standing water in the center of the basin) from date of first filling following 1 September to the date of first drying. Second hydroperiod was defined as the number of days from second filling to second drying. Total hydroperiod was defined as the sum of first and second hydroperiods. We used Pearson product-moment correlations to examine the relationships between rainfall and first, second, and total hydroperiods. Correlations were also performed between total hydroperiod and the total number of metamorphosing juveniles as well as their species richness. Data used in the Pearson product-moment calculations were log-transformed.

The influence of rainfall during a species breeding season and the number of metamorphosed juveniles in previous years on the number of breeding females was analyzed using simple and partial Spearman's rank correlations. For most species, the number of metamorphs in the previous two years was used in these correlations because most individuals reached reproductive maturity and could participate in breeding activities at two years of age. However, if most metamorphs matured at one year of age, then only metamorphs in the previous year were used. Kendall's partial rank correlation between female breeding population size and year was calculated to test for trends in breeding population sizes over time for each species, after correcting for rainfall during the species' breeding season. Significance levels for the Kendall's partial correlations were calculated from the quantile estimates of Maghsoodloo (1975).

Potential predictors of the number of metamorphosed juveniles produced per breeding female were analyzed separately for each species using a Tobit regression model for left-censored data (Tobin, 1958; SAS Institute, Inc., 1990). The Tobit model was used because the number of juveniles produced per female was zero (left-censored) for many observations. For salamanders, both simple and partial regression coefficients were calculated using first hydroperiod, initial larval density of the species, and pooled initial larval density of all other salamanders (potential competitors or predators) as predictor variables. For anurans, simple and partial regression coefficients were calculated using first hydroperiod, initial larval density of the species, pooled initial larval density of all other anurans (potential competitors), and pooled initial larval density of all salamanders (potential predators) as predictor variables.

In 1988 and 1989, first hydroperiod at Rainbow Bay was only 3 days and 20 days, respectively. Little or no breeding occurred during these 2 years; therefore they were deleted from the Tobit regression analyses. We assumed that in other years all females present at the pond oviposited. For anurans, the initial density of larvae could only be approximated by the density of breeding females (total number divided by the volume of the pond) because average fecundity was not available for each species. For salamanders, larval density was calculated from the number of breeding females of each species multiplied by their average estimated egg number and then divided by the volume of the pond. We used the volume of the pond on the same date each year (1 March) because of the difficulty of standardizing larval density at the same stage of development for all species. This procedure may have under- or over-estimated the initial larval densities of some species and hence the impact of density in our analyses of species interactions. *Gastrophryne carolinensis* was excluded from Tobit analyses because its larvae were not consistently present in the pond at the same time as those of the other anurans and salamanders analyzed. The number of juvenile *B. terrestris, P. nigrita,* and *S. holbrooki* produced per breeding female was not analyzed by Tobit regression because these species produced juveniles in too few years (2–3) for the regressions to be informative. These three species, however, were included in the calculation of heterospecific anuran density for analyses of other species.

III. HYDROPERIOD

A. Annual Dynamics

The dates of filling and drying of Rainbow Bay, and hence hydroperiod, varied greatly from year to year (Table II). First filling most often occurred in the winter between November and February and corresponded to the peak reproductive season for many species of amphibians. The earliest date of filling during the study was 6 September (1979, following a hurricane) and the latest was 24 April (in 1988). There was a strong positive relationship between cumulative annual rainfall (1 September–31 August) and first hydroperiod ($r = 0.76, P = 0.0006, n = 16$) as well as total hydroperiod ($r = 0.57, P = 0.02, n = 16$). The second hydroperiod was not significantly correlated with cumulative annual rainfall ($r = -0.16, P = 0.5594, n = 16$). Second fillings occurred when heavy rains fell during the summer (e.g., 1990), or when moderate rains fell shortly after the pond dried for the first time (e.g., 1982).

B. Relationship to Species Diversity and Reproductive Success

The number of species that produced metamorphosed juveniles increased significantly each year with the total hydroperiod ($r = 0.84, P < 0.0001, n = 16$,

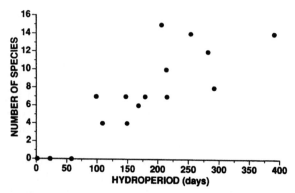

FIGURE 1. Relationship between the diversity of metamorphosing juvenile amphibians and total hydroperiod at Rainbow Bay. Each data point represents 1 year from 1979 to 1994.

Fig. 1). In addition, the total number of larvae metamorphosing (pooling all species) increased directly with total hydroperiod at Rainbow Bay ($r = 0.83, P < 0.0001, n = 16$, Fig. 2). Longer hydroperiods allowed a greater diversity of seasonally reproducing species to breed in the pond. Longer hydroperiods also permitted a greater number of larvae the opportunity to reach the critical size to initiate metamorphosis (Wilbur and Collins, 1973) and hence allowed more juveniles of more species to be recruited into terrestrial adult populations.

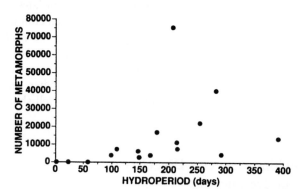

FIGURE 2. Relationship between the number of metamorphs of 13 amphibian species and total hydroperiod at Rainbow Bay. Each data point represents 1 year from 1979 to 1994.

IV. ANNUAL VARIATION IN
BREEDING POPULATIONS

For each species, the number of breeding females varied by two or three orders of magnitude among the 16 years (Figs. 3 and 4). It is important to note that the drift fence with pitfall traps only captures those adults migrating to the pond to breed, which may not represent the entire adult population. Variation in breeding population sizes can result from the effects of climatic conditions on migratory activity, as well as from demographic factors (and fence trespass for some species). For some species, variation in breeding population size was related more to rainfall (a climatic factor) than to juvenile recruitment in the previous years (a demographic factor), whereas variation for other species was more related to juvenile recruitment (Table III). Breeding population size was significantly positively correlated with rainfall during the breeding season for two species of amphibians: the salamander *A. talpoideum* and the anuran *B. terrestris*. One additional species of salamander, *A. opacum*, also showed the same relationship after partial correlations were performed to remove the effects of the number of metamorphs produced in previous years (Table III). The partial correlation of rainfall and breeding population size was nearly significant ($P = 0.06$) for *A. tigrinum* and *S. holbrooki*.

We assume that rainfall affects the conditions available for some species to migrate successfully from the terrestrial environment to the pond and that more rainfall allows more individuals to reach the pond (Packer, 1960; Hurlbert, 1969; Gibbons and Bennett, 1974; Sinsch, 1990). For example, adult *A. talpoideum* migrate only at night and during or shortly after periods of rain; individuals do not migrate on nights without rain (Semlitsch, 1985a; Semlitsch and Pechmann, 1985). In addition, the positive correlation between rainfall and pond hydroperiod (Section III.A.) suggests that rainfall could provide amphibians with a predictive cue to the presence of water in the pond and to how long it will remain. Adults may forego breeding migrations if it is likely that there is no water in the pond or that the pond may dry before larvae are able to metamorphose. Climatic limitation was reflected more generally for other species by the observation that in the driest years (1981, 1985, 1988, and 1989) almost all species had few or no breeding females due to drought conditions (Table II, Figs. 3 and 4). The reduction of breeding females was unlikely an exclusive effect of mortality during these dry periods because large numbers of previously marked females were recaptured during the year or two following such droughts. This suggests that the reduction of breeding females is an effect of climatic limitation on migratory activity, or on food resources and the acquisition of energy for reproduction, or both.

The breeding population size of a species may vary widely due to fluctua-

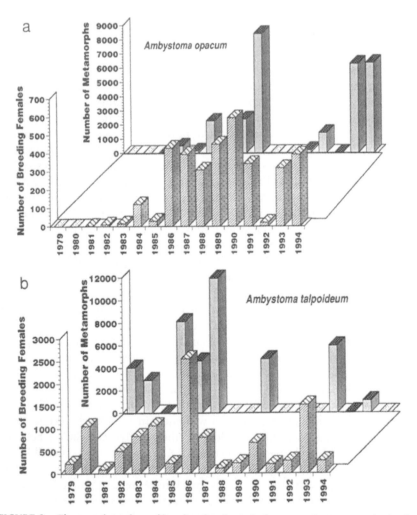

FIGURE 3. The annual numbers of breeding females and of metamorphosing juveniles for five species of salamanders ((a) *A. opacum*, (b) *A. talpoideum*, (c) *A. tigrinum*, (d) *E. quadridigitata*, and (e) *N. viridescens*) at Rainbow Bay. Bars represent the total numbers collected in pitfall traps at the drift fence for each year from 1979 to 1994.

tions in juvenile recruitment; if recruitment rate is high and variable relative to adult survival rate, terrestrial density dependence is weak, and population dynamics are determined at the local pond or are synchronous at nearby ponds from which individuals may immigrate. Past juvenile recruitment and breeding population sizes also may be strongly correlated because of long-term

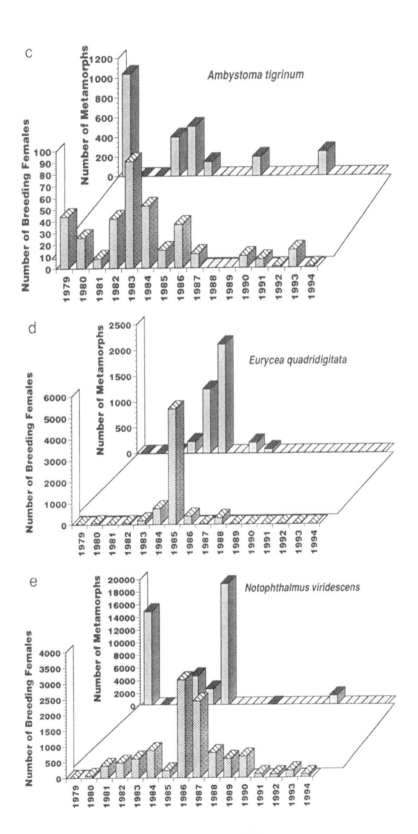

c   *Ambystoma tigrinum*

d   *Eurycea quadridigitata*

e   *Notophthalmus viridescens*

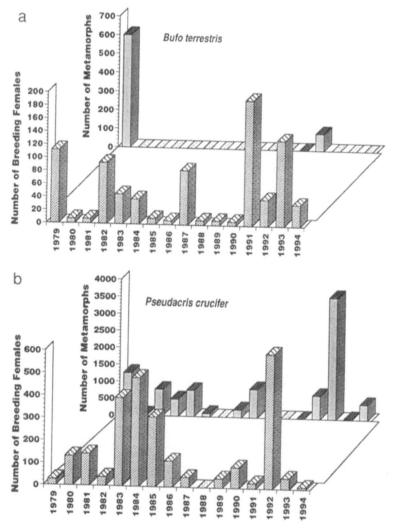

FIGURE 4.   The annual numbers of breeding females and of metamorphosing juveniles for eight species of anurans ((a) *B. terrestris*, (b) *P. crucifer*, (c) *P. nigrita*, (d) *P. ornata*, (e) *R. clamitans*, (f) *R. utricularia*, (g) *S. holbrooki*, and (h) *G. carolinensis*) at Rainbow Bay. Bars represent the total numbers collected in pitfall traps at the drift fence for each year from 1979 to 1994.

increases or decreases in total population size. There were four species that showed a significant positive correlation between the number of breeding females and metamorphs produced in previous years: *A. opacum*, *E. quadri-digitata*, *P. nigrita*, and *P. ornata*. The correlation was nearly significant ($P = 0.06$) for *A. tigrinum* (Table III). Correlations for all five of these species

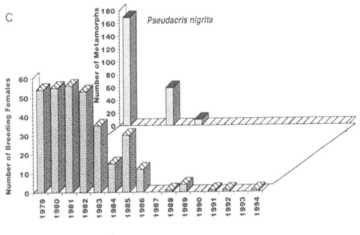

c

Pseudacris nigrita

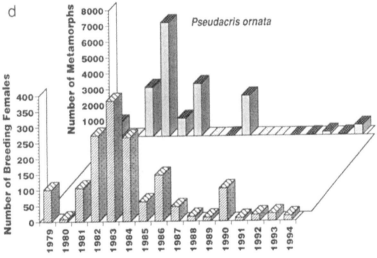

d

Pseudacris ornata

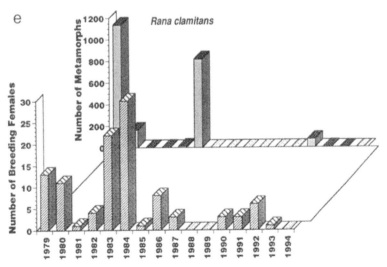

e

Rana clamitans

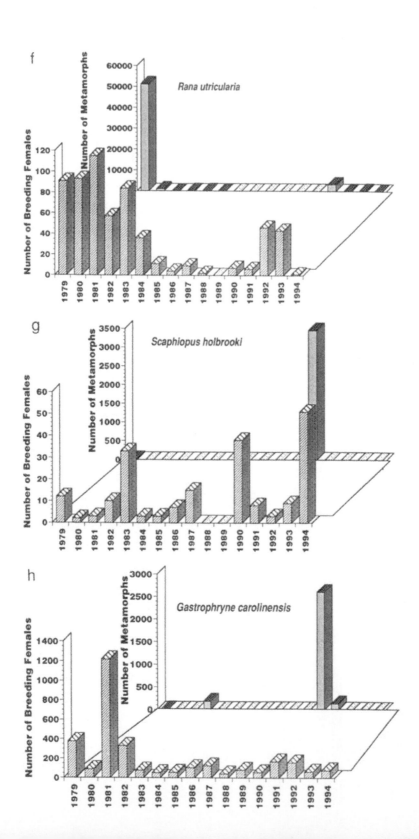

f   Rana utricularia

g   Scaphiopus holbrooki

h   Gastrophryne carolinensis

TABLE III   Summary of the Spearman Partial (and Simple) Rank Correlations of the Number of Breeding Females with Cumulative Rainfall during the Species' Breeding Season and the Number of Metamorphs in Previous Years (See Methods)

Species	Cumulative rainfall	Number of metamorphs
Caudata		
A. opacum	0.57*	0.86*
	(0.11)	(0.80*)
A. talpoideum	0.63*	0.43
	(0.58*)	(0.20)
A. tigrinum	0.55	0.63*
	(0.37)	(0.53)
E. quadridigitata	−0.20	0.63*
	(−0.25)	(0.66*)
N. viridescens	−0.16	0.45
	(−0.10)	(0.43)
Anura		
B. terrestris	0.76*	0.48
	(0.57*)	(0.44)
P. crucifer	−0.05	0.04
	(0.03)	(0.04)
P. nigrita	−0.32	0.73*
	(−0.27)	(0.72*)
P. ornata	0.30	0.62*
	(0.03)	(0.57*)
R. clamitans	0.24	0.24
	(0.37)	(0.32)
R. utricularia	0.21	0.33
	(0.11)	(0.40)
S. holbrooki	0.55	−0.10
	(0.48)	(−0.02)

Note. Significant correlation coefficients are indicated by an asterisk ($P < 0.05$). The two predictor variables were not significantly correlated for any species (all $P > 0.20$).

were significant after the effect of rainfall was removed by partial correlation (Table III). For A. opacum, A. tigrinum, and P. nigrita, these correlations may reflect significant trends in their population sizes over the course of the study (see following discussion). Caldwell (1987) suggested from mortality data that both P. ornata and P. nigrita were short-lived, which could increase their sensitivity to variation in juvenile recruitment. For some species the lack of a significant correlation between breeding population sizes and prior juvenile recruitment may have resulted from migration between Rainbow Bay and other nearby ponds. The presence of a breeding population of females at Rainbow Bay during most years despite infrequent juvenile recruitment, e.g., for B. terrestris and S. holbrooki, is consistent with this hypothesis. Three of the four lowest correlations between breeding population sizes and juvenile recruitment were for species whose adults frequently trespass the drift fence: *P. crucifer, R. clamitans,* and *R. utricularia.*

The breeding population sizes of five species exhibited significant correlations with year, i.e., significant trends over the time period of the study (Table IV). Correlations for these same five species were significant after adjusting for the effect of breeding season rainfall (Table IV). Both the simple and partial correlations with year were positive and significant for *A. opacum,* indicating an increase over time. No *A. opacum* bred at Rainbow Bay during the first two years of our study, but several hundred females bred during eight of the last nine years (Fig. 3). This suggests that *A. opacum* may have recently colonized or recolonized Rainbow Bay. Alternatively, its population may have fluctuated to a small size at the time our study began and then recovered without the necessity for recolonization.

Breeding population sizes of the other four species, *A. tigrinum, P. nigrita, R. clamitans,* and *R. utricularia,* decreased significantly during the period of our study (Table IV; Figs. 3 and 4). We attribute these decreases primarily to the lack of juvenile recruitment associated with extended drought conditions from 1985 to 1990 (Table II; Figs. 3 and 4). For *A. tigrinum* and *P. nigrita,* there was

TABLE IV Summary of the Trend Analysis Using Kendall's (Tau^{-b}) Rank Correlations of the Number of Breeding Females with Year, Cumulative Rainfall during the Breeding Season, and Partial Rank Correlations with Year after Removing the Effect of Rainfall

Species	Year	Rainfall	Year (correcting for rainfall)
Caudata			
A. opacum	0.58*	0.08	0.57*
A. talpoideum	0.03	0.47*	−0.03
A. tigrinum	−0.46*	0.28	−0.53*
E. quadridigitata	−0.23	−0.18	−0.21
N. viridescens	−0.15	−0.08	−0.14
Anura			
B. terrestris	−0.01	0.44*	−0.08
G. carolinensis	−0.22	0.20	−0.22
P. crucifer	−0.17	0.01	−0.17
P. nigrita	−0.72*	−0.18	−0.71*
P. ornata	−0.33	0.07	−0.35
R. clamitans	−0.39*	0.27	−0.45*
R. utricularia	−0.53*	0.03	−0.54*
S. holbrooki	0.15	0.33	0.11

Note. Significant correlation coefficients are indicated by an asterisk ($P < 0.05$). The two predictor variables were not significantly correlated for any species (all $P > 0.40$).

a significant partial correlation of breeding population size with past juvenile recruitment (Table III). It is also possible that the increase in *A. opacum* contributed to the decrease in the other four species, due to predation by its larvae on the other species larvae. There was a negative association between the density of heterospecific salamander larvae (all species pooled) and per-capita juvenile recruitment for three of the species that declined (see section V.; one species, *P. nigrita*, was not analyzed). Although the total density of salamander larvae (all species pooled) did not increase significantly over time (Kendall correlation with year = 0.27, $P = 0.17$, $n = 14$), *A. opacum* may be a particularly important predator because its larvae are usually the first present in the pond and initially larger in body size.

V. LARVAL SUCCESS: INTERACTIONS AMONG SPECIES AND WITH HYDROPERIOD

A. Salamanders

The simple Tobit regressions between the number of metamorphosing juveniles per breeding female and length of the first hydroperiod were significant for three of the five species of salamanders (Table V). After correcting for any linear effects of conspecific and heterospecific larval density, first hydroperiod was positively related to increased reproductive success only for *A. talpoideum*. For *A. tigrinum* and *N. viridescens*, the partial regression coefficient for hydroperiod was not significantly different from zero even though the simple regression coefficient was. This indicates that any effects of hydroperiod on these two species could not be separated from the effects of the other predictor variables. Hydroperiod was significantly correlated with one or both of the other predictor variables for all the salamander species (Table V). This may be, in part, because the other variables were larval densities calculated using 1 March pond volume, which was itself significantly correlated with pond hydroperiod (Spearman's rank correlation $r = 0.78$, $P < 0.001$).

A positive relationship of first hydroperiod with the number of metamorphs is most likely attributable to more time available to complete larval development. The species for which this relationship was weakest were the two autumn breeders, *A. opacum* and *E. quadridigitata* (Table V), both of which usually migrate to the pond and breed early in the season (September–November). Their larvae metamorphose earlier and are less affected by a short hydroperiod than those of species (*A. talpoideum*, *A. tigrinum*, and *N. viridescens*) that breed later in the season (i.e., winter or early spring; Table V).

The simple and partial Tobit regressions of the total number of metamorphs per female of a species with its initial larval density were significant only for

TABLE V Summary of the Tobit Regression Analyses for the Number of Metamorphosed Juveniles Produced Per Breeding Female of Five Salamander Species at Rainbow Bay

| | | | First pond hydroperiod | Initial density | |
	Non-censored (N)	Left-censored (N)		Same species	All other species
A. opacum	11	1	0.02 ± 0.05 (0.04 ± 0.05)	−38.5 ± 178.8 (−139.7 ± 150.7)	−9.9 ± 12.2 (−12.8 ± 9.9)
A. talpoideum	10	4	0.07 ± 0.02* (0.10 ± 0.03*)	−205.1 ± 37.7* (−222.3 ± 65.9*)	−17.2 ± 38.8 (−94.6 ± 74.9)
A. tigrinum	8	6	0.06 ± 0.04 (0.11 ± 0.04*)	−533.8 ± 1122.8 (−1490.2 ± 923.0)	−140.2 ± 81.3 (−185.1 ± 83.6*)
E. quadridigitata	8	6	−0.02 ± 0.02 (−0.01 ± 0.02)	−48.3 ± 58.7 (−50.2 ± 103.9)	−2.8 ± 3.3 (−1.9 ± 3.5)
N. viridescens	7	7	0.08 ± 0.05 (0.11 ± 0.05*)	17.2 ± 123.1 (−114.4 ± 103.1)	−193.9 ± 115.2 (−207.4 ± 118.9)

Note. Partial regression coefficients are indicated ±1 SE. Corresponding simple regression coefficients are given in parentheses. Significance ($P < 0.05$) is indicated by an asterisk. There were significant ($P < 0.05$) Spearman rank correlations between the following pairs of predictor variables: hydroperiod and density of the same species (negative) for *A. talpoideum* and *N. viridescens*, hydroperiod and density of other species (negative) for all species, and density of the same species and of other species (positive) for *A. talpoideum, A. opacum, N. viridescens,* and *E. quadridigitata.*

A. talpoideum (Table V). The regression coefficients were negative and likely indicate density-dependent reproductive success. This species is abundant at Rainbow Bay (Table I), and it is likely to numerically and behaviorally dominate competitive relationships with other species as well as with itself (Semlitsch and Walls, 1993). Although the additional density effect of the other species also influenced *A. talpoideum* negatively, the regressions were not significant (Table V). The reproductive success of the other four species, *A. opacum, A. tigrinum, E. quadridigitata,* and *N. viridescens,* was not significantly affected by their own initial larval densities (Table V).

The Tobit analyses detected only weak evidence for negative interactions (competition or predation) among salamander species. The only significant Tobit regression for the number of metamorphosing juveniles per breeding female with the initial density of heterospecific larvae was the simple regression for *A. tigrinum* (Table V). The corresponding partial regression coefficient was smaller and not quite significant ($P = 0.08$). This suggests that any effects of heterospecific density on *A. tigrinum* were partially confounded with the effects of hydroperiod, with which heterospecific density was significantly correlated (Table V). Both the simple and partial Tobit regressions for heterospecific density effects on *N. viridescens* were nearly significant ($P = 0.08$ and $P = 0.09$, respectively). The weakest effects of heterospecific salamander density were again for the two species that breed the earliest in the season, *A. opacum* and *E. quadridigitata.* Early breeding reduces the temporal overlap between their larvae and those of other salamanders. In addition, it provides *A. opacum* and *E. quadridigitata* with a size advantage over other larvae.

B. Anurans

The simple coefficient for *R. utricularia* was the only positive, significant Tobit regression coefficient with first hydroperiod as a predictor of the number of metamorphosing juveniles per breeding female (Table VI). The simple coefficient was also nearly significant for *R. clamitans* ($P = 0.08$), but the corresponding partial regression coefficient was not significant for either species (Table VI). This suggests that any effect of hydroperiod on *R. utricularia* and *R. clamitans* could not be distinguished from the effects of other variables, two of which were significantly correlated with hydroperiod (Table VI). It is likely that longer hydroperiods would increase the number of metamorphs for these two species, because they have relatively long larval periods. Longer hydroperiods may have allowed more individuals to reach the critical size to initiate metamorphosis. Species such as *P. crucifer* have a shorter larval period (<45 days) and were unlikely constrained by hydroperiod in most years (i.e., except in 1985, 1988, and 1989; Fig. 4 and Table II).

TABLE VI Summary of the Tobit Regression Analyses for the Number of Metamorphosed Juveniles Produced Per Breeding Female of Four Anuran Species at Rainbow Bay

			First pond hydroperiod	Initial density		
	Non-censored (N)	Left-censored (N)		Same species	All other anurans	Salamanders
P. crucifer	13	1	-0.08 ± 0.06	-62297 ± 28593*	-46630 ± 53743	-8.7 ± 10.8
			(0.02 ± 0.05)	(-61299 ± 28729*)	(-65481 ± 43749)	(-14.7 ± 10.8)
P. ornata	12	2	-0.17 ± 0.05*	-29794 ± 85937	-62925 ± 25230*	-16.2 ± 13.5
			(-0.04 ± 0.06)	(-76312 ± 71277)	(-36154 ± 27176)	(-14.1 ± 12.1)
R. clamitans	8	5	-0.17 ± 0.10	9538474 ± 3675740*	-164589 ± 60104*	-574 ± 179*
			(0.21 ± 0.12)	(-1620982 ± 3934530)	(-96752 ± 51292)	(-456 ± 224*)
R. utricularia	10	4	-0.24 ± 0.68	-393941 ± 1665789	-500097 ± 375056	-3999 ± 1140*
			(1.52 ± 0.69*)	(-684877 ± 2278149)	(-734259 ± 379757)	(3980 ± 1149*)

Note. Partial regression coefficients are indicated ±1 SE. Corresponding simple regression coefficients are given in parentheses. Significance ($P < 0.05$) is indicated by an asterisk. There were significant ($P < 0.05$) Spearman rank correlations between the following pairs of predictor variables for all species: hydroperiod and density of other anurans (negative), hydroperiod and density of salamanders (negative), and density of other anurans and of salamanders (positive). Also, density of same species was significantly correlated with density of other anurans (positive) for R. utricularia and P. ornata, and with hydroperiod (negative) and density of salamanders (positive) for P. ornata.

There was a significant negative relationship between hydroperiod and per-capita juvenile recruitment of *P. ornata,* after correcting for the other predictor variables (Table VI). This means that *P. ornata* tended to do more poorly in years with long first hydroperiods. Rainbow Bay was more likely to fill early in these years (e.g., 1980, 1991), before the breeding season of *P. ornata.* Its tadpoles may therefore have been exposed to greater levels of competition and predation (from invertebrates and vertebrates) than in years when *P. ornata* bred shortly after the pond filled. It should be noted that *P. ornata* also did poorly in short-hydroperiod years when the pond dried early (e.g., 1985, 1990), suggesting that intermediate hydroperiods were the most favorable.

Second hydroperiod also may have affected juvenile recruitment of some species, but this was not formally analyzed here. The only large cohort of *S. holbrooki* metamorphs was produced in 1990 after Rainbow Bay dried and then refilled (Fig. 4; Table II). *Gastrophryne carolinensis* produced metamorphs only when Rainbow Bay refilled for at least several weeks during the summer after drying completely (1979, 1982, and 1990) or nearly drying (1991). In most other years Rainbow Bay did not hold water for an adequate period during the summer breeding season of *G. carolinensis,* or did so but did not dry first, and no metamorphs of this species were produced (Fig. 4; Table II).

Conspecific density was a significant predictor of the number of juveniles produced per breeding female for *P. crucifer* and *R. clamitans* but not for *P. ornata* or *R. utricularia* (Table VI). Both the simple and the partial regression coefficients were significant and negative for *P. crucifer,* suggesting that intra-specific competition reduced survival to metamorphosis in years that larval densities were high. Intraspecific density effects may have been more detect-able for *P. crucifer* because its breeding population sizes at Rainbow Bay aver-aged larger than those for all other anurans except *G. carolinensis* (Table I). For *R. clamitans,* the simple regression coefficient for conspecific density was nega-tive and not significant, whereas the partial coefficient was positive and signifi-cant (Table VI). This suggests that survival to metamorphosis of *R. clamitans* may have been facilitated by higher densities of conspecific larvae, after cor-recting for other factors.

The partial Tobit regression coefficients for heterospecific anuran density were negative and significant for *P. ornata* and *R. clamitans* (Table VI), while the corresponding simple coefficients were negative and not significant (but nearly so for *R. clamitans;* $P = 0.06$). This suggests that interspecific competi-tion measurably reduced per-capita juvenile recruitment of these two species at Rainbow Bay, after accounting for the effects of other variables. Tobit regres-sion coefficients for heterospecific anuran density were negative but not signif-icant for *P. crucifer* and *R. utricularia* (Table VI), although the simple coeffi-cient for *R. utricularia* was nearly significant ($P = 0.053$).

The simple and partial Tobit regressions for the initial density of salamander larvae with the number of metamorphosed juveniles per breeding female were significant for *R. clamitans* and *R. utricularia* (Table VI). The regression coefficients were negative and likely indicated predation by salamander larvae on the tadpoles. Per-capita juvenile recruitment of *P. crucifer* and *P. ornata* was not significantly affected by the density of salamander larvae (Table VI). These results are consistent with Morin's (1983a) experimental finding that salamander predation was greater on *R. utricularia* tadpoles than on *P. crucifer* tadpoles. Salamander predation on *P. ornata* eggs and tadpoles may be reduced because *P. ornata* usually breeds just after the pond fills, before larger salamander larvae are present in the pond. The negative relationship between salamander density and juvenile recruitment of *R. clamitans* that we observed is not consistent with Walters' (1975) observations that eggs and tadpoles of this species were unpalatable to adult *N. viridescens* and larval *Ambystoma*.

The relationship between second hydroperiod and juvenile recruitment observed for some anuran species at Rainbow Bay may be interpreted as evidence for the importance of interspecific interactions. The fact that *G. carolinensis* metamorphs were produced only after Rainbow Bay dried, then refilled, suggests that other species have negative effects on this species and that drying is necessary to eliminate predators or competitors (either amphibians or other taxa). Similarly, *S. holbrooki* produced a large number of metamorphs only in 1990, when Rainbow Bay refilled after drying.

VI. DISCUSSION AND CONCLUSIONS

Our 16-year study found that pond hydroperiod (disturbance) is a primary source of variation in community structure for this natural community of pond-breeding amphibians (*sensu* Menge and Sutherland, 1987). Larval competition and predation form other axes that are positioned along a continuum of hydroperiod, and the strength of their influence on the success of species is mediated by pond hydroperiod. Although hydroperiod, competition, and predation all had a detectable influence on the amphibian community at Rainbow Bay, the effects of these factors were often difficult to separate. Correlative analyses are less powerful for sorting out confounding factors (see a discussion of field data in Petranka and Sih, 1986) than manipulative experiments in replicated artificial ponds (e.g., Morin, 1983a; Wilbur, 1984, 1987). The predictor variables used were often themselves correlated. Furthermore, we lacked sufficient replication to test for statistical interactions among the variables, which may play an important role (Wilbur, 1987).

Juvenile production for all species was virtually episodic, with large numbers of metamorphs being produced in only a small number (1–7 years) of the 16 years. Our data show that temporal variation may favor the reproductive success of different species in different years. Nevertheless, in many cases a good year for a species was also a good year for its competitors and predators. Our third prediction, that hydroperiod is a highly variable but still significant predictor of the number and diversity of metamorphosing amphibians at Rainbow Bay, was supported by the data. Relationships between species diversity and hydroperiod were suggested previously by Heyer *et al.* (1975) and Wilbur (1980, 1984). Their emphasis, however, was on spatial variation among ponds rather than temporal variation. Rapid pond drying often causes complete reproductive failure in amphibian species (e.g., Blair, 1957; Tevis, 1966; Shoop, 1974; Newman, 1987; Semlitsch, 1987). Early pond drying is sometimes viewed as an unusual or catastrophic event with little or no influence on community structure because of its seemingly random occurrence. Yet, early pond drying was not rare but rather a pervasive risk at Rainbow Bay. Although hydroperiod is systematically linked to rainfall, and rainfall can be viewed as random, its effects on community structure and dynamics are predictable. Over the 16 years of our study, 4 (25%) of the years (1981, 1985, 1988, and 1989) had short total hydroperiods (≤ 100 days), and complete or nearly complete reproductive failure occurred for most species. Although short hydroperiods were catastrophic for most species in these years, other species actually benefitted from pond drying followed by refilling in other years, presumably because it eliminated predators (both salamander and insect larvae), competitors, or both (e.g., *G. carolinensis, S. holbrooki* in Fig. 4; and *Hyla femoralis* in Pechmann *et al.,* 1989). Seven (44%) of the years (1979, 1980, 1982, 1983, 1984, 1991, and 1993) had long total hydroperiods (>200 days), which allowed species that breed late or have long larval periods to produce metamorphosed juveniles. Numbers and species diversity of metamorphosing juveniles were, on average, highest in these years. The remaining 5 years were intermediate in total hydroperiod (between 100 and 200 days). There was no clear indication that species diversity was maximized at intermediate hydroperiods as suggested by Heyer *et al.* (1975) and Wilbur (1980, 1984), probably because the extreme case of no pond drying did not occur at Rainbow Bay for comparison.

Juvenile recruitment of all species was limited by hydroperiod in the driest years. The Tobit regressions tested for a relationship between hydroperiod and per-capita juvenile recruitment above the estimated threshold hydroperiod for non-zero recruitment. Longer hydroperiods were significantly associated with increased recruitment for three of five salamanders and one of four anurans examined by the Tobit analyses. The positive effect of longer hydroperiods, however, could be separated from the effects of predation and competition only for the salamander *A. talpoideum.* Longer hydroperiods were associated with decreased recruitment of one anuran, *P. ornata,* after correcting for other variables.

Our fourth prediction, that the density of competitors affects the number of metamorphosing juveniles per breeding female, was supported for some species. There was evidence for intraspecific competition for one abundant salamander and one abundant anuran. There was evidence for facilitation in another anuran species. The number of metamorphosing juveniles was significantly affected by interspecific competition in two of four anuran species tested. The density of heterospecific salamander larvae had a significant effect on juvenile recruitment in only one of five salamander species, and this effect could not be statistically separated from that of hydroperiod.

Data from Rainbow Bay also supported our fifth prediction, that the density of salamander larvae is a significant predictor of per-capita juvenile recruitment for anurans. A significant negative relationship was detected for two of four anuran species analyzed, most likely due to predation by the salamander larvae on the anuran tadpoles. Thus, predation, competition, and hydroperiod all had detectable effects on the numbers of metamorphosing juvenile amphibians at Rainbow Bay.

Fluctuations in juvenile recruitment, the effects of rainfall on breeding migrations, and perhaps other factors resulted in wide fluctuations in breeding population sizes, affirming our first prediction. Annual variation, short-term trends, and long-term trends (our second prediction) all were observed during the course of our 16-year study. The result was that community structure, at least in terms of relative abundances, varied continuously. The breeding population size of *A. opacum* began at 0 and increased significantly during this period. No species were lost from Rainbow Bay during our study, but numbers of four species declined significantly, in some cases to very low levels. We view these long-term trends as prolonged natural fluctuations related to climatic variation, predation, competition, and other natural interacting factors. Some amphibian populations are reported to have declined or disappeared in isolated, protected areas around the world since the 1970s (Wake and Morowitz, 1990; Blaustein *et al.*, 1994). The concern is that these losses have resulted from undetected human impacts. Data from Rainbow Bay illustrated the difficulty of distinguishing declines due to human activities from natural fluctuations (Pechmann *et al.*, 1991).

Our analyses in this paper concentrated on the larval stage of the complex life cycle of amphibians. Population sizes and community diversity may also be affected by processes acting upon the terrestrial stages (Istock, 1967; Wilbur, 1980, 1996; Werner and Gilliam, 1984; Loreau and Ebenhoh, 1994; Pechmann, 1994). Local dynamics may also be influenced by immigration and emigration among ponds, i.e., by processes occurring at a metapopulational or regional level (Hanski and Gilpin, 1991; Ricklefs and Schluter, 1993). For example, *Ambystoma opacum* may have colonized or recolonized Rainbow Bay during our study. For some species the low frequency of successful reproduction observed at Rainbow Bay, coupled with a lack of correlation between juvenile recruitment and breeding population sizes, suggests that their persistence at this site may be aided by metapopulation "rescue effects" or "source-sink" processes (Brown and Kodric-Brown, 1977; Pulliam, 1988).

In conclusion, environmental variation resulted in striking changes in the relative abundances of amphibian species at Rainbow Bay, both from year to year and over the 16 years of the study. This community cannot be viewed as being in equilibrium; nevertheless, species interactions and density dependence play a significant role. We suggest that long-term results from Rainbow Bay can be generalized to other amphibian communities. Variation in breeding population sizes and larval success among years similar to that observed at Rainbow Bay has been observed at other ponds in our geographic region in short-term studies (Gibbons and Bennett, 1974; Semlitsch, 1983, 1985b; Pechmann et al., 1989). Similarly high levels of variability are well-documented in natural amphibian communities across a wide range of geographic regions (e.g., Bannikov, 1948; Blair, 1961; Tevis, 1966; Heyer, 1973, 1979; Shoop, 1974; Gill, 1978; Wiest, 1982; Berven and Grudzien, 1990; Dodd, 1992). Thus, there is no reason to believe that our results are unique because of geographic region, pond type, or species diversity.

The critical step in equating the dynamics of various pond-breeding amphibian communities is to first position ponds along a hydroperiod gradient ranging from extremely ephemeral (e.g., desert pools) to extremely permanent (e.g., spring-fed ponds or bogs). The importance of competition among species, and predation by fish, salamanders, and aquatic insects, will then vary with hydroperiod. In addition, it is important to consider whether a pond's position on the gradient of hydroperiod is static or dynamic. Our long-term data on the hydroperiod of Rainbow Bay indicates that some ponds change position along this gradient annually, and thus, the effects of competition and predation on species are also dynamic.

A critical question is whether these processes affecting the persistence and relative abundance of species are predictable or random. Although rainfall and pond hydroperiod may be viewed as stochastic influences, our data show that they produced a predictable pattern, interacting with competition and predation. Species unable to cope with high frequencies of years unfavorable for reproduction may persist for short periods of time but, barring immigration, would certainly become extinct locally as mortality exceeded reproduction. Thus, regulation of community structure within a pond occurs through the predictable interaction of rainfall, hydroperiod, competition, and predation.

VII. SUMMARY

Our 16-year study of amphibians in a natural temporary pond found that hydroperiod is a primary source of variation in community structure. Larval competition and predation also have a detectable influence, but the strength of their influence on the success of species is mediated by pond hydroperiod. Although hydroperiod, competition, and predation all had a detectable influence on the amphibian community at Rainbow Bay, the effects of these factors were often difficult to separate. Juvenile production for all species was virtually episodic, with large numbers of metamorphs being produced in only a small number (1–7) of the 16 years. Although hydroperiod was highly variable, it was still a significant predictor of the number and diversity of metamorphosing amphibians at Rainbow Bay. Our data show that temporal variation may favor the reproductive success of different species in different years. Nevertheless, in many cases a good year for a species was also a good year for its competitors and predators. Juvenile recruitment of all species was limited by a short hydroperiod in the driest years. In years with longer hydroperiods, the density of competitors affected the number of metamorphosing juveniles per breeding female for some species. The density of salamander larvae was also a significant predictor of per-capita juvenile recruitment for anurans. A significant negative relationship was detected for two of four anuran species analyzed, most likely due to predation by the salamander larvae on the anuran tadpoles. These results indicate that community structure, at least in terms of relative abundances, varied continuously. However, regulation of community structure within a pond occurs through the predictable interaction of rainfall, hydroperiod, competition, and predation.

Acknowledgments

We thank the many people at SREL who helped construct the drift fence and check pitfall traps at Rainbow Bay but especially K. L. Brown, J. P. Caldwell, A. C. Chazal, A. M. Dancewicz, R. A. Estes, J. Garvin, P. E. Johns, J. Kemp, G. C. Knight, T. K. Lynch, J. H. McGregor, G. B. Moran, M. K. Nungesser, K. K. Patterson, R. A. Seigel, C. A. Shoemaker, and L. J. Vitt. We also thank C. M. Bridges, B. W. Buchanan, M. L. Crump, D. E. Gill, C. G. Murphy, M. J. Parris, A. M. Welch, and S. Wise for their critical review of the manuscript and P. M. Dixon and K. A. Garrett for statistical advice. This research was supported by U.S. Department of Energy contract DE-AC09-76SR00819 with the University of Georgia.

References

Bannikov, A. G. (1948). On the fluctuation of anuran populations. *Tr.—Akad. Nauk SSSR* **61**, 131–134.

Berven, K. A. (1988). Factors affecting variation in reproductive traits within a population of wood frogs (*Rana sylvatica*). *Copeia*, pp. 605–615.

Berven, K. A. (1990). Factors affecting population fluctuations in larval and adult stages of the wood frog (*Rana sylvatica*). *Ecology* **71**, 1599–1608.

Berven, K. A., and Gill, D. E. (1983). Interpreting geographic variation in life-history traits. *Am. Zool.* **23**, 85–97.

Berven, K. A., and Grudzien, T. A. (1990). Dispersal in the wood frog (*Rana sylvatica*): Implications for genetic population structure. *Evolution (Lawrence, Kans.)* **44**, 2047–2056.

Blair, W. F. (1957). Changes in vertebrate populations under conditions of drought. *Cold Spring Harbor Symp. Quant. Biol.* **22**, 273–275.

Blair, W. F. (1961). Calling and spawning seasons in a mixed population of anurans. *Ecology* **42**, 99–110.

Blaustein, A. R., Wake, D. B., and Sousa, W. P. (1994). Amphibian declines: Judging stability, persistence, and susceptibility of populations to local and global extinctions. *Conserv. Biol.* **8**, 60–71.

Brockelman, W. Y. (1969). An analysis of density effects and predation in *Bufo americanus* tadpoles. *Ecology* **50**, 632–644.

Brown, J. H., and Kodric-Brown, A. (1977). Turnover rates in insular biogeography: Effect of immigration on extinction. *Ecology* **58**, 445–449.

Caldwell, J. P. (1987). Demography and life history of two species of chorus frogs (Anura: Hylidae) in South Carolina. *Copeia*, pp. 114–127.

Collins, J. P., and Cheek, J. E. (1983). Effect of food and density on development of typical and cannibalistic salamander larvae in *Ambystoma tigrinum nebulosum*. *Am. Zool.* **23**, 77–84.

Connell, J. H. (1978). Diversity in tropical rain forests and coral reefs. *Science* **199**, 1302–1310.

Dodd, C. K., Jr. (1991). Drift fence associated sampling bias of amphibians at a Florida sandhills temporary pond. *J. Herpetol.* **25**, 296–301.

Dodd, C. K., Jr. (1992). Biological diversity of a temporary pond herpetofauna in north Florida sandhills. *Biodiversity Conserv.* **1**, 125–142.

Gibbons, J. W. (1976). Thermal alteration and the enhancement of species populations. *In* " Thermal Ecology. II. Energy Research and Development Administration Symposium Series (CONF-750425)" (G. W. Esch and R. W. McFarlane, eds.), pp. 27–31. Natl. Tech. Inf. Serv., Springfield, VA.

Gibbons, J. W., and Bennett, D. H. (1974). Determination of anuran terrestrial activity patterns by a drift fence method. *Copeia*, pp. 236–243.

Gibbons, J. W., and Semlitsch, R. D. (1982). Terrestrial drift fence with pitfall traps: An effective technique for quantitative sampling of animal populations. *Brimleyana* **7**, 1–16.

Gill, D. E. (1978). The metapopulation ecology of the red-spotted newt, *Notophthalmus viridescens* (Rafinesque). *Ecol. Monogr.* **48**, 145–166.

Hairston, N. G., Sr. (1986). Species packing in *Desmognathus* salamanders: Experimental demonstration of predation and competition. *Am. Nat.* **127**, 266–291.

Hanski, I., and Gilpin, M. E. (1991). "Metapopulation Dynamics." Academic Press, London.

Heyer, W. R. (1973). Ecological interactions of frog larvae at a seasonal tropical location in Thailand. *J. Herpetol.* **7**, 337–361.

Heyer, W. R. (1979). Annual variation in larval amphibian populations within a temperate pond. *J. Wash. Acad. Sci.* **69**, 65–74.

Heyer, W. R., McDiarmid, R. W., and Weigmann, D. L. (1975). Tadpoles, predation, and pond habitats in the tropics. *Biotropica* **7**, 100–111.

Hurlbert, S. H. (1969). The breeding migrations and interhabitat wandering of the vermilion-spotted newt *Notopthalmus viridescens* (Rafinesque). *Ecol. Monogr.* **39**, 465–488.

Istock, C. A. (1967). The evolution of complex life cycle phenomena: An ecological perspective. *Evolution (Lawrence, Kans.)* **21**, 592–605.

Jaeger, R. G. (1971). Competitive exclusion as a factor influencing the distributions of two species of terrestrial salamanders. *Ecology* **52**, 632–637.

Kats, L. B., Petranka, J. W., and Sih, A. (1988). Antipredator defenses and the persistence of amphibian larvae with fishes. *Ecology* 69, 1865–1870.

Lawler, S. P. (1989). Behavioural responses to predators and predation risk in four species of larval anurans. *Anim. Behav.* 38, 1039–1047.

Levin, S. A., and Paine, R. T. (1974). Disturbance, patch formation and community structure. *Proc. Natl. Acad. Sci. U.S.A.* 71, 2744–2747.

Loreau, M., and Ebenhoh, W. (1994). Competitive exclusion and coexistence of species with complex life cycles. *Theor. Popul. Biol.* 46, 58–77.

Maghsoodloo, S. (1975). Estimates of the quantiles of Kendall's partial rank correlation coefficient. *J. Stat. Comput. Simul.* 4, 155–164.

Menge, B. A., and Sutherland, J. P. (1987). Community regulation: Variation in disturbance, competition, and predation in relation to environmental stress and recruitment. *Am. Nat.* 130, 730–757.

Morin, P. J. (1981). Predatory salamanders reverse the outcome of competition among three species of anuran tadpoles. *Science* 212, 1284–1286.

Morin, P. J. (1983a). Predation, competition, and the composition of larval anuran guilds. *Ecol. Monogr.* 53, 119–138.

Morin, P. J. (1983b). Competitive and predatory interactions in natural and experimental populations of *Notophthalmus viridescens dorsalis* and *Ambystoma tigrinum*. *Copeia*, pp. 628–639.

Morin, P. J. (1986). Interactions between intraspecific competition and predation in an amphibian predator-prey system. *Ecology* 67, 713–720.

Morin, P. J. (1987). Predation, breeding asynchrony, and the outcome of competition among treefrog tadpoles. *Ecology* 68, 675–683.

Newman, R. A. (1987). Effects of density and predation on *Scaphiopus couchi* tadpoles in desert ponds. *Oecologia* 71, 301–307.

Newman, R. A. (1989). Developmental plasticity of *Scaphiopus couchii* tadpoles in an unpredictable environment. *Ecology* 70, 1775–1787.

Odum, E. P., Finn, J. T., and Franz, E. H. (1979). Perturbation theory and the subsidy-stress gradient. *BioScience* 29, 349–352.

Packer, W. C. (1960). Bioclimatic influences on the breeding migration of *Taricha rivularis*. *Ecology* 41, 509–517.

Pechmann, J. H. K. (1994). Population regulation in complex life cycles: Aquatic and terrestrial density-dependence in pond-breeding amphibians. Ph.D. Dissertation, Duke University, Durham, NC.

Pechmann, J. H. K., Scott, D. E., Gibbons, J. W., and Semlitsch, R. D. (1989). Influence of wetland hydroperiod on diversity and abundance of metamorphosing juvenile amphibians. *Wetlands Ecol. Manage.* 1, 3–11.

Pechmann, J. H. K., Scott, D. E., Semlitsch, R. D., Caldwell, J. P., Vitt, L. J., and Gibbons, J. W. (1991). Declining amphibian populations: The problem of separating human impacts from natural fluctuations. *Science* 253, 892–895.

Petranka, J. W. (1989). Density-dependent growth and survival of larval *Ambystoma*: Evidence from whole-pond manipulations. *Ecology* 70, 1752–1767.

Petranka, J. W., and Sih, A. (1986). Environmental instability, competition, and density-dependent growth and survivorship of a stream-dwelling salamander. *Ecology* 67, 729–736.

Pfennig, D. W. (1990). The adaptive significance of an environmentally-cued developmental switch in an anuran tadpole. *Oecologia* 85, 101–107.

Pulliam, H. R. (1988). Sources, sinks, and population regulation. *Am. Nat.* 132, 652–661.

Ricklefs, R. E., and Schluter, D. (1993). "Species Diversity in Ecological Communities: Historical and Geographical Perspectives." Univ. of Chicago Press, Chicago.

Ross, T. E. (1987). A comprehensive bibliography of the Carolina bays literature. *J. Elisha Mitchell Sci. Soc.* 103, 28–42.

SAS Institute, Inc. (1990). "SAS/STAT User's Guide," Version 6.03, 4th ed. SAS Institute Inc., Cary, NC.

Scott, D. E. (1990). Effects of larval density in *Ambystoma opacum*: An experiment in large-scale field enclosures. *Ecology* 71, 296–306.

Scott, D. E. (1994). The effect of larval density on adult demographic traits in *Ambystoma opacum*. *Ecology* 75, 1383–1396.

Semlitsch, R. D. (1983). Structure and dynamics of two breeding populations of the eastern tiger salamander, *Ambystoma tigrinum*. *Copeia*, pp. 608–616.

Semlitsch, R. D. (1985a). Analysis of climatic factors influencing migrations of the salamander *Ambystoma talpoideum*. *Copeia*, pp. 477–489.

Semlitsch, R. D. (1985b). Reproductive strategy of a facultatively paedomorphic salamander *Ambystoma talpoideum*. *Oecologia* 65, 305–313.

Semlitsch, R. D. (1987). Relationship of pond drying to the reproductive success of the salamander *Ambystoma talpoideum*. *Copeia*, pp. 61–69.

Semlitsch, R. D., and Pechmann, J. H. K. (1985). Diel pattern of migratory activity for several species of pond-breeding salamanders. *Copeia*, pp. 86–91.

Semlitsch, R. D., and Walls, S. C. (1993). Competition in two species of larval salamanders: A test of geographic variation in competitive ability. *Copeia*, pp. 587–595.

Semlitsch, R. D., Scott, D. E., and Pechmann, J. H. K. (1988). Time and size at metamorphosis related to adult fitness in *Ambystoma talpoideum*. *Ecology* 69, 184–192.

Sharitz, R. R., and Gibbons, J. W. (1982). "The Ecology of Southeastern Shrub Bogs (Pocosins) and Carolina Bays: A Community Profile," FWS/OBS-82/04. U.S. Dept. of the Interior, Fish Wildl. Serv., Washington, DC.

Shoop, C. R. (1974). Yearly variation in larval survival of *Ambystoma maculatum*. *Ecology* 55, 440–444.

Sinsch, U. (1990). Migration and orientation in anuran amphibians. *Ethol. Ecol. Evol.* 2, 65–79.

Skelly, D. K. (1995). A behavioral trade-off and its consequence for the distribution of *Pseudacris* treefrog larvae. *Ecology* 76, 150–164.

Smith, D. C. (1983). Factors controlling tadpole populations of the chorus frog (*Pseudacris triseriata*) on Isle Royale, Michigan. *Ecology* 64, 501–510.

Smith, D. C. (1987). Adult recruitment in chorus frogs: Effects of size and date at metamorphosis. *Ecology* 68, 344–350.

Southerland, M. T. (1986). Coexistence of three congeneric salamanders: The importance of habitat and body size. *Ecology* 67, 721–728.

Tevis, L., Jr. (1966). Unsuccessful breeding by desert toads (*Bufo punctatus*) at the limit of their ecological tolerance. *Ecology* 47, 766–775.

Tobin, J. (1958). Estimation of relationships for limited dependent variables. *Econometrica* 26, 24–36.

Travis, J., Keen, W. H., and Juiliana, J. (1985a). The effects of multiple factors on viability selection in *Hyla gratiosa* tadpoles. *Evolution (Lawrence, Kans.)* 39, 1087–1099.

Travis, J., Keen, W. H., and Juiliana, J. (1985b). The role of relative body size in a predator-prey relationship between dragonfly naiads and larval anurans. *Oikos* 45, 59–65.

Van Buskirk, J. (1988). Interactive effects of dragonfly predation in experimental pond communities. *Ecology* 69, 857–867.

Van Buskirk, J., and Smith, D. C. (1991). Density-dependent population regulation in a salamander. *Ecology* 72, 1747–1756.

Wake, D. B., and Morowitz, H. J. (1990). "Declining Amphibian Populations—A Global Phenomenon?" Report to Board on Biology, National Research Council on Workshop in Irvine, CA, 19–20 February 1990; reprinted in *Alytes* 9, 33–42 (1991).

Walters, B. (1975). Studies of interspecific predation within an amphibian community. *J. Herpetol.* 9, 267–279.

Werner, E. E., and Gilliam, J. F. (1984). The ontogenetic niche and species interactions in size structured populations. *Annu. Rev. Ecol. Syst.* 15, 393–425.

Werner, E. E., and McPeek, M. A. (1994). Direct and indirect effects of predators on two anuran species along an environmental gradient. *Ecology* 75, 1368–1382.

Wiest, J. A., Jr. (1982). Anuran succession at temporary ponds in a post oak-savanna region of Texas. *In* "Herpetological Communities" (N. J. Scott, Jr., ed.), pp. 39–47. U.S. Fish Wildl. Serv., Washington, DC.

Wilbur, H. M. (1972). Competition, predation, and the structure of the *Ambystoma-Rana sylvatica* community. *Ecology* 53, 3–21.

Wilbur, H. M. (1980). Complex life cycles. *Annu. Rev. Ecol. Syst.* 11, 67–93.

Wilbur, H. M. (1984). Complex life cycles and community organization in amphibians. *In* "A New Ecology: Novel Approaches to Interactive Systems" (P. W. Price, C. N. Slobodchikoff, and W. S. Gaud, eds.), pp. 195–224. Wiley, New York.

Wilbur, H. M. (1987). Regulation of structure in complex systems: Experimental temporary pond communities. *Ecology* 68, 1437–1452.

Wilbur, H. M. (1988). Interactions between growing predators and growing prey. *In* "Size-structured Populations" (B. Ebenman and L. Persson, eds.), pp. 157–172. Springer, Berlin.

Wilbur, H. M. (1996). Multi-stage life cycles. *In* "Spatial and Temporal Aspects of Population Processes" (O. E. Rhodes, Jr., R. K. Chesser, and M. H. Smith, eds.), pp. 75–108. Univ. of Chicago Press, Chicago.

Wilbur, H. M., and Collins, J. P. (1973). Ecological aspects of amphibian metamorphosis. *Science* 182, 1305–1314.

Wilbur, H. M., and Fauth, J. E. (1990). Experimental aquatic food webs: Interactions between two predators and two prey. *Am. Nat.* 135, 176–204.

Bird Communities

Role of the Sibling Species in the Dynamics of the Forest-Bird Communities in M'Passa (Northeastern Gabon)

A. BROSSET

Muséum National d'Histoire Naturelle, URA1183/CNRS, 91800 Brunoy, France

I. INTRODUCTION

Where the mature tropical rainforest is cleared, the original and relatively stable bird community is replaced by a dynamic new one. Studies of the ecological and evolutionary mechanisms involved in this replacement are few (Driscoll and Kikkawa, 1989; Lovejoy *et al.*, 1986; Terborgh and Weske, 1969; Terborgh *et al.*, 1990; Thiollay, 1986). This chapter is an attempt to describe one of these mechanisms, namely, the role of the sibling species in the repopulation by a new community of birds of a small deforested patch included within a huge block of primary forest in eastern Gabon.

One characteristic of the communities of organisms in lowland rainforests is the existence of series of closely related sympatric species. Species packing has been described in plants (Rogstad, 1990) and animals, especially in birds (Cody, 1980; Kikkawa *et al.*, 1980; Haffer, 1992). Tropical Africa seems to be a region where groups of monophyletic sympatric species are particularly numerous; for example, Haffer (1992) lists 152 parapatric groups of Afrotropical birds that encompass some 375 species. During our 20-year study in the 2-km² M'Passa site, we recorded 55 pairs of sibling species of birds; most of which were not cited in Haffer's list.

However, birds are not the only vertebrates forming sibling species pairs in M'Passa; the phenomenon is also recorded in rodents (squirrels of the genus *Funisciurus*), insectivores (shrews of the genus *Crocidura*), bats (genera *Nycteris* and *Hipposideros*), and even forest-dwelling cyprinodont fishes of the genus *Diapteron* (Brosset, 1982a, 1988; ECOTROP, 1979). Clearly M'Passa is a highly favorable site for the study of closely related sympatric species, considered here as sibling species.

The main questions approached in this paper are the following: What do we mean by the term "sibling species," and how do we ascertain whether two allied species are sibling species? Are the species of sibling pairs spatially isolated in M'Passa? What is the ecological niche of each member of a sibling species pair in the primitive rainforest and in the nearby modified habitats? What are the ecological and behavioral isolating mechanisms that might reduce interspecific aggression and prevent hybridization between sibling species in their contact zone? Are there consistent differences between the species of sibling pairs that show general trends in the divergent evolution of sibling species? And finally, what are the historical aspects of sibling species formation in West Africa? The consequences of patchy deforestation of the rainforest, by permitting contact of sibling species that were formerly isolated, are discussed in the conclusion from the point of view of species richness and conservation.

II. STUDY SITE, AVIFAUNA, METHODOLOGY, AND SOME DEFINITIONS

A detailed description of the Ivindo Basin and of the M'Passa study area may be found in Brosset and Erard (1986) and in UNESCO (1987). The birds of the study area, which includes the M'Passa campus and the surrounding forest, were first investigated by myself between 1964 and 1968, when the site was entirely covered with mature rainforest. Between 1969 and 1970, a plot of about 10 ha, on the border of the large Ivindo River, was deforested to establish the campus; partial deforestation covered another approximately 20 ha. A dirt road 6 km long with deforested borders about 40 m wide was established between M'Passa and the large deforested area of the Makokou village; this road was an access route for birds of secondary habitat that invaded cleared areas of M'Passa after 1970.

From 1970 to 1983, the site was investigated by a dozen professional ornithologists of different nationalities. Four of them spent more than 1 year in the area, and between 1963 and 1982 I spent 47 months at M'Passa, spread over 22 visits. Special mention must be made of C. Erard, who spent 3–4 consecutive months each year from 1972 to 1985 in the Makokou–M'Passa area and who has contributed a considerable amount of original data to the present work.

The accuracy of our observations on M'Passa birds was easily validated, thanks to the utilization of different methods by different observers. These methods and their results are given in Brosset and Erard (1986), with a summary in Brosset (1990). Methods used to measure changes in the avifauna composition included the following: daily records of the birds observed in the surveyed habitats; collection of 1400 specimens of 314 species, 1963 to 1966; ringing (banding) in M'Passa of 4017 birds, 1971 to 1982; and identification and observation of 1623 nests of 214 noncolonial species. Further, films and photos of 123 species were obtained by A. Devez in the field, and most of the vocalizations of the birds present in M'Passa were recorded by C. Chappuis and C. Erard (Chappuis, 1974, 1981; Brosset and Erard, 1986).

A. What Are Sibling Species?

Before going further, it is necessary to define precisely the term sibling species. Considering the fact that many pairs or groups of closely allied species are represented at the regional scale, several authors have discussed this issue, mostly from the perspective of speciation processes. Due to the complexity of the subject, great difficulties have arisen in terminology. Groups of closely allied, monophyletic species were termed superspecies by Mayr (1942), following the "Artenkreis" used by Rensch (1929) for similar assemblages. Genermont and Lamotte (1980) later introduced the term supraspecies, Amadon (1956) the term allospecies, Smith (1955, 1965) the term parapatric species, Prigogine (1984a,b) the term paraspecies, Cain (1954) the term species-group, Short (1969, 1972) the term semispecies, and Amadon and Short (1976) the term megaspecies (see revision of Haffer, 1992, and Prigogine, 1984a,b).

All these terms apply only in part to the phenomenon we describe here. Some of the definitions include comparison with species that are too distantly related or so closely related that no one can be sure that they are really distinct species;

some include the existence of hybrids at the contact zone, which we have not observed in the several thousand specimens examined by my colleagues and myself at M'Passa. Most studies of species groups consider pattern on a regional scale broader than that of our study, which was localized within a very small area. The definition of sibling species most suitable in the present situation is closest to the definition of Mayr (1963, 1980): sibling species are morphologically similar or identical populations that are reproductively isolated. Nevertheless, Mayr's definition, based as it is on the biological species concept, does not take into account the allopatric, parapatric, and sympatric components of spatial pattern (Hall and Moreau, 1970). Here we define sibling species as morphologically similar populations that are reproductively isolated and that are sympatric and sometimes syntopic, in contact well within the limits of their respective geographic ranges.

B. How Are Two Allied Species Ascertained to Be Sibling?

Species were considered to be sibling pairs on the basis of strong morphological similarity. But despite this similarity, the two species forming a pair are not conspecific, and differences in morphology, ecology, and behavior remain as constant and reliable indications of species' identity within the M'Passa populations. The specific status of each sibling species is unanimously recognized by taxonomists (e.g., Brown et al., 1982; Urban et al., 1986; Fry et al., 1988; Keith et al., 1992); the single exception concerns two described closely allied species of Criniger (White, 1956; Rand et al., 1959), for which specific status is not clearly established in the M'Passa populations (Brosset and Erard, 1986).

We are aware of the limits and shortcomings of a sibling species criterion that takes into consideration only visually apparent aspects of morphology. However, since congruence of morphological similarity and molecular relatedness is a matter of conjecture for these species pairs, comparative morphology is the only method available to classify pairs of closely allied species as sibling species.

To eliminate arbitrariness from the morphological similarity concept (Johnson, 1963), morphological comparisons must use characters that lend themselves to quantification. In this chapter, such characters are the following: average weight and wing length (from our files or, when lacking there, from The Birds of Africa handbook); shape, especially of the tail, feet, and bill (from collected specimens); plumage patterns of males (with percentage similarity estimated by measurement of similar vs dissimilar areas from a side view of specimens); and distinctive shade of coloration (with percentage similarity obtained by the same method as above). The average of these five similarity percentages gives an overall percentage similarity of morphology, which is rounded to the nearest 5%.

C. Are Sibling Species Spatially Isolated?
Parapatry, Sympatry, and Syntopy

The methods utilized to ascertain the locations of individual birds in M'Passa during this 20-year study have produced sufficient numbers of observations to permit a precise mapping of the spatial distributions of sibling species. Comparisons between species pairs show that their spatial disposition is complex and varies with species and habitat. No general pattern may be discerned from our data; only broad categories may be classified, one of which applies in the majority of cases.

In these categories, some pairs of sibling species consist of purely mature-forest dwellers and others of purely secondary vegetation dwellers, in each case with or without vertical or horizontal spatial segregation. These species are here considered as more or less syntopic. In the other species pairs—in fact in the majority—one species occupies the mature rainforest and the other the deforested area. Such species meet along the ecotone in a narrow transition zone and are considered sympatric but not syntopic. This difference is fundamental to understanding the historical and ecological relationships between the species forming sibling species pairs, and the issue is examined in detail.

III. THE M'PASSA AVIFAUNA AND SIBLING SPECIES PAIRS

According to the long-term results of this study, the composition of the M'Passa avifauna was as follows. The resident species of the primary forest numbered 175; in the deforested part of the area, they were replaced by 138 resident newcomers and 51 seasonal migrants, mostly Palaearctic. Thus a total of 364 bird species were seen in and around the campus between 1964 and 1985. Within the 50,000-km^2 Ivindo Basin, the biogeographical region in which M'Passa is situated, the number of recorded bird species was 424.

Sibling species pairs numbered 15 (30 species) in the mature forest and increased to 55 after various cleared and successional habitats were represented in and around the M'Passa campus after deforestation. Table I lists 36 pairs of sibling species at M'Passa in which morphological similarity lies between 80 and 95%. Table II lists another 18 species pairs of which 7 have a percentage similarity of only 75%; these pairs are listed separately because, being migratory or

TABLE I Morphological Similarity in Sibling Species

Sibling species	Weight		Wing length		General morphology	Patterns of males	Coloration	General % of similarity
Sarothura rufa	76	90%	34	90%	95%	80%	80%	85
Sarothura boehmi	81		37					
Tauracus persa	185	90%	270	95%	95%	90%	75%	90
Tauracus macrorynchos	174		268					
Cercococcyx mechowi	137	80%	270	95%	95%	100%	95%	95
Cercococcyx olivinus	145		268					
Centropus anselli	188	80%	270	80%	90%	80%	80%	80
Centropus monachus	177		256					
Apaloderma narina	132	80%	63	85%	95%	95%	100%	90
Apaloderma aequatorialis	121		71					
Alcedo cristata	55	100%	15	80%	95%	95%	95%	95
Alcedo leucogaster	55		17					
Merops gularis	92	85%	31	70%	90%	75%	70%	80
Merops muelleri	85		23					
Gymnobucco calvus	88	90%	60	85%	90%	90%	85%	90
Gymnobucco peli	82		53					
Pogoniulus bilineatus	52	95%	13	80%	95%	95%	95%	95
Pogoniulus subsulfureus	49		11					
Indicator exilis	70	95%	19	95%	90%	85%	90%	90
Indicator willcoksi	72		20					
Smithornis capensis	73	90%	21	95%	95%	80%	80%	90
Smithornis rufolateralis	67		22					
Psalidoprocne nitens	96	95%	10	90%	70%	90%	65%	80
Psalidoprocne pristoptera	98		11					

Taxon							
Dryoscopus gambensis	78	95%		70%	85%	90%	80
Dryoscopus sabini	80						
Oriolus brachyrhynchus	110	90%		95%	90%	90%	90
Oriolus nigripennis	121						
Dicrurus atripennis	114	90%	41	75%	80%	95%	80%	80
Dicrurus adsimilis	126		53				
Andropadus gracilis	72	95%	23	85%	95%	95%	95%	95
Andropadus ansorgei	75		20				
Andropadus virens	81	95%	22	80%	90%	85%	95%	90
Andropadus latirostris	84		27				
Andropadus gracilirostris	76	100%	30	90%	95%	95%	95%	95
Andropadus curvirostris	77		26				
Baepogon indicator	100	100%		95%	90%	90%	95
Baepogon clamans	101						
Phyllastrephus icterinus	75	90%	23	80%	100%	95%	95%	90
Phyllastrephus xavieri	79		28				
Alethe castanea	88	90%	32	85%	95%	85%	85%	90
Alethe poliocephala	93		37				
Neocossyphus rufus	117	90%	63	75%	95%	75%	75%	80
Neocossyphus poensis	105		50				
Geokichla princei	104	90%	61	70%	95%	95%	80%	85
Geokichla cameronensis	95		45				
Malacocincha fulvescens	73	95%	28	95%	95%	95%	95%	95
Malacocincha rufipennis	72		27				
Camaroptera chloronota	52	100%	13	95%	90%	90%	80%	90
Camaroptera brevicauda	54		12				

continues

TABLE I (Continued)

Sibling species	Weight	Wing length	General Morphology	Patterns of males	Coloration	General % of similarity
Macrosphenus concolor	59	14	95%	95%	80%	95
Macrosphenus flavicans	59	14				
	100%	100%				
Muscicapa seth-smithi	57		95%	95%	85%	95
Muscicapa epulata	57					
	100%					
Batis poensis	50		90%	90%	95%	95
Batis minima	48					
	95%					
Myoparus griseigularis	66	14	90%	90%	90%	95
Myoparus plumbeum	67	14				
	95%	100%				
Diaphorophia castanea	59	13	90%	95%	95%	90
Diaphorophia tonsa	54	12				
	90%	90%				
Tchitrea viridis	78	14	80%	70%	60%	80
Tchitrea batesi	75	15				
	95%	95%				
Anthreptes seimundi	54		90%	95%	90%	90
Anthrepes batesi	51					
	95%					
Nectarinia chloropygia	49	7	95%	90%	95%	90
Nectarinia minulla	46	6				
	95%	85%				
Nicator chloris	100	45	95%	95%	95%	80
Nicator vireo	75	28				
	75%	50%				
Malimbus racheliae	80	32%	90%	75%	80%	90
Malimbus cassini	91	36				
	90%	90%				
Lonchura poensis	49		95%	75%	80%	90
Lonchura cucullata	49					
	100%					

TABLE II   Percentage of Similarity in Migratory or Rare Sibling Species

Sibling species		% similarity
Lampribis rara	L. olivacea	75
Accipiter tousseneli	A. castanilius	75
Falco naummani PM	F. tinnunculus PM	80
Sorathrura elegans	S. pulchra	85
Turtur tympanistria	T. afer	75
Apus apus PM	A. pallidus PM	90
Ceyx picta	C. lecontei	80
Halycon senegalensis	H. malimbicus	80
Bycanistes cylindricus	B. subcylindricus	75
Hirundo semirufa	H. senegalensis	80
Dryoscopus senegalensis	Chlorophoneus bocagei	75
Acrocephalus scirpaceus PM	A. baeticatus	90
Phylloscopus trochilus PM	P. sibilatrix PM	95
Fraseria ocreata	F. cinerascens	85
Malimbus rubricollis	M. malimbicus	75
Pirenestes ostrinus	P. rothschildi	90
Nigrita canicapilla	N. luteifrons	75
Ploceus preussi	P. dorsomaculatus	90

Note: PM, Palaearctic migrant.

rare, they were less intensively studied. One species pair, *Criniger calurus* and *C. ndussumensis*, is not included in these tables because the separate specific status of *C. ndussumenis* is not clearly apparent in the M'Passa population (see previous discussion).

A. Pairs of Partially or Totally Syntopic Species

Syntopic sibling species pairs occur both in the rainforest and in modified habitats. For some pairs, the syntopy is complete, and no spatial segregation between the two species has been recorded. Other pairs—the majority—exhibit horizontal or vertical segregation inside the rainforest or within the deforested area.

1. Sibling Species Inhabiting the Mature Rainforest in Complete Syntopy

Examples of syntopic rainforest species are illustrated in Fig. 1; three pairs of sibling species of the genera *Trichastoma, Malimbus,* and *Neocossyphus* belong to this category. No spatial segregation was noticed in two pairs of very closely

FIGURE 1. Pairs of syntopic sibling species with no spatial segregation. (Top right) *Trichastoma fulvescens* and *T. rufipennis*; (top left) *Malimbus cassini* and *M. racheliae*; (bottom) *Neocossyphus rufus* and *N. poensis*.

allied species of Akalas living in the undergrowth of the mature forest. These babblers, *Trichastoma rufipennis* and *T. fulvescens,* have such similar morphology, coloration patterns, and songs that the species are difficult to separate in the field and even in the hand. No definite spatial or ecological segregation between the two was detected, and both were encountered together in the same mixed parties following swarms of army ants (Brosset, 1969). The only obvious differences concern the nests, which have different structures and positions, and the egg colors, which are bright blue in *T. rufipennis* and white blotched reddish in *T. fulvescens*. Differences in egg coloration may be a response to heavy parasitism by several species of cuckoo (Brosset, 1976; Brosset and Erard, 1974).

In sibling species of thrushes *Neocossyphus*, the habitats, general ecology, and behavior also seem to be the same; both are regular components of mixed parties following army ants, as are the sibling species of *Trichastoma*.

Malimbus cassini and *M. racheliae* are similar species, distinguished by the color of the male breast: red for *M. cassini* and mostly yellow for *M. racheliae*.

Both are regular components of the mixed species associations that exploit the foliage of the high canopy in mature forest. The foraging behavior and the composition of the social group (one female with several males) is the same in both species. The two species differ in the location of their nests, which are built in different microhabitats. The nests of *M. cassini* (25 observed) invariably hang from the tops of palm trees, *Calamus* or *Ancistrophylum,* in a swampy area, while the nests of *M. racheliae* (3 observed) hang from the underside of the high canopy in the driest parts of the M'Passa plateau (Brosset, 1978).

2. Pairs of Strictly Rainforest Species with Vertical Segregation

In four pairs of sibling species common in the rainforest around the M'Passa station, one species is restricted to the canopy, and the other lives below in the undergrowth or in the middle stratum (see Fig. 2). This vertical stratification of spatial niches was observed in sibling species of forest bulbuls of the genera *Andropadus* and *Phyllastrephus.* Subtle differences in coloration and size separate *Phyllastrephus icterinus* from *P. xavieri,* but as male *P. icterinus* are the same size as female *P. xavieri,* only vocalizations allow one to distinguish the two species in the field. Localization by song shows *P. icterinus* to be an undergrowth dweller, usually encountered between 0 and 6 m high, with *P. xavieri* living above, between 8 and 25 m high. *Andropadus gracilirostris* and *A. curvirostris,* morphologically very similar, share the mature forest habitat in the same way as the *Phyllastrephus* siblings, with *A. curvirostris* below and *A. gracilirostris* above. The same vertical stratification of spatial niches also exists in the warblers *Macrosphenus flavicans* and *M. concolor.* These sibling species inhabit separate strata of vegetation with *M. concolor* in the canopy and *M. flavicans* below (Brosset and Erard, 1986). One of the most remarkable cases of vertical stratification in a pair of sibling species has been described in detail by Erard (1987, 1990) in forest flycatchers of the genus *Dyaphorophyia. Dyaphorophyia castanea* and *D. tonsa* are very closely allied species, both of which are abundant and regularly distributed in the primary forest; their vertical segregation is clear-cut, with *D. tonsa* in the high canopy and *D. castanea* in the undergrowth and middle strata.

3. Pairs of Purely Rainforest Dwellers with Horizontal Segregation

This category refers to sibling species pairs in which species are spatially segregated in different habitats inside the mature forest. The main axis of segregation is between habitats on the M'Passa plateau and those on the slopes and riverine habitats along the Ivindo River. This is the case for the four pairs

FIGURE 2. Pairs of syntopic sibling species vertically segregated in the canopy (G) and in the undergrowth (H). (Middle right) *Macrosphenus flavicans* (G) and *M. concolor* (H); (middle left) *Andropadus gracilis* (G) and *A. ansorgei* (H); (bottom right) *Dyaphorophyia tonsa* (G) and *D. castanea* (H); (bottom left) *Phyllastrephus xavieri* (G) and *P. icterinus* (H).

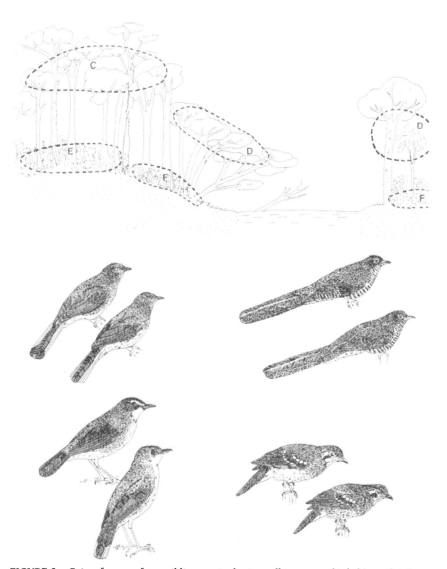

FIGURE 3. Pairs of mature forest sibling species horizontally segregated inhabiting the plateau forest canopy (C) or the undergrowth (E) or the slopes and riverine habitat in the canopy (D) or the undergrowth (F). (Middle right) *Cercococcyx olivinus* (C) and *C. mechowi* (D); (middle left) *Baepogon indicator* (C) and *B. clamans* (D); (bottom right) *Geokichla princei* (E) and *G. cameronensis* (F); (bottom left) *Alethe castanea* (E) and *A. poliocephala* (F).

of sibling species in the genera *Geokichla, Alethe, Cercococcyx,* and *Baepogon* (Fig. 3).

The pair of ground thrushes *Geokichla princei* and *G. cameronensis* closely resemble each other in morphology, pattern of plumage, and behavior, but *G. princei* is invariably found on the plateau (where 15 nests were located) and the rarer *G. cameronensis* (8 captures) was observed moving about in pairs only in riverine habitat (the nest of this species is unknown). The thrushes *Alethe castanea* (283 captures) and *A. poliocephala* (50 captures) show a similar type of spatial segregation. *Alethe castanea* is an ant follower and inhabits the plateau where its territorial songs are heard and where eight nests were found. Occasionally *A. poliocephala* was observed following the same army ant swarms as *A. castanea,* but its song perches and nests were exclusively in riverine habitats. The honeyguide bulbuls *Baepogon indicator* and *B. clamans* show the same type of localization; the former is a plateau bird, the latter a species of the riverine habitats where it nests (Erard, 1977). These two species are separable only by the pattern of the tail and by the color of the males' eyes (white in *B. indicator* and red in *B. clamans*). The two cuckoos *Cercococcyx mechowi* and *C. olivinus* are so similar that even in the hand their specific identities are difficult to establish. They are shy and secretive, and they cannot be separated in the field except by the songs, which are far-reaching and frequently heard at night. Each cuckoo has two types of songs, one similar in both species (of the *Cuculus solitarius* type) and the other quite different and species specific (Chappuis, 1974). *Cercococcyx olivinus* were heard on the plateau and *C. mechowi* in riverine forest.

Some species pairs of closely allied flycatchers, *Muscicapa cassini* and *M. sethsmithi,* and *Fraseria ocreata* and *F. cinerascens,* show a similar type of spatial segregation, with one in the plateau and the other in the riverine habitat (Erard, 1987, 1990).

4. Pairs of Syntopic Sibling Species of Exclusively Modified Habitats

Species pairs in this category are newcomers and few, the residents numbering only three (Fig. 4). Two pairs are strictly syntopic in a few hectares of the grassy area regularly cleared in the center of the campus. They are the flufftails *Sorathrura rufa* and *S. boehmi* and the mannikins *Spermestes* (*Lonchura*) *cucullata* and *S. poensis.* In these two pairs the morphology, pattern of coloration, and general way of life are very similar; they are equally abundant in the campus, where they live side by side without visible spatial segregation.

The two sunbirds *Nectarinia chloropygia* and *N. minulla* are differentiated only by a narrow bar of blue on the lower breast and a slightly smaller size in *N. minulla.* Both are found in the cleared campus area, where they display interspecific territoriality. Although they live side by side, their nest locations are

FIGURE 4. Pairs of syntopic sibling species, both spatially restricted to anthropogenic habitats. (Top) *Nectarinia chloropygia* (B) and *N. minulla* (E); (bottom left) *Sorathrura rufa* (A) and *S. boehmi* (A); (bottom right) *Spermestes (Lonchura) poensis* (B) and *S. cucullata* (B).

nevertheless distinct, with those of *N. chloropygia* built within the campus, mostly in the ornamental vegetation, and those of *N. minulla* in the natural regrowths along the ecotone forest open area (Brosset and Erard, 1986).

B. Pairs of Sibling Species in Which One Is a Strictly Rainforest Species and the Other a Species Purely of Secondary Habitats

Twenty pairs of sibling species belong to this category. In these pairs, the rainforest species (species A) is the original inhabitant of the whole M'Passa area prior to deforestation, and the second (species B) is a new arrival which has replaced species A in cleared areas. In nine cases the territories of the newcomer species B abut those of species A at the limit of deforestation around the campus. But in 11 other cases a more or less deep penetration of species B exists along the border into the territories of species A.

1. Cases in Which Species A and B Show Abutting Territories without Overlap

In these sibling pairs the ecological segregation is complete, and the interspecific territorial limits are clear-cut; to this category belong sibling species of cuckoos, kingfishers, bee-eaters, barbets, broadbills, orioles, shrikes, and flycatchers.

Two *Centropus* cuckoos are found at M'Passa (Fig. 5), *C. anselli* and *C. monachus.* The two species show some differences in size and pattern of coloration, although their general morphology, vocalizations, and behavior are similar. The two species are spatially segregated. *Centropus anselli* is common and widespread in the rainforest with a patchy distribution in dense, low undergrowth. *Centropus monachus* is common in open secondary vegetation, especially in bushes scattered throughout recently deforested areas. The blue bee-eaters *Merops gularis* and *M. muelleri* are two species closely related by morphology, color pattern, and behavior; the former typifies old plantations and partially deforested areas with scattered big trees, while the latter is characteristic of the pristine mature rainforest. The broadbills *Smithornis capensis* and *S. rufogularis* are sibling species that are very close to each other in all respects; *S. capensis* is a resident of secondary habitats and is strictly segregated spatially from *S. rufogularis,* which is common only in the rainforest. The small and closely allied barbets *Pogoniulus subsulfureus* and *P. bilineatus* sing monotonous songs of similar quality but differing in rhythm, a feature that permits their differentiation in the field. *Pogoniulus bilineatus* was exclusively identified in the large patches of deforestation, especially around the villages, with *P. subsulfureus* common in mature rainforest and old secondary forests (see Fig. 5).

The kingfishers *Alcedo cristata* and *A. leucogaster* are very similar in all morphological aspects; the former remains localized on the border of the rivers and pools, while the latter is an inhabitant of low vegetation in both primary and old secondary forest. The two species were never seen, or captured, side by side (Fig. 6).

Two local orioles, *Oriolus brachyrhynchus* and *O. nigripennis,* morphologically very similar but easily separable by the song, are also spatially segregated in M'Passa, *O. brachyrhynchus* being in the mature forest and *O. nigripennis* in the high secondary trees (*Musanga* spp.) bordering the dirt roads and deforested areas (Fig. 7). In the two sibling shrikes *Dryoscopus gambiensis* and *D.*

FIGURE 5.   Examples of pairs of sibling species in which one species is restricted to the mature forest (1 on the map of M'Passa) and the other restricted to the deforested area (2 on the map). These species are cleanly segregated by habitat without spatial interdigitation. (Middle right) *Merops gularis* (2) and *M. muelleri* (1); (middle left) *Centropus anselli* (1) and *C. monachus* (2); (bottom right) *Smithornis rufogularis* (1) and *S. capensis* (2); (bottom left) *Pogoniulus subsulfureus* (1) and *P. bilineatus* (2).

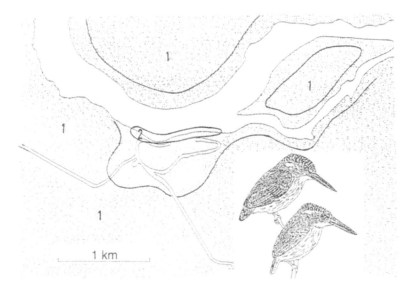

FIGURE 6. Sibling species of kingfishers in which one, *Alcedo cristata* (top), is spatially restricted to the riverine open habitat (2 on the map), and the other, *Alcedo leucogaster* (bottom), extends into the forest (1 on the map).

sabini, the males are similarly sized and have convergently similar patterns of coloration. However, the bill is heavier in *D. sabini*, and the female coloration patterns differ between species. *Dryoscopus gambiensis* is a purely secondary habitat dweller, while *D. sabini* is exclusively met in the canopy of the high mature rainforest (Fig. 7).

In flycatchers, a difficult group but well known thanks to the detailed study of Erard (1987, 1990), the same type of ecological segregation has been described in the pairs of closely allied species: *Batis poensis* and *B. minima*, *Muscicapa epulata* and *M. seth-smithi* (Fig. 7). *Myoparus griseigularis* and *M. plumbeum* along with *Tchitrea viridis* and *T. batesi* are further examples of similar species pairs, with one species restricted to the primary forest and the other to secondary vegetation.

2. Cases in Which Territories of Species A and B Abut with Some Range Overlap along the Primary Forest/Secondary Forest Ecotone

In another 10 pairs of sibling species, territorial exclusion and a clear-cut segregation of spatial ranges are less pronounced. At the limits of species' habitat ranges, territories may be located within the habitat typical of the second species to an extent that varies with the species involved. These incursions are more usually of the type that species B invades the habitat of species A, but the opposite does occur in certain pairs. This category of spatial organization has been observed in sibling species of tauracos, trogons, barbets, honeyguides, swallows, drongos, bulbuls, warblers, and sunbirds.

The pair of green tauracos *Tauracus persa* and *T. macrorynchos* have similar appearance and behavior, but are distinguished by shades of coloration, differently disposed patches of white and red on the head, and subtle differences in

FIGURE 7. Examples of sibling species pairs in which one species occupies rainforest (A), the other deforested areas (B). The territories abut, without spatial overlap. (Middle right) *Oriolus nigripennis* (B) and *O. brachyrynchus* (A); (middle left) *Dryoscopus sabini* (A) and *D. gambiensis* (B); (bottom right) *Batis poensis* (B) and *B. minima* (A); (bottom left) *Muscicapa epulata* (B) and *M. seth-smithi* (A).

the first notes of the series "ko-ko-ko" composing the song. *Tauracus persa* is a bird of the secondary vegetation, most common wherever the old parasol trees *Musanga* are dominant, the nest being built in dense bushes along the ecotones at the contact between clearing and forest. *Tauracus macrorynchos* is a rainforest touraco, nesting in the dense foliage of trees within the forest. *Tauracus persa* is a dynamic species, entering the forest more or less deeply along the ecotone where it meets its sibling *T. macrorynchos* (see Fig. 8). Strong interspecific competition occurs between the two, each invariably and immediately responding to the song of the other; details concerning the territorial conflicts between these sibling species may be found in Decoux and Erard (1988).

Like in the tauracos, the sibling species pair of trogons *Apaloderma* comprises one species (*A. narina*) living in modified habitats and another (*A. aequatorialis*) living in the darker and thicker part of middle strata in the mature primary forest (Fig. 8). The latter is not met outside its specific habitat, but *A. narina* is an eclectic bird, common in fragmented landscapes and in natural or artificial openings within forests, on summits of hills, along dirt roads, on the borders of man-made deforested areas, and also wherever the canopy is broken by tornadoes. Both species are morphologically similar, distinguished by patches of naked skin in the faces that are green in *A. narina* and yellow in *A. aequatorialis*. The songs are species characteristic and very different from one another. During the breeding season, displaying males form unisexual groups in which both species may be represented, singing in the same group at the limit of their respective range (Brosset, 1983).

In barbets of the genus *Gymnobucco*, the three species common at M'Passa share very similar morphology and pattern of coloration. We consider here *G. calvus* and *G. peli* to be sibling species, possibly forming a sibling species trio with *G. bonapartei*. They differ from one another in minute details: the nasal bristly tufts, characteristics of the *Gymnobucco*, are up-oriented in *G. peli* and down-oriented in *G. calvus*. *Gymnobucco peli* penetrated into deep, mature forest, where *G. calvus* was not found. Yet all three *Gymnobucco* species could be seen nesting side by side in the same stumps along the mature forest/secondary forest ecotone, and so the spatial segregation between these sibling species is neither continuous nor complete (Fig. 8).

The honeyguides of the genus *Indicator* constitute a trio of closely allied sympatric species at M'Passa (Fig. 8). We consider here *I. exilis* and *I. willcocksi* to be sibling species distinguished only by details of the facial pattern; the third species, *I. minor*, is also quite similar. The biology of the *Indicator* species remains poorly known. The species *I. exilis* has been observed mostly in secondary habitats and *I. willcocksi* almost exclusively in the rainforest. The territories of honeyguides, if any, are probably not associated with spatial divisions of the landscape, nor with a particular type of forest, but rather with the presence of bee colonies.

FIGURE 8.   Examples of sibling species pairs in which one species is a primary forest inhabitant (1 on the map) and the other an inhabitant of modified forest (2 on the map). There is spatial overlap of territories at the limits of the respective habitat ranges. (Middle right) *Apaloderma narina* (2) and *A. aequatorialis* (1); (middle left) *Tauracus persa* (2) and *T. macrorynchos* (1); (bottom right) *Indicator willcocksi* (1) and *I. exillis* (2); (bottom left) *Gymnobucco peli* (1) and *G. calvus* (2).

FIGURE 9. Sibling species pairs involving primary forest versus modified forest inhabitants. (Middle right) *Andropadus gracilirostris* (A) and *A. curvirostris* (B); (middle left) *Dicrurus atripennis* (A) and *D. adsimilis* (B); (top right) *Anthreptes seimundi* (B) and *A. batesi* (A); (top left) *Camaroptera brevicauda* (B) and *C. chloronota* (A).

The two local species of saw-wing swallows *Psalidoprocne* at M'Passa appear different in the field due to the shape of the tail (short and square in *P. nitens*, forked in *P. pristoptera*). Otherwise, *P. nitens* and the local subspecies of *P. pristoptera*, *P. p. petiti*, are very similar in all respects, including morphologically and behaviorally. Generally their spatial niches are complementary, with *P. nitens* in the mature forest area and *P. pristoptera* in and around the secondary vegetation. Yet both species are sometimes found side by side, with some range overlap along the contact zone between the primary and secondary habitats.

Three species of drongos occur at M'Passa, and two of these, *Dicrurus atripennis* and *D. adsimilis*, may be considered sibling species. They are similar in most characters, with *D. adsimilis* being slightly larger with a more forked tail. *Dicrurus atripennis* is the classical drongo of the mature rainforest; *D. adsimilis* is a species of modified habitats. They meet in narrow zones along the dirt roads crossing the rainforest and in the ecotone bordering the clearings (Fig. 9). *Andropadus ansorgei* and *A. gracilis* are likewise sibling species that are morphologically very similar. In most areas they are spatially segregated, *A. ansorgei* in the canopy of primary rainforest and *A. gracilis* in the higher secondary vegetation. They have been observed singing close to each other in high canopy trees where the two habitats are adjacent.

Two sympatric and closely allied species of warblers, *Camaroptera chloronota* and *C. brevicauda*, are differentiated only by the shade of the plumage and by the longer tail of *C. brevicauda* (Fig. 9). Their loud songs are distinctive, and their nests are both of the "tailor bird" type. At M'Passa, *C. chloronota* nests in the large leaves of the Marantaceae in the forest undergrowth, while *C. brevicauda* nests typically in *Solanum* bushes in the clearings. They meet only on the border of the forest, where their songs can be heard at short distances from each other (Brosset and Erard, 1986).

The sunbirds *Anthreptes seimundi* and *A. batesi* are two similarly very small and dull species, separable in the field only by subtle differences in plumage shades and attitudes (Fig. 9). *Anthreptes batesi* is a bird of the primary forests, hunting in groups in the canopies of the highest trees, while *A. seimundi* is an inhabitant of the secondary forest. These sibling species are spatially segregated according to the classical scheme, with one species in the mature rainforest and the other in the modified habitats.

C. Some Special Cases of Spatial Segregation in Sibling Species

The case of the two extremely common forest bulbuls *Andropadus virens* and *A. latirostris* is a special one. Both species have been the subject of long-term, detailed studies (Brosset, 1981a,b,c, 1982b). Do these birds qualify as sibling species? The morphological similarities are great between the two. Young *A. latirostris* are distinguishable from *A. virens* only in that they are slightly larger. The adults are separable by the malar yellow "moustaches" of *A. latirostris*, which the birds can hide or exhibit according to the circumstances; this ornament is lacking in *virens* (Fig. 10). Both nesting and social behaviors are similar in the two species. It is the spatial distribution of these two birds that is unusual, in that it varies with the geographic locality. For example, in Bengoué 130 km east of M'Passa, *A. virens* is numerically dominant in the local bird community, occupying the whole area at high density, whereas *A. latirostris* is far less numerous and more local (Brosset, 1981a). The opposite occurs at M'Passa, where 791 *A. latirostris* were mist-netted in the whole area vs 114 *A. virens* mist-netted mostly in the ecotone, in riverine habitats, and in dense secondary vegetation. The problem is that both species are found side by side in all types of habitats close to or within the forest. As the numerical and spatial dominance of the two species varies according to the locality, at M'Passa *A. latirostris* would be species A and *A. virens* species B, but the contrary is true in Bengoué (Brosset and Erard, 1986). Their spatial segregation, if any, appears to be variable and dynamic.

Another atypical situation is that of two species of nicator, *Nicator chloris* and *N. vireo*. These two species are morphologically similar (Fig. 10), with *N. chloris* being larger. The songs, loud and musical, are different, but the rudimentary, flat nests are both of the same type (Brosset and Erard, 1976). Among the M'Passa avifauna, *N. chloris* is the most eclectic of all species: the bird and its nests have been observed in all types of habitats, from the middle of the

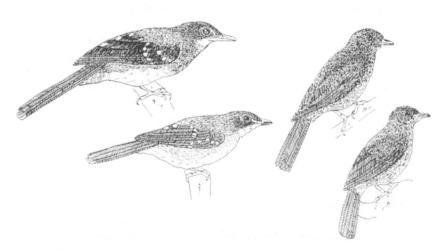

FIGURE 10. Special cases of questionable spatial segregation in sibling species involve (right) *Andropadus latirostris* and *A. virens* and (left) *Nicator chloris* and *N. vireo*.

mature forest to the ornamental bushes in the clearing of the M'Passa campus. The species *N. vireo* is restricted to the primary/secondary forest ecotone or to the thick vegetation of the galleries along the rivers; in these habitats, the two *Nicator* species live and nest side by side. No spatial segregation has been noticed in the areas where the two species co-occur.

IV. DISCUSSION

A frame of reference for the discussion that follows can be found in Haffer (1992), although Haffer's main concern was parapatric species. However, several questions raised by Haffer apply to this study: (a) What is the local distribution of the representative species forming a sibling pair? (b) In what manner is ecological and reproductive isolation maintained? (c) Why and how did the contact zone originate? (d) When and where did the sibling species originate?

A. What Are the Local Spatial Distributions of the Representative Species Forming a Pair?

In answer to this question, data from the M'Passa area show that the spatial disposition of the members of sibling species pairs is complex and varied, ranging from pairs of species living side by side in complete syntopy to others spatially segregated in distinct habitats. Yet other sibling species may be seen in very close spatial proximity, but occupying different subhabitats or compartments of the ecosystem and never coming into contact. Clear spatial subdivisions exist in some rainforest sibling species, which may be vertically or horizontally segregated. In a majority of the sibling pairs, one species is a rainforest dweller and the other a recent immigrant inhabiting only the deforested area.

B. How Is Isolation Maintained in Each Zone?

It is a matter of fact that reproductive isolation, especially sexual isolation, is firmly maintained regardless of the category of spatial overlap exhibited by the sibling species. During the course of the present study, several thousands of birds belonging to sibling species pairs were examined in the hand by competent ornithologists; no hybrid birds were detected. In the several hundred nests that were observed and photographed, all were attended by conspecific parents. Our data strongly and consistently suggest that sibling species do not hybridize in M'Passa; implicitly there exist efficient mechanisms for reproductive isolation, whether spatial, sexual, or behavioral. In what manner is the isolation between sibling species maintained? No doubt some ecological factors are involved, but environmental factors are probably not the most important proximal mechanism. Observations of interspecific territoriality strongly suggest that competition for space and food may exist between the sibling species. Nevertheless, examination and comparison of the use of visual signals exhibited in sexual displays suggest that spatial isolation may result from behavioral mechanisms important in the prevention of hybridization between sibling species. That is, reproductive isolation is assured by behavioral displays and by the morphological/behavioral attributes involved in such displays, even though ecological isolation may be indistinct and competition between sibling species severe.

1. Spatial Isolation by Interspecific Territorial Exclusion

Interspecific territoriality results in spatial segregation, but with a sharing of trophic niches, the advantage presumably going to the best-adapted species. This type of isolation may be detected by observations of interspecific territoriality (Murray, 1981). The territories of each species in the sibling pair may have common borders, along which territorial contests occur between the two; discussion of some well-documented contests at M'Passa may be found in Chappuis (1974, 1981), Brosset and Erard (1986), and Erard (1987). In most of the cases, the song of one species attracts the other, which responds by aggressive postures and vocalizations (Decoux and Erard, 1988).

2. Spatial Exclusion Resulting from Premating Isolation Mechanisms

Most of the distinctive morphological characters humans use to visually identify the two entities forming a sibling pair may be used in the same way by the birds for species recognition. In M'Passa, these morphological traits are various. Some of the characters that concern the patterns of the tail or the wings are visible from afar when exhibited. They are typical of species that display in relatively open places, such as the thrushes *Neocossyphus*, the drongos, the orioles, the bulbuls *Baepogon*, the honeyguides, and the mannakins *Spermestes*. A more common type of character concerns ornamentations of the head, which are highly species specific and probably utilized to recognize the correct partner at close quarters, especially useful in species displaying in thickets. In some pairs, both species have on the face patches of bare skin, brightly and differently colored according to the species. Such patches are exhibited in, e.g., ibises, tauracos, and trogons, and a most remarkable example is that of the four species of flycatchers *Dyaphorophyia* sympatric in the M'Passa area. The orbital wattles of bare skin are bright green in *D. concreta*, turquoise blue in *D. blisseti chalybea*, red in *D. castanea*, and purplish in *D. tonsa*. The brightness of these colored wattles increases when the birds are excited (Erard, 1987; A. Brosset, unpublished). In other sibling species, it is the plumage of the head that shows different colored patches. In this way *Andropadus gracilis* has a white orbital ring that its sibling *A. ansorgei* lacks. The thrush *Alethe castanea* possesses a white supercilium that is absent in *A. poliocephala*. In the species pair of *Baeopogon*, only the male *B. indicator* exhibits a white eye; the eyes of its female and of both sexes of *B. clamans* are dull red. This special characteristic of the male *B. indicator* may indicate that, in these two sibling syntopic species, female choice prevails. One can identify the very closely allied barbets *Gymnobucco calvus* and *G. peli* by examination of their bristly nasal tufts, which are up-oriented in *G. peli* and down-oriented in *G. calvus*. In *Andropadus latirostris*, the adult possesses bright yellow "moustaches" that the bird can reveal or conceal at will; not visible in nesting birds, the "moustaches" are exhibited in the lek by displaying birds (Brosset, 1982b). This ornament is absent in young *A. latirostris*, which is similar to *A. virens*.

In interspecific encounters, vocalizations are certainly important in specific recognition. The results of Chappuis' (1974, 1981) research suggest that the divergent evolution of vocalizations in sibling species is less pronounced in tropical birds than in those of the temperate areas. In the M'Passa sibling species, the response of one species to the call of the other seems to indicate that the acoustic signal is not sufficient to prevent aggression between the two species. At close quarters, visual signals seem of prime importance in the regulation of reproductive isolation.

It is generally assumed that morphological traits related to premating isolation mechanisms must be the result of intraspecific selection. These traits are used to recognise conspecific sexual partners. In the 110 sibling species recorded in M'Passa, 98 exhibit no sexual dimorphism: shape and color pattern are the same in the male and female. This fact suggests that intraspecific sexual recognition was not the only factor in the evolution of traits related to recognition. Their use by sibling species, to recognize each other as different species, suggests some form of interspecific pressure acting as reinforcement of these traits by the selection for morphological divergence where sibling species, formerly isolated, come into contact (Cody, 1969; Butlin, 1989).

3. Spatial Isolation Due to Early Ecological Imprinting

Many species of birds, when dispersing or migrating, choose stopping places with vegetation physiognomically similar to that of the area in which they were raised. A similar mechanism may explain also the localization of nests which, for a given species, may vary among allopatric populations having different social traditions. A similar attachment to natal habitat may result in the spatial segregation of allopatric populations in secondary contact. Early learning of locality, habitat, and socially acquired ecological traditions, in and around the nest, may contribute to the spatial isolation of incipient sibling species. This is particularly relevant to pairs that include migratory species, for instance, the kingfishers *Halcyon senegalensis* and *H. malimbicus*, the sedge-warblers *Acrocephalus scirpaceus* and *A. baeticatus,* and the leaf-warblers *Phylloscopus trochilus, P. sibilatrix,* and *P. bonelli.* Spatial isolation is achieved by these recent imigrants to M'Passa by their selecting habitats physiognomically closest to those of the natal habitats. Indeed, there is a conspicuous avoidance of the mature rainforest by migrants, which were never seen in M'Passa before the partial deforestation of the site (Brosset and Erard, 1986).

4. Other Morphological Differences Possibly Enhancing Spatial Segregation

Most of the mechanisms which maintain spatial segregation in sibling species are most likely related to reproductive isolation. As the structure of the habitats and their biological components are distinct in most of the cases, this spatial segregation would provide different ecological pressures and foster further divergent morphological adaptation in sibling species pairs. To test this notion, we used quantitative characters, average weights and wing lengths, to reveal potentially repetitive patterns in morphological differences between the sibling species, patterns indicative of constant trends in the divergent evolution of both (see Table I). Comparisons permit an assessment of the degree and trend of morphological differences and their possible relation to the spatial niches of each species and of whether general rules may be deduced from the results.

In pairs of species sharing the same primary forest habitat (genera *Neocossyphus* and *Malachocynchla*) and the same secondary vegetation (genera *Sorathrura, Spermestes,* and *Nectarinia*), three species pairs show no size differences and the other shows minimal differences. In species pairs inhabiting the primary forest with vertical segregation, similar size is observed in the sibling species of flycatchers (*Dyaphorophyia, Myoparum,* and *Tchitrea*). In the bulbuls, the canopy species *Andropadus gracilis* is smaller than the undergrowth species *Andropadus ansorgei;* but in *Phyllastrephus* it is the canopy species *P. xavieri* that is bigger and the undergrowth species *P. icterinus* that is smaller. In the primary forest sibling species showing horizontal spatial segregation (one species in the plateau, the other in riverine habitat), the cuckoos *Cercococcyx* both are similarly sized, in *Alethe* the riverine species *poliocephala* is bigger and the plateau species *castanea* is smaller, and in *Geockichla* the plateau species *princei* is bigger and the riverine species *Cameronensis* is smaller.

The category of sibling species pairs of species A, the mature forest inhabitant, and species B, restricted to modified habitats, is of special interest because such pairs probably have evolved in allopatry under different ecological regimes. The question then can be asked whether their different origins have resulted in divergent evolution in both size and weight. Our data rule out this possibility, as comparisons between such pairs show that within-pair differences are variously oriented among pairs. In most species pairs, these differences appear hardly significant (e.g., in *Tauracus, Apaloderma, Psalidoprocne, Pogonolius,* and *Alcedo*). In other cases, species A (forest species) is bigger (e.g., in *Centropus*), but species of *Merops* and *Dicrurus* show the opposite trend, with species A being the smaller. Further, our initial prediction that species evolving in dense vegetation might have shorter wings and lower weights than species evolving in relatively open habitat is not supported by the M'Passa data. Morphological differences in the sibling species appear primarily related to secondary sexual characters rather than to morphological adaptations to different habitats.

C. Why Does Contact between Sibling Species Occur at M'Passa?

The proximate historical cause of contact between sibling species at our site is the creation of the M'Passa campus and the more or less complete deforestation of a limited area of primary forest. This deforestation was followed by successional regrowth by plant species typical of cleared or anthropogenically disturbed areas. New contact zones were thus created between the birds of mature forest and birds colonizing from nearby modified habitats.

Some more distal causes are due to the relative proximity of the several square kilometers of deforested area around the Makokou village settlement, 12 km from M'Passa. Most of the secondary habitat species that rapidly invaded the M'Passa campus probably originated in the Makokou area, from which the newly created dirt road Makokou–M'Passa and the Ivindo River provide access routes to M'Passa. In fact colonization progress along the road from Makokou to M'Passa for several species of birds and mammals has been consistently substantiated (Brosset and Erard, 1986).

The deforestation of the Makokou area itself is not very old; the development of the village dates from about 1940. The question is then where did the birds of the Makokou area originate? The closest natural openings in the rainforest are the Boué savannahs, about 110 km distant. But the Boué avifauna differs from that of Makokou and M'Passa (A. Brosset, unpublished). Evidence has been provided that birds ecologically restricted to secondary habitats may fly over mature rainforest canopy and find and rapidly populate new and isolated deforested patches far inside continuous forest blocks (e.g., the Belinga campus, 110 km northeast of M'Passa, Brosset and Erard, 1986). Likely the bird species that populated the Makokou clearing colonized from afar, but the problem of the geographical origins of these birds remains unsolved.

D. When and Where Do Sibling Species Originate?

The answers to these questions remain largely hypothetical; just as in Neotropical birds (Cracraft and Prum, 1988), the phylogenetic patterns of avian differentiation within the Afrotropics is poorly understood. In Africa, the history of the paleovegetation and paleoclimate during the last three million years is characterized by the alternation of humid periods favorable to the extension of rainforest species and dry periods favorable to the extension of steppe and savannah species. Most probably such alternation promoted differentiation among populations comprising the original taxonomic unit from which the sibling pairs arose. But in the absence of fossil or molecular data, the "age" of sibling species must remain a subject of pure speculation. They are probably older than the last major glaciation, when it seems that eastern Gabon was a part of the rainforest refuge of southern Cameroun (Maley, 1987). This refuge would not have played a role in differentiation but only in the conservation and subsequent dispersal of "old" species. In M'Passa, the recently increased number of sibling species is not linked to the process of speciation at the global level, but is mostly due to recent colonization of the area by propagules from nearby or more distant successional habitats.

The second question, also difficult to answer, addresses where the sibling species originated. Have sibling species diverged from a common ancestor, or is one species of the pair derived from the other, for instance, by way of a founder effect? In both hypotheses the question remains as to whether the ancestor or ancestral population inhabited open areas or rainforest (or both). Some inferences are possible by comparing the present-day geographic ranges of the member species of those sibling pairs sympatric but not syntopic. This comparison is highly informative: of 19 pairs of sibling species chosen because their geographical ranges are quite well known, the B species of modified habitats have larger ranges than the A species, whose ranges typically are included within those of B. Within the pairs, the B species inhabiting modified

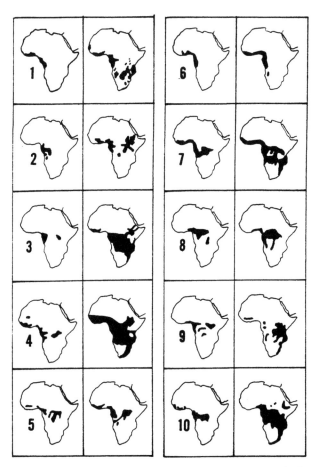

FIGURE 11. Comparative geographic ranges of pairs of sibling species in which A is a rainforest species, and B a species of modified habitats. On each map the rainforest species is mapped on the left and is the first-named species of each pair: (1) *Tauracus macrorynchos* and *T. persa;* (2) *Centropus anselli* and *C. monachus;* (3) *Apaloderma aequatorialis* and *A. narina;* (4) *Alcedo leucogaster* and *A. cristata;* (5) *Merops muelleri* and *M. gularis;* (6) *Gymnobucco peli* and *G. calvus;* (7) *Pogoniulus subsulfureus* and *P. bilineatus;* (8) *Indicator willcocksi* and *I. exilis;* (9) *Smithornis rufolateralis* and *S. capensis;* (10) *Psalidoprocne nitens* and *P. pristoptera.*

habitats seem much more adaptable and dynamic than the A species, which are ecologically restricted to rainforest. Range maps in Figs. 11–12 (largely interpreted from the *Birds of Africa* handbook) show no propensity for species A to adapt to new habitats; these reflect the conservative characteristics of old taxa, and their ranges are probably decreasing with deforestation. Species B, on the

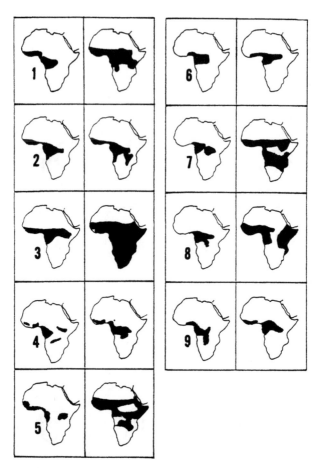

FIGURE 12.   Comparative geographic ranges of pairs of sibling species in which A is a rainforest species, and B a species of modified habitats. On each map the rainforest species is mapped on the left and is the first-named species of each pair: (1) *Dryoscopus sabini* and *D. gambiensis*; (2) *Oriolus brachyrynchus* and *O. atripennis*; (3) *Dicrurus atripennis* and *D. adsimilis*; (4) *Andropadus ansorgei* and *A. gracilis*; (5) *Camaroptera chloronota* and *C. brevicauda*; (6) *Muscicapa seth-smithi* and *M. epulata*; (7) *Myoparus griseigularis* and *M. plumbeum*; (8) *Tchitrea batesi* and *T. viridis*; (9) *Cyanomitra batesi* and *C. seimundi*.

other hand, show the dynamic and behavioral characteristics of "young" species. Unlike sibling species pairs A and B, syntopic sibling species have geographical ranges which are more often quite similar (Figs. 13–15).

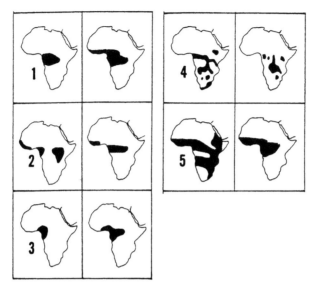

FIGURE 13. Comparative geographic ranges of sibling pairs in which member species share the same spatial niche. The first-named species is depicted to the left in each map. (1) *Neocossyphus rufus* and *N. poensis*; (2) *Trichastoma fulvescens* and *T. rufipennis*; (3) *Malimbus racheliae* and *M. cassini*; (4) *Sorathrura rufa* and *S. boehmi*; (5) *Lonchura cucculata* and *L. poensis*.

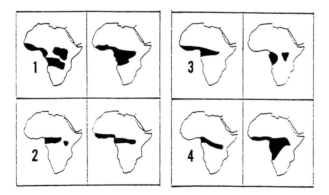

FIGURE 14. Comparative geographic ranges of sibling pairs that are syntopic but in which member species inhabit different strata of the vegetation in the rainforest. The first-named species is depicted to the left in each map. (1) *Andropadus gracilirostris* and *A. curvirostris*; (2) *Phyllastrephus xavieri* and *P. icterinus*; (3) *Macrosphenus concolor* and *M. flavicans*; (4) *Dyaphorophia tonsa* and *D. castanea*.

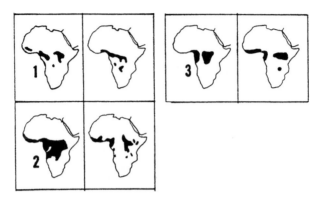

FIGURE 15. Comparative geographic ranges of sympatric sibling species inhabiting different subhabitats within the rainforest. The first-named species is depicted to the left in each map. (1) *Cercococcyx olivinus* and *C. mechowi*; (2) *Alethe castanea* and *A. poliocephala*; (3) *Geokichla princei* and *G. cameronensis*.

V. CONCLUSIONS

Prior to deforestation in 1970, 2 km² of the M'Passa area were covered by primary mature rainforest. The number of local bird species, 175, was fairly characteristic of tropical lowland rainforests on different continents in the Old World (Brosset, 1990). The deforestation of the M'Passa campus created a mosaic of successional habitats that were rapidly invaded by 189 species of new settlers, increasing the total species number to 364 for this small area. Especially striking is the increase in the relative number of sibling species pairs in the new situation; in the primeval rainforest the number of sibling species was initially 30 (22% of the avifauna), but between 1970 and 1982 the number of sibling species recorded in M'Passa increased to 110 (30% of the enhanced avifauna). The increased representation of siblings in the avifauna is statistically significant ($\chi^2 = 6.22$, $df = 1$, $P < 0.02$), showing that sibling species contribute more to the increase of species richness at the local scale than do nonsiblings.

The increase of species richness in M'Passa does not argue against the views of most conservationists. Note that the described process is local and artificial, being the result of human encroachment and disturbance. Placing into contact taxonomic units that are otherwise allopatric or parapatric produces quickly and locally what evolution might accomplish in thousands of years. The increased species richness locally does not result from speciation, nor does the local invasion by species of secondary habitats from nearby anthropogenic areas change the specific richness at the level of the Afrotropical avifauna. The β-diversity component is obviously enhanced since clearing introduces new habitats, but the α-diversity of the rainforest birds will most likely decline, at least ultimately.

According to Ricklefs (1987, 1989), the local and regional processes can have distinct dynamics; this is true for the bird communities in the M'Passa region. The impact of small patches of deforestation included within large blocks of untouched forest, as at M'Passa, is probably minimal at the levels of regional and global specific richness. On the other hand, in the inverse situation, that is, of fragmented small patches of mature rainforest included within large areas of modified habitats, it is generally true that species richness decreases dramatically at the regional level. The mature rainforest species are the first to disappear, especially those having large ranges and specialized niches and those inhabiting the lower strata of the rainforest vegetation. The comparison of sibling species pairs in Figs. 11 and 12, one inhabiting the primitive forest and the other secondary vegetation, is most informative. Invariably the species of modified habitats show larger geographical ranges than the rainforest species. Deforestation would thus increase the global range of the former while decreasing that of the latter, a process to be observed whenever and wherever deforestation occurs. On a large scale, deforestation would result in declining numbers of sibling species by enhancing extinction proneness and the likelihood of eradication in the rainforest species of sibling pairs, the species which are likely the oldest and the most interesting from the conservation point of view.

VI. SUMMARY

This chapter discusses the results of a long-term study on changes in rainforest birds in Gabon, West Africa, following local deforestation. A 2-km^2 study site at M'Passa in northeastern Gabon originally was entirely covered by mature rainforest and supported 175 bird species, including 30 sibling species. After an approximately 30-ha deforestation of the area in 1979, the cumulative number of species recorded on the site gradually increased to 364 by 1983, mostly through the addition of successional species. Statistical analysis shows that sibling species, which increased to 110, contributed more to the enhancement of local species richness than nonsibling species.

Special attention was paid to the spatial organization of the species forming the sibling pairs. In the mature rainforest, two pairs of sibling species were completely syntopic, without any visible spatial segregation; in four pairs, the sibling species were vertically segregated, with one in the canopy and one in the undergrowth; and in five pairs, the species were horizontally segregated, with one in the plateau and one in the slope and riverine habitats. After the deforestation, three pairs of new immigrants were found exclusively in syntopy in the cleared areas, but the majority—some 30 pairs—was composed of one species restricted to the rain forest and its sibling invading the deforested area. Interspecific isolation mechanisms that prevent territorial contests and hybridization were studied at the contact zone; most were related to the use of head coloration patterns in species displaying in dense vegetation and to wing or tail patterns in those displaying in more open places.

In the absence of fossil and molecular information, the history of the evolution of the sibling species remains poorly known, and relations with past climatic changes in Africa are impossible to establish with certitude. Nevertheless, comparison of the range maps of the rainforest versus modified habitat sibling species invariably shows that the range of the former is smaller. Consequently, there is consistent indirect evidence that among sibling species pairs the rainforest species is the older and more conservative while the species of modified habitats is the younger and more dynamic. Finally, the role of the sibling species is discussed from the point of view of conservation on local, regional, and global scales.

Acknowledgments

I am particularly grateful to Professor C. Erard, Dr. D. Lachaise, and Professor R. Ricklefs for their useful comments. In its final stages, the manuscript was thoroughly reviewed, and substantially improved in both style and content, by Professor M. Cody in his capacity as book editor. An important source of information for this chapter and synthesis is *The Birds of Africa*. Great thanks are expressed to its editors, L. H. Fry, S. Keith, and E. J. Urban, and to the illustrator, M. Woodcock. The preparation and illustration of the manuscript were executed by S. Jouard and M. Charles-Dominique. Fieldwork was supported by the Centre National de la Recherche Scientifique, France.

References

Amadon, D. (1956). The superspecies concept. *Syst. Zool.* **15**, 245–249.

Amadon, D., and Short, L. L. (1976). Treatment of subspecies approaching species status. *Syst. Zool.* **25**, 161–167.

Brosset, A. (1969). La vie sociale des oiseaux dans une forêt équatoriale du Gabon. *Biol. Gabonica* **5**, 29–69.

Brosset, A. (1976). Observations sur le parasitisme de la reproduction du Coucou émeraude *Chrysococcyx cupreus* au Gabon. *Oiseau R.F.O.* **46**, 201–208.

Brosset, A. (1978). Social organization and nest-building in the forest weaver birds of the genus *Malimbus* (Ploceinae). *Ibis* **120**, 27–37.

Brosset, A. (1981a). Occupation du milieu et structure d'une population du bulbul forestier *Andropadus latirostris*. *Oiseau R.F.O.* **51**, 115–126.

Brosset, A. (1981b). La périodicité de la reproduction chez un bulbul de forêt équatoriale africaine *Andropadus latirostris*. Ses incidences démographiques. *Rev. Ecol.* (*Terre Vie*) **35**, 109–129.

Brosset, A. (1981c). Evolution divergente chez deux bulbuls sympatriques (Pycnonotidae). *Alauda* 49, 94–111.

Brosset, A. (1982a). Le peuplement des Cyprinodontes du bassin de l'Ivindo, Gabon. *Rev. Ecol. (Terre Vie)* 36, 233–292.

Brosset, A. (1982b). The social life of the African forest Yellow-whiskered Greenbul *Andropadus latirostris*. *Z. Tierspsychol.* 60, 239–255.

Brosset, A. (1983). Parades et chants collectifs chez les Coucourous du genre *Apaloderma*. *Alauda* 51, 1–10.

Brosset, A. (1988). Le peuplement de mammifères insectivores des forêts du Nord-Est du Gabon. *Rev. Ecol. (Terre Vie)* 43, 23–46.

Brosset, A. (1990). A longterm study of the rainforest birds in M'Passa (Gabon). *In* "Biogeography and Ecology of Forest Bird Communities" (A. Keast, ed.), pp. 259–274. Academic Publ., The Hague, The Netherlands.

Brosset, A., and Erard, C. (1974). Note sur la reproduction des *Illadopsis* de la forêt gabonaise. *Alauda* 42, 385–395.

Brosset, A., and Erard, C. (1976). Première description de la nidification de quatre espèces en forêt gabonaise. *Alauda* 44, 205–235.

Brosset, A., and Erard, C. (1986). Les oiseaux des régions forestières du Nord-Est du Gabon. Vol. 1. Ecologie et comportement des espèces. *Rev. Ecol. (Terre Vie), Suppl.* 3, 1–297.

Brown, L. H., Urban, E. J., and Newman, K. (1982). "The Birds of Africa," Vol. 1. Academic Press, New York.

Butlin, R. (1989). Reinforcement of premating isolation. *In* "Speciation and its Consequences" (D. Otte and J. Endler, eds), pp. 158–179. Sinauer Assoc., Sunderland, MA.

Cain, A. J. (1954). "Animal Species and their Evolution." Hutchinson's University Library, London.

Cody, M. L. (1969). Convergent characteristics in sympatric populations: A possible relation to interspecific competition and aggression. *Condor* 71, 222–239.

Cody, M. L. (1980). Species packing in insectivorous birds communities: Density, diversity and productivity. *Proc. Ornithol. Congr., 17th, 1978*, Berlin, pp. 1071–1077.

Chappuis, C. (1974). Illustration sonore de problèmes bioacoustiques posés par les oiseaux de la zone éthiopienne. *Alauda* 42, 467–500.

Chappuis, C. (1981). Illustration sonore de problèmes bioacoustiques posés par les oiseaux de la zone éthiopienne. *Alauda* 49, 35–58.

Cracraft, J., and Prum, R. (1988). Patterns and processes of diversification: Speciation and historical congruence in some tropical birds. *Evolution (Lawrence, Kans.)* 42, 603–620.

Decoux, J. P., and Erard, C. (1992). *Tauraco macrorhynchus* and *T. persa* in northeastern Gabon: Behavioral and ecological aspects of their coexistence. *Proc. Pan-Afr. Ornithol. Congr., 7th*, Nairobi, Kenya, 1988, pp. 369–380.

Driscoll, P. V., and Kikkawa, J. (1989). Bird species diversity of lowland tropical rainforests in the New Guinea and Northern Australia. *Ecol. Stud.* 69, 123–152.

ECOTROP (1979). "Liste des Vertébrés de la région de Makokou," Mimeogr. rep. Laboratoire ECOTROP/CNRS, Paris.

Erard, C. (1977). Découverte du nid de *Baepogon clamans*. *Alauda* 45, 271–277.

Erard, C. (1987). Ecologie et comportement des Gobemouches (Aves: Muscicapinae, Platysteirinae, Monarchinae) du Nord-Est du Gabon. Vol. 1. Morphologie des espèces et organisation du peuplement. *Mém. Mus. Natl. Hist. Nat., Ser. A (Paris)* 138, 1–256.

Erard, C. (1990). Ecologie et comportement des Gobemouches (Aves: Muscicapinae, Plastysteirinae, Monarchinae) du Nord-Est du Gabon" Vol. II. Organisation sociale et reproduction. *Mém. Mus. Natl. Hist. Nat., Ser. A (Paris)* 146, 1–233.

Fry, C. H., Keith, S., and Urban, E. K. (1988). "The Birds of Africa," Vol. 3. Academic Press, San Diego, CA.

Genermont, J., and Lamotte, M. (1980). Le concept biologique de l'espèce dans la zoologie contemporaine. *In* "Les problèmes de l'espèce dans le monde animal" (C. Boquet, J. Genermont, and M. Lamotte, eds.), Vol. 3. Soc. Zool. Fr., Paris.

Haffer, J. (1992). Parapatric species of birds. *Bull. Br. Ornithol. Club* 113, 250–264.

Hall, B. P., and Moreau, R. E. (1970). "An Atlas of Speciation in African Passerine Birds." Br. Mus. (Nat. Hist.), London.

Johnson, N. K. (1963). Biosystematics of sibling species of flycatchers in the *Empidonax hammondii-oberholseri-wrightii* complex. *Univ. Calif., Berkeley, Publ. Zool.* 66, 79–238.

Keith, S., Urban, E. K., and Fry, C. H. (1992). "The Birds of Africa," Vol. 4. Academic Press, London.

Kikkawa, J., Lovejoy, T. E., and Humphrey, P. S. (1980). Structural complexity and species clustering of birds in tropical rainforest. *Proc. Ornothol. Congr., 17th,* Berlin, 1978, pp. 1071–1077.

Lovejoy, T. E., Bierregaard, R. O., Jr., Rylands, A. B., Malcolm, C. E., Quintela, L. H., Harper, L. H., Brown, K. S., Jr., Powell, A. H., Powel, G.V.N., Schubart, H.O.R., and Hays, M. B. (1986). Edge and other effects of isolation on Amazon forest fragments. *In* "Conservation Biology: The Science of Scarcity and Diversity" (M. E. Soulé, ed.), pp. 257–285. Sinauer Assoc., Sunderlands, MA.

Maley, J. (1987). Fragmentation de la forêt dense humide africaine et extension des biotopes montagnards au quaternaire récent: Nouvelles données polliniques et chorologiques. Implications paléoclimatiques et biogéographiques. *In* "Paleoecology of Africa" (A. A. Balkema, ed.), pp. 307–333. Brookfield, Rotterdam.

Mayr, E. (1942). "Systematics and the Origin of Species." Columbia Univ. Press, New York.

Mayr, E. (1963). "Animal Species and Evolution." Belknap Press of Harvard, Cambridge, MA.

Mayr, E. (1980). Problems of the classification of birds, a progress report. Ervin Streseman Memoria Lecture. *Acta Congr. Int. Ornithol., 17th,* Berlin, 1978, pp. 95–112.

Murray, B. G., Jr. (1981). The origins of adaptative interspecific territorialism. *Biol. Rev. Cambridge Philos. Soc.* 56, 1–22.

Prigogine, A. (1984a). Speciation problems in birds with special reference to the Afrotropical regions. *Mitt. Zool. Mus. Berl.* 60, Suppl. Ann. Ornithol., 8, 3–27.

Prigogine, A. (1984b). Secondary contact zone in central Africa. *Proc. Pan-Afr. Ornithol. Congr., 5th,* 1980, pp. 81–96.

Rand, A. L., Friedmann, H., and Taylor, M. A. (1959). Birds from Gabon and Moyen Congo. *Fieldiana, Zool.* 41, 221–411.

Rensch, B. (1929). "Das Prinzip geographischer Rassenkreise und das Problem der Artbildung." Borntràger, Berlin.

Ricklefs, R. E. (1987). Community diversity: Relative roles of local and regional processes. *Science* 235, 167–171.

Ricklefs, R. E. (1989). Speciation and diversity: The integration of local and regional processes. *In* "Speciation and its Consequences" (D. Otte and J. Endler, eds.), Vol. 24, pp. 599–622. Sinauer Assoc., Sunderland, MA.

Rogstad, S. H. (1990). The biosystematics and evolution of the *Polyalthia hypoleuca* species complex (Annonaceae) of Malesia. II. Comparative distributional ecology. *J. Trop. Ecol.* 6, 387–408.

Selander, R. K. (1971). Systematics and speciation in birds. *Avian Biol.* 1, 57–147.

Short, L. L. (1969). Taxonomic aspects of avian hybridization. *Auk* 86, 84–105.

Short, L. L. (1972). Hybridization, taxonomy and avian evolution. *Ann. MO. Bot. Gard.* 59, 447–453.

Smith, H. M. (1955). The perspective of species. *Turtox News* 33, 74–77.

Smith, H. M. (1965). More evolutionary terms. *Syst. Zool.* 14, 57–58.

Terborgh, J., and Weske, J. S. (1969). Colonisation of secondary habitats by peruvian birds. *Ecology* 50, 765–782.

Terborgh, J., Robinson, S. K., Parker, T. A., III, Munn, C. A., and Pierpont, N. (1990). Structure and organization of an Amazonian bird community. *Ecol. Monogr.* **60**, 213–238.

Thiollay, J. M. (1986). Alteration of raptor communities along a succession from primary rainforest to secondary habitats. Proc. Symp. British Ecological Society (Cambridge 1985): Tropical forest birds: Ecology and conservation. *Ibis* **128**, 172.

UNESCO-MAB-ECOTROP-IRET (1987). "Makokou, Gabon: A research station in tropical forest ecology: Review and publications," Publ. No. 7, UNESCO, Paris.

Urban, E. K., Fry, C. H., and Keith, S. (1986). "The Birds of Africa," Vol. 2. Academic Press, London.

White, C.M.N. (1956). Notes on the systematics of African bulbuls. *Bull. Br. Ornithol. Club* **76**, 155–158.

CHAPTER 11

Bird Communities in the Central Rocky Mountains

MARTIN L. CODY

Department of Biology, University of California, Los Angeles, California 90095

I. INTRODUCTION

This chapter will review research from long-term studies of bird communities in the central Rocky Mountains. The majority of such studies are censuses of breeding bird communities, repeated over a series of different years at the same site, but some pertain to wintering bird communities. In some cases censuses from related habitats at nearby sites are contrasted; in others different habitats, which may share species, are compared within and among years.

The main part of the chapter concerns research at two sites within Grand Teton National Park (GTNP), Wyoming, in adjacent but very different habitats: a semi-desert grass–sagebrush site of low, dry, and open vegetation and a marshy site dominated by scrubby willows. For these two sites not only are census results available for a series of years spanning more than a quarter of a century but also information on characteristics of species' habitat selection within sites, on foraging ecology, and on interactions among the breeding species. After an introductory section describing the physical attributes of the central Rockies region is a section in which general census studies are reviewed, after which the results of the GTNP work are covered. The chapter concludes with an overview of the lessons learned from these long-term bird studies.

A. The Central Rockies: Geography, Topography, and Climate

The central Rocky Mountains, from Colorado to Montana, are part of a chain of mountains extending from New Mexico into northwestern Canada, over nearly 30° of latitude. This range clearly deserves its epithet as the "backbone of North America," and is broadly coincident with the continental divide. While generally a product of crustal folding, the mountains are geologically complex. They include some of the oldest rocks on the continent (Precambrian schists and gneiss) and some of the youngest, tallest, and most active mountains (Teton Range, rising west of the Teton Fault through Jackson Hole), where sedimentary limestone can be found at 4000 m elevation next to impressive glaciers. In this region synclinal basins overlay vast coal reserves, and the volcanic and hydrothermal features of the nation's first national park, Yellowstone (YNP), are appreciated by millions of visitors annually. Several popular and informative books are available for regional and more local geological history and overviews (e.g., McPhee, 1986; Lageson and Spearing, 1988; Knight, 1994).

A real and symbolic barrier since pioneer times, the Rockies separate the largely arid and mountainous west from the central plains and eastern lowlands. As a major topographical feature, this cordillera not only constitutes a barrier of biogeographical significance for the biota east-to-west, but a habitat inroad to the southern part of the continent from the north, and to a lesser extent as a northerly avenue for the southern biota of the mountains of northwest Mexico (Sierra Occidental). Between the mile-high plains grasslands on the east and the desert river basins of the Colorado and Columbia drainages on the west, the mountain topography supports a wide range of vegetation types. Woodland and forested habitats occur in the foothills, valleys, and canyons, and extend up to treeline (ca. 3300 m at 45° N), with arctic-alpine tundra above.

Mountains of this stature have a marked influence on climate and intercept both winter storms from the northwest originating in the Gulf of Alaska and summer storms from the southeast out of the Gulf of Mexico. The former attenuate to the south, and the latter to the north, each providing for strong seasonal precipitation gradients. As is expected for sites at high elevation and locations over 1000 km from the nearest coast, the climate is "continental," with extreme seasonality. Climograms for three locations are shown in Fig. 1. At Boulder (1620 m; lat. 40° N) in the foothills of northeastern Colorado near

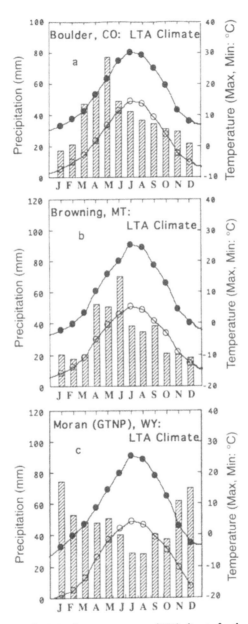

FIGURE 1. Climograms depicting long-term average (LTA) climate for three sites in the eastern foothills of the central and northern Rocky Mountains: (a) Boulder, Colorado, (b) Browning, Montana, and (c) Moran (near Grand Teton National Park), Wyoming. Mean monthly precipitation (mm) is represented by bars; mean monthly maximum and minimum temperatures (°C) are represented by solid and open circles.

Rocky Mountain National Park (RMNP), annual precipitation averages 457 mm and peaks in spring to early summer; monthly mean temperature minima exceed zero 7 months of the year. Browning, Montana (1320 m elevation, latitude 49° 20′ N), at the edge of the plains east of Glacier National Park (GNP), has a similar precipitation pattern (average 407 mm annually), but temperatures are lower, with mean monthly minima below zero October–April and mean monthly maxima below zero December–February. In GTNP, at Jackson Lake dam near Moran, Wyoming, at an intermediate latitude (44° N) and a higher elevation (2000 m), summer days are warm and relatively dry (one-sixth of the annual 584 mm precipitation falls in the three summer months, June–August), but nights are cold, and daily temperatures might range over 25°C. Only these three summer months have mean monthly temperature minima above zero, and, with a scant average of 35 frost-free days per year, the season is short; temperatures from late fall to spring are typically subzero accompanied by heavy snowfall.

Besides the extreme seasonality of the midcontinental montane climate, year-to-year weather variations are also conspicuous. At Boulder, the coefficient of variation (CV) in yearly precipitation is around 0.25 with over one-third of the annual totals falling beyond one standard deviation from the mean (Fig. 2a). In GTNP, over the 33-year period, CV of annual precipitation is somewhat lower (0.18), but still more than one-third of the years are similarly very wet or very dry. Even annual mean temperatures vary widely, with figures beyond one SD of the mean (2.26°C) in 9 of 33 years (Fig. 2b). The region might then be characterized by short, warm summers and long, severe winters, with considerable variability among years in precipitation and temperatures and further unpredictability in the timing or onset of the short growing season.

B. Life Zones and Bird Habitats

The abrupt topography of the central Rockies provides a wide range of local physical conditions that vary with elevation and to a lesser extent with latitude. Variations in vegetation follow those in climate and insolation, precipitation, and drainage and correspond also with changes from the deep soils of alluvial basins to rocky crags and scree slopes, from desert to alpine and xeric to mesic. An east–west transect across the central Rockies would begin in the short-grass plains with cottonwoods (*Populus* spp.) along drainage courses, strike sagebrush (*Artemisia* spp.) in drier valleys or pinyon–juniper woodland in the lower foothills, ascend through drier, open forests of ponderosa pine (*Pinus ponderosa*) or wetter forests of lodgepole pine (*P. contorta*), and encounter denser and taller spruce-fir forests in the sheltered canyons with subalpine fir at higher elevations up to tundra. Virtually all of Merriam's life zones might be encountered within a few dozen kilometers. A variety of habitat types occur, with a dependence on local conditions and the incidence or recency of perturbations such as drainage

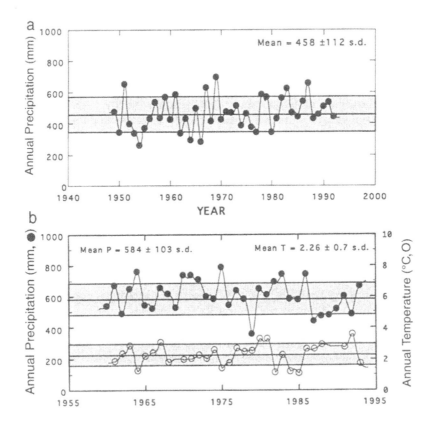

FIGURE 2.   Year-to-year variation in weather at two sites in the central Rocky Mountains. (a) Boulder, Colorado: yearly precipitation totals (mm), mean (458 mm) ± standard deviation (shaded area). About 30% of the annual totals lie >1 SD from the overall mean. (b) Moran, Wyoming: yearly precipitation data (mm) and annual mean temperatures (°C), with mean (2.26°C) and standard deviation (shaded area). Thirty-six percent of precipitation totals and 27% of annual temperatures are similarly deviant.

patterns, rock slides, glacial moraines, or fires. These habitats include riparian vegetation, lakes, marsh and meadowland, willows, and aspen woodland. Correspondingly, the Rockies are renowned for their wildlife in general and for their many species of birds and larger mammals in particular.

C. The Birds: Residents vs Migrants

There are over 200 breeding bird species in the central Rockies from foothills to peaks, and the diversity of waterbirds and raptors is particularly high. The four national parks from south to north record around 140 (RMNP) to 165 (GTNP,

YNP, GNP) species that are more or less regular breeders; see Johnsgard (1986) for an excellent synopsis of the avifauna of this region. The marked climatic seasonality of the area renders many habitats that are productive for breeding birds inhospitable in the off-season, as they are often under meters of snow. Thus the more open habitats, from grassland and marsh to sagebrush and scrub, support a breeding bird community that is nearly wholly migrant, whereas woodland and forest retain some species in winter as year-round residents; these may be joined by others that winter in these sites but breed elsewhere. Typically in the coniferous forests, around 45% of the species seen are year-round residents, 40% summer (breeding) visitors, and 15% winter visitors. Summer visitors are numerically dominant in the breeding community and, while the number of wintering species often approaches that in the breeding community, densities are much lower, even of residents, many of which undertake a partial withdrawal in the off-season. The overwintering birds are the usual mix common to most north-temperate woodlands and forests in the non-breeding season, including corvids, woodpeckers, nuthatches, chickadees and finches; specifically lacking are the foliage-gleaning and flycatching insectivores that comprise the larger part of the bird biomass in the breeding community.

The summer visitors to the central Rockies display a variety of migration strategies, from species that make local, often elevational, habitat shifts in winter (American robin, rufous-sided towhee, song sparrow; N.B. scientific names are found in the Appendix); to species that undertake short-distance migrations south, in many cases to grassland, desert, or scrub habitats in the southern United States (mourning doves, wrens, blackbirds, meadowlarks, and many emberizid sparrows); to species that are long-distance migrants to forest, woodland, or edge habitats in Central and South America (swallows, flycatchers, warblers, vireos, and tanagers). For example, three of five species listed in Table IV, and seven of the 15 in Table V, have been found wintering in western Mexico where they occupy either cloud forest, tropical deciduous forest, thornforest, or pine-oak-fir forest, exclusively or in some combination (Hutto, 1980, 1985, 1986, 1992). Thus breeding bird communities are largely assembled anew each spring, with components that are gathered from a wide variety of overwintering sources, including distant, local, and on-site.

II. LONG-TERM CENSUSES IN THE ROCKY MOUNTAINS

In this section I review some of the long-term census data published for several habitats in the central Rockies of Colorado. The overview gives a perspective on changes in breeding birds and their densities within sites and among years and relates such changes to habitat type, alterations over time in vegetation within habitats, and degree of variation in bird populations that can be considered normal in this region.

A. Riparian Cottonwoods

Within extended census series are embodied two sorts of potential changes and trends: those associated with long-term changes of the bird populations independent of the habitat or site, and those associated with change in the habitat itself. A census series from riparian cottonwoods (e.g., Kingery et al., 1971, and later) near Denver, CO, illustrates the latter aspect. Censuses were initiated at the 9.7 ha site after the South Platte River was dammed to form a reservoir in 1971 and continued for a decade as the cottonwoods gradually succumbed. Initially some 25 bird species were recorded, with total densities of 15.8 pairs/ha dominated by house wrens, yellow warblers, western wood pewees, and American robins. The forest was fully flooded by 1980 with 2–3 m of water; by 1981 the species count was reduced to 40% and total density to 30% of their former values.

Figure 3 shows the populational responses of various species to changes in the habitat. Clearly different species responded to the habitat change in different ways. The dominant foliage insectivore, yellow warbler, declined steadily (Fig. 3a) as trees died and lost their foliage, while the lower-foraging common yellowthroat enjoyed a mid-period boom associated with early stages of flooding and a preference for wetter habitat. The numbers of both vireo species likewise declined (Fig. 3a), with those of the red-eyed vireo preceding the warbling vireo by a few years; the earlier demise of the former perhaps is associated with mid-period increases in the latter species. Figure 3b shows a precipitous decline in western wood pewee, a flycatcher of the forest canopy and subcanopy, while the low-foraging flycatcher typical of more open sites, eastern kingbird, was little affected by flooding, and the purely aerial tree swallow increased markedly as more cottonwoods died and provided abundant nest sites.

Ground-foraging American robins and mourning doves disappeared within the decade, but the standing cottonwoods provided nesting habitat for ground-foraging but more widely-ranging grackles and starlings, permitting their increase (Fig. 3c). House wrens (Fig. 3d) were initially favored, apparently by dead trees and the additional nest sites thereby provided, and trunk-and-branch searchers were less affected by foliage loss. They succumbed late (but convincingly), whereas the two woodpecker species reached the same endpoint in a more gradual fashion (Fig. 3d). These data clearly show a dominance of long-term trends associated with profound habitat change and also, in no less compelling a fashion, the variety of ways in which the bird populations may respond to such change.

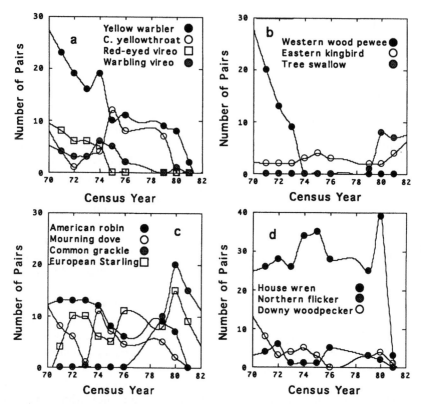

FIGURE 3. Changes in breeding bird species and densities after flooding and tree die-off in a cottonwood forest after Chatfield Dam construction, South Platte River, Colorado. The responses of different species vary in kind and in magnitude. Data are from Kingery *et al.* (1971, and later); see text for discussion.

B. Scrub-Oak Communities

Scrub-oak (*Quercus gambelii*) is a dominant component of a distinctive brushy vegetation in south-central Colorado and adjacent areas of Utah and New Mexico; the habitat occurs at around 2000 m elevation, often adjacent to pinyon pine (*Pinus edulis*) woodland and sagebrush. Two scrub-oak sites in El Paso County, Colorado, were censused independently over a partially overlapping series of years, 1979–1982 (e.g., Jackson and St. Helens, 1980; site 1) and 1980–1984 (e.g., Abbott *et al.*, 1985; site 2). A comparison of their results serves to illustrate similarities and differences in census results derived from the same habitat at sites only about 12 km apart and to highlight the critical influence of site-specific factors even though habitat type and the general locality are constant.

Site 1 supported an average 12.5 ($\pm$ 1.5 SD) breeding species over the census interval, compared to an average 19.0 ($\pm$ 2.0 SD) at site 2, even though the second site, at 8 ha, was less than half the area of the first. Total densities differed by a similar degree in the same direction. Fifteen species were recorded in common at the two sites, with 7 additional species restricted to site 1 and 15 to site 2. However, the same 7 species comprise 66–91% of the total bird numbers in both sites (in order of overall density, rufous-sided towhee, scrub jay, American robin, broad-tailed hummingbird, black-billed magpie, mourning dove, and house finch), and thus the sites share their numerically dominant species and differ largely in uncommon (low-density) species.

The most common species in the habitat, rufous-sided towhee, shows a parallel trend in density between sites (Fig. 4), indicating that factors responsible for its interyear variation are shared between sites. In the next ranked species, scrub jay (Fig. 4a) and American robin (Fig. 4b), there is no particular correspondence in trends between sites over years. Note, however, that density trends of the broad-tailed hummingbird, common only at site 1, parallel those of the robin at site 1; the black-billed magpie, common only at site 2, shows good correspondence with the American robin at that site (Fig. 4b). Presumably site-specific factors account for the correspondences, and such factors apparently differ between sites and years; numerical trends among years seem to be controlled by both regional and local factors and affect different species to different degrees.

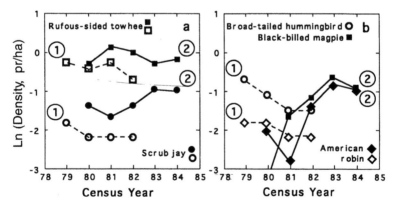

FIGURE 4.   Breeding densities of several of the more common species in a scrub-oak habitat at two sites (labeled 1 and 2) in El Paso County, Colorado, in a partially overlapping series of yearly censuses. Symbols for site 1 are open; those for site 2 are solid. Species may show similar trends over time within sites or between sites, or they may show no apparent similarity. Data are from Jackson and St. Helens (1980, and later) and Abbott et al. (1985, and earlier); see text for discussion.

C. Ponderosa Pine Forest

Ponderosa pine (*Pinus ponderosa*) is a dominant forest tree of the drier and more open foothill slopes in the southern Rocky Mountains; a long-standing census series was initiated in 1954 in this habitat on Boulder City parkland at 1800 m (e.g., Hering, 1984, 1985, 1991, and previous) and still continues. Over a 40-year period 33 bird species bred at the 8-ha site, but the number in any one year ranges fourfold (5–21) and averages 14.6 ± 4.7 SD. However, the most common species are predictably present nearly every year, and one-third of the total species occur in approximately 75% of the yearly breeding bird censuses. Absolute abundances vary among years around a mean of 4.4 pairs/ha (SD 2.08), with CVs of all species <1 and averaging 0.72 ± 0.21. Most of the 11 most common species, in terms of density and incidence, covary positively (Fig. 5) in density [42/50 correlations positive, 18 significantly so ($P < 0.05$), with one significant negative correlation]. In contrast, in the 11 taxa that constitute the middle tercile of species ranked by density × incidence, species are generally not positively covariant (24/50 correlations positive, 4/24 significantly so), and cross-correlations between species in the first and species in the second terciles are positive (57 of 99, 10 significant) or negative (42 of 99, 5 significant) with about equal frequency. Thus, the relative abundances of the common species remain quite constant, and interyear variations in species and densities appear to be due to factors at the site that birds recognize and respond to more or less in unison: good years are seen by most species as good, and poor years as poor.

   Some contributing factors to interyear variation at the site can be tentatively identified as intrinsic and biotic. Thus, densities in the three most common species, mourning dove, western wood pewee, and chipping sparrow, are all correlated significantly and negatively with black-billed magpie numbers (present in 5/18 years), as is the density of a fourth common species, pygmy nuthatch, with Steller's jay density. Both of these corvids are significant nest predators. However, migratory status does not seem to be a factor since residents such as pygmy nuthatch, pine siskin, mountain chickadee, and Steller's jay; short-distance migrants such as American robin, chipping sparrow, and mourning dove; and long-distance migrants such as western wood pewee, western tanager, solitary vireo and broad-tailed hummingbird are all represented in the first tercile. Similarly, resident hairy woodpecker, northern flicker, white-breasted nuthatch, and red crossbill; short-distance migrants such as house wren, house finch, lesser goldfinch, and dark-eyed junco; and longer-distance migrants such as violet-green swallow compose the second tercile.

   It seems likely that variations between years in the site's resources or productivity are responsible for a major part of interyear variation in bird species composition and density. Most obviously, the forest at this site was thinned in 1982, following which several species declined conspicuously (Fig. 5) and have

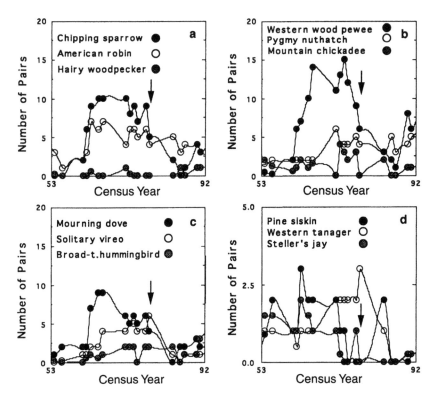

FIGURE 5. Densities of some of the more common breeding bird species at a ponderosa pine site in Boulder, Colorado, over a nearly 40-year interval. Note that species groupings here are for graphical convenience only. A partial thinning of the forest site occurred in 1982 (see arrow), a disturbance to which some, but not all, species appear to have responded by declines. Data are from Hering (1991, and prior); see text for discussion.

since recovered to some extent. However, such an explanation apparently does not explain overall increases in bird densities from the 1950s to the 1970s (L. Hering, personal communication). One possibility is that increased bird densities from the mid 1950s were a response to recovery from the drought years 1950–1954 (Fig. 2a) and to increased food resources in years of more normal precipitation. Long-term, unidirectional trends in species composition seem absent from the data, with the possible exception of the brown-headed cowbird, which became regular in the censuses of the early 1990s.

In addition to the breeding bird censuses, winter bird censuses have been conducted at the same site and reported for 14 intermittent years between 1955 and 1991 (e.g., Hering, 1992, and earlier). A total of some 23 wintering species were recorded, averaging 13.5 ± 2.8 (SD) species in any given year, and two-thirds of these (15 species) have an incidence of 50% or more in the winter counts (i.e., occur in at least half). The variations in numbers of these regular wintering species are shown in Fig. 6.

Are the numbers of wintering birds more or less variable than the breeding birds? Ten of the regular species are relatively constant and show CV values averaging 0.62 ± 0.10 SD; this is comparable to interyear variation in the breeding species. In the other five species CVs all exceed one: American crow (1.04), American robin (1.19), pine siskin (1.54), red crossbill (1.77), and red-breasted nuthatch (2.21). Thus, among the corvids the common Steller's jay and black-billed magpie are more constant than the sporadic crow (Fig. 6b); the more generalized ground-forager (gray-headed junco) is more regular than the two more-specialized seed-eaters (Fig. 6c); and one nuthatch species is rela-

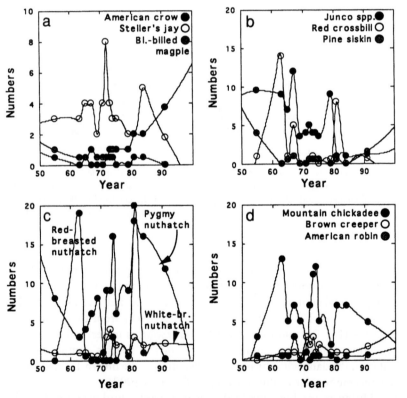

FIGURE 6. Densities of some of the more common wintering bird species over a nearly 40-year interval at a ponderosa pine site in Boulder, Colorado (see Fig. 5). Species are grouped in a–c by taxonomic affinity: (a) corvids, (b) finches, and (c) sittids. Other common species are included in d.

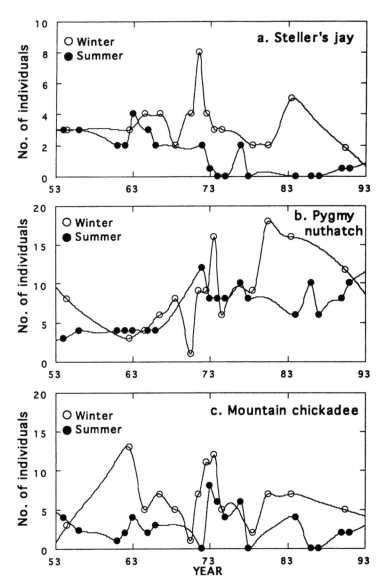

FIGURE 7.   Covariance between breeding and wintering bird numbers in a ponderosa pine forest in Boulder, Colorado, varies among species and among different time intervals. A correspondence in numbers between seasons presumably reflects a mutual interdependence between seasons, whereas noncorrespondence may indicate an independence of regulatory factors that are season specific.

tively constant, a second is less so, and a third is wildly erratic (Fig. 6d). The picture that emerges is that, with similar species numbers present in both breeding and wintering seasons, and a similar proportion of the species totals recorded as regulars in either season, the breeding species numbers are more variable than wintering species numbers (CV 0.32 versus 0.21), but variation in the total breeding density is similar to that in winter bird counts (CV 0.46 versus 0.43).

Are breeding and wintering densities closely related in resident birds? Data for three common year-round residents are shown in Fig. 7. The patchiness of the data over time precludes detailed statistical tests for congruence. However, intercensus slopes (positive or negative) in breeding versus wintering data covary positively for mountain chickadees (χ^2 = 4.44, df = 1, P < 0.05), suggesting parallel trends over time between seasons. There is no such covariation for Steller's jay or pygmy nuthatch data and no obvious relation between numbers in adjacent seasons. The former result is perhaps that of simple expectations, a carryover from one season to the next. The latter result might occur for at least three reasons: (a) the site is not a closed system, and either breeding or wintering densities are influenced by populations or factors off-site, (b) breeding and wintering numbers are regulated by wholly different factors that operate independently from each other, and (c) density dependence between seasons is weak, masking the influence of winter–summer and subsequent summer–winter inputs.

Weather is the most likely density-independent factor to influence bird numbers, directly or indirectly via resource production. Weather variation potentially affects summer reproduction and recruitment, overwinter survival, and carrying capacities in both seasons. A complete census data set, identification of the relevant weather variables, and, even better, direct measurement of resources on the site would be required to resolve these possibilities.

III. STUDIES IN GRAND TETON NATIONAL PARK, WYOMING

A. Topography, Habitats, and Birds

Grand Teton National Park (GTNP) in northeastern Wyoming epitomizes the wide range of habitats in the central Rockies, from the Great Basin Desert sagebrush dominated by *Artemisia tridentata* and *A. cana* in the floor of Jackson Hole (ca. 2000 m) to the arctic-alpine tundra of the ca. 4000-m Teton Range along its western edge; from the Snake River bottomlands (1900 m) of meadow, marsh, willow (*Salix* spp.), and cottonwoods (*Populus angustifolia* and *P. trichocarpa*) through the aspen (*Populus tremuloides*) and lodgepole pine (*Pinus contorta*) woodlands and forest of the foothills to spruce-fir forest (*Pseudotsuga menziesii* and *Picea engelmannii*) in the western canyons. The evolution of this unique landscape is well described in Love and Reed (1968), and a recent treatment by Knight (1994) gives a detailed ecosystem perspective on the region. The climate in GTNP (see Figs. 1c and 2c) is representative of its high elevation and mid-continental location, with moderate seasonality and year-to-year variation in precipitation, but strong seasonality and high year-to-year variation in monthly mean temperatures (Cody, 1974). Around 120 landbird species are known to breed in GTNP on a regular basis, and the avifauna includes species representative of the various migration strategies discussed in Sect. I.

The geographical position of GTNP, at the western edge of the Great Plains, the northeastern edge of the Great Basin Desert, the eastern edge of the western coniferous forests, and near the southern edge of the spruce-fir dominated forests of the northern latitudes, coincides with the range extremities of a large number of bird species. It is the western limit of a number of breeding birds that are plains species (e.g., brown thrasher, lark bunting, clay-colored sparrow, and bobolink), the southwestern limit of species chiefly eastern in distribution (e.g., red-headed woodpecker, least flycatcher, red-eyed vireo, American redstart, and common grackle), the eastern or northeastern limit of species that are largely southwestern in distribution (e.g., broad-tailed hummingbird, pinyon jay, canyon wren, sage thrasher, green-tailed towhee, and sage sparrow), and the southern limit of northern species that are near their lower latitude boundaries, at least with respect to the Rocky Mountains (e.g., harlequin and other ducks, black-backed and three-toed woodpeckers, varied thrush, northern waterthrush, Townsend's warbler, and winter wren).

B. Bird Population Changes since GTNP Establishment

Bird status surveys were conducted in 1933–1937 soon after the establishment of the park in 1929 (USDI, Washington). The early survey, while clearly of a preliminary nature, provides useful qualitative data for comparison with present-day measures of species occurrence and abundance, giving a half-century perspective on changes in the GTNP avifauna. The comparison is necessarily somewhat one-sided, as the early survey was clearly incomplete, and omits mention of a number of species now common; they were presumably common then in GTNP as they were in similar habitats in adjacent Yellowstone National Park during the same period (USDI, Washington). It is likely that species rare in GTNP in the 1930s were also omitted from the early list. Thus, the most reliable comparisons are among earlier listed species, particularly those that were regarded as common.

Some of the more apparent changes in status are summarized in Table I. Several species in the first part of the table, previously listed, are almost certainly erroneous omissions, as they are now abundant and were common in the 1930s in similar habitats in adjacent Yellowstone National Park (e.g., white-breasted nuthatch, Swainson's thrush, orange-crowned warbler, and Wilson's warbler). But several of the species previously unlisted and now recorded are likely to reflect genuine increases in abundance rather than omissions; these include American crow, European starling, brown-headed cowbird, and house finch; all but the latter are associated with disturbed grasslands, and all but the first-mentioned are involved in more recent range increases. Two of these species (European starling and brown-headed cowbird) have increased conspicuously in numbers since the 1960s and are now common. One of the species unlisted earlier, clay-colored sparrow, was relatively common in several GTNP sites in the late 1960s when it could be found in dry, brushy sites throughout Jackson Hole and the associated Gros Ventre valley. However, it could not be found in the park from 1991 to 1993, but it reappeared as a single territorial male in 1994. This easily overlooked species can best be described as rare, and perhaps it always has been rare in GTNP; but perhaps, as with several other birds near their range margins in this area, there are years or intervals when they are in fact common. Indeed, a substantial proportion of these unlisted species are near the limits of their geographic distribution in this region.

Species previously listed and now apparently more common include barn and cliff swallows and the house wren, each of which is favored by buildings and habitation. The third section of Table I lists birds which were rare breeding species in the 1930s and seem to have the same status presently.

Last, a number of species listed in Table I were previously considered common or abundant and appear presently to be much less so. In this category are some I have not found breeding in GTNP since my studies began there. Notably, many apparent changes in abundance pertain to species typical of grasslands and open or brushy country. Crows, cowbirds, and starlings are now common, and a rather large number of species associated with these habitats appear to have declined. These include sharp-tailed grouse, common nighthawk, eastern and western kingbirds, Say's phoebe, bobolink, American goldfinch, lark sparrow, gray catbird, western meadowlark, and grasshopper sparrow. The more extensive grasslands in the southern and northeastern parts of Jackson Hole are currently heavily grazed by cattle and (especially in winter) by the Greater Yellowstone Ecosystem's elk herd. Perhaps a combination of habitat changes and expanding populations of the parasitic cowbirds and introduced starlings can account for the current scarcity of the native grassland and brush species.

TABLE I   Some Changes in Status of Breeding Birds in Grand Teton National Park and Adjacent Areas of Jackson Hole in the Last 55 Years[a]

Species	1930s	Mid-1990s[b]
Species unlisted in 1930s survey		
Sandhill crane	—	Fairly common; YNP[c]
Three-toed woodpecker	—	Uncommon; rare YNP
Bank swallow	—	Fairly common; common YNP
Rough-winged swallow	—	Uncommon; rare YNP
American crow	—	Fairly common; common YNP
White-breasted nuthatch	—	Common; uncommon YNP
Winter wren	—	One recent record[d]
Marsh wren	—	Occasional, most years; rare YNP
Golden-crowned kinglet	—	Common in high-elevation spruce; YNP
European starling	—	Rare 1960s; common 1990s
Swainson's thrush	—	Abundant; common YNP
Veery	—	NS (a few recent records); YNP
Townsend's warbler	—	NS (occasional breeding); common YNP
Orange-crowned warbler	—	Fairly common; YNP
Wilson's warbler	—	Abundant; common YNP
American redstart	—	NS; reported breeding 1994
Northern waterthrush	—	Occasional breeding
Common grackle	—	Reported south end of JH[d]; not in GTNP
Northern oriole	—	NS (reported from SP); rare YNP
Bobolink	—	Bred in NER since 1980s; occasional YNP
Brown-headed cowbird	—	Occasional 1960s, common 1990s; common YNP
Lark sparrow	—	Rare (dwarf aspen, 1994); rare YNP
Clay-colored sparrow	—	Common late 1960s–early 1970s; rare 1990s
House finch	—	Fairly common
Species listed in 1930s that have apparently increased since		
Callope hummingbird	± Rare	Fairly common, especially willow flats
Empidonax spp.	Uncommon	Only willow flycatcher listed; in addition dusky flycatcher is common, cordilleran flycatcher uncommon, and least flycatcher rare in GTNP
Barn swallow	Fairly common	Common
Cliff swallow	Fairly common	Common
Black-billed magpie	Uncommon	Common
Brown creeper	Rare	Fairly common
House wren	Fairly common	Abundant
Common yellowthroat	Uncommon	Abundant
Green-tailed towhee	Rare	Common
Lincoln's sparrow	Uncommon	Common
White-crowned sparrow	Uncommon	Abundant
Rarer species that have apparently maintained their status		
Eastern kingbird	Rare	Rare; Snake River bottoms
Gray catbird	Rare	Rare; a few seen most years

continues

TABLE I   (Continued)

Species	1930s	Mid-1990s[b]
Sage thrasher	"Listed"	Uncommon
Western meadowlark	Uncommon	Uncommon
Lazuli bunting	Rare	Rare
Evening grosbeak	Occasional	Reported[d] winter only
Species that have apparently declined since the 1930s		
Sharp-tailed grouse	Uncommon	NS
Common nighthawk	Abundant	Uncommon to rare
Western kingbird	Uncommon	NS
Say's phoebe	Common	Rare; occurs at Jackson
Stellar's jay	Uncommon	Restricted to canyon bottoms
American goldfinch	Common	Uncommon
Grasshopper sparrow	Uncommon	NS

[a]Waterbirds are not considered. Species are listed in taxonomic order in the table. Abbreviations used: GTNP, Grand Teton National Park; JH, Jackson Hole; NER, National Elk Refuge; SP, South Park.
[b]Observations on current status are my own. NS, not seen, 1960s–1993.
[c]YNP (in first section of table) refers to status in 1930s survey from Yellowstone National Park; otherwise not recorded.
[d]fide Steven Cain, GTNP Nat. Res. Manager; other reports by sundry observers.

C. Birds of the Aspen Woodlands

Aspen woodlands are found scattered throughout GTNP, but are especially prominent on the lower hills and moraines along the southern and eastern reaches of Jackson Hole. Flack (1976) incorporated several sites within GTNP into his treatment of aspen bird communities ranging from Canada to Arizona. Three of his sites, Elk Ranch East (site 22; Flack, 1976, Table 1), Elk Ranch West (site 24), and Cow Lake/Signal Mountain (site 25), were recensused from 1992 to 1994 (M. L. Cody, unpublished data). Both original and recent census sites were 5 ha in size. In the data summary given in Table II, nocturnal and diurnal raptors, other wide-foraging species such as ruffed grouse, northern raven, and cowbirds, and irruptive and/or late-breeding finches such as pine siskin, red crossbill, and American goldfinch have been omitted (as in Flack, 1976).

Some 32 bird species were censused at the three sites; of which the 15 most common, those with densities in excess of one pair per 5 ha, average ca. 90% of the total bird density (averaging over all sites and years). There are differences in vegetation structure among sites, and from Elk Ranch West to Elk Ranch East to Cow Lake aspen stem-density declines and tree stature in-

creases, both currently and in 1966, and concomitantly the incidence of con-
ifers increases, from near zero to ca. 3/ha. Many birds show more or less
constant density across the aspen sites (e.g., house wren, northern flicker, tree
swallow, western wood pewee), but in others there are trends associated with
changes in vegetation composition and structure, producing turnovers in the
bird community. For example, several species are most common at the lower,
brushier site (white-crowned sparrow, *Empidonax* spp., yellow warbler, orange-
crowned warbler, and house finch); others at the tallest and most open site
(dark-eyed junco, American robin, chipping sparrow, and yellow-rumped
warbler) in which some 6–8 additional (and incidental) species more typical
of open conifer forests occur exclusively. There appear to be few obvious
systematic differences in species composition between 1966 and the 1990s,
and in general the same species in similar rank order dominate the early and
the later communities; year-to-year variation within the 1990s is similar to that
between these censuses and those taken 25 years earlier. Nevertheless, of the
10 most common species, house wren was significantly more common and
Empidonax spp., warbling vireo and dark-eyed junco significantly less common in the 1990s censuses (*t* tests; $P < 0.05$).

The total bird densities vary modestly ($CV = 0.1$) over sites and among
years, with no discernible difference between earlier and later totals; the same
10 species comprise two-thirds to five-sixths of the total densities at all sites
and years (Table II). One view of the relative similarities of the aspen bird
communities among years is given in Fig. 8, in which the proportion of total
bird density is accumulated on the ordinate as a function of the same first-
ranked 15 species listed in Table II. Note that the first 4 to 5 species are very
similar in relative proportions within sites regardless of the census year and
generally account for about 50% of the total bird density. The density accumulation curves of the 1990s span the 1966 data at two of the three sites, but
at Elk Ranch East there appear to be consistent differences; this is due to the
currently higher density of house wren and lower densities of the next four
ranked species at this location. Note that the 1994 data are the most divergent
at all three sites; this was a very early breeding season in a dry year, and an
unusually high proportion of atypical species occurred in the aspens.

D. Birds of the Sagebrush Flats

Much of the floor of Jackson Hole is covered by sagebrush; where the sagebrush is tall and dense, as around the airport, sage grouse and sage thrashers
are present, but where the vegetation is lower, more grassy, and open, the bird
community lacks these habitat specialists and is composed mostly of emberizid sparrows. A 5-ha site in open sagebrush was established in 1966 adjacent

TABLE II Bird Communities at Three Aspen Woodland Sites in GTNP, 1966–1994[a]

	Elk Ranch West				Elk Ranch East				Cow Lake/Signal Mtn						
	JADF	MLC	MLC	MLC	JADF	MLC	MLC	MLC	JADF	MLC	MLC	MLC			
Species	1966	1992	1993	1994	1966	1992	1993	1994	1966	1992	1993	1994	Sum	Ave	Rank
House wren	2	3	8	4.5	2	6	8	7.5	3	4	6	5	59.0	4.917	1
Empidonax spp.	7	3	1	4	6	2	4	3	4	4	1	2.5	41.5	3.458	2
Tree swallow	3	3	6	4	5	3	3	2	3	3	4	2	41.0	3.417	3
Warbling vireo	4	2	2	3.5	6	2	3	3.5	3	2.5	3	3	37.5	3.125	4
Western wood pewee	4	3	3	2	4	2	2	4.5	2	2.5	3	4	36.0	3.000	5
White-crowned sparrow	4	6	4	4	2	3	3	4					30.0	2.500	6
Dark-eyed junco	4	3	1	1	3		1	2	4	1	3	2.5	25.5	2.125	7
Mountain bluebird	2	1	4	2	2	2	2	1.5	2	1	3	1.5	24.0	2.000	8
Robin		2	1	3	1	1	2	2	3	4	2	1.5	22.5	1.875	9
Yellow warbler	3		2	1	2	3	4	3			3		21.0	1.750	10
Northern flicker	1	1	2	2	2	1	1	2	2	1	2	1	18.0	1.500	11
Black-capped chickadee	2			2	2	1	1	2		3	1	2	16.0	1.333	12
Downy woodpecker	1	2		2		1	1	1		3	2	1	14.0	1.167	13
Chipping sparrow									2	4	3	4	13.0	1.083	14
Yellow-rumped warbler	2	2		1	1	1	1		1	2	1	0.5	12.5	1.042	15
Red-naped sapsucker				2		1				2		2	7.0	0.583	16
Orange-crowned warbler			3	1			1	1				1	6.0	0.500	17

Species												Average abundance	Average density	No.
House finch	1		2	2		1	1				0.5	5.5	0.458	18
White-breasted nuthatch		0.5	1	1.5		1	1				1	4.5	0.375	19
Green-tailed towhee					1		1		1		1.5	4.0	0.333	20
MacGillivray's warbler	1				1		1					3.0	0.250	21
Broad-tailed hummingbird			0.5		1		1	1				2.5	0.208	22
Cassin's finch								2		2	2.0	2.0	0.167	23
Calliope hummingbird			0.5			1					1.5	1.5	0.125	24
Clark nutcracker			0.5	0.5	0.1		0.2				0.5	1.3	0.108	25
Hairy woodpecker							1					1.0	0.083	26
Western tanager							1	1				1.0	0.083	27
Wilson's warbler						1						1.0	0.083	28
Lazuli bunting						1						1.0	0.083	29
Mountain chickadee											1	1.0	0.083	30
Cedar waxwing											0.5	0.5	0.042	31
Gray jay											0.5	0.5	0.042	32
Total species:	14	13	14	21	17	18	18	14		15	21			
Total density:	40	32	40	44	31	40	42	33		38	38			
Average species:	16			16		18	18				17			
Average density: (1992–1994)	39			39		38	38				38			

[a] Census sites were 5 ha.

Note. JADF = data from Flack.

MLC = data from Cody.

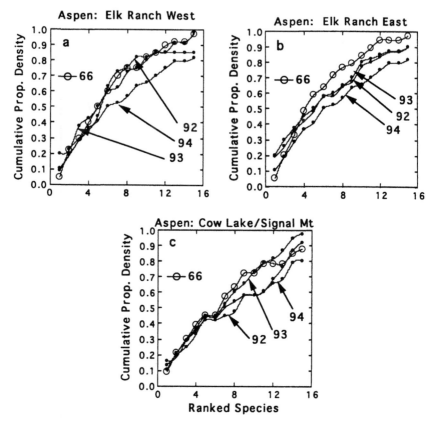

FIGURE 8. Bird communities at three aspen woodland sites in Grand Teton National Park, Wyoming. The proportion of the total bird density at each site is accumulated (ordinate) as a function of the (same) rank order of species, from most (left) to least (right) common, along the abscissa. Note that data from the 1990s straddle the 1966 data in two cases and that 1994 data are in general more different than those of other years. Data from 1966 are from Flack (1976).

to the Snake River 1 km east of Jackson Lake Dam, as reported and analysed in Cody (1974); the breeding birds were censused and territories mapped on a 15 × 15-m grid, and observations of foraging height and behavior recorded. The site comprises perhaps one-fifth of the drier river bench flat at this location (although much similar habitat is located within a few kilometers). Adjacent slopes are aspen or lodgepole pine woodland, and adjacent wetlands are marshy meadowland or willow flats. The same plot was used to assess territorial interactions among species in several subsequent years (1966–1970; Cody 1974), although it was not systematically re-surveyed. This site was reestablished in 1991, located as exactly as possible on the same coordinates (based on memory, photographs, and maps of vegetation features from the previous study), and has been censused each year since (1991–1994).

Vegetation structure at the site was measured in 1966 as a foliage profile, a plot of vegetation density versus height assessed separately for each grid quadrat. Similar vegetation measurements were made in 1991. All evidence suggests that the vegetation at the plot has not changed in any detectable fashion over the last 28 years; the same plant species in similar proportions occur in similar locations within the plot as they did originally, and the same peripheral willow clumps, similarly sized, occur in the same locations at the plot's edge.

By standing in this plot for hundreds of hours over many years, one of course sees many bird species, and most of those listed in the five sections of Table III can be seen on a typical morning's visit. However, just four species listed in the first category (vesper, savannah, Brewer's, and white-crowned sparrows) defend territories, breed, and feed regularly within the site, and a fifth species, chipping sparrow, does so occasionally. Species composition within this group has been stable over the years, and it is this group for which accurate density data exist and are comparable among years.

The densities of the emberizid sparrows are given in Table IV. Total densities vary some 25% (CV = 0.08) between the highest (2.78 pairs/ha; 1992) and lowest (2.24 pairs/ha; 1991) figures, which occur in adjacent years and span the 1966 data (2.40 pairs/ha). Further, the species order of mean abundance at the site in 1966 is maintained over the census period. However, densities of each species do vary among years; white-crowned sparrow is relatively constant (CV = 0.12), Brewer's sparrow is most variable (CV = 0.40), and vesper and savannah sparrows are intermediate (CV = 0.23 and CV = 0.36, respectively). Between-year variations are more pronounced in the densities of individual species than in the community as a whole, indicating either density compensation or reciprocal reactions to local conditions within years. Savannah sparrow density appears to respond predominantly to year-to-year variation in precipitation; the species is more common in wet years (1992) and less so in dry years (1994). The opposite trend best describes Brewer's sparrow variation.

The most common species at the grass–sage site occupy scarcely more than half of the plot, even in their good years. Clearly not all of the plot is equally suitable for the different sparrows; different species view different parts of the plot as best suited to their habitat requirements. Mean profiles of the vegetation within territories of the four regular species are shown in Fig. 9, with white-crowned sparrows in the densest tall vegetation, savannah sparrows preferring the more open, grassier sites, and Brewer's and vesper sparrows with similar and intermediate habitat preferences.

Differences in habitat preferences among species can be used to qualify the

TABLE III Birds Recorded in Breeding Season at Grass–Sage Site, GTNP, 1966–1994, Indicating Status, Frequency, Site Use, and Habitat Association

Species breeding on site, territories mapped

Vesper sparrow	Regular, 1966–1994
Savannah sparrow	Regular, 1966–1994
Brewer's sparrow	Regular, 1966–1994
Chipping sparrow	Irregular, 1990s; spillover from pines
White-crowned sparrow	Regular, 1966–1994

Species nesting on site but foraging elsewhere

Red-breasted merganser	1966–1994; site adjacent to Snake River
Gadwall	1966–1994; site adjacent to Snake River

Species regularly foarging over or in site

Cliff swallow	1966–1994
Brewer's blackbird	1966–1994; breeds nearby
Brown-headed cowbird	First seen 1990s
European starling	First seen 1991–1992; irregular 1993–94

Species foraging occasionally in or over site

Common nighthawk	1966–1994; wide forager
Northern flicker	1966–1994; from adjacent lodgepole pines
Tree swallow	1966–1994; from adjacent aspens
Bank swallow	1966–1994; from adjacent river
Barn swallow	1966–1994; wide forager
Common raven	1966–1994; wide forager
American robin	1966–1994; from adjacent aspens and pines
House finch	1990s; from nearby aspen and cottonwoods

Species foraging nearby and/or seen flying over

Sandhill crane	1966–1994; from adjacent meadows and wetlands
Kildeer	1966–1994; from adjacent river banks
Violet-green swallow	1966–1994; from adjacent lodgepole pines
Black-billed magpie	1966–1994; from adjacent willows and aspens
American crow	1966–1994; from meadows and rangeland
Mountain bluebird	1966–1994; from adjacent aspens
Black-headed grosbeak	1966–1994; from adjacent aspens

individual quadrats within the study site and generate the spatial distribution of preferred versus marginal habitat for each species. I used a factor analysis of the amount of vegetation (under the foliage profile) at heights of 0–0.075, 0.075–0.15, 0.15–0.225, 0.225–0.3, 0.3–0.45, 0.45–0.6, 0.6–0.75, 0.75–0.9, 0.9–1.05, 1.05–1.2, and taller than 1.2 m to describe the interquadrat variation in vegetation structure within the study plot; the first two factors account for two-thirds of the total variation. For each species, I identified subsets of the 208 quadrats used (located within territories) in at least four years, 1991–1994, and classified their distributions by confidence ellipses. Quadrats within

TABLE IV Site Characteristics and Bird Community Data for a Grass–Sagebrush Site

Study year:	1966	1991	1992	1993	1994
Site area (ha):	4.97	4.97	4.68	4.68	4.68
15 × 15-m quads:	221	221	208	208	208
Vesper sparrow					
Quads occupied	129	52	83	123	110
Breeding pairs	4	5	4	7	6
Total breeding density	0.78	0.62	0.62	1.06	0.82
Savannah sparrow					
Quads occupied	67	45	74	57	22
Breeding pairs	4	5	6	5	3
Total breeding density	0.44	0.50	0.75	0.64	0.26
Brewer's sparrow					
Quads occupied	108	55	101	31	95
Breeding pairs	6	5	9	5	8
Total breeding density	0.92	0.82	1.15	0.35	1.30
Chipping sparrow					
Quads occupied	0	12	0	26	0
Breeding pairs	—	1	—	1	—
Total breeding density	—	0.04	—	0.21	0
White-crowned sparrow					
Quads occupied	45	15	31	35	19
Breeding pairs	2	2	2	2	2
Total breeding density	0.26	0.26	0.26	0.32	0.23
Summary					
Total breeding species	4	5	4	5	4
Total quads × species occupying quads	349	183	289	267	246
Total bird density	2.40	2.24	2.78	2.58	2.61

Note. For each species, "quads occupied" indicates the number of 15 × 15 quadrats within the species' territories, and "breeding pairs" indicates the total number of pairs with territories completely within or partially overlapping the study site. Breeding density has units of pairs per hectare.

the 33% confidence ellipses were termed "core habitat" for the species in question, those between the 33 and 67% ellipses were called "secondary habitat," and those beyond the 67% ellipses were called "marginal habitat." In this way each quadrat was classified with respect to each species' habitat preference. Species' distributions in the habitat plane are shown in Fig. 10; the distinction of Savannah sparrow habitat is shown by the upper-left distribution, and the relative similarity in prime habitat for Brewer's and vesper sparrows is seen by the near coincidence of the ellipses describing their core habitat.

The consistency over years of habitat choice by the sparrows within this plot is illustrated by Fig. 11. Savannah sparrow, for example, exhibits a strong

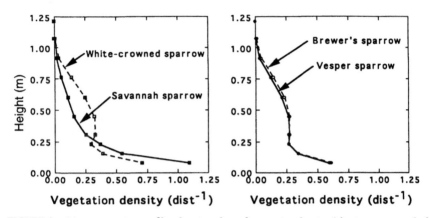

FIGURE 9.   Mean vegetation profiles showing plots of vegetation density (abscissa; reciprocal of mean horizontal distance to half-cover) vs height above the ground (ordinate) for four species of emberizid sparrows at a grass–sagebrush plot in Grand Teton National Park. Savannah sparrows (with a preference for open, grassy habitat) and white-crowned sparrows (with a preference for taller and denser sage) are most distinct, but Brewer's and vesper sparrows have very similar preferences.

preference for the grassier quadrats, which are relatively uncommon within the site. Just 29/208 quadrats are classified as core savannah sparrow habitat, with 26 others secondary and 153 marginal. This species is seen consistently to occupy similar parts of the study area over subsequent years, and clearly its choice of territories is strongly tied to areas suited to its preference for grassy locations. Habitat at this site is even more marginal for white-crowned sparrows (about 5% core habitat), but the species is similarly consistent in habitat choice among years (Fig. 11), with territories positioned each year to coincide most closely with quadrats of taller sagebrush adjacent to off-site willow patches that are often used as song-perches.

Both Brewer's and vesper sparrows are common in the study area, in accordance with the availability of suitable habitat; 46 and 56% of the site, respectively, is either core or secondary habitat for these species. However, core habitat is scattered throughout the site for both species (Fig. 11) and interspersed with less suitable quadrats. Accordingly, both species vary in the disposition of their territories among years. There is a tendency (cf. last two columns of the figure) for the two species to utilize different parts of the site in any one year, despite their similar habitat preferences.

Variations in site occupancy among years will depend on several factors. At the broadest resolution, more individuals may contest territorial habitat if overwintering success, and therefore spring population sizes, is high. Second, conditions at the site in any given year may elevate or depress habitat quality

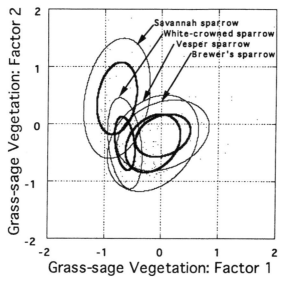

FIGURE 10. The Grand Teton National Park grass–sagebrush site is composed of 208 15 × 15-m quadrats, which are characterized by factor analysis based on vegetation density within stacked horizontal layers. Axes F(1) and F(2) account for two-thirds of the total variation in vegetation at the site. For each species, the distributions of those quadrats that were occupied in at least 3 of 4 years, 1991–1994, are summarized by the congruent ellipses encompassing one-third and two-thirds of the quadrats. Quadrats within the smaller ellipses are termed "core" habitat for the species; between the two ellipses, "secondary" habitat; and beyond the larger ellipses, "marginal" habitat.

and provide an override on the static assessment of habitat suitability provided by measures of vegetation structure. Third, the cooccupancy of quadrats by other species may add another factor to be evaluated in the quest for a suitable territory. Last, the spatial disposition of high-quality quadrats is important. Quadrats are not used individually, and may be acceptable only if surrounded by others of like quality; isolated quadrats which individually rank highly are relatively useless, whereas even the best territories may perforce include poorer habitat within territory-sized areas. This is termed an "adjacency constraint" by Haila *et al.* (1995) who considered this factor together with "habitat constraints," i.e., vegetation structure characteristics, in evaluating the occupancy of heterogeneous pine-spruce forest in Finland by a variety of bird species over successive years.

The interaction at various spatial scales of these diverse influences on habitat selection contributes largely to the extensive literature on this subject; see, e.g., Cody (1985) for an overview. The list might be extended by other factors such as those of history, tradition, or philopatry (e.g., Wiens, 1985),

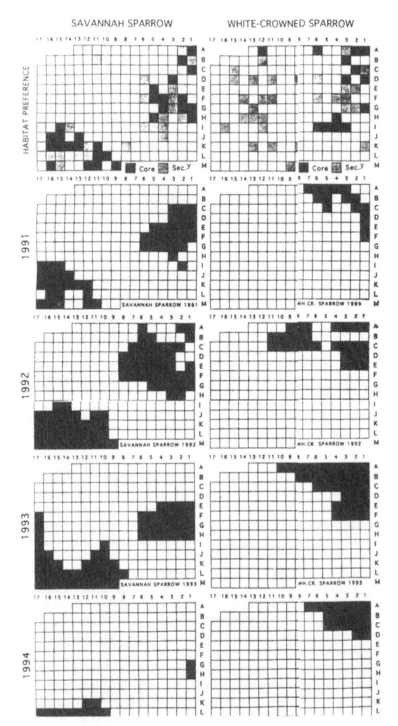

FIGURE 11. Each quadrat at the Grand Teton National Park grass–sagebrush site is classified as core (heavy shading), secondary (light shading), or marginal (unshaded) for each of four sparrow species. The occupancy of quadrats over 4 years is shown for each species in rows 2–5. Note the

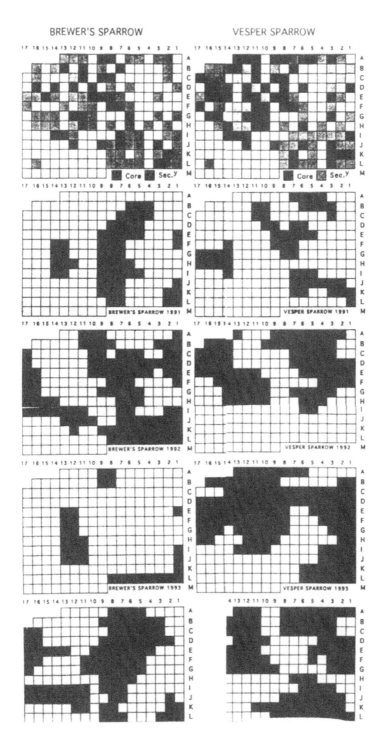

consistency of habitat choice over years for savannah and white-crowned sparrows and, in contrast, the variation in site occupancy for Brewer's and vesper sparrows, which in fact show some tendency to avoid quadrat co-occupancy.

factors that exacerbate the correct interpretation of habitat use from data generated on-site within a single season.

Some of these influences on site utilization are illustrated by savannah sparrow data in Fig. 12. In 1992, a wet year, savannah sparrow density was high, and $74/208 = 0.356$ quadrats were occupied. Of these, significantly

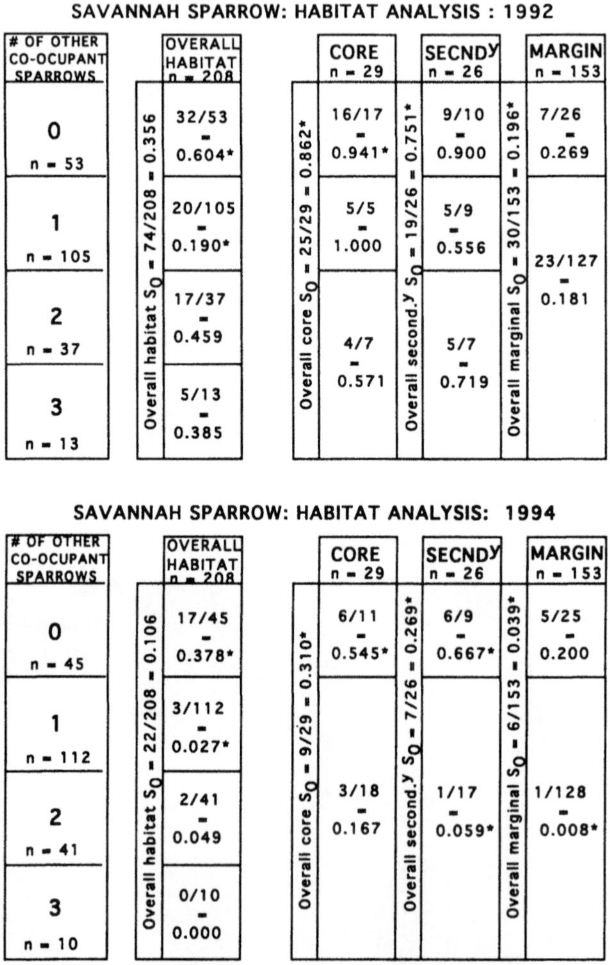

SAVANNAH SPARROW: HABITAT ANALYSIS : 1992

# OF OTHER CO-OCUPANT SPARROWS	OVERALL HABITAT n = 208	CORE n = 29	SECNDY n = 26	MARGIN n = 153
0 (n = 53)	32/53 — 0.604*	16/17 — 0.941*	9/10 — 0.900	7/26 — 0.269
1 (n = 105)	20/105 — 0.190*	5/5 — 1.000	5/9 — 0.556	23/127 — 0.181
2 (n = 37)	17/37 — 0.459	4/7 — 0.571	5/7 — 0.719	
3 (n = 13)	5/13 — 0.385			

Overall habitat $S_Q = 74/208 = 0.356$ Overall core $S_Q = 25/29 = 0.862^*$ Overall second.y $S_Q = 19/26 = 0.751^*$ Overall marginal $S_Q = 30/153 = 0.196^*$

SAVANNAH SPARROW: HABITAT ANALYSIS: 1994

# OF OTHER CO-OCUPANT SPARROWS	OVERALL HABITAT n = 208	CORE n = 29	SECNDY n = 26	MARGIN n = 153
0 (n = 45)	17/45 — 0.378*	6/11 — 0.545*	6/9 — 0.667*	5/25 — 0.200
1 (n = 112)	3/112 — 0.027*			
2 (n = 41)	2/41 — 0.049	3/18 — 0.167	1/17 — 0.059*	1/128 — 0.008*
3 (n = 10)	0/10 — 0.000			

Overall habitat $S_Q = 22/208 = 0.106$ Overall core $S_Q = 9/29 = 0.310^*$ Overall second.y $S_Q = 7/26 = 0.269^*$ Overall marginal $S_Q = 6/153 = 0.039^*$

FIGURE 12. Analysis of site occupancy for savannah sparrows in the grass–sagebrush site. In 1992 (top) the species was present at high density (74 quadrats occupied), and in 1994 (bottom) at low density (22 quadrats occupied). In both years the species occupied significantly more quadrats where no other sparrows were present (second column), an effect that was magnified in its core and secondary habitats (cols. 3 and 4). See text for further discussion.

more were monopolized by the sparrow, and significantly fewer than expected by chance were cooccupied by other species (by χ^2; see col. 2). However, this pattern has at least two reasonable explanations (see, e.g., Cody, 1981); the first, that it is a consequence of differential habitat selection, has already been established, while the second, that savannah sparrows avoid habitat that is (or would be) cooccupied by other species, remains to be tested. This is done in columns three through five. First note that savannah sparrows occupy significantly more core habitat (occupancy rate: 0.862) and secondary habitat (occupancy: 0.751), based on predictions from overall habitat occupancy of 0.356, than is expected by chance, and significantly less marginal habitat (occupancy: 0.196). Further, within their core habitat (col. 3), savannah sparrows have a significantly higher occupancy rate where the habitat is not shared with other species (occupancy: 0.941); good habitat appears to be avoided where other species are present. Just the same interaction was described in 1966 (Cody, 1974), where the discrepancy between overlap in habitat preference and overlap in space within the site was ascribed to interspecific interactions.

In 1994, a dry year, savannah sparrows were much less common at the site, and their overall occupancy rate dropped to 0.106 (Fig. 12, bottom). The same preferences for core and secondary habitat over marginal habitat are apparent; even stronger is the aversion to shared territory, with significantly higher occupancy rates in monopolized core and secondary habitat, and significantly lower occupancy rates in shared marginal habitat. It seems that the effects of interspecific interactions on Savannah sparrows are more pronounced in a poor year for this species.

The same sort of analysis may be repeated for the other sparrows at the site; Brewer's and vesper sparrows are particularly interesting, because of the similarity in their habitat requirements, as well as the fact that variations in their abundances are not synchronized (e.g., 1993 had the lowest Brewer's sparrow density and very high vesper sparrow density). In 1992, both occupy core habitat to a significantly greater extent when no other sparrows are cooccupants, but the two species show a variable interaction over habitat among years (see Table V). Brewer's and vesper sparrows show a significant aversion to shared territories when overall sparrow density in the field is high (1992) and a significant tendency to cooccupy their shared core habitats when overall density is low (1991); in other years of intermediate sparrow density (e.g., 1966), there is neither a positive nor a negative association of the species (Cody, 1974, 1993, 1994). Note, finally, that this result is not what one would expect if the sparrow densities measured at the grass–sage site were the product of local factors that varied among years. If they were, 1992 would be a good year because local conditions were good and sparrow densities tracked local conditions. In such circumstances, we might expect higher territorial overlap in 1992 because of the good local conditions and low overlap when densities

TABLE V   Analysis of Brewer's and Vesper Sparrow Habitat Cooccupancy, Based on the 65 Quadrats in the Grass–Sage Site That Are either Core or Secondary Habitat for Both Species

	Year							
	1991		1992		1993		1994	
Species	Quads occupied	P	Quads occupied	P	Quads occupied	P	Quads occupied	P
B	17	0.262	35	0.538	12	0.185	34	0.523
V	22	0.338	30	0.462	49	0.754	41	0.631
	Obs	Exp	Obs	Exp	Obs	Exp	Obs	Exp
BV'	5	11.3	27	18.8	2	3.0	16	12.5
B'V	10	16.2	22	13.9	39	39.9	23	19.6
BV	12	5.8	8	16.2	10	9.1	18	21.5
B'V'	38	31.8	8	19.2	12	13.0	8	11.4
χ^2	13.9*		16.6*		0.5		3.14	

Note. The upper part of the table gives the numbers of quadrats and proportions of the overall site occupied by the two species. The lower part of the table contrasts the observed (Obs) and expected (Exp) numbers of quadrats that have just Brewer's sparrow (BV'), just Vesper sparrow (B'V), both species (BV), and neither (B'V').
*$P < 0.05$ that the two species are nonrandomly disposed with respect to each other.

were low (indicating poor local conditions). The opposite is seen, leading us to suspect that variations in densities among years are more the product of extrinsic rather than local factors; thus the sparrows apparently make local adjustments to externally driven higher or lower densities, reflected by variation among years in the levels of territorial overlap that are tolerated.

E. Birds of the Wet Willows

Salt (1957a) reported breeding bird composition and species' densities in a willows habitat in Jackson Hole and discussed (Salt, 1957b) the habitat selection and relative abundances of Lincoln's, fox, and song sparrows at the same site. Subsequently I have studied his approximate site intensively, from the 1960s to the early 1970s (Cody, 1974), and again from 1991 through 1994 (Cody, 1991, 1992, 1993). The site, which is below the Willow Flats overlook in GTNP, looks superficially homogeneous and appears as a large patch of uniform willow scrub up to 3–4 m high bordered by wet meadows or sagebrush.

I established my site in 1966 to coincide in part with Salt's (1957a) 55 × 800-m belt transect. Between 1966 and the 1990s the location of the site was changed somewhat because of beaver activity; several small streams that run through the site to the Snake River were found in 1991 to be dammed by beavers, flooding a part of the previous study area and promoting its partial relocation (with approximately 60% overlap with the new site) and subsequent expansion in 1992. The flooding remained high through 1993, but water levels were down in 1994, when beavers were no longer occupying the ponds and maintaining the dams.

The study site appears much more heterogeneous from within than from without, and encompasses considerable variation in vegetation at the resolution of 15 × 15-m quadrats. This variation was indicated in Cody (1974, Fig. 7), and is shown updated in this chapter in Fig. 13. The major components of this variation are the extent of open water, of grassy areas without shrub cover, and of cover by willows of different heights. Variation in vegetation height best describes this heterogeneity (without resorting to multivariate statistics), and I use the proportion of vegetation of different heights to characterize habitats chosen for territories in the different breeding birds.

The site affords breeding habitat to a diverse suite of birds, including six emberizid finches and five paruline warblers, together with a flycatcher, magpie, wren, thrush, and a hummingbird. The species recorded here over the years of study are summarized in Table VI, where the first third of the 43 listed species is the regular breeding bird community. The species for which density and territory data were collected are listed in Table VII, which compares species composition and densities at the site in 1952–1954, 1966, and 1991–1994. Three paruline warblers (yellow and Wilson's warblers and common yellowthroat) are regular breeders, as are the willow flycatcher and four species of emberizid finches (song, fox, Lincoln's and white-crowned sparrows). Year-to-year variations in density among these species are reflected in CV values ranging from 0.14 to 0.54; CVs average 0.16 for the total breeding bird density and the summed warbler and sparrow densities.

Among the species regularly censused at the site since the 1960s, only one species was not found by Salt (1957a) in the early 1950s, namely Wilson's warbler. Notably, this is a species that was not listed in the infant GTNP survey data of USDI (1937), and conceivably the species has become much more common in the region from the 1960s to the 1990s.

Before discussing the ecology of the regular breeding species further, some remarks on the irregular breeding species are warranted. Some changes in the bird community are clearly attributable to the changes within the site, specifically with regard to water levels. With increased flooding at the site, several water birds bred or foraged there in the 1990s (see Table V); among the passerines marsh wren and northern waterthrush were added as breeding species. More subtle changes also resulted from higher water levels; ground-foraging Swainson's thrush occupied the taller willows along watercourses in

FIGURE 13. Variation in vegetation height in the wet willows site, Grand Teton National Park. Such variation is relevant to the bird species inhabiting the site, especially to the four more common species of paruline warblers that forage at different heights in the vegetation and concomitantly select and defend territories corresponding to differences in vegetation height.

TABLE VI   Birds of the Wet Willows Site and Their Status, 1966–1994

Breeding species, territories mapped routinely

Parulines
1. Yellow warbler	Regular breeder, 1966–1994
2. Common yellowthroat	Regular breeder, 1966–1994
3. Wilson's warbler	Regular breeder, 1966–1994
4. MacGillivray's warbler	Irregular breeder, 1966–1994
5. Northern waterthrush	Irregular breeder, 1990s

Emberizines
6. Song sparrow	Regular breeder, 1966–1994
7. Fox sparrow	Regular breeder, 1966–1994
8. Lincoln's sparrow	Regular breeder, 1966–1994
9. White-crowned sparrow	Regular breeder, 1966–1994
10. Clay-colored sparrow	Regular breeder, 1966–early 1970s
11. Savannah sparrow	Irregular breeder, 1990s

Others
12. Calliope hummingbird	Irregular breeder, 1966–1994
13. Willow flycatcher	Regular breeder, 1966–1994
14. Marsh wren	Irregular breeder, 1990s
15. Swainson's thrush	Irregular breeder, 1966–1994

Other breeding species, territories not mapped

Dryland species
16. Black-billed magpie	Regular breeder, 1966–1994
17. Warbling vireo	Bred on site, 1993

Wetland-associated species
18. Virginia rail	Bred on site, 1992, 1994
19. Green-winged teal	Bred on site, 1992
20. Gadwall	Bred on site, 1994
21. Ruddy duck	Bred on site, 1992, 1993
22. American coot	Bred on site, 1992, 1993
23. Common snipe	Bred on site, 1992–1994

Species breeding nearby, and using site in late season with juveniles

24. Red-naped sapsucker	Regular forager, 1966–1994
25. Black-capped chickadee	Regular forager, 1966–1994

Species foraging regularly in or over site

26. Cliff swallow	Regular forager, 1966–1994
27. Barn swallow	Regular forager, 1966–1994
28. American robin	Regular forager, 1966–1994

Species foraging occasionally in or over site

29. Great blue heron	1994
30. Black-crowned nightheron	1994
31. Sandhill crane	Occasional, 1966–1994
32. Broad-tailed hummingbird	1993
33. Northern flicker	Occasional, 1966–1994; wide forager
34. Violet-green swallow	Occasional, 1966–1994; wide forager

continues

TABLE VI   *(Continued)*

35. Rough-winged swallow	Infrequent, 1966–1994; wide forager
36. Cedar waxwing	1992
37. Red-winged blackbird	1991
38. Brown-headed cowbird	Occasional, 1990s
39. Black-headed grosbeak	Occasional, 1993–1994

Species foraging adjacent to plot or flying over

40. Northern harrier	Regular, 1966–1994; from adjacent meadows
41. Common raven	1966–1994; wide forager
42. American crow	1990s; wide forager
43. American goldfinch	1993; casual

1966 but has been essentially absent since their flooding. Calliope humming-birds also were common in 1966, depending on stream-bank nectar-producing plants such as *Castilleja,* but this species was much reduced at the site in the 1990s, recovering somewhat in the dry summer of 1994.

A further notable change, but one less readily attributable to within-site changes, has been in the clay-colored sparrow population. This species was common in the 1960s within the willows in the drier and more open glades but has not bred there in the 1990s; indeed, the species was located within GTNP only once, in 1994, a dry year. Perhaps associated with the clay-colored sparrow's absence, savannah sparrows have occurred within the willows site sporadically in the 1990s, utilizing the same open and grassy plots previously used by clay-colored sparrows.

There are considerable differences in habitat preferences among the breeding bird species. These differences were described previously (Cody, 1974, Fig. 9) and are illustrated for the six more common sparrows in Fig. 14. Although there is considerable habitat overlap among them, the species consistently select the same or similar parts of the site for location of territories. Song sparrow territories are generally located in the wettest sites (Fig. 14; height "0" indicates open water); Lincoln's sparrows in areas with significantly lower and denser willows (heights 2–6 ft or 0.6–1.8 m); fox sparrows where the willow cover is closed and tall (8–14 ft or 2.5–4.2 m); and white-crowned sparrows preferentially in sites where open, grassy habitat is well represented (±1 ft, or 0.3 m).

In terms of foraging ecology and body sizes, the two most similar species are song and Lincoln's sparrows, and these species show habitat preferences that are most diametrically opposed: wet sites with taller willows and dry sites with low willows, respectively (Fig. 14). Accordingly, the changes in densities of these two species, relative to changes within-site, are revealing. Densities of the two species are significantly negatively correlated ($r = -0.94$, $P < 0.05$), and the change of numerical dominance from Lincoln's to song sparrow from

TABLE VII Species and Densities of the Wet Willows Bird Community in GTNP

Year of census:	1952–1954[a]	1966	1991	1992	1993	1994	CV
Site area (ha):	4.0	2.21	2.20	3.26	3.26	3.26	
Yellow warbler							
Quads occupied		61	66	113	114	107	
Breeding pairs		8	8	11	13	11	
Total breeding density	2.53	1.99	2.00	2.61	2.91	2.67	0.17
Wilson's warbler							
Quads occupied		48	53	19	52	36	
Breeding pairs		6	6	2	4	3	
Total breeding density	—	1.81	1.55	0.40	1.10	0.64	0.54
Common yellowthroat							
Quads occupied		85	51	85	83	86	
Breeding pairs		11	7	10	12	11	
Total breeding density	1.28	3.08	1.45	2.06	2.55	2.50	0.26
MacGillivray's warbler							
Quads occupied		5	0	0	19	8	
Breeding pairs		1	—	—	2	1	
Total breeding density	0.25	0.09	—	—	0.26	0.08	
Northern waterthrush							
Quads occupied		0	0	8	0	6	
Breeding pairs		—	—	1	—	1	
Total breeding density	—	—	—	0.18	—	0.15	
Willow flycatcher							
Quads occupied		34	35	34	49	44	
Breeding pairs		2	2	2	3	2	
Total breeding density	1.28	0.36	0.50	0.37	0.41	0.49	0.15
Lincoln's sparrow							
Quads occupied		80	41	48	40	49	
Breeding pairs		6	6	5	6	6	
Total breeding density	1.61	1.99	1.27	0.92	0.89	1.05	0.37
Fox sparrow							
Quads occupied		58	47	55	67	74	
Breeding pairs		5	5	6	6	7	
Total breeding density	0.50	1.00	0.82	1.04	1.21	1.01	0.14
Song sparrow							
Quads occupied		27	33	62	77	74	
Breeding pairs		3	5	8	9	8	
Total breeding density	0.74	0.90	1.36	1.44	1.72	1.55	0.22
White-crowned sparrow							
Quads occupied		71	27	32	38	36	

continues

TABLE VII (*Continued*)

Year of census: Site area (ha):	1952–1954[a] 4.0	1966 2.21	1991 2.20	1992 3.26	1993 3.26	1994 3.26	CV
Breeding pairs		6	3	3	4	3	
Total breeding density	0.37	0.36	0.82	0.64	0.86	0.77	0.29
Clay-colored sparrow							
Quads occupied		28	0	0	0	0	
Breeding pairs		3	—	—	—	—	
Total breeding density	—	0.72	—	—	—	—	
Savannah sparrow							
Quads occupied		0	0	9	7	4	
Breeding pairs		—	—	1	2	1	
Total breeding density	—	—	—	0.06	0.20	0.05	
Swainson's thrush							
Quads occupied		15	0	4	0	0	
Breeding pairs		2	—	1	—	—	
Total breeding density	0.17	0.27	—	0.03	—	—	
Calliope hummingbird							
Quads occupied		33	0	7	6	37	
Breeding pairs		3	—	2	1	2	
Total breeding density	1.07	1.36	—	0.09	0.07	0.58	1.37
Marsh wren							
Quads occupied		0	0	10	0	12	
Breeding pairs		—	—	1	—	1	
Total breeding density	—	—	—	0.25	—	0.22	
Summary totals for above 15 mapped species							
Species breeding	10	12	8	13	11	13	
Total breeding density	9.80	14.93	9.77	10.09	12.18	11.76	0.18
Paruline spp.	3	4	3	4	4	5	
Total breeding density	4.06	6.97	5.00	5.25	6.82	6.04	0.15
Emberizine spp.	4	5	4	5	5	5	
Total breeding density	3.22	5.97	4.27	4.10	4.88	4.43	0.16

[a]Data from Salt (1957a), who employed a belt transect 18 × 267 m and reported densities in terms of individuals/acre. Salt's densities, according to his own calibration with studies mapping territories, translate from 1 indivudal/area to 1 pair/area, indicating belt transect counts of largely singing males. Salt's larger area included some taller vegetation, accounting for four additional passerine species adding 0.83 pair/ha and increasing his overall bird density from 9.80 to 10.63 pair/ha.

Note. For each species, "quads occupied" indicates the number of 15 × 15-m quadrats within the species territories, and "breeding pairs" indicates the total number of pairs with territories completely within or partially overlapping the study site. Breeding density units are pairs per hectare. CV, coefficient of variation.

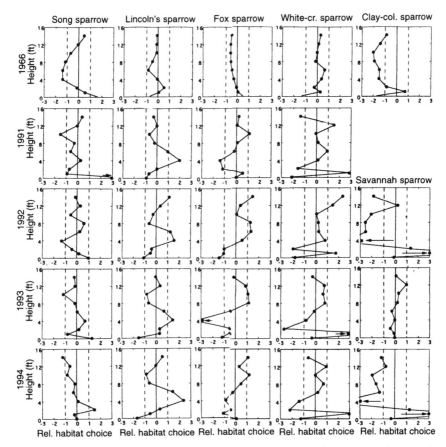

FIGURE 14. Habitat preference diagrams for six species of emberizid sparrows in the wet wil-lows site, Grand Teton National Park. Preferences are shown relative to the vegetation at the site as a whole, where the proportions of vegetation by height in each 15 × 15-m quadrat are stan-dardized (mean = 0, SD = 1), and the vegetation within the territories of each species is summarized in different years as shown. Species differ in habitat preferences and are consistent in habitat choice from year to year. Note that song and Lincoln's sparrows, closely related species of very similar body size, exhibit habitat choices that are near mirror images of each other, indicating their use of different portions of the site.

1966 to the 1990s parallels the change from drier to wetter conditions within the site. Notably, of all of the breeding species within the willows, these species are the only two between which direct aggressive interactions have been observed.

The paruline warblers, being foliage insectivores, are presumably less af-fected by within-site changes such as ground-level flooding; the three regular breeding species are joined sporadically by northern waterthrush, which cen-ters its territory on the large beaver pond, and by MacGillivray's warbler, which is present within the site in most years but is never common. Differences in habitat choice among the parulines are clearcut, with mean vegetation heights within territories increasing from common yellowthroat to yellow warbler, Wilson's warbler, and MacGillivray's warbler (Fig. 15). These differences are consistent among years, with preferences of yellowthroats for quadrats with vegetation in the 2- to 6-ft (0.6–1.8 m) range, of yellow warbler for 6- to 10-ft (1.8–3 m) vegetation, of Wilson's warbler for 8- to 12-ft (2.4–3.6 m) willows, and of MacGillivray's warbler for generally defending territories around the tallest vegetation present.

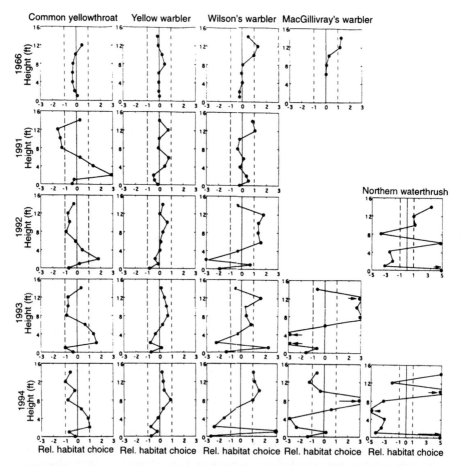

FIGURE 15. Habitat preferences of five species of paruline warblers at the wet willows site, Grand Teton National Park (see Fig. 14 for format). Preferences for increasingly tall habitat are apparent from common yellowthroat through yellow warbler and Wilson's warbler to Mac-Gillivray's warbler. Note that the ubiquitous yellow warbler exhibits scarcely any preference, as it inhabits nearly all quadrats at the site.

Habitat selection patterns provide an explanation for the relative differences in density among species, as the site does not supply suitable habitat to all species equally. The frequency distribution of vegetation by height within the site is shown in Fig. 16, in which vegetation 2–3 m high is clearly the most common, vegetation 0.8–2 m is less common, while vegetation taller than 3 m is quite scarce. These availabilities mirror warbler densities (see Table VII) with yellow warbler the most common, followed in rank order by common yellowthroat, Wilson's warbler, and MacGillivray's warbler.

Preferences among the warblers for territory locations based on vegetation height correspond to interspecific differences in foraging heights. All species are similarly sized foliage insectivores and, with the exception of Wilson's warbler which indulges in some hover-gleaning, foraging behavior is similar among species. The warblers, however, are conspicuously different in foraging height distribution, with foraging heights matching the upper levels of the predominant vegetation within their territories. The match is seen in Fig. 17 (using 1993 data), where different species exploit vegetation at different heights (left hand column), corresponding to the vegetation that is preferentially incorporated into their territories (right hand column).

I return now to further consideration of density variations in the common warbler species (e.g., those of Fig. 17) among years. Such variations are apparently driven by external factors such as differences in overwintering success, rather than by interyear differences within the site, where at least over a few

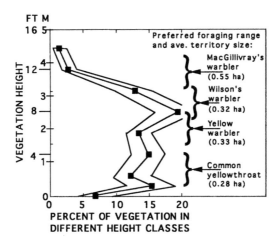

FIGURE 16.   The frequency distribution of vegetation of different heights at the wet willows site shows that habitat availability varies by vegetation height. The relative densities of the warbler species corresponds to the relative availability of their preferred vegetation at the site. Taller vegetation (>3 m) is particularly scarce, and MacGillivray's warbler, which prefers the tallest willows, defends correspondingly larger territories.

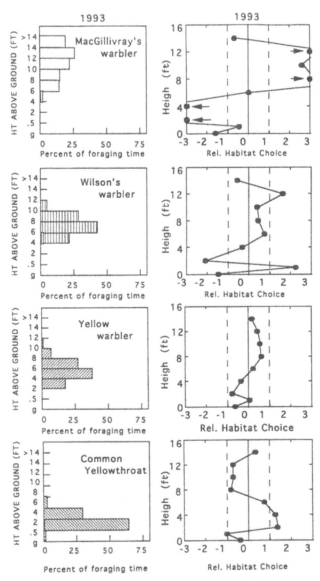

FIGURE 17. Habitat preferences, in terms of vegetation height, differ among species (right-hand column) and correspond to the interspecific differences in foraging height distribution (left-hand column). Across species, territories are located within the site where vegetation height matches foraging height.

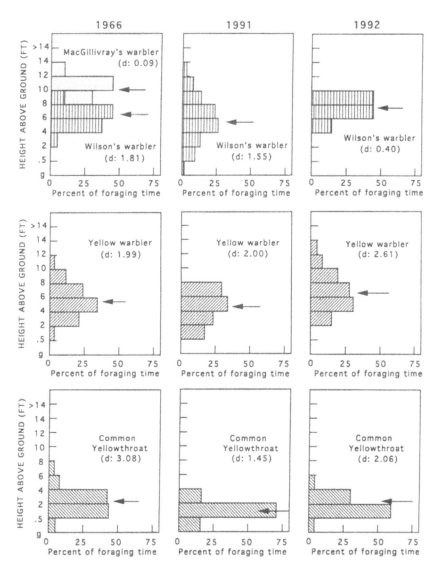

FIGURE 18. Frequency distributions of foraging heights vary conspicuously among species and also vary somewhat between years as a function of interyear differences in species' densities. See text for discussion.

decades the willow vegetation has remained quite constant. Accepting provisionally their likely extrinsic origin, the density variations are reflected on-site in two ways: by differences in species' habitat utilization among years, and by differences in species' foraging height distributions. In 1966 common yellow-throat was abundant at the site, outnumbering even yellow warbler, and consequently it exhibited no specific habitat preferences relative to the site as a whole (Fig. 15). In 1991 common yellowthroats were just half as common; their habitat preferences for low willows are clear and foraging heights are lower (Fig. 18). In 1992 Wilson's warbler was especially uncommon (as it was elsewhere in GTNP), and the repercussions were several: (a) the species was restricted to its most favorable habitat within the study site in 1992, and its habitat preferences appeared sharper than in other years; (b) its foraging activities were higher in 1992, corresponding to its restriction to taller, preferred willows; (c) yellow warblers were common in 1992, and their foraging heights were shifted up in the vegetation to parallel the shift in Wilson's warbler; and (d) in turn, common yellowthroats foraged higher in that year (Fig. 18). Note that there are no significant correlations among warbler densities, there are no direct interspecific interactions, and there are extensive overlaps among species pairs in both territorial habitat and foraging height. Yet there remains the potential for density compensation and niche shifts in habitat and foraging height following interyear density variation. While density variations might be generated by off-site effects, they have consequences for the organization of the bird community within the site.

An overall assessment of the bird community at the willows site balances change with constancy. There was the advent of Wilson's warbler to the site apparently some time prior to the early 1960s, with no plausible explanation for its earlier absence. There was a turnover in sparrows, clay-colored to savannah, that might be attributable to longer term climate cycles. There were within-site changes related to beavers and water levels, probably enhancing the habitat for marsh wrens and northern waterthrushes, changing it for the worse from the perspective of ground-foraging thrushes and forb-dependent hummingbirds, and apparently altering the balance in two common sparrows, favoring song sparrow and disfavoring Lincoln's. Other density variations were apparently unrelated to either long-term abiotic factors or local, within-site habitat changes; both Wilson's and MacGillivray's warblers varied in density among years in a fashion that was certainly not site- or locality-specific (as their variations were at least park-wide), and such variation was probably a regional phenomenon related to off-season (overwinter) events. Some of this variation was accommodated by niche shifts within the community.

IV. SYNOPSIS

A. Intrinsic vs Extrinsic Factors

Temporal constancy versus change in community composition and structure is a topic at the heart of community studies, and many aspects of it are illustrated by the birds of the central Rockies. Temporal aspects may be contrasted to spatial aspects of community constancy; the replicability of bird communities within habitats, even within years at sites not too distant, is far from perfect (see "Scrub Oak Communities," Sect. II.B., in this chapter; also Cody, 1994). This presumably was owing at one level to variation in the details of vegetation structure and at another to factors such as habitat area, access, and composition of the adjacent landscape with its concomitant provisions for colonization sources or spillover effects.

Another source of poor replicability will be sampling effects, and this applies to both spatial and temporal replication; bird communities that are composed wholly of nonresidents and some of these long-distance migrants are especially vulnerable to species turnover and community recomposition among years. Given the wealth of potential for variation in both local and distantly operating effects, it is notable that both the grass–sage flats and the wet willows sites have been remarkably constant in species composition over at least a 25-year interval. In both sites some 80–85% of the bird biomass is composed of the same species each year; the same species occur, in comparable numbers, exhibiting the same habitat preferences and foraging habits and participating in the same interactions among species. That whatever variation does occur is as readily seen in adjacent years as between years a quarter-century apart attests to the lack of any directional or longer-term change in these communities.

The wet willows site exemplifies most of the factors that can influence bird communities and their stability or variation over time. Long-term community studies are sensitive to, and may document, changes due to long-term changes in climate and changes in bird ranges and population sizes that might be associated with climatic change. Clearly studies conducted over a longer term are more likely to accrue within-site (directional) changes that are due to climatic change and may span or sample a wider range of local climatic variation due to differences among years in abiotic conditions.

A useful classification of the cause of variation in populations and the communities they constitute recognizes extrinsic and intrinsic factors relative to the study site. Clearly major habitat changes within-site are intrinsic; those produced by flooding and tree deaths at the cottonwood site are obvious examples, responsible for predictable and understandable directional changes in the breeding bird community. Within-site habitat change, by forest thinning, was also reflected in changes in the bird community at the ponderosa pine site, and flooding within the wet willows contributed predictable changes there, in both species composition and density.

Extrinsic factors are more difficult to account for and are *de facto* a category that is held accountable when no clear intrinsic factors can be identified. In some cases, such as the variation in warblers among years in the wet willows, "extrinsic" seems an appropriate term. When Wilson's warbler is common in

GTNP, it is common within this site; when MacGillivray's warbler is common within the park, it usually breeds in the wet willows, and otherwise not. Thus the phenomenon can be established as a broader, regional effect (by monitoring other habitats in GTNP), and it is clearly not related to conditions intrinsic to the site. Presumably the occasional use of the willows by warbling vireos, or of the grass–sage by chipping sparrows, has a similar explanation and is not explicable by intrinsic factors.

Judging by variations in the warbler species over a 25-year interval, such changes are not directional, but simply mirror year-to-year variation in extrinsic factors. Other changes may be indeed directional and extrinsic to the site where they are observed, though regionally generated. The apparent increases in GTNP of American crows, European starlings, cliff swallows, and house wrens may be readily attributable to factors within the park (e.g., enhanced foraging opportunities in overgrazed lands or additional nest site options on buildings and other structures), but the increased incidence of house wrens in aspen woodland or of starlings in the grass–sage site are not easily related to within-site factors. There is, in effect, a continuum from extrinsic to intrinsic factors; they are clearly identical for resident species, but for species that are short-distance migrants, and for species that are enabled by changes in one habitat to expand their use of another, they are indistinct.

Another view of this puzzling aspect of interyear variation is obtained by consideration of the many species that are at or near their range margins in GTNP. Clay-colored sparrows are an example, as are gray catbirds, least flycatchers, American redstarts, and many others. There seems to be an association between drier conditions within Jackson Hole and the presence of clay-colored sparrows and others; of wetter conditions and the presence of northern waterthrush, gray catbird, and others. Given these associations, it seems that intrinsic factors must play a large role; yet, unless local factors covary with factors beyond Jackson Hole and enhance the availability of birds to breed at their range margins, they will not be detectable.

B. Migrants vs Residents

Variation in extrinsic factors will apply least to resident birds, most to long-distance migrants. Migration distance might be assumed *a priori* to be correlated with the importance of extrinsic influences; in this respect, it seems relevant to year-to-year variation in grass–sage sparrow densities, as Brewer's sparrow varies the most and winters largely south of the United States–Mexico boundary, white-crowned sparrow varies the least and winters within or close to the Rockies, and vesper and savannah sparrows are intermediate in both respects. However, the CVs of emberizids and parulines in the wet willows are comparable in magnitude, even though the former winter more locally and the latter are largely Neotropical migrants. Empirically, it seems that overwinter survival can be just as unpredictable at sites nearer breeding habitats as it is farther away, and the ponderosa pine census data tend to support this contention; neither residents nor summer visitors are more predictable or constant among years, and variation in both overwinter success and in breeding season resources appears to contribute to year-to-year differences in breeding density.

C. Advantages of Long-Term Studies

Both advantages and disadvantages characterize long-term community studies (with the tacit assumption that some combination of time, energy, and funding is limited). The ecology of a breeding bird community can be understood to a large extent within the context of a single breeding season: species composition and densities can be measured; habitat choice characterized and related to variation in, and availability of, vegetation features; interspecific differences in foraging ecology in turn related to preferred vegetation; and interspecific interactions over territory clarified. Repeating these measurements over a few adjacent years, given the considerable and often dramatic differences among years in local weather patterns in the central Rockies, will provide an indication of how such variations affect different species to different degrees and generate minor rearrangements in community structure. But both potential fallacies that may be supported by short-term study can be dispelled in the longer term; the constancy of species, in incidence, density, habitat use, and interspecific associations, might be either more or less apparent in the shorter term than is verified in the longer. For example, 1992 and 1993 were quite different years in many ways for the grass–sage sparrows, but by considering other years over a 25-year span they are in fact both somewhat extreme, and most years will be intermediate. The same 2 years in the wet willows produced different warblers (northern waterthrush in 1992 and MacGillivray's warbler in 1993); Swainson's thrush and marsh wren were found in just one year; and threefold differences existed in savannah sparrow and Wilson's warbler densities. However, in the longer term, the first five of these species are seen as marginal in all years at the site, and the poor year for Wilson's warbler in 1992 appears as the anomaly; the relative stability of the core species is convincing, as the longer-term data set exemplifies.

Finally, of course, there do exist longer-term trends that can be identified only with long-term monitoring. The dynamics of species at the edges of their geographical ranges can be studied only over longer time periods encompassing broader variations in climate pattern and population cycles. Concern over the declining trends in some Neotropical songbirds is another example requiring a longer-term monitoring effort (Keast and Morton, 1980; Smithsonian Institution, 1991; Terborgh, 1989), and increasing trends in alien species (e.g., European starlings) or others undergoing range extensions with detrimental effects on native birds (e.g., brown-headed cowbirds) are another.

V. SUMMARY

The central Rocky Mountains lie at or near the edge of many bird species' distributions. The higher elevation continental climate produces severe winters and short, unpredictable summers, and a major proportion of the avifauna is migratory. In these circumstances, the year-to-year consistency of bird communities, in terms of species composition and density, is of particular interest and is considered for a variety of foothill to montane habitats.

Published census data are discussed for several habitats in Colorado to address questions of census replicability within a habitat (scrub-oak); of avian responses to consistent, unidirectional habitat change (in flooded cottonwood forest); and of long-term breeding and wintering bird community stability and variation (in ponderosa pine forest). In Grand Teton National Park, Wyoming, census data covering (noncontinuously) a 30-year period are available for aspen and grass–sage habitats, and for a wet willows site the bird community data reach back over 40 years.

In very few cases discussed, the bird communities occupy habitats that have undergone substantial and systematic changes; more commonly there has been minor or no habitat change over several decades, although variations in abiotic conditions from year to year may have been conspicuous. Factors affecting bird community variations are discussed relative to their extrinsic (off-site) or intrinsic (on-site) origins. Examples are given of both sorts of influences, with attention paid to the prominence of community stability in the face of the diversity of such factors mitigating against it. While some examples are cited of population changes at the 20- to 40-year time scales, most community variation is minor in scope, observable between adjacent years; it may be readily ascribed to either particular intrinsic or to poorly understood extrinsic effects. The consistency of other community attributes, habitat preference and utilization patterns, and interspecific interactions is also reported for two sites (grass–sage and wet willows) for which extensive and detailed community data exist.

Acknowledgments

The author's research in Grand Teton National Park was generously supported by the New York Zoological Society, the University of Wyoming, the National Park Service, and the National Science Foundation.

Appendix
Scientific Names of Birds Mentioned in Text

American coot *Fulica americana*
American crow *Corvus brachyrhynchos*
American goldfinch *Carduelis tristis*
American redstart *Setophaga ruticilla*
American robin *Turdus migratorius*
Bank swallow *Riparia riparia*
Barn swallow *Hirundo rustica*
Black-backed woodpecker *Picoides arcticus*
Black-billed magpie *Pica pica*
Black-capped chickadee *Parus atricapilla*
Black-crowned nightheron *Nycticorax nycticorax*
Black-headed grosbeak *Pheuticus melanocephalus*
Bobolink *Dolichonyx oryzivorus*
Brewer's blackbird *Euphagus cyanocephalus*
Brewer's sparrow *Spizella breweri*
Broad-tailed hummingbird *Selasphorus platycercus*
Brown-headed cowbird *Molothrus ater*
Brown creeper *Certhia americana*
Calliope hummingbird *Stellula calliope*
Cassin's finch *Carpodacus cassinii*
Cedar waxwing *Bombycilla cedrorum*
Chipping sparrow *Spizella passerina*
Clark's nutcracker *Nucifraga columbiana*
Clay-colored sparrow *Spizella pallida*
Cliff swallow *Petrochelidon pyrrhonota*
Common grackle *Quiscalus quiscula*
Common nighthawk *Chordeiles minor*
Common raven *Corvus corax*
Common snipe *Gallinago gallinago*
Common yellowthroat *Geothlypis trichas*
Cordilleran flycatcher *Empidonax difficilis*
Dark-eyed junco *Junco hyemalis*
Downy woodpecker *Picoides pubescens*
Dusky flycatcher *Empidonax oberholseri*
Eastern kingbird *Tyrannus tyrannus*
Eurasian starling *Strunus vulgaris*
Evening grosbeak *Coccothraustes verpertinus*
Fox sparrow *Melospiza iliaca*
Gadwall *Anas strepera*
Golden-crowned kinglet *Regulus satrapa*

Grasshopper sparrow   *Ammodramus savannarum*
Gray catbird   *Dumetella carolinensis*
Gray jay   *Perisoreus canadensis*
Great blue heron   *Ardea herodias*
Green-tailed towhee   *Pipilo chlorurus*
Green-winged teal   *Anas crecca*
Hairy woodpecker   *Picoides villosus*
House finch   *Carpodacus mexicanus*
House wren   *Troglodytes aedon*
Kildeer   *Charadrius vociferus*
Lark bunting   *Calamospiza melanocorys*
Lark sparrow   *Chondestes grammacus*
Lazuli bunting   *Passerina amoena*
Least flycatcher   *Empidonax minimus*
Lesser goldfinch   *Carduelis psaltria*
Lincoln's sparrow   *Melospiza lincolni*
Loggerhead shrike   *Lanius ludovicianus*
MacGillivray's warbler   *Oporornis tolmiei*
Marsh wren   *Cistothorus palustrus*
Mountain bluebird   *Sialia currucoides*
Mountain chickadee   *Parus gambeli*
Mourning dove   *Zenaida macroura*
Northern flicker   *Colaptes auratus*
Northern harrier   *Circus cyaneus*
Northern oriole   *Icterus galbula*
Northern waterthrush   *Seiurus novaboracensis*
Orange-crowned warbler   *Vermivora celata*
Pine siskin   *Spinus pinus*
Pygmy nuthatch   *Sitta pygmaeus*
Red crossbill   *Loxia curvirostra*
Red-breasted merganser   *Mergus serrator*
Red-breasted nuthatch   *Sitta canadensis*
Red-eyed vireo   *Vireo olivaceus*
Red-naped sapsucker   *Sphyrapicus nuchalis*
Red-winged blackbird   *Agelaius phoeniceus*
Rough-winged swallow   *Stelgidopteryx serripennis*
Ruddy duck   *Oxyura jamaicensis*
Rufous-sided towhee   *Pipilo erythrophthalmus*
Sage grouse   *Centrocercus urophasianus*
Sage thrasher   *Oreoscoptes montanus*
Sandhill crane   *Grus canadensis*
Savannah sparrow   *Passerculus sandwichensis*
Say's phoebe   *Sayornis saya*

Scrub jay   *Aphelocoma caerulescens*
Sharp-tailed grouse   *Tympanuchus phasianellus*
Solitary vireo   *Vireo solitarius*
Song sparrow   *Melospiza melodia*
Steller's jay   *Cyanositta stelleri*
Swainson's thrush   *Catharus ustulatus*
Tennessee warbler   *Vermivora peregrina*
Three-toed woodpecker   *Picoides tridactylus*
Townsend's warbler   *Dendroica townsendi*
Tree swallow   *Tachycineta bicolor*
Varied thrush   *Ixoreus naevius*
Veery   *Catharus fuscesens*
Vesper sparrow   *Poocetes grammineus*
Violet-green swallow   *Tachycineta thalassina*
Virginia rail   *Rallus limicola*
Warbling vireo   *Vireo gilvus*
Western kingbird   *Tyrannus verticalis*
Western meadowlark   *Sturnella neglecta*
Western tanager   *Piranga ludoviciana*
Western wood pewee   *Contopus sordidulus*
White-breasted nuthatch   *Sitta carolinensis*
White-crowned sparrow   *Zonotrichia leucophrys*
Willow flycatcher   *Empidonax trailii*
Wilson's warbler   *Wilsonia pusilla*
Winter wren   *Troglodytes troglodytes*
Yellow warbler   *Dendroica petechia*
Yellow-rumped warbler   *Dendroica coronata*

References

Abbott, F., Byers, D., and van Horn, D. (1985). Census #89: Scrub oak-mountain mahogany woodland. *Am. Birds* **39**, 113.

Cody, M. L. (1974). "Competition and the Structure of Bird Communities," Monogr. Popul. Biol. No. 7. Princeton Univ. Press, Princeton, NJ.

Cody, M. L. (1981). Habitat selection in birds: The roles of habitat structure, competitors, and productivity. *BioScience* **31**, 107–113.

Cody, M. L. (1985). "Habitat Selection in Birds." Academic Press, Orlando, FL.

Cody, M. L. (1991). Population densities and community structure of birds in Jackson Hole: A reassessment after 25 years. *UW/NPS Res. Cent., 15th Annu. Rep.*, pp. 144–147.

Cody, M. L. (1992). Population densities and community structure of birds in Jackson Hole: A reassessment after 25 years. *UW/NPS Res. Cent., 16th Annu. Rep.*, pp. 3–30.

Cody, M. L. (1993). Population densities and community structure of birds in Jackson Hole: A reassessment after twenty-five years. *Bull. Ecol. Soc.* **74**, Suppl., 195.

Cody, M. L. (1994). Mulga bird communities: Species composition and predictability across Australia. *Aust. J. Ecol.* **19**, 206–219.

Flack, D. (1976). "Bird Populations of Aspen Forests of Western North America," *Ornithol. Monogr.* No. 19. American Ornithologists Union, Anchorage, KY.

Haila, Y., Nicholls, A. O., Hanski, I. K., and Raivo, S. (1995). Stochasticity in bird habitat selection: Year-to-year changes in territory locations in a boreal forest bird assemblage. *Oikos* (in press).

Hering, L. (1984). Census # 135: Ponderosa pine forest. *Am. Birds* **38**, 66.

Hering, L. (1985). Census # 93: Ponderosa pine forest. *Am. Birds* **39**, 113.

Hering, L. (1991). Breeding bird counts: Ponderosa pine forest. *J. Field Ornithol.* **62**, 30.

Hering, L. (1992). Winter bird counts: Ponderosa pine forest. *J. Field Ornithol.* **63**, 115.

Hutto, R. L. (1980). Winter habitat distribution of migrant landbirds in western Mexico, with specific reference to foliage-gleaning insectivores. *In* "Migrant Landbirds in the Neotropics: Ecology, Behavior, Distribution and Conservation" (A. Keast and M. Morton, eds.). Smithsonian Institution, Washington, DC.

Hutto, R. L. (1985). Habitat selection by nonbreeding, migratory landbirds. *In* "Habitat Selection in Birds" (M. L. Cody, ed.), pp. 455–476. Academic Press, Orlando, FL.

Hutto, R. L. (1986). Migratory landbirds in western Mexico: A vanishing habitat. *West. Wildlands* **11**, 12–16.

Hutto, R. L. (1992). Habitat distributions of migratory landbird species in western Mexico. *In* "Ecology and Conservation of Neotropical Migrant Landbirds" (J. M. Hagan, III and D. W. Johnston, eds.), pp. 221–239. Smithsonian Inst. Press, Washington and London.

Jackson, E., and St. Helens, S. D. (1980). Census # 119: Scrub oak-mountain mahogany woodland. *Am. Birds* **34**, 73.

Johnsgard, P. A. (1986). "Birds of the Rocky Mountains." Univ. of Nebraska Press, Lincoln, NE.

Keast, A., and Morton, M., eds. (1980). "Migrant Landbirds in the Neotropics: Ecology, Behavior, Distribution and Conservation." Smithsonian Institution, Washington, DC.

Kingery, H., Hurley, N., Trainer, J., and Bottoroff, R. (1971). Census # 5: Flooded cottonwood forest. *Am. Birds* **25**, 966–967.

Knight, D. H. (1994). "Mountains and Plains: The Ecology of Wyoming Landscapes." Yale Univ. Press, New Haven, CT.

Lageson, D. R., and Spearing, D. R. (1988). "Roadside Geology of Wyoming." Mountain Press, Missoula, MT.

Love, J. D., and Reed, J. C., Jr. (1968). "Creation of the Teton Landscape." Grand Teton Nat. Hist. Assoc., Moose, WY.

McPhee, J. (1986). "Rising from the Plains." Farrar, Straus, & Giroux, New York.

Salt, G. W. (1957a). An analysis of avifaunas in the Teton Mountains and Jackson Hole, Wyoming. *Condor* **59**, 373–393.

Salt, G. W. (1957b). Song, Lincoln's and fox sparrows in a Tetons willow thicket. *Auk* **74**, 258.

Smithsonian Institution (1991). "Birds over Troubled Forests." Smithsonian Migratory Bird Center, National Zoological Park, Washington, DC.

Terborgh, J. (1989). "Where Have All the Birds Gone?" Princeton Univ. Press, Princeton, NJ.

Wiens, J. (1985). Habitat selection in variable environments: Shrub-steppe birds. *In* "Habitat Selection in Birds" (M. L. Cody, ed.), pp. 227–251. Academic Press, Orlando, FL.

CHAPTER 12

Finch Communities in a Climatically Fluctuating Environment

PETER R. GRANT AND B. ROSEMARY GRANT
*Department of Ecology and Evolutionary Biology, Princeton University,
Princeton, New Jersey 08544*

I. INTRODUCTION

Space has been referred to as the final frontier for ecological theory (Kareiva, 1993). Without an understanding of the effects of spatial variation in properties of organisms and their environments no theory of population or community ecology will be complete (Bell, 1992). Given the current attention being paid to modeling spatial effects upon ecological systems (Levin, 1992; Kareiva, 1993), we have chosen to emphasize the other dimension in this chapter: time. Long after the final frontier of space has been breached, the importance of time will remain incompletely known, if only because we can never know as much as we want to know about the past. We can go anywhere on the globe in space and study communities directly by observation or experiment. We cannot do the same in time.

Long-Term Studies of Vertebrate Communities

Our concern in this chapter is primarily with change in communities caused by environmental fluctuations, and secondarily with the connection between observable change and change over longer periods of time in the past. Seasonal and annual fluctuations in environmental conditions over periods on the order of decades are well known from direct observation and measurement. Probably the best known for the longest period of time are fluctuations in climate, especially temperature. In recent years the records have been extended backward in time through analysis of tree rings, alluvial deposits, ice cores, coral cores, etc. (Anderson et al., 1992; Bush et al., 1992; Diaz and Markgraf, 1992; Broecker, 1994; Rahmstorf, 1994; Schlesinger and Ramankutty, 1994), and it has become apparent that temperature fluctuations display two patterns: periodic or quasi-periodic oscillations and state shifts.

Biotic consequences of environmental change are expected to depend partly on the magnitude and rate of change and partly on the duration of altered conditions relative to the life span of affected organisms (Ricklefs and Schluter, 1993; Jackson, 1994; Mangel and Tier, 1994). Changes in environmental conditions about a fixed or moving mean may result in communities that are not in equilibrium with their environment at any time (Davis, 1986; Graham, 1986; Campbell and McAndrews, 1993). Rather, the composition of a community may constantly lag behind what current conditions permit, better adapted to the past than to the present.

This is the background against which contemporary communities should be viewed. The most important lesson to be drawn from the past is that no contemporary study can expect to encompass the full range of conditions experienced by the community. Nevertheless, awareness of changes in the past should help us to understand the present. If the current trend in global warming continues, it may provide us with invaluable data on how communities respond to fairly rapid, sustained, directional change in the environment (Kareiva et al., 1993), responses that mimic those that occurred in the past. A more dependable source of information is the performance of communities living in highly varying environments. In this chapter we explore the responses of one small community living in such an environment, taking advantage of a once-in-a-century environmental event.

We have three goals. The first is to illustrate what may be learned from extending a short-term study of several years' duration to one lasting for 2 decades. Material in the chapter is organized in such a way as to achieve this goal. The second is to discuss the relevance of new knowledge and insights gained over this period to theories of community structure. The third is to consider the connection between results of contemporary studies and changes in environmental conditions over much longer time periods in the past. These last two topics are pursued in the Discussion.

II. BACKGROUND

A. Origins of the Study

Charles Elton used to encourage the study of animal populations on small islands, the idea being that more can be learned from simple communities than complex ones (Crowcroft, 1990). Islands are valuable for studying the ecology of diversity because their boundaries are discrete, and the populations and communities they support can be characterized unequivocally. Twenty years ago, the major questions about diversity on islands were:

1. What determines the pattern of variation in number of species among islands?
2. What determines which particular species coexist?
3. What determines the relative abundance of each species?

The framework for answering these questions was the theory of island biogeography developed by MacArthur and Wilson (1963, 1967). They suggested that the answer to the first question could be found in a balance between colonization and extinction. One then had to know what factors determined each of these processes, and since they could only rarely be studied as dynamic processes through time, except when speeded up experimentally (e.g., Simberloff and Wilson, 1969; Schoener, 1974) or cataclysmically (Fridriksson, 1975; Thornton et al., 1990), the hope was entertained that a study of several differing communities at a single point in time would reveal those factors by statistical means. The second and third questions were to be answered in the somewhat different framework of interspecific competition theory (e.g., Schoener, 1982; Chesson and Case, 1986).

To answer these questions we began a study of Galápagos ground finch communities in 1973, with a special focus on testing Lack's (1945, 1947, 1969) ideas about the importance of interspecific competition (Abbott et al., 1977; Abbott, 1980; Grant, 1986a). Two-week visits to several islands were made to measure food supply, finch numbers, and diets. Food supply was found to vary among islands. It determined to a large extent how many species and which particular ones occurred on an island. Interspecific competition was indicated by the absence or rarity of particular combinations of ecologically similar species (Abbott et al., 1977; Simberloff and Connor, 1981; Grant and Schluter, 1984). The role of competition was confirmed by complementary analyses later (Grant and Grant, 1982; Grant and Schluter, 1984; Schluter and Grant, 1984; Schluter et al., 1985). The overall conclusion from this first study was that food supply variation among islands and interspecific competition for food jointly determined the interisland variation in composition of the community and diets and numbers of its members (Abbott *et al.*, 1977).

The initial study was comparative and static. It was followed by a return to several of the islands to compare community features at different seasons but at the same site: i.e., comparative and dynamic. In accordance with expectations from competition theory, diets of coexisting species generally diverged when food abundance declined from wet season to dry season (Smith *et al.*, 1978).

In view of the pronounced annual variation in rainfall, the next question was whether these results were repeatable or specific to the particular year and particular conditions under which the observations were made. In 1975 we sacrificed breadth for depth and began an intensive investigation of finch species and their food supply on one island, Daphne Major. It was still continuing 20 years later. This particular island was chosen because of its small size (0.34 km²) and manageable population sizes of a few hundred individuals and because, alone among the islands visited twice in 1973, its populations were apparently fully resident. Local persistence of banded birds from April to December was 90% or higher on Daphne and less than 50% at all the other sites (Grant *et al.*, 1975; Smith *et al.*, 1978). We first describe the island and its inhabitants before proceeding to the study itself.

B. Daphne Major and Its Inhabitants

Daphne Major is located 8 km north of the much larger Santa Cruz island (904 km²) in the center of the Galápagos archipelago. The island is a volcanic or pyroclastic cone (de Paepe, 1966) close to the equator. Its age is not known. The nearest dated locality is Baltra island, 1.37 ± 0.16 (SD) million years old, but the nearby north shore of Santa Cruz is half that age or less (Cox, 1983; see also Hickman and Lipps, 1985). During maximum glacial times, when the sea level was 120 m lower than at present (Bard *et al.*, 1990), Daphne was connected to the neighboring islands of Santa Cruz, Seymour, and Baltra (Simpson, 1974). It became an island about 15,000 years ago.

The climate is strongly seasonal. A hot and wet season of variable length and wetness in the first few months of the year is followed by a longer and cooler dry season (Grant, 1985).

The simplicity of the community is shown by the small number of plant species. Forty species were recorded on the island in our first few years (Boag and Grant, 1984a), and only 6 more (rare) species have been added since: *Tournefortia psilostachya, Heliotropium curassavicum, Alternanthera filifolia, Rhynchosia minima, Galactia striata,* and *Cyperus anderssonii.* The last 3 of these have disappeared, possibly without breeding, as have 5 more from the earlier list: *Scalesia crockeri, Cleome viscosa, Gossypium barbadense, Porophyllum ruder-*

ale, and *Trianthema portulacastrum.* If one considers the 8 "extinct" species as transients, the list stands at 38, only half of which are moderately common (Boag and Grant, 1984a). The invertebrate fauna is entirely terrestrial, as the island lacks surface water. Following the advent of rains and the production of leaves on trees (*Bursera graveolens* and *Croton scouleri*) and various shrubs and annual herbs, larvae of several moth species appear and numbers of grasshoppers (*Schistocerca migratoria*) multiply. These are the dominant herbivorous members of the arthropod fauna. A few species of spiders, beetles, and flies, as well as scorpions, are also common.

Darwin's finches are the prime focus of study. Two species are common residents. They are the medium ground finch, *Geospiza fortis* (~17 g), and the cactus finch, *Geospiza scandens* (~21 g). Two other rarer species are the small ground finch, *G. fuliginosa* (~12 g), and the large ground finch, *G. magnirostris* (~30 g). A population of about half a dozen pairs of Galápagos martins (*Progne galapagoensis*), an occasional breeding pair of yellow warblers (*Dendroica petechia*), 10–50 pairs of Galápagos doves (*Nesopelia galapagoensis*), and one or two pairs of black-crowned night herons (*Nyctanassa violacea*) complete the resident landbird fauna. Six species of seabirds breed on the island. Significant immigrants are short-eared owls (*Asio flammeus*), egrets (*Casmerodias albus* and *Bubulcus ibis*), and herons (*Ardea herodias, Butorides sundevalli* and *B. striatus*), which occasionally visit the island and kill finches. Two species of lizards (*Microlophus [Tropidurus] albemarlensis* and *Phyllodactylus galapagoensis*) are the only other resident vertebrates; there are no mammals.

Finches interact directly with many of the arthropods, plants, egrets, and owls. Finches eat spiders, lepidoptera and diptera larvae, seeds, fruits, nectar, and pollen, and they are eaten, though not often and never as nestlings, by egrets and owls. All other interactions are indirect. For example, finches occasionally glean fragments of fish brought to the island and dropped by seabirds, and they may compete occasionally with doves for some of the seeds.

The island is in a completely natural state. This was an important factor in our choice of the island for study. No animals or plants have been introduced by humans as far as we know, and the only disturbance to the habitat has been the creation of paths by walking. The island is small enough that birds can fly the length in a minute, and habitat heterogeneity is minor (Gibbs and Grant, 1987a; Price, 1987).

C. The Framework

Based on our early experience, we postulated a simple causal chain to account for community dynamics of finches at a given site such as Daphne (Grant and Boag, 1980; Grant and Grant, 1980a). In the seasonally arid Galápagos environment rainfall varies annually to a marked degree. Annual rainfall determines plant and arthropod production, which in turn determines finch production, largely through the arthropod component, and subsequent survival of finches through the dry season, largely through the plant component (seeds, fruit, nectar, and pollen). We could write this as RAINFALL → VEGETATION → FOOD → FINCH NUMBERS. The relative success of each species is determined by the availability of the particular components of the total food supply that constitute its diet.

While self-evidently correct to some degree, this scheme could nevertheless miss some other essential factors, and as a result it could be a poor guide to an understanding of fluctuations in numbers of finches. For example, food production might be determined more by temperature conditions, which vary seasonally and annually, than by rainfall. Finch numbers might be determined more directly by predators or parasites than by food supply. Reasons for considering these to be of minor importance have been given elsewhere (Grant and Boag, 1980; Boag and Grant, 1984a,b; Grant, 1986a).

The framework allows us to assess a possible role of interspecific competition in determining finch numbers. Competition is plausible, not just because diets diverged when food supply declined but because diets overlapped even at the lower food density on Daphne (Smith *et al.*, 1978).

We need to make it clear that experimental methods to demonstrate the causal connections we have postulated are not available to us on this or other Galápagos islands. Instead we have inferred causality from the results of statistical testing of biological hypotheses, using natural, uncontrolled, quantitative observations and measurements of environmental and biotic variables. In principle it would be possible to test the causal hypotheses by experiments in similar systems elsewhere (B. R. Grant and P. R. Grant, 1989, p.xviii; e.g., see Smith *et al.*, 1980; Arcese and Smith, 1988).

D. A Medium-Term Study on Daphne

We used the framework to structure our observations and measurements. Over the period 1976 to 1984 several associates, and to a lesser extent ourselves, determined rainfall, food supply, diets, finch natality and mortality, and hence numbers in every year. By the end of the period we had witnessed a drought in 1977 and the wettest year of the century (late 1982 to mid-1983). Finches had fluctuated markedly in numbers, mortality had varied from heavy to light, and natality had varied from none to as many as 39 eggs (and 25 fledglings) produced by one female. The driving force was apparently rainfall, and the link between rain and finches was food supply. There were good (wet) years for the food supply and finches, and there were bad (dry) years for both (Grant and Grant, 1980b; Boag and Grant, 1984a,b; Grant, 1986b; Gibbs and Grant,

1987a,b). These results will be discussed in more detail in a following discussion.

At this point it appeared we had seen everything; we wrote of the value of long-term studies for an understanding of communities in variable environments so as to encompass potentially important rare events (Boag and Grant, 1984a,b; Grant, 1986b; Gibbs and Grant, 1987b; see also Wiens, 1977, 1989). But when does a study become long-term? When it includes climatic or demographic extremes, however defined? Surely not if they occur in successive years. In our view a long-term study should cover a number of generations, determined by the severity and frequency of environmental fluctuations, in order to give some measure of the variation in conditions to which different generations of the study organisms are exposed. Yet by 1984 we did not have a precise estimate of generation length, and we did not know how long birds were capable of living. Motivated in part by this ignorance we continued the study. We now know that generation length is on the order of five years, but to find this out we had to follow a cohort (hatched in 1978) to virtual extinction which took 13 years (Grant and Grant, 1992a). One member was still alive and breeding in 1994 at 16 years of age!

Since the interval between successive El Niño events or droughts may be 5 years or longer, we double the generation length and obtain 10 years as a minimum for what we now call a long-term study of Darwin's finches. By this criterion our study up to 1984 did not qualify as long-term. It is best described as medium-term. Although these designations are merely a convention, they allow us to ask the question, what does a long-term study tell us that is not revealed in the short or medium term? Specifically, what if anything did observations and measurements in the decade after 1984 reveal that was not known or was incorrectly known in the nine years up to and including 1984?

To answer that question we use the framework of causal links in the chain connecting environmental input (rainfall) to community output (finch numbers). We adopt three analytical procedures. Taking 1983, the year of exceptional rainfall, as the pivotal point, we (a) compare means before and after this event, (b) compare relationships between variables before and after the event, or (c) sequentially add data from successive years to the combined data accumulated before to see if the new data produce a change or not. From this approach we stand to gain confirmation and possibly strengthening of repeatable patterns; new patterns, insights, and understanding; or contradiction and confusion!

III. LONG-TERM STUDY

A. Rainfall

Annual rainfall is highly variable and strongly correlated among islands (Grant, 1985). Figure 1 shows the variation on Daphne compared with variation at the

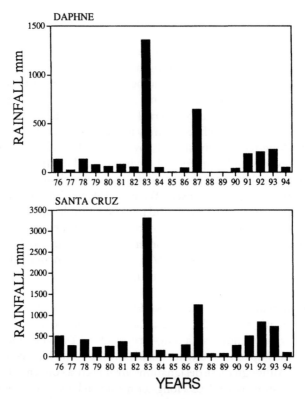

FIGURE 1. Annual rainfall on Daphne Major and Santa Cruz. The exceptional El Niño event of 1983 began in November 1982; rainfall records for November and December 1982 have been added to the totals for 1983 on both islands. Santa Cruz records are from the Charles Darwin Research Station on the south coast. No records are available for the north coast. It is in the rain shadow of the central highlands (Itow, 1975) and has a more arid vegetation than the area around the Research Station. Daphne is 8 km north of Santa Cruz.

Charles Darwin Research Station on the south side of Santa Cruz island. The records are dominated by the exceptional rainfall associated with the 1982–1983 El Niño event. In addition, very wet years contrast with years of virtually no rain. For example the minimum and maximum on Daphne were 1 mm in 1989 and 1359 mm in 1982–1983.

There are three noteworthy patterns in the Daphne rainfall. The first is the repeated occurrence of extensive rain associated with ocean warming called El Niño, or ENSO (El Niño Southern Oscillation) in its wider manifestation (Philander, 1990; Diaz and Markgraf, 1992); these occurred in 1982–1983, 1987, and 1991–1993. El Niño events occur at roughly 3- to 5-year intervals (Graham and White, 1988; Anderson, 1992; Enfield, 1992; Quinn, 1992; Dunbar *et al.*,

1994). The rainfall associated with them varies in amount (Fig. 1) and duration. Not all El Niño events, classified by sea surface temperature anomalies, are associated with heavy rainfall on the Galápagos. For example, an El Niño of moderate strength was identified in 1976 on the continental coast (Quinn, 1992). Sea surface temperatures at Galápagos were not especially high that year, nor was rainfall, whereas at the start of the following year sea surface temperatures were actually higher (Dunbar et al., 1994), yet little rain fell that year (Fig. 1).

The second pattern is the repeated occurrence of droughts in 1977, 1985, and 1988–1989. These typically occurred in association with unusually low sea surface temperatures (La Niña) at the opposite pole of the Southern Oscillation (Philander, 1990). The third is the occurrence of successive years of similar conditions: 1978–1981 (moderate), 1988–1989 (dry), and 1991–1993 (wet). This last pattern implies that conditions for the biota persist in being favorable or unfavorable for 2 years or more. We needed 2 decades to witness this.

Returning to the main theme, our knowledge of rainfall patterns underwent a large change from the end of the medium-term study (1984) onward. By 1984 we thought we had seen the full range of variation in annual rainfall. The following year, however, was a drought of exceptional severity (only 4 mm of rain), and we witnessed two more droughts later. These changes had an effect upon our estimates of both the mean and the variance in annual rainfall. Addition of each successive year's data to the total accumulated beforehand shows how means and variances change with increasing time span (e.g., Fig. 2). For example, cumulative mean annual rainfall on an arithmetic plot against time rose rapidly in 1983, then descended at a decreasing rate to an approximately constant value of 180 mm/year. The same pattern is evident over the same number of years on Santa Cruz, and it gives a long-term average of about 550 mm/year, which is approximately three times larger.

Figure 2 illustrates the patterns on a logarithmic scale, which emphasizes the effect of low values (e.g., droughts) more than high ones. On a logarithmic scale cumulative standard deviations, used as a measure of variation, approach asymptotic values. Daphne and Santa Cruz differ in two important respects with regard to cumulative standard deviations on this scale; the maximum is reached much earlier on Santa Cruz and is much lower. Thus Daphne is drier and more variable than Santa Cruz. The temporal patterns and differences in variation are confirmed by a comparison of coefficients of variation.

To what extent are the results on Daphne specific to the particular times of observation? We cannot answer this directly, but we can answer the same question applied to Santa Cruz by extending the records there backward in time and examining their effects on previous patterns. Addition of rainfall records for the preceding 11 years, 1965–1975 (Fig. 3), does not alter the

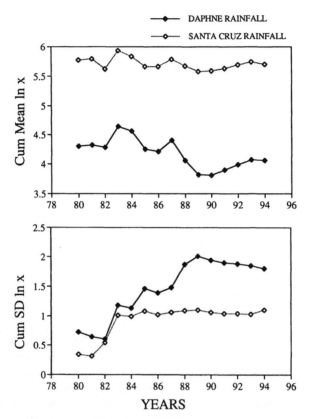

FIGURE 2. Cumulative means and standard deviations of annual rainfall on a natural log scale for Daphne and Santa Cruz. Points for 1980 represent the cumulative records for the years 1976–1980 inclusive. All subsequent means and standard deviations are recalculated with the successive addition of ln rainfall for 1 more year.

temporal pattern after 1976, but it (a) lowers the long-term annual mean to about 480 mm (i.e., by about 12%), (b) reduces the effect of the 1983 total on the rise in cumulative mean on the log scale, and (c) reduces the cumulative standard deviation by about 25%. These effects probably apply to Daphne as well. Over an even longer period, say 50 years, they are likely to be stronger as the influence of rare, extreme events like the 1982–1983 El Niño become progressively more diluted by values closer to the mean.

To summarize, one decade after 1984 the long-term estimate of mean annual rainfall on Daphne had fallen, which is not surprising since we did not expect another El Niño event of the same severity experienced in 1982–1983

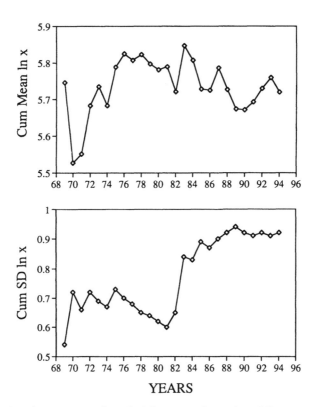

FIGURE 3.   Cumulative means and standard deviations of annual rainfall on a natural log scale for Santa Cruz in the years 1965–1994. Points for 1969 represent the cumulative records for the years 1965–1969 inclusive. All subsequent means and standard deviations are recalculated with the successive addition of ln rainfall for 1 more year. The exceptional rainfall of 1983 had little effect upon the long-term mean but a large effect upon the standard deviation.

to occur. What is surprising is a doubling of variation in annual rainfall (Fig. 2).

B. Finch Numbers

It is easy to see that finches are affected by rainfall. In wet years a rich food supply of arthropods and seeds is produced, finches breed repeatedly (Fig. 4), and natality exceeds mortality. In dry years finches do not breed, or breed with low success, and survival is low, especially for juveniles (Fig. 5). *Geospiza fortis*

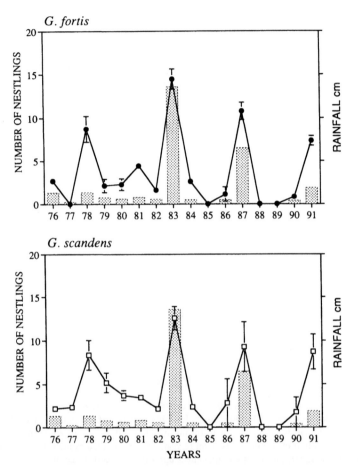

FIGURE 4. Annual variation in the mean number of nestlings produced by *G. fortis* (●) and *G. scandens* (□) females in relation to rainfall (histogram). Vertical bars show 95% confidence intervals. Only those pairs breeding in the first month of the breeding season after the first rains are included. Other birds, usually young birds, began breeding later in some years, or, as in the dry years of 1977, 1982, 1986, and 1990, did not breed at all. Pre-rains breeding by a few pairs of *G. scandens* has been ignored. Mean number of nestlings is highly correlated with rainfall on a logarithmic scale for both *G. fortis* ($r = 0.946$, $P < 0.0001$) and *G. scandens* ($r = 0.928$, $P < 0.0001$); so is mean number of clutches per female ($P < 0.001$ in each case). Usual clutch sizes are three to five eggs.

and *G. scandens* respond to rainfall similarly, and as a result their numbers tend to covary (Fig. 6). Changes in numbers from year t to $t + 1$ are correlated with rainfall in year t for both *G. fortis* ($r^2 = 0.237$, $P < 0.05$) and *G. scandens* ($r^2 = 0.518$, $P < 0.01$). The correlations are higher when changes are expressed as a

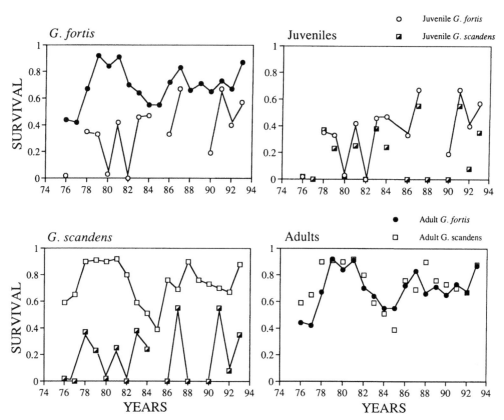

FIGURE 5.   Annual survival of *G. fortis* (●) and *G. scandens* (□). The left panels show that adult survival is consistently higher than juvenile (first year) survival for both species. The right panels show that *G. fortis* juveniles survive better than *G. scandens* juveniles (with one exception, 1978), whereas there is no tendency for adults of one species to survive better than adults of the other. Averages of the annual survival values ± one standard deviation are 0.693 ± 0.145 for adult *G. fortis*, 0.736 ± 0.153 for adult *G. scandens*, 0.351 ± 0.223 for juvenile *G. fortis*, and 0.190 ± 0.202 for juvenile *G. scandens*. Survival of juveniles was taken from fledging, usually February or March, to the end of the year. Data for juveniles in 1992 and 1993 are restricted to those fledging before the middle of March. Note that in some years there was no breeding.

proportion of starting numbers: for *G. fortis* $r^2 = 0.471$ and for *G. scandens* $r^2 = 0.520$.

Cumulative means and standard deviations of the two species differ strikingly on a logarithmic scale (Fig. 7). Cumulative mean *G. fortis* numbers rose to a maximum in 1994, whereas the equivalent for *G. scandens* fell substantially from a maximum in 1984. Cumulative variation in *G. fortis* numbers exhibited stability for a decade, then rose sharply in the subsequent 2 years. In

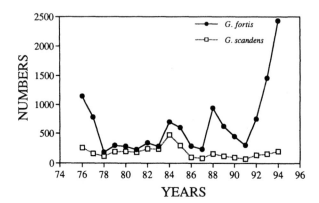

FIGURE 6.   Total numbers of *G. fortis* (●) and *G. scandens* (□) alive on Daphne on Jan. 1 each year; extended and revised from Grant and Grant (1993). The fraction of the total without bands was estimated by mark-recapture methods and enumeration of breeding birds (Boag and Grant, 1984a; Grant and Grant, 1992a). Hybrids (F_1) have been excluded, but backcrosses to each of the parental species (see Grant, 1993) have been included in the species totals.

contrast *G. scandens* achieved and maintained approximately constant variation after a decade. None of these trends was evident at the end of the medium-term study in 1984.

A phase diagram (Fig. 8) pinpoints the changes in the numerical relationship between the species. The "community" of finches would be stable (but not stationary) if densities fluctuated in phase and tended to return to the same point. If this were the case, joint densities of *G. fortis* and *G. scandens* would tend to move up and down a trend line whose slope would measure the unchanging proportions of the species. Although there is evidence for this from 1987 to 1992, at other times the trend lines differ and are offset. There are two decreases, 1976–1978 and 1984–1987, and two increases, 1978–1984 and 1992–1994. None of them coincide. Expressed quantitatively, the proportion of *G. scandens* was initially 0.23 in 1976, rose to a maximum of 0.46 in 1983, and then fell to a minimum of 0.08 in 1994. Transitions from one stable state to another, i.e., from one set of proportions to another, occurred at two times of decline, in 1977–1978 and 1984–1986, and one period of increase, 1992–1993. As discussed later in this chapter, the first two transitions were also times of change in the food supply.

C. Food Supply

The link between rainfall and finches is made by food supply via the vegetation. Seeds have been sampled much better than arthropods. Since most mor-

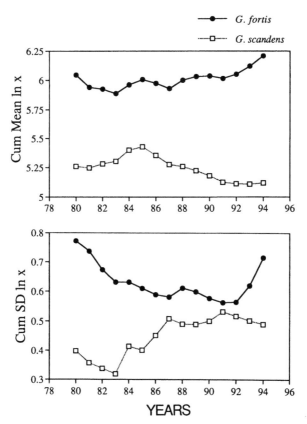

FIGURE 7. Cumulative means and standard deviations of population numbers (Fig. 6) of *G. fortis* (●) and *G. scandens* (□) on a natural log scale. Initial values represent computations for the years 1976–1980. Subsequent values were computed by serially adding annual data.

tality occurs in the dry season, when arthropods are scarce and seeds form the majority of the diets, we concentrate on annual variation in seed supply. Figure 9 shows that profound changes took place in the seed community at the end of the medium-term study. These changes persisted until our regular sampling finished in 1991. We know the major trends continued to 1993 from samples obtained that year in part of the seed sampling grid (unpublished data).

We have used an index of size and hardness of seeds to characterize the difficulty experienced by finches when attempting to extract the kernels (Abbott et al., 1977). It is $(DH)^{1/2}$, where D is depth in mm measured as the second largest dimension orthogonal to the first, and H is hardness measured in kgf and converted to newtons by multiplying by 9.81. Seeds fall conveniently into three size-hardness classes. The small-soft seed category [$(DH)^{1/2}$

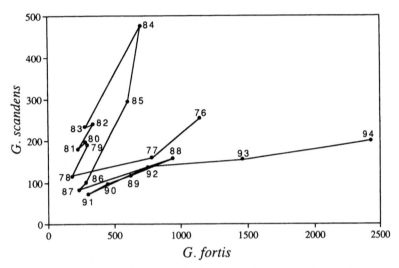

FIGURE 8. Phase diagram of the numerical relationship of the two species showing the temporal pattern of change in proportions. Data are from Fig. 6. Numbers signify years. Members of *G. fortis* always outnumber *G. scandens*.

< 4] comprises 22 species, a medium category [$(DH)^{1/2} = 9-14$] comprises 3 species, only 1 of which (*Opuntia echios*) is common in the dry season, and a large-hard category [$(DH)^{1/2} = 27$] comprises just 1 species, *Tribulus cistoides*.

Cast into these three categories, the seed supply shows a profound and sustained change (Fig. 10) that occurred in 1983 as a result of the exceptional rainfall and exceptional growth of herbaceous vegetation. The biomass of annual and perennial plants producing small-soft seeds increased enormously. The growth of these, and of vines in particular, smothered the prostrate perennial *Tribulus cistoides* and many of the *Opuntia echios* bushes. As a result seed production of these two species fell dramatically.

Daphne was converted from a large-seed island to a small-seed island in 1983. Before 1983 the biomass of large seeds was consistently higher than the biomass of small seeds. This was reversed in 1983, and the reversal was sustained for the rest of the study (including 1993). Biomass of medium seeds was intermediate. It was consistently higher than biomass of small seeds prior to 1983, and lower in all years afterward except for 1991. Analyses of variance contrasting the seed supply before and after 1983 demonstrated statistically strong differences in these seed categories (Grant and Grant, 1993).

The rise in small seed biomass in 1983, followed by a period of stability at an elevated level, was due partly to a storage effect. Effects of past conditions

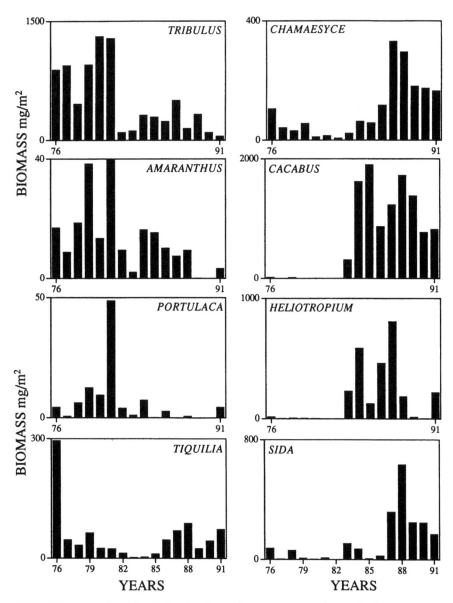

FIGURE 9.   Annual variation in the abundance of eight common seed types. Seeds were counted in 50 randomly chosen 1-m² quadrats in a permanent grid in January and again in April or May, summed, and converted to biomass by multiplying by mean wet weight masses in Boag and Grant (1984a,b). The panels are arranged to contrast the tall plants and vines (*Cacabus*) that prospered after the El Niño year of 1983, on the right, with those prostrate growing ones that were smothered to some extent by overgrowth from others, on the left.

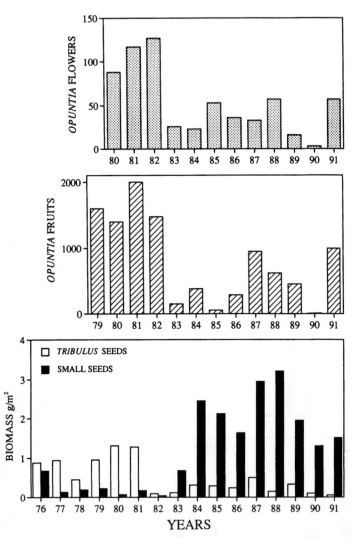

FIGURE 10. Annual variation in numbers of *Opuntia* flowers and fruits on 10 regularly censused bushes, and the biomass of seeds in the small and large (*Tribulus*) categories sampled in the permanent grid (see Fig. 9). Note the switch-over point in seed proportions in 1983. In 1982, before the El Niño event, *Tribulus* seed production was affected by a rust. From Grant and Grant (1993).

were stored in the perennial plant biomass and hence seed production potential (in addition to the seed bank in the soil).

These results show why there is more than one trend line in Fig. 8. The 1984–1986 transition is accounted for by the reversal of *Opuntia* and small seed biomasses. The earlier 1977–1978 transition is explained by the control of *Opuntia* seeds by *G. scandens* through dominance behavior (Boag and Grant, 1984a), which forced *G. fortis* to concentrate on a decreasing supply of small and large seeds.

D. Diets

How did the changing food supply, caused by fluctuations in rainfall, bring about changes in finch numbers? We address this question by examining annual variation in dry season diets and annual variation in dietary overlap between species.

1. Annual Heterogeneity in Dry Season Diets

Individually recognizable birds were recorded feeding on a variable number of days in each of the dry seasons. The observations were cast into five dietary categories: small seeds, medium seeds (*Opuntia echios*), large seeds (*Tribulus cistoides*), *Opuntia* flowers (nectar and pollen), and all other items, including arthropods, combined. This last category was a minor one for all species at all times. For an analysis of annual variation in the diet of each species only one observation was used for each of the many individuals observed each year. This was chosen as the most frequent type of observation or, with tied frequencies, the first record. Observations were made in every year from 1976 to 1994 before breeding had started. We have excluded observations in the dry seasons of the El Niño years of 1983, 1987, and 1993 because they were sparse or lacking, and observations of flower feeding in 1976–1978 because they were recorded slightly differently then. Records of *G. magnirostris* were sufficient for inclusion after 1984.

Annual variation in diets is clearly demonstrated by all three species. Proportional use of the five dietary categories by each of the species varied significantly among years (Table I). To guard against the unlikely possibility that this result was caused by lack of independence among years (some birds were recorded in 2 or more years), the analysis was repeated with a restricted set of data from three years that maximized the number of birds: 1985, 1988, and 1994. No individual entered the analysis in more than one year. The results are the same (Table I).

TABLE I   Annual Heterogeneity in Dry-Season Diets of *Geospiza*

	N^a	χ^2	df^b	P
		All years		
G. fortis	3405	1154.83	48	0.0001
G. scandens	1134	342.49	36	0.0001
G. magnirostris	174	34.31	21	0.0335
		Restricted data: 1985, 1988, 1994		
G. fortis	792	115.28	8	0.0001
G. scandens	265	28.67	6	0.0001
G. magnirostris	79	17.44	6	0.0078

Note: Hybrids (Grant, 1993) not included.

[a] Number of individuals each year, summed over all years. Some individuals were recorded in more than 1 year. These multiple records are included in the analysis of data from all 16 years above but not in the analysis of the restricted data below.

[b] These differ between species because G. *scandens* did not feed on large seeds, and G. *magnirostris* did not feed on flowers. G. *magnirostris* were observed in the years 1985–1994 only.

The three finch species have clearly separable diets on two niche axes (Fig. 11). The spread of values for each species is a representation of the annual variation demonstrated in Table I. Even G. *magnirostris*, a specialist on seeds, varied annually in the degree to which *Tribulus* seeds monopolized the diet. A fourth species, G. *fuliginosa*, was observed feeding in all years, but is not included in Fig. 11 because in most years records of banded and hence individually identifiable birds were scarce. The position of this species would be in the top left corner of Fig. 11, i.e., 100% small seeds and no flowers. In the 3 years of most numerous observations of banded individuals there was no heterogeneity. All birds observed in 1985 ($N = 8$), 1989 ($N = 9$), and 1994 ($N = 8$) fed on small seeds.

The annual variation displayed in Fig. 11 is translated into a temporal pattern in Fig. 12. This shows that, despite annual fluctuations in dietary proportions within species, differences between species were maintained. Thus G. *scandens* fed on *Opuntia* seeds more than G. *fortis* in all 16 years, and more than G. *magnirostris* in all 8 years of joint records. Reciprocally, G. *fortis* fed on small seeds more than G. *scandens* in 15 of the 16 years, and both of them fed more than G. *magnirostris* on small seeds over the last 8 years. In addition, G. *magnirostris* consistently fed more on *Tribulus* seeds than G. *fortis*. The magnitude of this difference was large; proportions never fell below 0.50 for G. *magnirostris* and never rose above 0.35 for G. *fortis*. Finally, although G. *scan-*

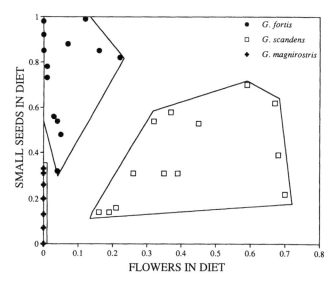

FIGURE 11. Dietary differences among species in the dry season on two niche axes: proportion of all feeding observations on *Opuntia* flowers and proportion of seed-feeding observations on small seeds. Each point represents the average for a species in one year; symbols: ● *G. fortis*, □ *G. scandens*, and + *G. magnirostris*.

dens and *G. fortis* co-varied in their proportional exploitation of *Opuntia* flowers across 13 years ($r = 0.686$, $P = 0.0096$), *G. scandens* exploited them consistently more than did *G. fortis* (Fig. 12).

The overall conclusion from these diet comparisons is that the long-term study contributed nothing new to knowledge gained from the short-term study. Instead, observations made after 1984 confirmed the earlier findings; they were repeatable. The pattern of feeding of *G. magnirostris* was an apparently new addition, but in fact it was known from a study of birds without bands before 1984 (Boag and Grant, 1984a,b).

There is one exception nonetheless to the overall conclusion. After 1983 *G. fortis* increased and maintained higher proportions of small seeds in the diet and lower proportions of *Opuntia* seeds in the diet (Fig. 12). A less pronounced trend toward a reduction of *Tribulus* seeds in the diet is also apparent. The trend away from *Opuntia* seeds contrasts with continuing exploitation of them by *G. scandens*. The trend away from *Tribulus* seeds contrasts with continuing exploitation of them by *G. magnirostris*. These contrasts suggest that *G. fortis* responded to changes in the food supply by shifting their diets whereas the other two did not. We next examine this possibility by contrasting the responses of *G. fortis* and *G. scandens* to a marked change in the availability of small seeds after 1983.

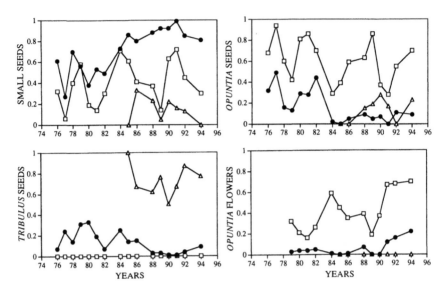

FIGURE 12. Consistent differences between species in four components of dry-season diets. As in Figure 11, proportions are expressed in relation to all food types for *Opuntia* flowers but in relation to total seeds only for *Opuntia*, *Tribulus*, and small seeds. Symbols: ● *G. fortis*, □ *G. scandens*, and △ *G. magnirostris*.

2. Dietary Responses to a Change in Food Supply

Considering just the seed supply alone, from 1976 onward the small seed component changed from a mean proportion of 0.03 before 1983 ($N = 7$ years) to a mean of 0.57 afterward ($N = 7$ years). The two species increased their proportional consumption of small seeds to a different extent and from different starting points (Fig. 13). *G. fortis* increased from a mean of 0.50 to a mean of 0.87, and *G. scandens* increased from a mean of 0.28 to a mean of 0.54.

Before 1983 both species consumed small seeds in significantly higher proportions than dictated by their occurrence in the environment; for *G. fortis* paired $t_6 = 8.14$, $P = 0.0002$; and for *G. scandens* paired $t_6 = 4.41$, $P = 0.0045$. After 1983 *G. fortis* still consumed small seeds to a disproportionate extent (paired $t_6 = 3.538$, $P = 0.0122$), but *G. scandens* did not (paired $t_6 = 0.615$, $P = 0.5612$). All tests were performed with arcsin-transformed data, and results were confirmed with Wilcoxon matched-pairs signed-rank tests. The difference between the species after 1983 is very clear; the proportion of small seeds in the diet of *G. fortis* was higher than the proportion available in 6 years out of 7, but was higher in only 1 year out of 7 for *G. scandens* (Fig. 13).

After 1984, when small seeds were relatively and absolutely abundant, *G.*

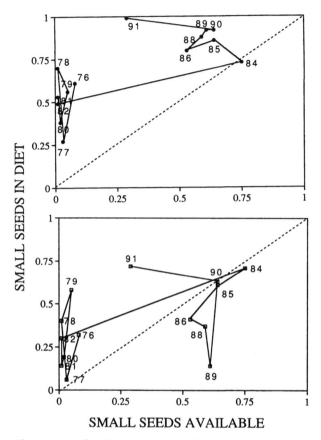

FIGURE 13. Changes in small seeds as a proportion of total seeds in the dry-season diets of *G. fortis* (●) and *G. scandens* (□) in relation to changes in the proportion of small seeds (numbers) available. Points represent annual averages, and those for successive years are connected by solid lines. If proportions in the diet and in the numbers available were equal, all points would fall along the broken lines. Note the difference between the species after 1983.

magnirostris differed from both of the other species in consistently feeding on large seeds more than small seeds and in feeding on large seeds in higher proportions than their occurrence in the pool of available seeds. They specialized on *Tribulus* seeds to a large degree before 1983 as well (Boag and Grant, 1984a,b) and were probably not affected by the food supply change in 1983. After 1983 the feeding niche of *G. fortis* was intermediate between the very different feeding niches of *G. scandens* and *G. magnirostris* (Fig. 14), in some years exactly so.

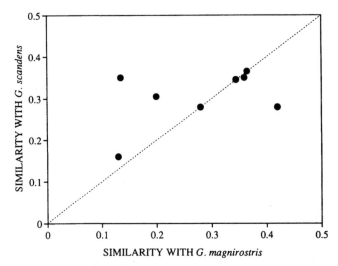

FIGURE 14. The intermediate position of the dry-season niche of *G. fortis*. In terms of similarity in five dietary categories (Table 1), the niche of this species is approximately equidistant from the niches of *G. scandens* and *G. magnirostris* in most years (1985–1994; excluding 1987 and 1993 for reasons of insufficient data). Similarity was measured by $\Sigma(\min p_{i1}, \min p_{j1})$, where p_1 is the proportion of diet item 1 in the diets of species *i* and *j*. The broken line indicates equality of the similarities. Three points fall on this line.

E. Interspecific Competition

Fluctuations in food supply influence finch numbers. Since diets of *G. fortis* and *G. scandens* overlap, do species compete for a limited food supply, and in so doing affect each other's numbers? We consider a possible competitive influence on diets first, then on survival.

1. Diets

Smith *et al.* (1978) suggested that competition for food would result in a divergence of diets as food supply declined from wet season to dry season. Thus comparing interspecific overlap in diets in the wet and dry season is a method for detecting a competitive influence on one or both species. The expectation was upheld in a comparison of wet and dry season diets of different ground finch species on several islands in 1973, including *G. fortis* and *G. scandens* on Daphne. Boag and Grant (1984a,b) extended the comparison on Daphne to the years 1976–1978 and found the same pattern when comparing diet overlaps immediately before and after the rains. We have extended the comparison for an additional 9 years. The pattern was shown in 10 of the 13 years, a result that was not expected by chance (Wilcoxon matched-pairs signed-rank test, $z = 1.818$, $P = 0.034$ one-tailed). Therefore it is highly repeatable. The three exceptional years were characterized by abundant *Opuntia* flowers and frequent feeding on them by both species. This short-term abundant resource can alleviate the effects of interspecific competition (Grant and Grant, 1980a). Overall, the data are consistent with an hypothesis of interspecific competition for food.

The change in food supply in 1983 might be expected to have altered interspecific dietary relations, although the direction is not easy to anticipate. On the one hand a decline in *Opuntia* seed abundance ought to have led to greater competition between *G. scandens,* an *Opuntia* specialist, and *G. fortis.* On the other hand the greater abundance of small seeds may have alleviated any competitive effects. We know that dry-season diet overlap declined after 1983 (Fig. 15), with a divergence in the exploitation of seeds being partly but not entirely responsible. Competition may have been a factor in the divergence. Competition with *G. magnirostris* may have been a factor also in the reduction of *Tribulus* seeds in the diet of *G. fortis.*

2. Survival

If species compete for food, we should expect to see an effect on survival as well as on diets. Gibbs and Grant (1987a) showed with multiple regression analysis that, from 1976 to 1983, adults of each species survived best in 4- to 6-month periods of high abundance of small seeds and low total finch density. A distinction between conspecific and heterospecific density effects was not made. Juveniles, which are more likely than adults to be affected by food shortage, were not included. We have performed similar multiple linear regression analyses with either (a) mortality of birds in their 1st year of life or (b) annual adult mortality, regressed on abundance of small, medium, and large seeds and numbers of *G. fortis* and *G. scandens* present at the beginning of each year (considered as adults). Mortalities were arcsin-transformed and all independent variables were ln-transformed.

For young birds there are three important results. First, for each species, mortality is significantly and negatively related to the abundance of small seeds. Second, for each species, mortality is significantly and positively related to the abundance of *G. fortis.* Third, for each species, mortality is not significantly related to the abundance of *G. scandens,* medium seeds, or large seeds. As indicated by the percentage variance in mortality explained, small seed abundance (47.8%) is a more important determinant of juvenile *G. fortis* mortality than are *G. fortis* numbers (18.7%). The opposite is true for *G.*

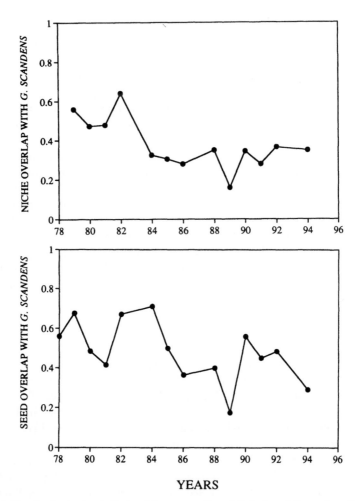

FIGURE 15.   Decline in the similarity of dry-season diets of *G. fortis* and *G. scandens*. Considering all dietary categories (upper) the decline is abrupt after 1983. For the seed components alone (lower), and over a slightly longer period, the decline is more gradual and is fitted by a regression line with negative slope; $b = -0.014 \pm 0.06$, $P = 0.0228$.

scandens juveniles; *G. fortis* numbers (38.9%) account for more of the variation in mortality than do small seeds (25.7%).

These results, which are similar to the earlier ones with adults, suggest that mortality of young *G. fortis* is influenced by intraspecific competition, whereas *G. scandens* mortality is more sensitive to interspecific competition from the more abundant adult *G. fortis*. Competition is for food. All juveniles and adult *G. fortis* feed predominantly on small seeds.

Interspecific competitive asymmetry was also suggested by the results of a previous study. During the drought of 1977, as food (seed) abundance declined the diets of *G. fortis* became broader while diets of *G. scandens* became narrower (Boag and Grant, 1984b). Nonetheless *G. scandens* probably have some effects on juvenile *G. fortis* survival, even though effects are not shown by statistical analyses here, as a result of their behavioral dominance to *G. fortis* of all ages at *Opuntia* feeding sites (Millington and Grant, 1983; Boag and Grant, 1984a,b; Price, 1987). Competitive effects of *G. scandens* on *G. fortis* numbers may be nonlinear, perhaps occurring only in years with a low supply of alternative foods.

The similar relationships for the two species between mortality and seed abundance and *G. fortis* numbers reflect the fact that there are good years and bad years, and they are the same for both species. *G. scandens* and *G. fortis* first-year mortalities covary ($r = 0.874$, $P = 0.0168$, $N = 11$ years).

The relationships for both species with small seed abundance and *G. fortis* numbers existed over the period 1976–1983. The partial regression coefficients exhibited virtually no change in either species when we sequentially lengthened the period by adding data from the four additional years (1984, 1987, 1990, and 1991) in which most adults bred. We conclude that the change in food supply after 1983 did not alter the influence of either food supply or competition on mortality of finches in their first year.

In contrast to these strong statistical patterns for juvenile mortalities there is only one significant relationship with adult mortality; *G. fortis* annual mortality is strongly associated with *G. fortis* numbers ($b = 0.298 \pm 0.081$, $P = 0.0043$, $N = 16$ years). This single factor statistically accounts for 54.9% of the variation in annual mortality. The association, and no other, is present in the data up to 1983 and in all other periods considered by adding one more year to the preceding ones until 1991 is reached. It thus appears to be a robust result. The corollary is that there is no statistical evidence for interspecific competition from annual adult mortalities. Competitive effects may be manifested over shorter periods of a few months, as suggested by the results prior to 1984 (Gibbs and Grant, 1987a), but this cannot be explored further because after 1984 we have only annual census data.

F. Are the Daphne Results Generalizable to Other Galápagos Islands?

This is the sort of question best answered by conducting the same type of study simultaneously at several sites over a large number of years. For example, a comparison of long-term studies (≥ 20 years) of tropical forests at various sites around the world enabled Phillips and Gentry (1994) to determine

that species turnover was generally increasing through time. We are less fortu-
nate. Only one other long-term study of Darwin's finches has ever been carried
out. It was conducted by our assistants and ourselves on the undisturbed
island of Genovesa in the northeast corner of the archipelago, from 1978 to
1988 (B. R. Grant and P. R. Grant, 1989) and hence at the same time as the
Daphne study. Food supply was not measured at the same time in every year,
so the links in the chain connecting rainfall and finch numbers are not as well
established as those on Daphne. Nevertheless there is enough evidence to
suggest that events on Daphne were repeated on at least one other island in the
archipelago. The evidence is as follows:

1. Annual rainfall totals on Daphne and Genovesa are strongly correlated,
whether the exceptional 1982–1983 and 1987 values are included ($r = 0.986$,
$N = 11, P = 0.0001$) or not ($r = 0.844, N = 9, P = 0.0042$).

2. Numbers of territorial males of the large cactus finch, *G. conirostris,* in a
study area on Genovesa varied annually in a correlated manner with numbers
of *G. scandens* territorial males on Daphne ($r = 0.628, N = 10, P = 0.052$) in
the years of breeding during 1978–1987 (B. R. Grant and P. R. Grant, 1989),
1990–1991, and 1994 (unpublished observation on Genovesa). Both species
are specialists on various parts and products of cactus. From mist-net captures
it is known that numbers of *G. conirostris* covaried with numbers of the other
two congeneric species on Genovesa, *G. magnirostris* and *G. difficilis* (B. R.
Grant and P. R. Grant, 1989).

3. Small seeds were produced in great abundance on both islands in 1983,
and *Opuntia* production declined drastically as a result of smothering of the
bushes by vines (illustrated in B. R. Grant and P. R. Grant, 1989).

The abundant rainfall in 1982–1983 had a long-term effect on the vegetation on both islands. This is difficult to quantify, but evident in photographs
and in the increased difficulty after 1983 of walking across Genovesa without
encountering a thicket of *Cordia lutea* vegetation. We have observed the same
long-term effects on the vegetation of the other islands studied before and after
1983: Española in 1979, revisited in 1988 (P. R. Grant and B. R. Grant, 1989);
Gardner by Floreana in 1980, revisited in 1991; and Champion in the years
1980–1991. This consistency is an indication that the community of finches
on Daphne was not alone in experiencing the effects of exceptional rainfall in
1982–1983 as well as droughts and the long-term consequences of an altered
composition of the vegetation.

G. The Number of Species in the Daphne
Finch Community

With a better understanding of community dynamics on Daphne we return to
the first of the questions discussed in Section II.A., the question of what

determines the number of species on an island. In particular, what governs the opposing processes of colonization and extinction?

As expected from island biogeographic theory, the number of species of Darwin's finches on each island varies as a function of island area and isolation (Hamilton and Rubinoff, 1967; Harris, 1973; Abbott et al., 1977). The relationship with area is strong and with isolation it is weak. On the basis of a simple regression of number of Geospiza species on ln area ($N = 22$ islands excluding Daphne, $r = 0.712$, $P = 0.0002$), an island the size of Daphne should support populations of two species. The logarithm of distance to nearest island makes no statistical contribution ($P = 0.9504$) to the relationship with ln area.

Daphne has more species than expected (Fig. 16). At the beginning of our study three species of ground finches were breeding on the island: G. fortis, G. scandens, and the much rarer G. fuliginosa. All three species were collected on the island for museums at the beginning of the century (Swarth, 1931) and observed breeding by Beebe (1924) and later by other visitors (Harris, 1973; Grant, 1984). As a breeding species G. fuliginosa has persisted on the island throughout our study, although always rare (<10 pairs).

Midway through our study Daphne was converted from a three-species island to a four-species island! The fourth species, G. magnirostris, visited the island in the dry season of every year, including 1973 and 1975. Sometimes as many as 50 or more were present in one season. At the beginning of each wet season their numbers declined quickly towards 0. We presumed they had returned to their island of origin, although a few were found dead on Daphne.

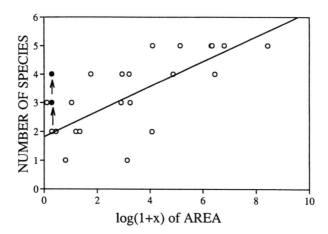

FIGURE 16.   Daphne has more Geospiza species than predicted from the regression of species number on the logarithm of area (km²) for 22 islands. The predicted number is 2; until recently it was three, and in 1994 it was 4. Predicted and the two observed numbers (solid circles) are connected by arrows. For the relationship without Daphne, $r = 0.712$, $P = 0.0002$; with Daphne included as a four-species island, $r = 0.641$, $P = 0.001$.

This pattern changed at the beginning of the El Niño event of 1982–1983 when three breeding pairs were formed from two males and three females (Gibbs and Grant, 1987a). Breeding was successful, and the population grew exponentially as a result of recruitment and to some extent further immigration (Grant and Grant, 1995a). Twenty-three pairs bred in 1993, and in the breeding season of 1994 there were at least 40 territorial males (Fig. 17).

Daphne is the smallest island supporting breeding populations of four *Geospiza* species (Fig. 16). Rábida, the next largest with four species (Abbott *et al.*, 1977), is approximately 15 times larger than Daphne (Wiggins and Porter, 1971). How and why did *G. magnirostris* colonize an island that was apparently full?

Colonization of an island by a new species is potentially influenced by several factors, including the size of the propagule, the presence of competitor species, conditions in the source area, conditions on the island, and, in various ways, human influences on all of them (MacArthur and Wilson, 1967; Lack, 1969, 1976; Williamson, 1981; O'Connor, 1986). Human influences are neither known nor suspected, and propagule size is not an issue in this particular colonization. The remaining three have been recently considered in detail (Grant and Grant, 1995a). We will simply summarize the conclusions.

The colonization of Daphne by *G. magnirostris* is not explained by any of the factors examined. First, it is not explained by a relaxation of competition.

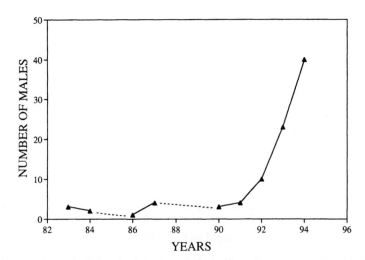

FIGURE 17. Exponential increase in number of breeding *G. magnirostris* males on Daphne. No breeding occurred in 1985, 1988, and 1989. The complete breeding was not observed in 1994, so all territorial males (40) observed in February of that year were assumed to breed.

Densities of *G. fortis* and *G. scandens* were not depressed at the time of the colonization in late 1982 (see Fig. 6). Moreover *G. magnirostris* are larger than the others and competitively superior when contesting ownership of nest sites or food resources. Second, it is not explained by changes taking place in the source areas of Santa Cruz and Santiago islands. It happened before the archipelago-wide buildup of finch numbers occurred as a result of extended breeding associated with prolonged rains. Third, the colonization is not explained by a change in food supply: neither seeds nor arthropods were unusually abundant. Fourth, it is not explained by an area effect: the island had more species than predicted before the colonization took place. Finally, it is not explained by isolation, at least not in a simple way: immigration took place repeatedly, often in large numbers, without resulting in colonization before the El Niño event began at the end of 1982.

With hindsight we cannot identify any ecological factor that prevented *G. magnirostris* from breeding before they did at the end of 1982. All we can suggest is the possibility that the rapid breeding response of other finches on the island to an exceptionally early and heavy rainfall may have induced singing and nest-building activity in the *G. magnirostris* males before they were in a physiological state of readiness to emigrate. Females, in turn, may have been stimulated by the males before they were ready to emigrate. Thus the colonization may be an example of a community catching up with the past. It took a rare event to nudge a species into joining a community whose environment was prepared to accept it.

The island now has more than the predicted number of breeding species partly because it is close to large source islands and receives new immigrants, some of which stay to breed, and partly because the food remains unusually abundant a decade after an exceptional El Niño event.

H. Extinction

It is easy to form the impression from the literature that natural extinctions of birds on islands are moderately common (Diamond and May, 1977; Williamson, 1981, 1983; Diamond, 1984; Pimm *et al.*, 1988). While it is true that many species have become extinct on islands around the world (Stattersfield, 1987; Fraser *et al.*, 1993; Smith *et al.*, 1993; Pimm *et al.*, 1994), a human influence can be discerned in most cases, or is strongly suspected (Caughley, 1994). Small populations of temperate zone migratory species appear to be especially prone to extinction (Abbott and Grant, 1976; Williamson, 1983; Pimm *et al.*, 1993). Natural extinctions of resident populations of birds well isolated on tropical islands are rare (Grant, 1986a; Mayer and Chipley, 1992). None are known from the Galápagos. Six populations of Darwin's finches, and possibly three more, disappeared in the century after Darwin's visit in 1835, but the causal factors are known or strongly suspected to be anthropogenic in all cases: habitat alteration and predation by introduced mammals (Grant, 1986a).

We have learned nothing about natural extinction of finch populations in modern times. A drought in 1988 and 1989, lasting for almost 1000 days (Grant and Grant, 1992a), did not eliminate any population on Daphne. No extinctions of finches were witnessed on Genovesa in the period 1973 to 1994. No changes in the composition of finch communities were revealed by visits to Champion, Gardner by Floreana, Gardner by Española, and Española itself before and after the El Niño event of 1982–1983 or by annual visits to the large island of Santa Cruz.

Thus Galápagos islands have stable communities of finch species: colonization and extinction appear to be very rare events. This conclusion must be tempered by the acknowledgment that most populations have not been observed often enough or long enough to know whether extinction (or colonization) has occurred or not (Grant, 1986a). For example in August 1983, just after the El Niño event had ceased, we counted an equal number of G. fortis and G. fuliginosa individuals in a small part of Seymour during a 2-hour visit, and estimated population sizes of each species to be over 100. The island (1.84 km²) is many times larger than Daphne Major and much less isolated (<1 km) from the nearest large island, Baltra (Wiggins and Porter, 1971). In 2½ days during the breeding seasons of 1990 we failed to find a single G. fortis individual. We estimated the breeding population of G. fuliginosa to be no more than 30 pairs. In addition, we observed two G. scandens. On a 1-day visit in 1991 we saw only G. fuliginosa. These fragmentary observations illustrate the difficulties. Although G. fortis were collected on the island at the turn of the century (Grant et al., 1985) and were observed commonly in 1983, we do not have breeding records on the island to know whether a breeding population of G. fortis became extinct in the drought of 1988–1989, or earlier, or whether those observed in 1983 (all in brown plumage and possibly immatures) were temporary immigrants from nearby Baltra and Santa Cruz islands.

IV. DISCUSSION

A. The Value of Long-Term Studies

Long-term studies of ecological communities are better than short ones, for at least five reasons. First, they allow better estimation of parameters of interest. Second, they permit improved statistical evaluation of hypotheses. Third, results obtained in the short term can be checked for reliability or repeatability.

Fourth, the longer the study, the greater the range of environmental conditions likely to be encountered, and for this reason long-term studies are more comprehensive than short-term ones. In particular, rare but important perturbations or subtle changes are more likely to be manifested over the long term. Finally, new knowledge may be gained that is impossible to obtain over short or medium time periods. For example, some processes are slow, like competitive exclusion, and some properties of populations and communities can be determined only after the lapse of many years, such as stability, equilibria, resilience, and the length of population cycles of long-lived animals.

Our medium-term study happened to include an extremely rare environmental perturbation, occurring perhaps no more often than once in 400 years or more (Dunbar et al., 1994). Even so, our understanding of rainfall patterns and their effects upon finches underwent a large change in the following decade. Estimates of mean annual rainfall fell and standard deviations increased. We encountered sequences of dry or wet years that had not occurred in the short term.

Similarly, numerical relationships of the two species changed profoundly over the long term. Cumulative mean G. fortis numbers reached a maximum at the end of the study, after 19 years, whereas cumulative mean numbers of G. scandens declined. Cumulative standard deviations of numbers of the two species also displayed contrasting patterns which were revealed only in the long term. These results clearly demonstrate the dependence of ecological dynamics and our understanding of them upon the temporal scale of study, just as some ecological processes are dependent on spatial scale (Levin, 1992).

Rare extreme events may have long-term effects on the community, and long-term studies are needed to include them, but there is inherent uncertainty about when such events will occur and how long the effects will last. This makes some long-term studies virtually unplannable, which is an argument for conducting such studies in an open-ended way.

B. Regulation of Finch Numbers By Food

Population numbers and community membership on Daphne are governed by the trophic level below and not by the trophic level above the finches (Grant, 1986a,b). There are three ways in which numbers of finches can be influenced by their resources. They may be (a) regulated by density-dependent factors (competition for food) about an equilibrium density determined by a fixed carrying capacity, (b) regulated about an equilibrium that fluctuates because the carrying capacity fluctuates, or (c) not regulated—i.e., numbers change always in a density-independent way.

Assessment of density dependence is simplified by assuming a fixed carrying capacity. Under these conditions density-dependent change in numbers from year to year results in a correlation between change in log density between each successive pair of years and the log of density in the first of each pair (Pollard *et al.*, 1987). A slope of -1 indicates a return to equilibrium following a perturbation (Williamson, 1987). There are statistical problems associated with testing the null hypothesis of no density dependence. They have been frequently discussed but not yet resolved (Pollard *et al.*, 1987; Williamson, 1987; Woiwood and Hanski, 1992; Murdoch, 1994).

Ignoring the statistical difficulties, we simply calculated the relationship between annual change in log numbers $(x_{i+1} - x_i)$ and starting log numbers (x_i) for the years 1976–1980, then serially added data from successive years to see how the correlation might change with changes in the environment. Figure 18 shows that the correlation starts very close to -1 for both species. Until 1981 populations of both species appear to be regulated. For *G. fortis* the

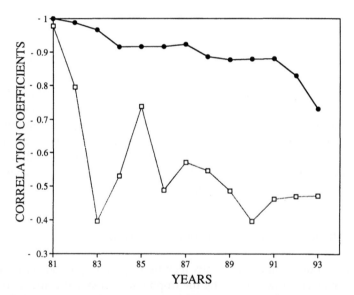

FIGURE 18. Correlations between density and changes in density as a cumulative function of time starting with 1976. A correlation of -1 indicates regulation by density-dependent processes. Over the first 6 years both species were apparently regulated. Serial addition of annual data from 1982 onward weakens the patterns, more rapidly for *G. scandens* ($\square$) than for *G. fortis* ($\bullet$). Correlation coefficients for *G. fortis* do not differ significantly ($P > 0.1$) from -1 at any time; they differ significantly ($P < 0.05$) from zero after 1984. Coefficients for *G. scandens* differ significantly from -1 after 1989 and from zero after 1990.

correlation gradually declines, then dips sharply at the end when density rises steeply in successive years (Fig. 6). This implies that density dependence weakened, especially in the final 3 wet years in a row. The rise and fall in *G. scandens* numbers during and after the 1982–1983 El Niño event resulted in a lower and fluctuating correlation, with relatively little change after 1986. We interpret this pattern to mean that numerical changes were either independent of density or dependent in a changing way as the food supply changed. In the absence of statistical evaluation we can conclude only that the evidence for density dependence is stronger earlier than later, and stronger for *G. fortis* than for *G. scandens*.

Boundedness in population fluctuations implies some degree of regulation through stabilizing density-dependent mechanisms acting some of the time (Royama, 1977; Pollard *et al.*, 1987; Murdoch, 1994). It does not imply regulation all of the time. Chesson and Case (1986) argued like Strong (1984) that density dependence is restricted to population extremes and may not occur at intermediate densities (see also Murray, 1979, 1994). However, even under this scheme, when carrying capacity fluctuates, density dependence may occur sometimes at densities that are, in the long term, intermediate.

If dry-season food supply can be taken as an imperfect index of carrying capacity, then carrying capacities do indeed change (Grant, 1986b). Therefore *G. fortis* numbers may be subject to regulation about a changing equilibrium caused by a fluctuating carrying capacity. *Geospiza scandens* numbers may be more tightly regulated by stronger territorial behavior (Millington and Grant, 1983; Boag and Grant, 1984a,b), but again about an equilibrium density that changes as a result of environmental change. In fact, the environmental change associated with the El Niño event was apparently more severe for *G. scandens* than for *G. fortis* (see Fig. 5) owing to a decline in *Opuntia* cactus abundance. Thus the carrying capacities of the two species are different because their adult diets are largely different: mainly small seeds in the case of *G. fortis*, mainly *Opuntia* products in the case of *G. scandens*.

The suggestion that populations of finch species are regulated by different food supplies is not new. It was considered by Lack (1947) in his pioneering studies of Darwin's finches and extended to related species in continental regions by Newton (1967), Pulliam and Parker (1979), Schluter (1988), Schluter and Repasky (1991), and others. It is an integral part of general theories of niche partitioning (Schoener, 1982). Our study has provided quantitative support over a long period of time for the importance of food as a population-regulating factor. The next step would be to experimentally test the hypothesis of food regulation in regions suitable for measuring the effects of manipulations (e.g., see Smith *et al.*, 1980; Arcese and Smith, 1988; Arcese *et al.*, 1992).

C. Relevance to Community Theories

There is no single theory for all communities. Different assemblages of different organisms living in different environments and having different histories display different trophic and taxonomic structures and very different dynamics (Diamond and Case, 1986; Ricklefs and Schluter, 1993). Chesson and Case (1986) reviewed the main theories of community structure and classified them into several types of classical (equilibrial) and nonequilibrial theories. Our study combines elements of both and fits closest to their scheme of discontinuous competition in a fluctuating environment. Their version of this does not allow a changing long-term environmental mean, which is probably wrong for our system on a scale of centuries (see next section), and predicts no limit to the similarity of resource use, which also appears to be wrong (Grant, 1986a).

The Daphne environment fluctuates strongly and to a large extent erratically despite quasi-periodic behavior in this region of the Pacific on the scales of 3 to 5 years (Graham and White, 1988; Enfield, 1992; Dunbar et al., 1994) and 20 years (Anderson, 1992; Anderson et al., 1992; Schlesinger and Ramankutty, 1994). These fluctuations cause fluctuations and occasionally major changes in the food supply. Changes in the food supply cause changes in finch numbers and proportions because dietary characteristics of the species are relatively fixed. The pulsed and variable nature of food production results in alternations between food limitation and relaxation of limitation; at times finch numbers are at a maximum set by the food supply, and at other times food supply increases faster than do finch numbers. The maximum itself varies. These alternations, combined with the fact that the diets of species overlap at all times, imply that competition between species is episodic.

The extent to which this model of community structure and dynamics is general is open to question. Terrestrial species that have been studied for many years in the temperate zone often display approximately constant population sizes for a period followed by sharp declines or increases (e.g., Lack, 1964; Enemar et al., 1984; Svensson et al., 1984; Williamson, 1984; Woolfenden and Fitzpatrick, 1984; Holmes et al., 1986, 1991; Arcese et al., 1992; Böhning-Gaese et al., 1994), which makes us think that the dynamics displayed by the Daphne community and their determination by climatic (weather) and food factors are not unique to Galápagos. They may simply be more pronounced. Moreover, in similar fluctuating environments in continental regions the dynamics of bird populations appear to be very similar to those on Daphne. Food production and supply vary dramatically among years of contrasting rainfall in arid regions (Pulliam and Brand, 1975), as do population sizes of related consumers such as emberizine sparrows (Pulliam and Parker, 1979) and reproductive performance (Rotenberry and Wiens, 1991). The similarities are striking in view of two major differences; continental species are probably subject to higher levels of predation, and they have the opportunity of avoiding unfavorable conditions by nomadic or regular movements (e.g., Schluter, 1988). To these should be added the possibility that disease plays a more important role in determining population numbers in continental regions than on islands, that is, until disease organisms are introduced to islands (Warner, 1968; van Riper *et al.*, 1986).

Steele (1985) and Steele and Henderson (1994) have characterized the dynamics of marine systems as being determined by large-scale processes, such as El Niño events, and contrasted them with those of terrestrial systems which are presumed to be determined more by local, biotic interactions (predator–prey, competitor–competitor, etc.) than by external perturbations. This generalization tends to break down in terrestrial systems affected by El Niño. These include not only Galápagos and the neighboring continental region but places far removed as a result of atmospheric teleconnections (e.g., Trenberth *et al.*, 1988; Diaz and Kiladis, 1992), for example, Brazil, Panama, various parts of North America, South Africa, Borneo, and the surrounding region of Australasia. Dynamics on the small island of Daphne, characterized by biotic interactions periodically superseded in importance by environmental perturbations, could be representative of community dynamics in a broad geographical range of similar terrestrial environments.

D. Extrapolation: Changes Over the Longer Term

Daphne finches, as a partially isolated community, have been in existence for approximately 15,000 years, equivalent to about 3000 generations. What has been the climatic history of the community, and what does the history imply about the fate of the species?

From sediment cores taken in lakes on two Galápagos islands (Colinvaux, 1972; Goodman, 1972), coral cores from Galápagos covering the last 400 years (Dunbar *et al.*, 1994; see Fig. 19), and evidence from various sources elsewhere (Anderson, 1992; Anderson *et al.*, 1992; McGlone *et al.*, 1992), we know that the history of Daphne and the other islands has been one of repeated and often rapid climatic change. Conditions were either drier than now or as warm and wet but no wetter (Colinvaux, 1972). Periods of one or more centuries in which El Niño events were common alternated with periods when they were apparently scarce or lacking.

Early in its history Daphne probably had no more species of finches than the four contemporary ones. Despite being larger, higher, and less isolated than now, the island's vegetation was probably sparser because the climate was drier. The community of finches is likely to have fluctuated between its current size and composition and a smaller one as the climate fluctuated between current and drier conditions. In periods of more frequent droughts than at present, extinctions are more likely, and the finch community would have become reduced to one or two species, which ones depending upon which particular plant species were able to persist. The generalist granivores *G. fortis* and *G. fuliginosa* would probably have the highest chance of surviving when the flora was impoverished. Recolonization by the others would then follow when wetter conditions returned. This dynamic view of the past, combined with modern observations of frequent dispersal to the island and one documented colonization event, implies that the Daphne community of finches has been "assembled" many times in the island's history; there have been many opportunities for finch species to combine to form the community. It also suggests that the community has a core of species, and other species are sometimes added or subtracted according to environmental circumstances.

These conclusions apply to the period from 15,000 to about 3000 years ago when an essentially modern climate replaced a drier one (Colinvaux, 1972). For the last 3000 years a general argument can be made that extinctions and replacements have remained likely as a result of strong fluctuations in the climate. How likely is debatable. The argument is as follows.

The longer the time interval between two points, the more different will the environment be, on average (Williamson, 1987; Bell, 1992). The longer the time series, the more long-term large amplitude changes are uncovered (Ariño and Pimm, 1995). Short-term fluctuations, the sort usually studied by ecologists, are imposed on the long-term ones. If this is generally true, it implies that density variation may continue to increase indefinitely because the environment is constantly changing in new ways and that all small populations are bound to go extinct sooner or later unless they can adapt to changing conditions fast enough or move to new areas (Pimm and Redfearn, 1988). Empirical support for this view consists of a tendency for the standard deviation of log density of various bird and insect populations to continue increasing with time (the length of census period) (Pimm and Redfearn, 1988, 1989; Ariño and Pimm, 1995; see also McArdle, 1989; McArdle *et al.*, 1990; McArdle and Gaston, 1992; Schoener and Spiller, 1992; Gaston and McArdle, 1994; Murdoch, 1994).

The question of whether density variation continues to increase is dependent on the time scale. For example, *G. scandens* appears to have approached an asymptote in less than 20 years. Until 1992 it could be claimed that *G. fortis* variation was also stable, but then three wet years in a row destroyed the stability. One must therefore ask whether in the near future some other unusual combination of climatic conditions might not arise and destroy the stability of the *G. scandens* variation. Nevertheless, as the period of time is lengthened, the effects of abrupt changes will become progressively smaller as they are dampened by an increasing number of preceding density values. Therefore, after a time, cumulative means and standard deviations will remain largely unaltered unless there is a substantial long-term environmental trend, such as gradual warming, or unless large amplitude changes occur. Our study is too short to detect very-long-term trends or variations, so we must resort to indirect information.

Coral ring data provide an indirect record of sea temperature fluctuations over a period of time much longer than that spanned by our observations of finches (Fig. 19). The record shows that it takes a long time for environmental variation, as measured in this manner, to approach an asymptote (~170 years). Even though conditions were warmer in some half-centuries than others, and more variable in some than others, there is no evidence of a long-term trend or of large amplitude changes preventing a close approach to an asymptote. If density variation displays a similar pattern, the risk of extinction may not be negligible over a century or more.

To be more specific, a once-in-a-millenium El Niño event of extraordinary severity, such as apparently occurred about 900 and 1400 years ago (Michaelsen and Thompson, 1992) and earlier (McGlone et al., 1992), could bring about a drastic reduction in cactus abundance and production and the demise of the cactus finches. Our study has shown that the effects of an exceptional El Niño event are not restricted to the year of occurrence: they last for more than a decade. In contrast, droughts, which would appear to be a more serious threat to finch populations, have relatively short-lived effects. According to our study, the severity of those effects depends upon preceding environmental conditions. Short-term reversals between El Niño events of abundant rainfall and La Niña droughts, brought about by the Southern Oscillation of atmospheric pressure differences across the tropical and southern Pacific (Nicholls, 1992), prevent severe climatic (drought) conditions from lasting long. We notice, for example, the occurrence of three successive years (1637–1639) of cool sea surface temperatures and hence low rainfall only once in the nearly 400-year record contained in Galápagos coral cores (Dunbar et al., 1994). Thus climatic fluctuations by themselves may not have the dramatic effects of causing extinction as often as the extreme values would indicate.

One final point deserves emphasis. Extinction may be inevitable in a strongly fluctuating environment unless species adapt to the changes. In this chapter we have ignored evolutionary change, yet evolutionary change, through natural selection and hybridization, has been repeatedly documented in the Daphne finches over the last 20 years (Grant et al., 1976; Boag and Grant, 1981; Price et al., 1984; Gibbs and Grant, 1987c; P. R. Grant and Grant, 1992b, 1994, 1995b; B. R. Grant and Grant, 1993). Long-term studies designed to

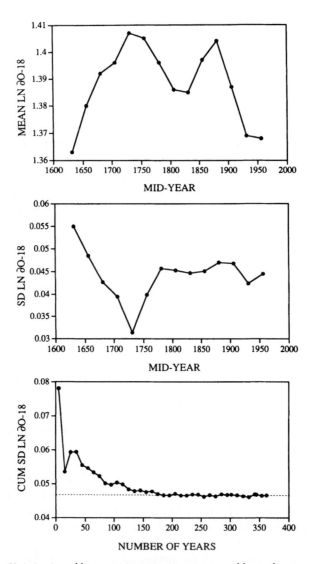

FIGURE 19. Variation in stable oxygen isotope ratios on a natural log scale as a measure of sea surface temperatures in the Galápagos corals *Pavonia clavus* and *P. gigantea*, 1607–1981 (data from Dunbar *et al.*, 1994). Negative signs of the original values have been ignored in the transformation; high values correspond to high sea surface temperatures (see Dunbar *et al.*, 1994, for further details). The upper panel shows 50-year annual means computed at 25-year intervals; owing to a few missing years the last three values are calculated from samples of 47, 44, and 40 years, respectively. The middle panel shows standard deviations calculated in the same way. The lower panel expresses cumulative standard deviations from the first five years onward; it took 170 years to reach a point (0.04713) below all subsequent fluctuations (maximum = 0.04719). Variation in isotope values may also reflect variation in rainfall as there is a moderate correlation ($r = 0.551$, $P = 0.0412$, $N = 14$ years) between them and rainfall on Santa Cruz island in the period 1965–1981 (1973–1974 missing, and 1975 omitted because the unusually late rains fell after the period of coral growth).

establish the link between environmental fluctuations and community dynamics should give explicit attention to the possibility of microevolution of the constituent species occurring during the course of the study.

V. SUMMARY

Previous studies of Darwin's finch communities on several Galápagos islands over a period of just under a decade established the importance of food supply in determining the number of species breeding on an island, which particular species they were, ecological differences between them, and their abundances. In this chapter we describe the main ecological results obtained from extending a medium-term study lasting less than 10 years (<2 generations) on the island of Daphne Major to a long-term study of 19 years (>3 generations).

Our medium-term study happened to include an extremely rare environmental perturbation, an El Niño event of exceptional severity and duration. Even so, our knowledge of rainfall patterns and their effects upon finches underwent a large change in the following decade. Estimates of mean annual rainfall fell and estimates of variance increased. We encountered sequences of dry or wet years that had not occurred in the medium term.

Similarly, numerical relationships of the two main species changed profoundly over the long term. Cumulative mean and cumulative standard deviation of *G. fortis* numbers on a log scale increased, whereas cumulative mean numbers of *G. scandens* decreased and the cumulative standard deviation stabilized. Changes in finch numbers were brought about by changes in their respective food supplies. A change from a preponderance of large seeds to a preponderance of small seeds, witnessed at the end of the medium-term study, persisted to the end. The abundance of *Opuntia* cactus also changed. In the face of these changes the diet of *G. scandens* remained basically unaltered, and as a result these finches were inflexibly at the mercy of a food supply that changed in composition and abundance. *Geospiza fortis* tracked the changing food supply to a greater extent and experienced higher survival.

The community of finches changed in one other respect. *Geospiza magnirostris*, a repeated immigrant to the island before the El Niño event of 1982–1983, became a member of the breeding community in December 1982. Numbers of breeders increased exponentially over the next decade as a result of local recruitment supplemented by additional immigration. No species went extinct.

These results show how consumers cope with erratically fluctuating food resources. The community of finches and the component species can only loosely be described as equilibrial, given the large and somewhat irregular fluctuations in annual rainfall upon which primary and secondary production depends, and the imperfect ability of consumer populations to track resources at all times. Dynamics on the small island of Daphne, characterized by biotic interactions periodically superseded in importance by environmental perturbations, could be representative of community dynamics over a broad geographical range of similar terrestrial environments affected directly or indirectly by El Niño events.

We use these results and interpretations to speculate about the ecology of finch communities over the last 15,000 years when Daphne became an island. Climatic conditions changed across millenia, and extinctions on Daphne probably occurred during relatively dry periods. This dynamic view of the past, combined with modern observations of frequent dispersal to the island and one documented colonization event, implies that the Daphne community of finches has been "assembled" many times in the island's history; there have been many opportunities for finch species to combine to form the community. It also suggests that the community has a core of species, and other species are sometimes added or subtracted according to environmental circumstances. Thus the biological consequences of extreme climatic conditions experienced during our long-term study provide us with a view of community dynamics in the historical and pre-historical past.

Acknowledgments

Our research has been supported by grants from NSERC (Canada) and NSF (USA), most recently DEB-9306753. We thank the Galápagos National Park authorities, the Charles Darwin Foundation, and the Charles Darwin Research Station for long-term support and the following assistants for their help with field research on Daphne: P. T. Boag, P. de Maynadier, H. L. Gibbs, J. P. Gibbs, K. T. Grant, N. Grant, G. C. Keys, S. Latta, I. Lovette, D. McCullough, S. J. Millington, D. Moore, K. Petit, R. Podolsky, T. D. Price, L. M. Ratcliffe, W. B. Rendell, G. Retzlaff, D. Rosenberg, K. Tarvin, A. Unal, C. Valle A., J. Weiland, and D. A. Wiggins. We thank M. L. Cody, T. D. Price, D. Schluter, and J. A. Smallwood for critical comments on the manuscript, and G. M. Wellington for supplying the coral isotope data.

References

Abbott, I. (1980). Theories dealing with the ecology of landbirds on islands. *Adv. Ecol. Res.* 11, 329–371.

Abbott, I., and Grant, P. R. (1976). Non-equilibrial bird faunas on islands. *Am. Nat.* 110, 507–528.

Abbott, I., Abbott, L. K., and Grant, P. R. (1977). Comparative ecology of Galápagos Ground Finches (*Geospiza* Gould): Evaluation of the importance of floristic diversity and interspecific competition. *Ecol. Monogr.* 47, 151–184.

Anderson, R. Y. (1992). Long-term changes in the frequency of El Niño events. *In* "El Niño: Historical and Paleoclimatic Aspects of the Southern Oscillation" (H. Diaz and V. Markgraf, eds.), pp. 193–200. Cambridge Univ. Press, Cambridge, UK.

Anderson, R. Y., Soutar, A., and Johnson, T. C. (1992). Long-term changes in El Niño/Southern Oscillation: Evidence from marine and lacustrine sediments. *In* "El Niño: Historical and Paleoclimatic Aspects of the Southern Oscillation" (H. F. Diaz and V. Margraf, eds.), pp. 419–433. Cambridge Univ. Press, Cambridge, UK.

Arcese, P., and Smith, J.N.M. (1988). Effects of population density and supplemental food on reproduction in song sparrows. *J. Anim. Ecol.* 57, 119–136.

Arcese, P., Smith, J.N.M., Hochachka, W. M., Rogers, C. M., and Ludwig, D. (1992). Stability, regulation, and the determination of abundance in an insular song sparrow population. *Ecology* 73, 805–822.

Ariño, A., and Pimm, S. L. (1995). On the nature of population extremes. *Evol. Ecol.* 9, 429–443.

Bard, E., Hamelin, B., Fairbanks, R. G., and Zindler, A. (1990). Calibration of the ^{14}C timescale over the past 30,000 years using mass spectrometric U-Th ages from Barbados corals. *Nature (London)* 345, 405–410.

Beebe, W. (1924). "Galápagos: World's End." Putnam, New York.

Bell, G. (1992). Five properties of environment. *In* "Molds, Molecules and Metazoa: Growing Points in Evolutionary Biology" (P. R. Grant and H. S. Horn, eds.), pp. 35–56. Princeton Univ. Press, Princeton, NJ.

Boag, P. T., and Grant, P. R. (1981). Intense natural selection in a population of Darwin's Finches (Geospizinae) in the Galápagos. *Science* 214, 82–85.

Boag, P. T., and Grant, P. R. (1984a). Darwin's Finches (*Geospiza*) on Isla Daphne Major, Galápagos: Breeding and feeding ecology in a climatically variable environment. *Ecol. Monogr.* 54, 463–489.

Boag, P. T., and Grant, P. R. (1984b). The classical case of character release: Darwin's finches (*Geospiza*) on Isla Daphne Major, Galápagos. *Biol. J. Linn. Soc.* 22, 243–287.

Böhning-Gaese, K., Taper, M. L., and Brown, J. H. (1994). Avian community dynamics are discordant in space and time. *Oikos* 70, 121–126.

Broecker, W. S. (1994). Massive iceberg discharges as triggers for global climate change. *Nature (London)* 372, 421–424.

Bush, M. B., Piperno, D. R., Colinvaux, P. A., DeOliveira, P. E., Krissek, L. A., Miller, M. C., and Rowe, W. E. (1992). A 14 300-yr paleoecological profile of a lowland tropical lake in Panama. *Ecol. Monogr.* 62, 251–275.

Campbell, I. D., and McAndrews, J. H. (1993). Forest disequilibrium caused by rapid Little Ice Age cooling. *Nature (London)* 366, 336–338.

Caughley, G. C. (1994). Directions in conservation biology. *J. Anim. Ecol.* 63, 215–244.

Chesson, P. L., and Case, T. J. (1986). Overview: Nonequilibrium community theories: Chance, variability, history, and coexistence. *In* "Community Ecology" (J. Diamond and T. J. Case, eds.), pp. 229–239. Harper & Row, New York.

Colinvaux, P. A. (1972). Climate and the Galápagos Islands. *Nature (London)* 240, 17–20.

Cox, A. (1983). Ages of the Galápagos Islands. *In* "Patterns of Evolution in Galápagos Organisms" (R. I. Bowman, M. Berson, and A. E. Leviton, eds.), pp. 11–23. Am. Assoc. Adv. Sci., Pacific Division, San Francisco.

Crowcroft, P. (1990). "Elton's Ecologists: A History of the Bureau of Animal Population." Univ. of Chicago Press, Chicago.

Davis, M. B. (1986). Invasions of forest communities during the holocene: Beech and hemlock in the Great Lakes Region. *In* "Community Ecology" (J. Diamond and T. J. Case, eds.), pp. 373–393. Harper & Row, New York.

de Paepe, P. (1966). Geologie van Isla Daphne Major (Islas Galápagos, Ecuador). *Natuurwet. Tijdschr. (Ghent)* 48, 67–80.

Diamond, J. M. (1984). 'Normal' extinctions of isolated populations. *In* "Extinctions" (M. H. Nitecki, ed.), pp. 191–246. Univ. of Chicago Press, Chicago.

Diamond, J. M., and Case, T. J., eds. (1986). "Community Ecology." Harper & Row, New York.

Diamond, J. M., and May, R. M. (1977). Species turnover rates on islands: dependence on census interval. *Science* 197, 266–270.

Diaz, H. F., and Kiladis, G. N. (1992). Atmospheric teleconnections associated with the extreme phase of the Southern Oscillation. *In* "El Niño: Historical and Paleoclimatic Aspects of the Southern Oscillation" (H. F. Diaz and V. Markgraf, eds.), pp. 7–28. Cambridge Univ. Press, Cambridge, UK.

Diaz, H. F., and Markgraf, V., eds. (1992). "El Niño: Historical and Paleoclimatic Aspects of the Southern Oscillation." Cambridge Univ. Press, Cambridge, UK.

Dunbar, R. B., Wellington, G. M., Colgan, M. W., and Glynn, P. W. (1994). Eastern Pacific sea surface temperature since 1600 A. D.: The O^{18} record of climate variability in Galápagos corals. *Paleoceanography* 9, 291–315.

Enemar, A., Nilsson, L., and Sjöstrand, B. (1984). The composition and dynamics of the passerine bird community in a subalpine birch forest, Swedish Lapland. A 20-year study. *Ann. Zool. Fenn.* 21, 321–338.

Enfield, D. B. (1992). Historical and prehistorical overview of El Niño/Southern Oscillation. *In* "El Niño: Historical and Paleoclimatic Aspects of the Southern Oscillation" (H. F. Diaz and V. Markgraf, eds.), pp. 95–117. Cambridge Univ. Press, Cambridge, UK.

Fraser, D. M., May, R. M., Pellew, R., Johnson, T. H., and Walter, K. S. (1993). Estimating extinction rates. *Nature (London)* 364, 494–496.

Fridriksson, S. (1975). "Surtsey." Wiley, New York.

Gaston, K. J., and McArdle, B. H. (1994). The temporal variability of animal abundances: Measures, methods and patterns. *Philos. Trans. R. Soc. London, Ser. B* 345, 335–358.

Gibbs, H. L., and Grant, P. R. (1987a). Adult survival in Darwin's ground finch (*Geospiza*) populations in a variable environment. *J. Anim. Ecol.* 56, 797–813.

Gibbs, H. L., and Grant, P. R. (1987b). Ecological consequences of an exceptionally strong El Niño event on Darwin's Finches. *Ecology* 68, 1735–1746.

Gibbs, H. L., and Grant, P. R. (1987c). Oscillating selection on Darwin's Finches. *Nature (London)* 327, 511–513.

Goodman, D. (1972). The paleoecology of the Tower island bird colony: A critical examination of the complexity-stability theory. Unpublished Ph.D Thesis, Ohio State University, Columbus.

Graham, N. E., and White, W. B. (1988). The El Niño cycle: A natural oscillator of the Pacific Ocean-Atmospheric system. *Science* 240, 1293–1302.

Graham, R. W. (1986). Response of mammalian communities to environmental changes during the late Quaternary. *In* "Community Ecology" (J. M. Diamond and T. J. Case, eds.), pp. 300–313. Harper & Row, New York.

Grant, B. R., and Grant, P. R. (1982). Niche shifts and competition in Darwin's Finches: *Geospiza conirostris* and congeners. *Evolution (Lawrence, Kans.)* 36, 637–657.

Grant, B. R., and Grant, P. R. (1989). "Evolutionary Dynamics of a Natural Population: The Large Cactus Finch of the Galápagos." Univ. of Chicago Press, Chicago.

Grant, B. R., and Grant, P. R. (1993). Evolution of Darwin's finches caused by a rare climatic event. *Proc. R. Soc. London, Ser. B* 251, 111–117.

Grant, P. R. (1984). The endemic land birds. *In* "Galápagos" (R. Perry, ed.), pp. 175–189. Pergamon, Oxford.

Grant, P. R. (1985). Climatic fluctuations on the Galápagos Islands and their influence on Darwin's Finches. *In* "Neotropical Ornithology" (P. A. Buckley, M. S. Foster, E. S. Morton, R. S. Ridgely, and F. G. Buckley, eds.), Ornithol. Monogr. No. 36, pp. 471–483. American Ornithologists' Union.

Grant, P. R. (1986a). "Ecology and Evolution of Darwin's Finches." Princeton Univ. Press, Princeton, NJ.

Grant, P. R. (1986b). Interspecific competition in fluctuating environments. *In* "Community Ecology" (J. M. Diamond and T. J. Case, eds.), pp. 173–191. Harper & Row, New York.

Grant, P. R. (1993). Hybridization of Darwin's Finches on Isla Daphne Major, Galápagos. *Philos. Trans. R. Soc. London, Ser. B* 340, 127–139.

Grant, P. R., and Boag, P. T. (1980). Rainfall on the Galápagos and the demography of Darwin's Finches. *Auk* 97, 227–244.

Grant, P. R., and Grant, B. R. (1980a). Annual variation in finch numbers, foraging and food supply on Isla Daphne Major, Galápagos. *Oecologia* 46, 55–62.

Grant, P. R., and Grant, B. R. (1980b). The breeding and feeding characteristics of Darwin's Finches on Isla Genovesa, Galápagos. *Ecol. Monogr.* 50, 381–410.

Grant, P. R., and Grant, B. R. (1989). The slow recovery of *Opuntia megasperma* on Española. *Not. Galápagos* 48, 13–15.

Grant, P. R., and Grant, B. R. (1992a). Demography and the genetically effective sizes of two populations of Darwin's Finches. *Ecology* 73, 766–784.

Grant, P. R., and Grant, B. R. (1992b). Hybridization of bird species. *Science* 256, 193–197.

Grant, P. R., and Grant, B. R. (1994). Phenotypic and genetic effects of hybridization in Darwin's Finches. *Evolution (Lawrence, Kans.)* 48, 297–316.

Grant, P. R., and Grant, B. R. (1995a). The founding of a new population of Darwin's Finches. *Evolution (Lawrence, Kans.)* 49, 229–240.

Grant, P. R., and Grant, B. R. (1995b). Predicting microevolutionary responses to directional natural selection on heritable variation. *Evolution (Lawrence, Kans.)* 49, 241–251.

Grant, P. R., and Schluter, D. (1984). Interspecific competition inferred from patterns of guild structure. *In* "Ecological Communities: Conceptual Issues and the Evidence" (D. R. Strong, D. Simberloff, L. G. Abele, and A. B. Thistle, eds.), pp. 201–233. Princeton Univ. Press, Princeton, NJ.

Grant, P. R., Smith, J. N. M., Grant, B. R., Abbott, I., and Abbott, L. K. (1975). Finch numbers, owl predation and plant dispersal on Isla Daphne Major, Galápagos. *Oecologia* 19, 239–257.

Grant, P. R., Grant, B. R., Smith, J.N.M., Abbott, I., and Abbott, L. K. (1976). Darwin's Finches: Population variation and natural selection. *Proc. Natl. Acad. Sci. U.S.A.* 73, 257–261.

Grant, P. R., Abbott, I. J., Schluter, D., Curry, R. L., and Abbott, L. K. (1985). Variation in the size and shape of Darwin's Finches. *Biol. J. Linn. Soc.* 25, 1–39.

Hamilton, T. H., and Rubinoff, I. (1967). On predicting insular variation in endemism and sympatry for the Darwin Finches in the Galápagos archipelago. *Am. Nat.* 101, 161–171.

Harris, M. P. (1973). The Galápagos avifauna. *Condor* 75, 265–278.

Hickman, C. S., and Lipps, J. H. (1985). Geologic youth of Galápagos islands confirmed by marine stratigraphy and paleontology. *Science* 227, 1578–1580.

Holmes, R. T., Sherry, T. W., and Sturgis, F. W. (1986). Bird community dynamics in a temperate deciduous forest: Long-term trends at Hubbard Brook. *Ecol. Monogr.* 56, 201–220.

Holmes, R. T., Sherry, T. W., and Sturgis, F. W. (1991). Numerical and demographic responses of temperate forest birds to annual fluctuations in their food resources. *Proc. Int. Ornithol. Congr., 20th,* Christchurch, *1990,* pp. 1559–1567.

Itow, S. (1975). A study of vegetation in Isla Santa Cruz, Galápagos islands. *Not. Galápagos* 17, 10–13.

Jackson, J.B.C. (1994). Constancy and change of life in the sea. *Philos. Trans. R. Soc. London, Ser. B* 344, 55–60.

Kareiva, P. (1993). Space: The final frontier for ecological theory. *Ecology* 75, 1.

Kareiva, P., Kingsolver, J. G., and Huey, R. B., eds. (1993). "Biotic Interactions and Global Change." Sinauer Assoc., Sunderland, MA.

Lack, D. (1945). The Galápagos finches (Geospizinae): A study in variation. *Occas. Pap. Calif. Acad. Sci.* 21, 1–159.

Lack, D. (1947). "Darwin's Finches." Cambridge Univ. Press, Cambridge, U.K.

Lack, D. (1964). A long-term study of the Great Tit (*Parus major*). *J. Anim. Ecol.* **34**, Suppl., 159–173.

Lack, D. (1969). The number of bird species on islands. *Bird Study* **16**, 193–209.

Lack, D. (1976). "Island Biology, Illustrated by the Land Birds of Jamaica." Univ. of California Press, Los Angeles.

Levin, S. A. (1992). The problem of pattern and scale in ecology. *Ecology* **73**, 1943–1983.

MacArthur, R. H., and Wilson, E. O. (1963). An equilibrium theory of island biogeography. *Evolution (Lawrence, Kans.)* **17**, 373–387.

MacArthur, R. H., and Wilson, E. O. (1967). "The Theory of Island Biogeography." Princeton Univ. Press, Princeton, NJ.

Mangel, M., and Tier, C. (1994). Four facts every conservation biologist should know about persistence. *Ecology* **75**, 607–614.

Mayer, G. C., and Chipley, R. M. (1992). Turnover in the avifauna of Guana Island, British Virgin Islands. *J. Anim. Ecol.* **61**, 561–566.

McArdle, B. H. (1989). Bird population densities. *Nature (London)* **338**, 628.

McArdle, B. H., and Gaston, K. J. (1992). Comparing population variabilities. *Oikos* **64**, 610–612.

McArdle, B. H., Gaston, K. J., and Lawton, J. H. (1990). Variation in the size of animal populations: Patterns, problems and artefacts. *J. Anim. Ecol.* **59**, 439–454.

McGlone, M. S., Kershaw, K. P., and Markgraf, V. (1992). El Niño/Southern Oscillation climatic variability in Australasian and South American paleoenvironmental records. *In* "El Niño: Historical and Paleoclimatic Aspects of the Southern Oscillation" (H. F. Diaz and V. Markgraf, eds.), pp. 435–462. Cambridge Univ. Press, Cambridge, U.K.

Michaelsen, J., and Thompson, L. G. (1992). A comparison of proxy records of El Niño/Southern Oscillation. *In* "El Niño: Historical and Paleoclimatic Aspects of the Southern Oscillation" (H. F. Diaz and V. Markgraf, eds.), pp. 323–348. Cambridge Univ. Press, Cambridge, UK.

Millington, S. J., and Grant, P. R. (1983). Feeding ecology and territoriality of the Cactus Finch *Geospiza scandens* on Isla Daphne Major, Galápagos. *Oecologia* **58**, 76–83.

Murdoch, W. W. (1994). Population regulation in theory and practice. *Ecology* **75**, 271–287.

Murray, B. G., Jr. (1979). "Population Dynamics: Alternative Models." Academic Press, New York.

Murray, B. G., Jr. (1994). On density dependence. *Oikos* **69**, 520–523.

Newton, I. (1967). The adaptive radiation and feeding ecology of some British finches. *Ibis* **109**, 33–98.

Nicholls, N. (1992). Historical El Niño/Southern Oscillation variability in the Australasian region. *In* "El Niño: Historical and Paleoclimatic Aspects of the Southern Oscillation" (H. F. Diaz and V. Markgraf, eds.), pp. 151–173. Cambridge Univ. Press, Cambridge, UK.

O'Connor, R. J. (1986). Biological characteristics of invaders. *Philos. Trans. R. Soc. London, Ser. B* **314**, 583–598.

Philander, S. G. (1990). "El Niño, La Niña and the Southern Oscillation." Academic Press, San Diego, CA.

Phillips, O. L., and Gentry, A. H. (1994). Increasing turnover through time in tropical forests. *Science* **263**, 954–958.

Pimm, S. L., and Redfearn, A. (1988). The variability of population densities. *Nature (London)* **334**, 613–614.

Pimm, S. L., and Redfearn, A. (1989). Reply. *Nature (London)* **338**, 628.

Pimm, S. L., Jones, H. L., and Diamond, J. (1988). On the risk of extinction. *Am. Nat.* **132**, 757–785.

Pimm, S. L., Diamond, J., Reed, T. M., Russell, G. J., and Verner, J. (1993). Times to extinction for small populations of large birds. *Proc. Natl. Acad. Sci. U.S.A.* **90**, 10871–10875.

Pimm, S. L., Moulton, M. P., and Justice, L. J. (1994). Bird extinctions in the central Pacific. *Philos. Trans. R. Soc. London, Ser. B* **344**, 27–33.

Pollard, E., Lakhani, K. H., and Rothery, P. (1987). The detection of density-dependence from a series of annual censuses. *Ecology* **68**, 2046–2055.

Price, T. D. (1987). Diet variation in a population of Darwin's Finches. *Ecology* **68**, 1015–1028.

Price, T. D., Grant, P. R., Gibbs, H. L., and Boag, P. T. (1984). Recurrent patterns of natural selection in a population of Darwin's finches. *Nature (London)* **309**, 787–789.

Pulliam, H. R., and Brand, M. (1975). The production and utilization of seeds in plains grasslands of southeastern Arizona. *Ecology* **56**, 1158–1166.

Pulliam, H. R., and Parker, T. A., III (1979). Population regulation of sparrows. *Fortschr. Zool.* **25**, 137–147.

Quinn, W. H. (1992). A study of Southern Oscillation-related climatic activity from A. D. 622–1900 incorporating Nile River flood data. *In* "El Niño: Historical and Paleoclimatic Aspects of the Southern Oscillation" (H. F. Diaz and V. Markgraf, eds.), pp. 119–149. Cambridge Univ. Press, Cambridge, UK.

Rahmstorf, S. (1994). Rapid climate transitions in a coupled ocean-atmosphere model. *Nature (London)* **372**, 82–85.

Ricklefs, R. E., and Schluter, D. (1993). Species diversity: Regional and historical influences. *In* "Species Diversity in Ecological Communities: Historical and Geographical Perspectives" (R. E. Ricklefs and D. Schluter, eds.), pp. 350–363. Univ. of Chicago Press, Chicago.

Rotenberry, J. T., and Wiens, J. A. (1991). Weather and reproductive variation in shrubsteppe sparrows: A hierarchical analysis. *Ecology* **72**, 1325–1335.

Royama, T. (1977). Population persistence and density-dependence. *Ecol. Monogr.* **47**, 1–35.

Schlesinger, M. E., and Ramankutty, N. (1994). An oscillation in the global climate system of period 65–70 years. *Nature (London)* **367**, 723–726.

Schluter, D. (1988). The evolution of finch communities on islands and continents: Kenya vs. Galápagos. *Ecol. Monogr.* **58**, 229–249.

Schluter, D., and Grant, P. R. (1984). Determinants of morphological patterns in communities of Darwin's Finches. *Am. Nat.* **123**, 175–196.

Schluter, D., and Repasky, R. (1991). Worldwide limitation of finch densities by food and other factors. *Ecology* **72**, 1763–1774.

Schluter, D., Price, T. D., and Grant, P. R. (1985). Ecological character displacement in Darwin's finches. *Science* **227**, 1056–1059.

Schoener, A. (1974). Colonization curves for planar marine islands. *Ecology* **55**, 818–827.

Schoener, T. W. (1982). Field experiments on interspecific competition. *Am. Sci.* **70**, 586–595.

Schoener, T. W., and Spiller, D. A. (1992). Is extinction rate related to temporal variability in population size? An empirical answer for orb spiders. *Am. Nat.* **139**, 1176–1207.

Simberloff, D., and Connor, E. F. (1981). Missing species combinations. *Am. Nat.* **118**, 215–239.

Simberloff, D., and Wilson, E. O. (1969). Experimental zoogeography of islands: The colonization of empty islands. *Ecology* **50**, 278–296.

Simpson, B. B. (1974). Glacial immigrations of plants: Island biogeographical evidence. *Science* **185**, 698–700.

Smith, F. D. M., May, R. M., Pellew, R., Johnson, T. H., and Walter, K. S. (1993). How much do we know about the current extinction rate? *Trends Ecol. Evol.* **8**, 375–378.

Smith, J. N. M., Grant, P. R., Grant, B. R., Abbott, I., and Abbott, L. K. (1978). Seasonal variation in feeding habits of Darwin's Ground Finches. *Ecology* **59**, 1137–1150.

Smith, J. N. M., Montgomerie, R. D., Tait, M. J., and Yom-Tov, Y. (1980). A winter feeding experiment on an island song sparrow population. *Oecologia* **47**, 164–170.

Stattersfield, A. (1987). A systematic list of birds presumed to have become extinct since 1600. *In* "Rare Birds of the World" (G. Mountford, ed.), pp. 241–246. Collins, London.

Steele, J. H. (1985). A comparison of terrestrial and marine ecological systems. *Nature (London)* **313**, 355–358.

Steele, J. H., and Henderson, E. W. (1994). Coupling between physical and biological scales. *Philos. Trans. R. Soc. London, Ser. B* **343**, 5–9.

Strong, D. (1984). Density-vague ecology and liberal population regulation in insects. *In* "A New Ecology: Novel Approaches to Interactive Systems" (P. W. Price, C. N. Slobodchikoff, and W. S. Gaud, eds.), pp. 313–327. Wiley, New York.

Svensson, S., Carlsson, U. T., and Liljedahl, G. (1984). Structure and dynamics of an alpine bird community, a 20 year study. *Ann. Zool. Fenn.* **21**, 339–350.

Swarth, H. S. (1931). The avifauna of the Galápagos Islands. *Occas. Pap. Calif. Acad. Sci.* **18**, 1–299.

Thornton, I.W.B., New, T. R., Zann, R. A., and Rawlinson, P. A. (1990). Colonization of the Krakatau Islands by animals: A perspective from the 1980s. *Philos. Trans. R. Soc. London, Ser. B* **328**, 131–165.

Trenberth, K. E., Branstator, G. W., and Arkin, P. A. (1988). Origins of the 1988 North American drought. *Science* **242**, 1640–1645.

van Riper, C., III, van Riper, S. G., Goff, M. L., and Laird, M. (1986). The epizootiology and ecological significance of malaria in Hawaiian land birds. *Ecol. Monogr.* **56**, 327–341.

Warner, R. E. (1968). The role of introduced diseases in the extinction of the endemic Hawaiian avifauna. *Condor* **70**, 101–120.

Wiens, J. A. (1977). On competition and variable environments. *Am. Sci.* **65**, 590–597.

Wiens, J. A. (1989). "The Ecology of Birds Communities," Vols. 1 and 2. Cambridge Univ. Press, Cambridge, UK.

Wiggins, I. L., and Porter, D. M. (1971). "Flora of the Galápagos Islands." Stanford Univ. Press, Stanford, CA.

Williamson, M. (1981). "Island Populations." Oxford Univ. Press, Oxford.

Williamson, M. (1983). The land-bird community of Stokholm: Ordination and turnover. *Oikos* **41**, 378–384.

Williamson, M. (1984). The measurement of population variability. *Ecol. Entomol.* **9**, 239–241.

Williamson, M. (1987). Are communities ever stable? *In* "Colonisation, Succession and Stability" (A. J. Gray, M. J. Crawley, and P. J. Edwards, eds.), pp. 353–371. Blackwell, Oxford.

Woiwood, I. P., and Hanski, I. (1992). Patterns of density dependence in moths and aphids. *J. Anim. Ecol.* **61**, 619–631.

Woolfenden, G. E., and Fitzpatrick, J. W. (1984). "The Florida Scrub Jay: Demography of a Cooperative-Breeding Bird." Princeton Univ. Press, Princeton, NJ.

CHAPTER 13

Waterfowl Communities in the Northern Plains

DOUGLAS H. JOHNSON

Northern Prairie Science Center, National Biological Service, Jamestown,
North Dakota 58401

I. INTRODUCTION

Features that determine the composition of avian communities have received extensive and enthusiastic attention, both empirically and theoretically (e.g., Cody, 1974; Strong *et al.,* 1984; Wiens, 1989a,b). Interspecific competition for limited resources is one influence widely regarded as critical, but others include species-specific responses to environmental conditions, predation, parasitism, commensal and mutualistic interactions, disturbance and chance, and historical events (Wiens, 1989b). Despite the attention they have received, the relative importance of these different features remains highly controversial.

Certainly the community of birds present in an area will be but a subset of the total population of each constituent species; thus, the total population size of each species is a potential influence on local communities. Since many bird communities in the temperate zone are reconstituted annually with migrants, the geographical location of a site within the breeding range is likely an important feature; clearly, an avian community outside the normal breeding range of a particular species is unlikely to include that species, irrespective of the suitability of vegetation and other habitat features. Further, if suitable breeding habitat exists in excess of its potential clientele, then suitable habitat encountered earlier in spring migration is more likely to be occupied than is habitat encountered later. And the breeding ranges of birds are known to shift from year to year, depending on climatic conditions, a phenomenon that represents a modification of the general breeding range at a finer scale of resolution.

The return of migrant birds to areas from which they fledged or where they previously bred, via philopatric behavior, can modify the occupancy of a particular habitat (Hildén, 1964); likewise, traditional use of breeding sites by birds can induce species to persist in unsuitable habitat (Wiens, 1985). Thus site fidelity in fact can weaken the relationship between habitat quality and number of birds on an area (Wiens, 1985).

Waterfowl have received less attention from community ecologists than have many other groups of birds, despite the importance attributed to waterfowl by biologists generally as well as by the public at large. Exceptions include work by Nudds and colleagues in Canada and by Pöysä and colleagues in Finland (see Nudds, 1992, and references contained therein). Further, waterfowl are important ecologically; in much of the prairie of the North American midcontinent, waterfowl are numerically among the most common bird species and are certainly dominant in terms of biomass.

This chapter addresses some of the influences on waterfowl communities in a mixed-grass prairie pothole habitat. It takes a temporal perspective, based on annual censuses of breeding ducks over 25 years on a specific study area at Woodworth, North Dakota. Any changes in the structural features of the habitat that may have occurred during this period at the census site were at most gradual. I examine how the waterfowl communities varied in response to influences that did change annually, such as climate, the conditions of the wetlands on the study area, the regional populations of birds from which the communities were constituted, and the population at Woodworth during the previous year. The results of the analyses are interpreted relative to individual characteristics of 11 waterfowl species: mallard (*Anas platyrhynchos*), gadwall (*Anas strepera*), green-winged teal (*Anas crecca*), blue-winged teal (*Anas discors*), northern pintail (*Anas acuta*), northern shoveler (*Anas clypeata*), American wigeon (*Anas americana*), canvasback (*Aythya valisineria*), redhead (*Aythya americana*), lesser scaup (*Aythya affinis*), and ruddy duck (*Oxyura jamaicensis*).

II. STUDY AREA

The Woodworth Study Area consists of 1231 ha of mixed-grass prairie pothole habitat in Stutsman County, east-central North Dakota (see, e.g., Higgins *et al.*,

1992). It is situated on the Missouri Coteau, a morainal belt extending across the state in a northwest to southeast direction. The rolling terrain contains numerous wetland basins, totaling 548 in the study area and encompassing 10% of the land area. About two-thirds of the basins are seasonally flooded (classification according to Stewart and Kantrud, 1971) and these account for about one-third of the wetland surface area. Twelve percent of the basins, and nearly half the surface area of wetlands, are semipermanently flooded. Except for a 53-ha permanent lake, other wetlands are temporary or ephemeral. All wetlands are of either fresh or slightly brackish water (Stewart and Kantrud, 1971).

Prior to its purchase by the U.S. Fish and Wildlife Service, land use in the study area was a mixture of cattle grazing and hay and crop production (Bayha, 1964). Those agricultural practices have continued in the privately owned portions of the study area. In the federally owned portions (87% of the area), management of the uplands since acquisition has emphasized restoration of the grasslands. Some formerly cropped fields were replanted to grasses or grass–legume mixtures, and unbroken grasslands have been managed mostly by prescribed burning. Release of the area from regular grazing and haying has resulted in increased areas of emergent vegetation in many wetland basins in the study area (H. A. Kantrud, Northern Prairie Science Center, personal communication).

III. METHODS

A. Duck Counts

Ducks were censused annually 1965–1989 according to the methods described in Higgins et al. (1992). Censuses were conducted while observers were walking or driving around the periphery of wetland basins. In large wetlands or those with extensive emergent vegetation, two individuals walked or waded in opposite directions around the basin and later compared notes to eliminate count duplications. If birds were flushed from a wetland during a survey, their flight to or descent into another wetland was observed to avoid counting the same birds on wetlands subsequently surveyed.

Breeding pair was the unit of record; the criteria used to indicate pairs, after Hammond (1969), were (1) observed pairs, (2) lone drakes, (3) lone hens only for diving ducks and only when males were not nearby, and (4) groups of up to five males or mixed sexes (except for northern shovelers and American wigeon, for which only pairs and lone males were counted).

Censuses were ordinarily conducted between 0700 and 1700 hours. In most years, censuses were repeated during the breeding season. Estimates of mallards, northern pintails, northern shovelers, and canvasbacks were based on censuses made in early May; estimates of gadwalls, blue-winged teal, redheads, lesser scaup, and ruddy ducks were based on early June censuses; estimates of American wigeon and green-winged teal were taken as the greater of the counts recorded on these two census dates.

B. Pond Counts

Wetlands in the Woodworth Study Area were surveyed several times each year (Higgins et al., 1992). The wetland basins were classified as containing water if at least 5% of the area was inundated by 25 mm or more of water. For the analyses reported here, the wetland surveys conducted during 1–15 May of each year are used, as they consist of measurements made closest to the time water-fowl settled in the area.

C. Weather Data

Monthly precipitation and monthly mean temperatures were obtained from weather stations near the study site. These included Steele and Pettibone, North Dakota, as well as the Woodworth Study Area itself.

D. Conserved Soil Moisture Index

The conserved soil moisture (CSM) index is a weighted average of precipitation during the 21 months preceding May of a particular year. It was developed by Williams and Robertson (1965) for agronomic purposes and popularized for waterfowl biologists by Boyd (1981), who suggested that it mirrored variation in wetlands.

E. Continental Population Indices

Annual indices of duck abundance were obtained by the U.S. Fish and Wildlife Service and the Canadian Wildlife Service from aerial surveys of the primary breeding ground of most duck species. The survey region encompasses about 3.03 million km^2 in the north-central United States, the prairie provinces, the Northwest Territories of Canada, and Alaska. In typical years more than 80% of the North American populations of the duck species considered in this chapter breed in this survey region (Johnson and Grier, 1988). The survey utilizes aerial and ground surveys conducted each May, with adjustments made for visibility; the survey methods were described by Martin et al. (1979).

F. Waterfowl Distributions and Philopatry

Information on the overall breeding distribution of each species, and on the tendency of species to return to the same breeding area in successive years, was taken from Bellrose (1980) and Johnson and Grier (1988), who summarized both published and unpublished information on those topics. Johnson and Grier (1988) also identified areas (strata as used for the continental surveys previously described) according to the average density of each species, from the highest to the lowest quartile.

G. Analytic Methods

The densities of the most common duck species were related to several explanatory variables, including the number of ponds in the Woodworth Study Area containing water during early May ("May ponds") and the conserved soil moisture (CSM) index, both of which reflect long-term precipitation patterns. Four more proximate climatic features were examined: mean temperature and total precipitation in April and in May. Other explanatory variables considered were the index of continental population size and the species count in the Woodworth Study Area during the previous year.

Statistical tools used [and SAS Institute, Inc. (1989, 1990) procedures] included Pearson product-moment correlation (Proc CORR), analysis of variance (Proc GLM), and stepwise regression (Proc REG), as well as descriptive methods (Proc UNIVARIATE).

In an effort to examine the suite of variables that might influence populations of ducks in the Woodworth Study Area, I attempted to model the numbers of each species there in relation to May ponds, CSM, mean temperature, and total precipitation in each of April and May, the count of each species in the previous year, and the continental population index. Five models were fitted for each species, which included, respectively: (1) only proximate habitat variables (May ponds, CSM, and temperature and precipitation values), (2) proximate habitat variables plus the previous year's count, (3) proximate habitat variables plus the continental index, (4) all variables mentioned above, and (5) the "best" model as determined by stepwise regression, with significance levels for including and deleting variables set at 0.15. Significance of the multiple correlation coefficient was determined for all but the "best" model, for which the stepwise selection of variables renders the R^2 values noncomparable.

We sought interactions among species above and beyond those related to similar responses by different species to abiotic or environmental features. This was done by calculating residuals from the best-fitting models for each species and relating them with regression to the counts of other species. Positive associations would be consistent with either the hypothesis that species actively associate with each other or the notion that the species respond similarly to certain environmental conditions that are not directly represented by the explanatory variables. Negative associations would be consistent with the hypothesis that the species compete with each other.

IV. RESULTS

A. Trends in and Relations among the Explanatory Variables

Both the number of May ponds and the index of conserved soil moisture varied dramatically during the 25-year period of study, but the CSM index less so (Fig. 1). Drought was especially pronounced during 1977, 1980–1981, and 1989. May ponds and CSM were strongly correlated ($r = 0.67, P < 0.001$; Table I). The number of May ponds was higher in years with cool April and May temperatures

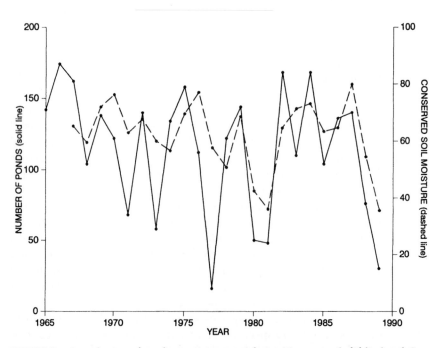

FIGURE 1. Annual count of ponds containing water during May survey (solid line) and Conserved Soil Moisture index (dashed line), Woodworth Study Area, 1965–1989.

TABLE 1   Correlations among Proximate Habitat and Weather Variables at Woodworth Study Area, North Dakota, 1965–1989

	Conserved soil moisture	Mean temperature		Precipitation	
		April	May	April	May
May ponds	0.67***	−0.62***	−0.50**	0.41**	0.12
Conserved soil moisture		−0.32	−0.21	0.36*	−0.11
Mean temperature in April			0.67***	−0.36*	0.02
Mean temperature in May				−0.39*	0.01
Precipitation in April					−0.23

*P < 0.10.
**P < 0.05.
***P < 0.01.

and wet Aprils; the CSM index was related to weather features in a similar manner, but less strongly (Table I).

Average temperatures during April and May tended to increase during the study period (Fig. 2). Temperatures in the two months were highly correlated (r

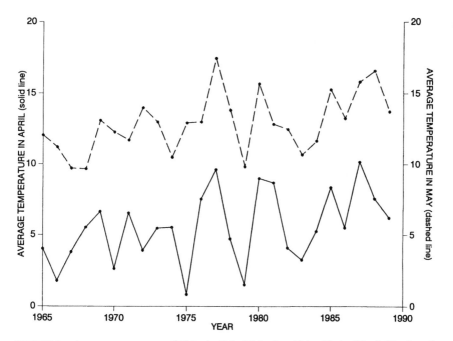

FIGURE 2.   Average temperatures (°C) in April (solid line) and May (dashed line), Woodworth Study Area, 1965–1989.

= 0.67, P < 0.001; Table I), indicating the persistence of seasonal temperature patterns within years. On the other hand, monthly precipitation totals were very erratic (Fig. 3). Unlike average temperatures, precipitation totals in April and in May were not significantly correlated, indicating that precipitation patterns were less enduring than temperature patterns.

B. Variations in Duck Populations

Seven dabbling duck species, three diving duck species, and one stiff-tailed duck were common in the Woodworth Study Area (Table II). In decreasing order of abundance, these were blue-winged teal (making up an average of 43% of the total waterfowl population), gadwall (12.9%), mallard (10.6%), northern shoveler (6.2%), lesser scaup (6.0%), northern pintail (5.6%), redhead (5.2%), ruddy duck (5.0%), canvasback (2.0%), American wigeon (1.9%), and green-winged teal (1.6%). Gadwall and mallard populations were the most consistent in size, as measured by the coefficient of variation; most variable in numbers were the canvasback, ruddy duck, and redhead (see Table II).

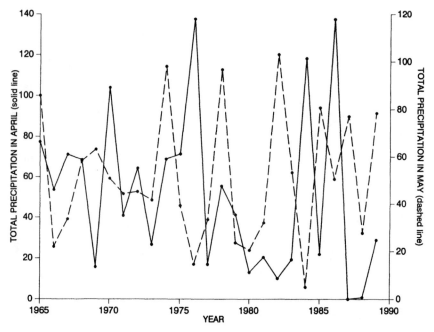

FIGURE 3. Total precipitation (mm) during April (solid line) and May (dashed line), Woodworth Study Area, 1965–1989.

TABLE II Average Number, Standard Deviation, and Coefficient of Variation of Breeding Pairs of Ducks in Woodworth Study Area, 1965–1989, and Trends (Correlation with Calendar Year) for Counts at Woodworth and at the Continental Scale

| | | Standard | | Trend | |
Species	Average	deviation	CV	Woodworth	Continent
Dabbling ducks					
Mallard	52.24	12.44	23.8	−0.09	−0.48**
Gadwall	63.56	17.30	27.2	−0.41**	−0.28
American wigeon	9.52	4.01	42.1	−0.59***	−0.30
Green-winged teal	7.72	4.08	52.8	−0.65***	0.26
Blue-winged teal	211.36	85.70	40.5	−0.32	−0.22
Northern shoveler	30.28	13.47	44.5	−0.02	−0.16
Northern pintail	27.44	12.09	44.1	−0.51***	−0.67***
Diving ducks					
Redhead	25.76	17.84	69.3	0.63***	−0.23
Canvasback	9.68	7.72	79.7	0.58***	−0.29
Lesser scaup	29.72	13.01	43.8	0.65***	0.10
Stiff-tail duck					
Ruddy duck	24.84	19.78	79.6	0.37*	0.41**

*$P < 0.10$.
**$P < 0.05$.
***$P < 0.01$.

C. Interactions among the Species

Among dabbling ducks in the Woodworth Study Area, populations of the different species generally positively correlated with each other (11 of 21 pairwise comparisons significant; Table III). Numbers of gadwalls and northern pintails especially were associated with other dabbling duck species, while at the other extreme, numbers of mallards were significantly related to those of only one other species, the gadwall. The three diving ducks plus ruddy ducks also tended to covary together (5/6 pairwise comparisons significant; Table III). Numbers of dabbling ducks and diving ducks tended to vary inversely, but just 3 pairwise comparisons were significant (with one positive), and $\chi^2 = 2.33$, $P > 0.05$ that the number of negative correlations (14/21) was higher than that expected by chance (10.5/21). It seems that the two groups varied independently.

Correlations between residuals from the best-fitting model for each species and counts of the other species were more frequently positive than negative (Table IV). Significant positive correlations ($P < 0.10$) with at least one other species were calculated for all modeled species except blue-winged teal. Some

TABLE III Correlation Coefficients between Annual Counts of Ducks at Woodworth Study Area, North Dakota, during 1965–1989

	Gadwall	American wigeon	Green-winged teal	Blue-winged teal	Northern shoveler	Northern pintail	Redhead	Canvasback	Lesser scaup	Ruddy duck
Mallard	0.52**	0.17	0.09	0.31	0.31	0.29	0.28	-0.01	0.02	0.05
Gadwall		0.40*	0.47*	0.52**	0.19	0.57**	0.01	-0.24	-0.31	0.08
American wigeon			0.73***	0.10	-0.01	0.59***	-0.50*	-0.37	-0.38	-0.38
Green-winged teal				0.03	-0.07	0.48*	-0.49*	-0.37	-0.45*	-0.26
Blue-winged teal					0.49*	0.40*	0.08	-0.19	-0.34	0.16
Northern shoveler						0.45*	0.46*	0.39	0.08	0.40*
Northern pintail							-0.06	-0.06	-0.21	0.15
Redhead								0.52**	0.54**	0.71***
Canvasback									0.57**	0.50*
Lesser scaup										0.34

*P < 0.05.
**P < 0.01.
***P < 0.001.

TABLE IV Correlations Coefficients and P Values between Residuals from Best-fitting Model for Each Species and Count of Other Species

Modeled species	Correlation coefficient with count of species										
	Mallard	Gadwall	American wigeon	Green-winged teal	Blue-winged teal	Northern shoveler	Northern pintail	Redhead	Canvasback	Lesser scaup	Ruddy duck
Mallard	—	0.41 / 0.05	-0.01 / 0.96	0.09 / 0.67	0.14 / 0.51	0.20 / 0.36	0.12 / 0.57	0.19 / 0.37	-0.03 / 0.90	-0.07 / 0.74	-0.05 / 0.83
Gadwall	0.46 / 0.02	—	0.40 / 0.05	0.41 / 0.04	0.37 / 0.07	0.05 / 0.80	0.47 / 0.02	-0.01 / 0.98	-0.22 / 0.30	-0.23 / 0.27	0.03 / 0.89
American wigeon	0.13 / 0.55	0.37 / 0.07	—	0.65 / 0.01	0.10 / 0.64	-0.06 / 0.79	0.46 / 0.02	-0.46 / 0.02	-0.31 / 0.13	-0.30 / 0.15	-0.38 / 0.06
Blue-winged teal	0.18 / 0.40	0.34 / 0.11	0.15 / 0.48	-0.01 / 0.96	—	0.03 / 0.90	-0.00 / 0.99	-0.04 / 0.84	-0.23 / 0.29	-0.48 / 0.02	-0.02 / 0.94
Northern shoveler	0.23 / 0.27	0.05 / 0.83	-0.04 / 0.85	-0.10 / 0.65	0.08 / 0.69	—	0.29 / 0.16	0.38 / 0.06	0.43 / 0.03	0.03 / 0.87	0.24 / 0.24
Northern pintail	0.28 / 0.17	0.48 / 0.02	0.41 / 0.04	0.34 / 0.10	0.25 / 0.23	0.50 / 0.01	—	0.19 / 0.36	0.36 / 0.08	0.03 / 0.90	0.33 / 0.10
Redhead	0.33 / 0.11	0.33 / 0.11	-0.27 / 0.21	-0.21 / 0.33	0.07 / 0.74	0.41 / 0.05	0.04 / 0.86	—	0.21 / 0.33	0.17 / 0.43	0.55 / 0.01
Canvasback	0.02 / 0.93	-0.35 / 0.09	-0.37 / 0.08	-0.53 / 0.01	-0.13 / 0.53	0.37 / 0.08	-0.07 / 0.74	0.40 / 0.05	—	0.37 / 0.08	0.41 / 0.05
Lesser scaup	0.08 / 0.72	-0.22 / 0.29	-0.29 / 0.16	-0.30 / 0.16	-0.42 / 0.04	0.11 / 0.62	-0.14 / 0.53	0.35 / 0.10	0.56 / 0.01	—	0.27 / 0.21
Ruddy duck	0.03 / 0.87	0.37 / 0.07	-0.23 / 0.28	-0.05 / 0.79	0.38 / 0.06	0.36 / 0.07	0.27 / 0.18	0.46 / 0.02	0.29 / 0.16	0.07 / 0.74	—

species pairs were symmetric in their significance; for example, residuals from the model for mallards were positively related to gadwall numbers, and residuals from the gadwall model were positively related to mallard counts. Other such species pairs with symmetric positive correlations were gadwall–American wigeon, gadwall–northern pintail, American wigeon–northern pintail, northern shoveler–redhead, northern shoveler–canvasback, redhead–ruddy duck, and canvasback–lesser scaup.

The only species pair with symmetrical negative correlation coefficients involved the blue-winged teal and lesser scaup. Other negative correlations were scattered, except that residuals from the model for canvasback were negatively related to counts of three species: gadwall, American wigeon, and green-winged teal.

D. Species Relations to Explanatory Variables

The populations of these duck species have fluctuated at the continent-wide level during the past quarter-century (Figs. 4–14), presumably in response to a variety of factors that include varying conditions of the wetland habitat, changes in land use, predator populations, hunting pressure, and possibly others. Correlations between numbers and calendar year indicate more-or-less linear trends over the period 1965–1989 that are negative at both local (Woodworth) and continental levels for all seven species of dabbling ducks except green-winged teal at the continental scale. Of these trends, decreases were significant at the local scale for green-winged teal, American wigeon, northern pintail, and gadwall (Table II). For mallard, the trend was significant at only the continental scale, whereas northern pintail declined significantly at both scales (Table II).

No diving duck had a significant continental trend (although two species showed a tendency to decline), whereas at the local scale all three species showed significant increases over the quarter-century. The stiff-tail, ruddy duck, increased significantly during this period at both local and continental scales.

Thus the population trends of ducks in the Woodworth Study Area did not always track continental changes closely, either in direction or in level of significance. Local counts are compared to the continental indexes in Figs. 4–14. Although correlations between counts at Woodworth and continental indexes were positive for virtually all species, they were significant for only ruddy duck, northern pintail, and blue-winged teal (Table V, last column).

The number of ducks in the study area was positively correlated with May ponds for each of the 11 species, but significantly so for only blue-winged teal ($r = 0.71$) and northern shoveler ($r = 0.58$), with northern pintail and ruddy duck

TABLE V Correlations between Counts of Ducks and Proximate Habitat and Weather Variables, Previous Year's Count, and Continental Index to Population Size at Woodworth Study Area, North Dakota, 1965–1989

	May ponds	Conserved soil moisture	Mean temperature		Precipitation		Previous count	Continental index
			April	May	April	May		
Mallard	0.21	0.24	−0.22	−0.01	0.23	−0.11	−0.48**	0.10
Gadwall	0.26	0.16	−0.42**	−0.11	0.36*	−0.19	−0.12	0.03
American wigeon	0.04	−0.11	−0.22	−0.09	−0.09	0.24	0.33	0.30
Green-winged teal	0.01	0.05	−0.22	−0.16	−0.05	−0.09	0.26	0.11
Blue-winged teal	0.71***	0.55***	−0.39*	−0.36*	0.44**	−0.12	−0.00	0.40**
Northern shoveler	0.58***	0.35	−0.41**	−0.20	0.13	0.16	0.04	0.20
Northern pintail	0.37*	0.17	−0.42**	−0.17	0.14	0.17	0.04	0.55***
Redhead	0.26	0.06	−0.12	0.04	−0.13	0.20	0.43**	0.14
Canvasback	0.06	−0.12	0.09	0.14	−0.02	0.05	0.66***	−0.03
Lesser scaup	0.10	0.10	0.15	0.39*	−0.16	0.25	0.51**	−0.00
Ruddy duck	0.35*	0.04	−0.22	−0.08	−0.09	0.17	0.38*	0.63***

*$P < 0.10$.
**$P < 0.05$.
***$P < 0.01$.

correlations indicating $0.10 > P > 0.05$ (Table V). Correlations with the Conserved Soil Moisture index, which serves as a proxy for wetland conditions, were generally similar to, but weaker than, those with May ponds (Table V).

Among the seven dabbling duck species, correlations with proximate weather variables, namely, April and May mean temperatures and precipitation, showed a general pattern of negative correlations with temperature in both months and generally positive correlations with precipitation, especially in April (Table V). These results are not surprising, as cooler and wetter conditions favor the maintenance of water in wetland basins. Neither of these relationships was expressed for the three diving ducks or ruddy duck, however.

The possibility that counts in any one year show carryover effects from the previous year is tested next. For two dabbling-duck species (American wigeon and green-winged teal), counts in successive years were weakly positively correlated (Table V), but not significantly. The mallard counts, inexplicably, were negatively correlated with those of the previous year. In contrast, counts of each diving duck species and the ruddy duck were strongly and positively related to those in the previous year.

Regression models provided different results for the various species. No model yielded a significant multiple correlation coefficient for the mallard, and R^2 reached only 22.7% for the best model. This result was consistent with the general lack of significance among simple correlations with explanatory variables (Table V). Similarly, models for the American wigeon and green-winged teal failed to produce significant R^2 values; for the former species, the R^2 value for the best model was only 14.0%, and in the latter species no best model was defined. Like the mallard, numbers of these two species were not strongly correlated with individual explanatory variables; their relatively low densities at Woodworth may explain the lack of identified relationships.

Regression models typically explained about a third of the variance in numbers of the gadwall. The "best" model, explaining about a fourth of the variance, included April temperature and May precipitation, both with negative coefficients. Stronger relationships were apparent in the blue-winged teal, with population numbers and proximate habitat variables giving $R^2 = 57.3\%$; here the best model explained two-thirds of the variation in numbers and included effects of May ponds, continental population, and the previous year's count. This last variable was included in the model despite the fact that its simple correlation coefficient with the species' numbers was 0 (Table V). A similar dependence of numbers on proximate habitat variables was shown in northern shoveler, where the best model included only May ponds and accounted for a third of the variation. Counts of the northern pintail were related both to proximate habitat features and to the continental index. The best model, explaining nearly half of the variation in numbers, included only the continental index, however.

Of the diving ducks, significant R^2 values for the redhead resulted from models including proximate habitat features and the previous year's count. The best model, with $R^2 = 44.6\%$, included the previous year's count, May precipitation, and May ponds. Numbers of canvasbacks were not related to proximate habitat variables, but were closely (and only) related to numbers in the previous year, with the best model explaining 43.8% of the variation. For the lesser scaup, models produced significant R^2 values when the previous year's count was included; the best model, accounting for half of the variation, included the previous year's count, May precipitation, May temperature, and May ponds.

The ruddy duck yielded models with significant multiple correlation coefficients whenever the continental population index was included. For this species the best model, incorporating only that variable, accounted for nearly half of the overall variance in census numbers.

V. OVERVIEW: THE SPECIES CONTRASTED

Mallard.   This species is the most widespread of the ducks, with a distribution centered in the southern Prairie Provinces of Canada; the stratum containing the Woodworth Study Area is in the second highest quartile of average densities (Johnson and Grier, 1988). Continental populations rose during 1965–1970, after which they declined (Fig. 4); however, numbers at Woodworth varied moderately without any evident trend.

Mallards at Woodworth were not strongly correlated with May ponds; Trauger and Stoudt (1978) and Leitch and Kaminski (1985) also noted no relations between mallard numbers and pond counts. In contrast, Johnson and Grier (1988) found that numbers of mallards correlated closely with pond counts in the Dakotas. The low correlation of mallard numbers with ponds at Woodworth may reflect the ability of this species to use a wide variety of wetland classes, including some of the more permanent wetlands, at Woodworth (Stewart and Kantrud, 1973). Its cosmopolitan habitat use conforms with the observation that proximate weather features did not seem to influence mallard numbers on the Woodworth Study Area.

Gadwall.   The breeding distribution of the Gadwall is centered in the Dakotas and the southern Prairie Provinces, and the Prairie Pothole Region of North Dakota consists of strata in the highest quartile of average densities. Populations declined somewhat during 1965–1989 at the continental scale, and the decline at Woodworth was more pronounced (Fig. 5).

Numbers of gadwalls in the Woodworth Study Area did not vary closely in response to numbers of wetland sites, although Johnson and Grier (1988)

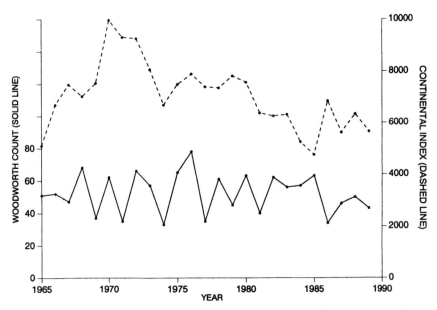

FIGURE 4. Counts of mallards at Woodworth Study Area and index to continental population, 1965–1989.

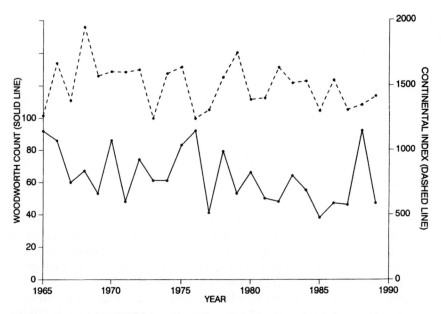

FIGURE 5. Counts of gadwalls at Woodworth Study Area and index to continental population, 1965–1989.

calculated r values of nearly 0.50 for that area in general. Gadwalls were more common in years with cool April temperatures; the negative coefficient of May precipitation in the best regression equation is difficult to explain.

American wigeon. The distribution of this species is centered farther north than most of the other dabbling ducks, with the area containing the Woodworth Study Area being in the second lowest quartile of densities. Continental populations declined after 1970, while populations at Woodworth fell erratically during the entire study period (Fig. 6).

There was no correlation between counts of American wigeon and wetland densities at Woodworth. This result was similar to that of Johnson and Grier (1988). The only weather variable that seemed to correspond to wigeon numbers was May precipitation, and only this factor was included in the best regression model.

Green-winged teal. This species is distributed much like the American wigeon, but with an even more northerly center. The stratum including the Woodworth Study Area is in the lowest quartile of densities identified by Johnson and Grier (1988). Although the continental population increased during 1965–1989, the Woodworth Study Area experienced a marked and significant decline (Fig. 7).

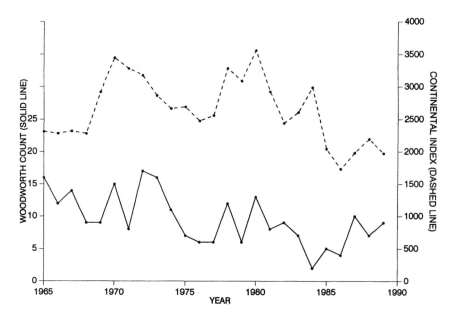

FIGURE 6.   Counts of American wigeon at Woodworth Study Area and index to continental population, 1965–1989.

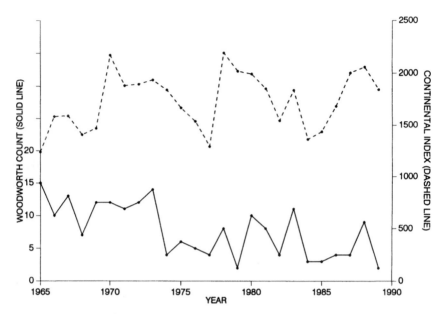

FIGURE 7.   Counts of green-winged teal at Woodworth Study Area and index to continental population, 1965–1989.

The lack of association between green-winged teal and wetland numbers in the Woodworth Study Area is consistent with the finding of Johnson and Grier (1988). No weather variable was significantly correlated with numbers of this species.

Blue-winged teal.   Blue-winged teal show a distribution similar to that of gadwall, with the Prairie Pothole Region of North Dakota containing densities in the highest quartile. Continental numbers of blue-winged teal declined from 1974 through 1989; while Woodworth populations generally declined over this period, there was much annual variation (Fig. 8).

The strongest association between duck numbers and wetland sites was recorded for the blue-winged teal, as befits a species with strong pioneering tendencies. The *r* value obtained for Woodworth was even higher than that found more generally by Johnson and Grier (1988). Numbers of this species also correlated negatively with temperatures in both April and May and positively with precipitation in April. The best regression model included none of these variables; however, it did include May ponds, continental population, and the previous year's count.

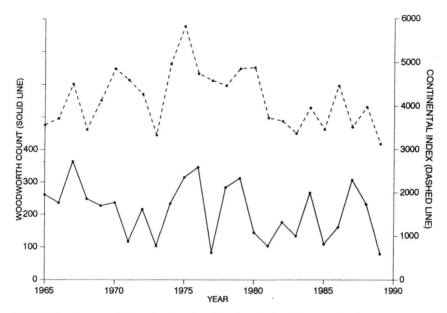

FIGURE 8. Counts of blue-winged teal at Woodworth Study Area and index to continental population, 1965–1989.

Northern shoveler. The shoveler is widely distributed, and the stratum containing the Woodworth Study Area is in the highest density quartile. Neither continental nor Woodworth populations experienced a decline during the study period (Fig. 9). As for the blue-winged teal, counts of the northern shoveler were strongly related to numbers of wetlands in the Woodworth Study Area. This finding was consistent with that of Johnson and Grier (1988). Numbers were also higher in years when April temperatures were low.

Northern pintail. Pintails are distributed somewhat more westerly than the other species, perhaps reflecting the fact that many of them winter in California. Highest density quartiles are in the southwestern portions of the Prairie Provinces and in Alaska. Pintails often are common in North Dakota, however, and the stratum containing the Woodworth Study Area is in the second highest quartile of densities. Continental populations of pintails fell dramatically between 1972 and 1989. Numbers at the Woodworth Study Area were correlated with continental numbers, but the declines were not in total synchrony (Fig. 10).

Although the correlation between numbers of pintails and wetlands in the

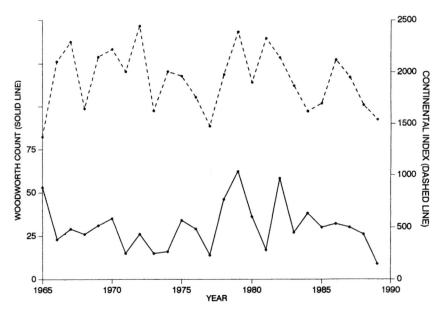

FIGURE 9. Counts of northern shovelers at Woodworth Study Area and index to continental population, 1965–1989.

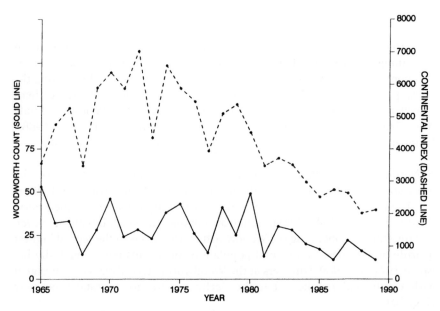

FIGURE 10. Counts of northern pintail at Woodworth Study Area and index to continental population, 1965–1989.

Woodworth Study Area was significant, its value ($r = 0.37$) was somewhat lower than those reported for strata in the Prairie Pothole Region in general. Pintail numbers in the Woodworth Study Area varied inversely with temperatures in April, but April temperature was not included in the best regression equation.

Redhead. The redhead achieves highest average densities in the parkland of the Prairie Provinces and in the Prairie Pothole Region of the Dakotas, and the stratum containing the Woodworth Study Area had average densities in the highest quartile. Continental populations of redheads showed no trend over the 25-year period, whereas those at Woodworth increased markedly after the mid-1970s (Fig. 11). The increased numbers of diving ducks and ruddy ducks in the Woodworth Study Area, in contrast to declines continentally, may reflect an increase in emergent vegetation in wetlands on the study area. These species most frequently nest over water in such vegetation, which increased after several years once grazing in the area had ceased. Dabbling ducks, on the other hand, ordinarily nest in upland vegetation and their numbers therefore would not be affected closely by the emergent vegetation.

Redhead numbers at Woodworth did not correspond very closely to those of wetlands and this was similar to what was reported by Johnson and Grier (1988).

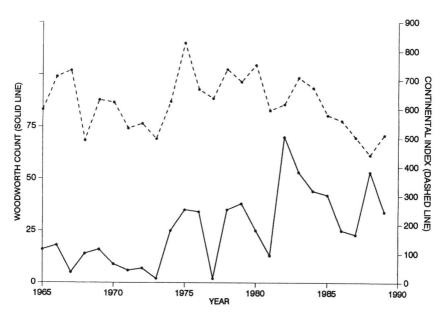

FIGURE 11.　Counts of redheads at Woodworth Study Area and index to continental population, 1965–1989.

Although simple correlation coefficients with weather variables were not significant, May precipitation entered the best regression equation.

Canvasback.   The highest density strata for canvasbacks are in the parkland region of Canada and in the far north; the Woodworth Study Area is included in the second highest quartile. Although continental numbers of canvasbacks tended to decline during the study period, the Woodworth population increased beginning in the late 1970s (Fig. 12).

Canvasback counts were not correlated to wetland counts in the Woodworth Study Area. Johnson and Grier (1988) reported *r* values in the range 0.20 to 0.40 for the Prairie Pothole Region of North Dakota. No weather variables were significantly related with counts of this species or were included in the best model.

Lesser scaup.   This species has a distribution centered farther north than any other considered in this chapter. Its highest densities are reported in Alaska and the Northwest Territories. The stratum containing the Woodworth Study Area is in the lowest quartile for average density. Continental popula-

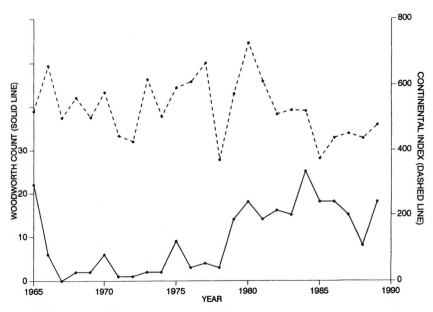

FIGURE 12.   Counts of canvasbacks at Woodworth Study Area and index to continental population, 1965–1989.

tions of scaup showed no linear trend; in contrast, the population in the Woodworth Study Area increased throughout the study period (Fig. 13).

At Woodworth, lesser scaup and wetland counts were not related. Johnson and Grier (1988) found moderately high correlation coefficients (0.20–0.40) for the general area. The best model for this species included both May weather variables, counts being highest in years with warm temperatures and much precipitation.

Ruddy duck. Ruddy ducks have a distribution concentrated in the southern Prairie Provinces and the Dakotas. Although the numbers of ruddy ducks continent-wide showed a significant upward trend during the study period, this result appears to be due to particularly high counts in a few years of the early 1980s (Fig. 14). Numbers at Woodworth also tended to increase and, in fact, mirrored the continental population ($r = 0.63$).

Ruddy ducks were fairly strongly related to wetland counts at Woodworth [$r = 0.35$; Johnson and Grier (1988) did not include this species in their study]. No weather variable appeared to influence numbers of ruddy duck on the study area.

FIGURE 13. Counts of lesser scaup at Woodworth Study Area and index to continental population, 1965–1989.

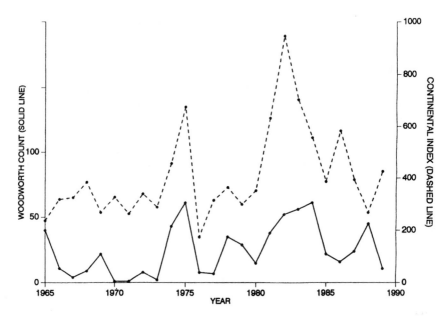

FIGURE 14.   Counts of ruddy ducks at Woodworth Study Area and index to continental population, 1965–1989.

VI. CONCLUSIONS

There is little published evidence that competition plays a strong role in
structuring waterfowl communities (Pöysä, 1983, 1984; Nudds, 1983), although interactions among species are difficult to detect since waterfowl populations are strongly affected by habitat quality and availability, and the species
exhibit generally similar responses to habitat changes (Pöysä, 1984). Waterfowl species do show differences in prey size preference (Nudds and Bowlby,
1984). Species do interact when feeding (Pöysä, 1986); their food habits differ
less in summer than in winter, probably because food resources are rarely
limiting during summer (DuBowy, 1988).

From the present analysis, the relative paucity of negative correlations
between residuals from the best-fitting models and counts of other species
does not support a strong role for interspecific competition among ducks at
Woodworth. The larger number of positive correlation coefficients and the
general symmetry between species pairs are consistent either with an active
association between those species or, more plausibly, with the hypothesis that
the species were responding similarly to changes in environmental conditions
that the measured variables did not adequately capture.

Wiens (1989b) observed that only rarely are community patterns the result of a single process, and this contention is consistent with results presented here. Ducks are widely recognized as responding immediately to dynamic wetland conditions by settling in areas where wetlands are favorable and deserting areas with dry basins, although philopatry also can influence where ducks settle (Johnson and Grier, 1988). The proximate habitat features that most influence the number of ducks using a particular breeding area are the number, type, and condition of wetlands in the area. This is especially true in an area such as Woodworth, where upland cover suitable for nesting dabbling ducks is abundant. The present analysis clearly shows opportunistic responses by waterfowl to wetland conditions, with such responses most pronounced in blue-winged teal, northern shoveler, northern pintail, and ruddy duck. The three dabbling ducks, but not the ruddy duck, make extensive use of seasonal wetlands (Stewart and Kantrud, 1973), i.e., those wetlands that are more variable from year to year than are semipermanent wetlands.

The lack of correlation between counts of a species in successive years is consistent with opportunistic settling in dynamically varying habitat, and it is prevalent among the dabbling ducks. In contrast, there are strong correlations in numbers between adjacent years in the diving ducks and ruddy duck, and these are species that make more use of semipermanent wetlands (Stewart and Kantrud, 1973) which are more persistent from one year to the next. In sum, the variability of a species' density clearly reflects the variability in its preferred habitat.

Philopatry also can contribute to a correlation between counts in successive years. Philopatry is an important characteristic of many species, particularly those that breed in stable habitats (e.g., Wiens, 1976). Among the waterfowl, diving ducks and possibly the ruddy duck tend to be most philopatric (Johnson and Grier, 1988). The present study found positive correlations between counts of all four of these species in successive years. Leitch and Kaminski (1985) reported similar findings for lesser scaup and canvasbacks but not for redheads and ruddy ducks. Although reproductively successful females of several dabbling-duck species will return to their previous breeding areas (Johnson and Grier, 1988; Lokemoen et al., 1990), the present study found few positive correlations between counts in successive years; these were for American wigeon and green-winged teal, and were not significant. Thus dabbling ducks seem to closely follow the habitat resources they select; there is little tracking inertia (sensu Hildén, 1964). Leitch and Kaminski (1985), in contrast, reported significant correlations between counts in successive years for five of seven dabbling ducks, including American wigeon but not green-winged teal.

Dabbling ducks as a group tended to respond to changes in habitat conditions in a similar fashion, but not identically. American wigeon and green-winged teal covaried most closely, as reflected by the correlation between numbers of the two species, their similar declines in the Woodworth Study Area, and changes in neither species relating to measured habitat conditions on the study area. The other five dabbling-duck species were positively correlated to one another and to wetland numbers, but differed in other ways. For example, the mallard and northern shoveler did not decline during the study period, as did the other species.

Numbers of the three diving ducks, and to a lesser degree the ruddy duck, were correlated with one another. All increased during the study period, regardless of continent-wide changes in populations. Increases at Woodworth were due possibly to increased emergent cover in wetlands, which produced more favorable conditions for nesting. Only the ruddy duck and possibly the redhead seemed to respond to local wetland conditions, as indexed by the May pond count.

The importance of the overall, continent-wide populations in influencing numbers at Woodworth varied by species. Counts in the Woodworth Study Area most closely tracked continental populations of the ruddy duck, northern pintail, and blue-winged teal. The ruddy duck and blue-winged teal do have somewhat more southerly breeding ranges, so that Woodworth is nearer the core of their ranges. For species that settle opportunistically, sites that the birds encounter early in their northward migration will, if habitat conditions are suitable, be occupied first. Accordingly, those sites should be occupied to their capacity and most likely will exhibit stronger relations between bird numbers and habitat conditions (Wiens, 1977).

VII. SUMMARY

Although waterfowl have been well studied generally, less work has been done on their community ecology. This paper addresses that topic using 25 years of counts of breeding ducks at a prairie-wetland site near Woodworth, North Dakota. Counts of 11 species at the site were related to the number of wetland basins containing water, an index of conserved soil moisture, temperature and precipitation measurements in April and May, an index of continental population sizes, and the counts at the census site during the previous year. All explanatory variables were related to the count of one or more waterfowl species, but the number of ponds and the continental population size were significant for more of the species than were the other variables.

Acknowledgments

Waterfowl were surveyed and ponds counted by leaders of the Woodworth Field Station, L. M. Kirsch (deceased), K. F. Higgins, and J. M. Callow, and numerous seasonal staff they supervised. I am grateful to all of them. B. R. Euliss prepared the graphics, and J. E. Austin, R. R. Cox, and R. R. Koford provided useful comments on earlier drafts of this chapter.

References

Bayha, K. D. (1964). History of the Woodworth study area with emphasis on land use. Unpublished report on file, Northern Prairie Science Center, Jamestown, ND.

Bellrose, F. C. (1980). "Ducks, Geese & Swans of North America," 3rd ed. Stackpole Books, Harrisburg, PA.

Boyd, H. (1981). "Prairie Dabbling Ducks, 1941–1990," Prog. Notes 119. Canadian Wildlife Service.

Cody, M. L. (1974). "Competition and the Structure of Bird Communities." Princeton Univ. Press, Princeton, NJ.

DuBowy, P. J. (1988). Waterfowl communities and seasonal environments: Temporal variability in interspecific competition. *Ecology* **69,** 1439–1453.

Hammond, M. C. (1969). Notes on conducting waterfowl breeding populations surveys in the north central states. *In* "Saskatoon Wetlands Seminar," *Rep. Ser.* 6, pp. 238–254. Canadian Wildlife Service.

Higgins, K. F., Kirsch, L. M., Klett, A. T., and Miller, H. W. (1992). Waterfowl production on the Woodworth Station in south-central North Dakota, 1965–1981. *U.S. Fish Wildl. Serv., Resour. Publ.* **180,** 1–79.

Hildén, O. (1964). Habitat selection in birds: A review. *Ann. Zool. Fenn.* **2,** 53–75.

Johnson, D. H., and Grier, J. W. (1988). Determinants of the breeding distributions of ducks. *Wildl. Monogr.* **100,** 1–37.

Leitch, W. G., and Kaminski, R. M. (1985). Long-term wetland-waterfowl trends in Saskatchewan grassland. *J. Wildl. Manage.* **49,** 212–222.

Lokemoen, J. T., Duebbert, H. F., and Sharp, D. E. (1990). Homing and reproductive habits of mallards, gadwalls, and blue-winged teal. *Wildl. Monogr.* **106,** 1–28.

Martin, F. W., Pospahala, R. S., and Nichols, J. D. (1979). Assessment and population management of North American migratory birds. *In* "Environmental Biomonitoring, Assessment, Prediction, and Management—Certain Case Studies and Related Quantitative Issues" (J. Cairns, Jr., G. P. Patil, and W. E. Waters, eds.), Stat. Ecol., Vol. 11, pp. 187–239. International Cooperative Publishing House, Fairland, MD.

Nudds, T. D. (1983). Niche dynamics and organization of waterfowl guilds in variable environments. *Ecology* **64,** 319–330.

Nudds, T. D. (1992). Patterns in breeding duck communities. *In* "Ecology and Management of Breeding Waterfowl" (B.D.J. Batt, A. D. Afton, M. G. Anderson, C. D. Ankney, D. H. Johnson, J. A. Kadlec, and G. L. Krapu, eds.), pp. 540–567. Univ. of Minnesota Press, Minneapolis.

Nudds, T. D., and Bowlby, J. N. (1984). Predator-prey relationships in North American dabbling ducks. *Can. J. Zool.* **62,** 2002–2008.

Pöysä, J. (1983). Resource utilization pattern and guild structure in a waterfowl community. *Oikos* **40,** 295–307.

Pöysä, H. (1984). Temporal and spatial dynamics of waterfowl populations in a wetland area—a community ecological approach. *Ornis Fenn.* **61,** 99–108.

Pöysä, H. (1986). Foraging niche shifts in multispecies dabbling duck (*Anas* spp.) feeding groups: Harmful and beneficial interactions between species. *Ornis Scand.* **17,** 333–346.

SAS Institute, Inc. (1989). "SAS/STAT User's Guide," Version 6, 4th ed., Vol. 2. SAS Institute, Inc., Cary, NC.

SAS Institute, Inc. (1990). "SAS Procedures Guide," Version 6, 3rd ed. SAS Institute, Inc., Cary, NC.

Stewart, R. E., and Kantrud, H. A. (1971). Classification of natural ponds and lakes in the glaciated prairie region. *U.S. Fish Wildl. Serv., Resour. Publ.* **92**, 1–57.

Stewart, R. E., and Kantrud, H. A. (1973). Ecological distribution of breeding waterfowl populations in North Dakota. *J. Wildl. Manage.* **37**, 39–50.

Strong, D. R., Jr., Simberloff, D., Abele, L. G., and Thistle, A. B., eds. (1984). "Ecological Communities: Conceptual Issues and the Evidence." Princeton Univ. Press, Princeton, NJ.

Trauger, D. L., and Stoudt, J. H. (1978). Trends in waterfowl populations and habitats on study areas in Canadian parklands. *Trans. North Am. Wildl. Nat. Resour. Conf.* **43**, 187–205.

Wiens, J. A. (1976). Population responses to patchy environments. *Annu. Rev. Ecol. Syst.* **7**, 81–120.

Wiens, J. A. (1977). On competition and variable environments. *Am. Sci.* **65**, 590–597.

Wiens, J. A. (1985). Habitat selection in variable environments: Shrub-steppe birds. *In* "Habitat Selection in Birds" (M. L. Cody, ed.), pp. 227–251. Academic Press, Orlando, FL.

Wiens, J. A. (1989a). "The Ecology of Bird Communities," Vol. 1. Cambridge Univ. Press, Cambridge, UK.

Wiens, J. A. (1989b). "The Ecology of Bird Communities," Vol. 2. Cambridge Univ. Press, Cambridge, UK.

Williams, G.D.V., and Robertson, G. W. (1965). Estimating most probable prairie wheat production from precipitation data. *Can. J. Plant Sci.* **45**, 34–47.

SECTION IV

Mammal Communities

CHAPTER 14

Small-Mammal Community Patterns in Old Fields

Distinguishing Site-Specific from Regional Processes

JAMES E. DIFFENDORFER[1], R. D. HOLT, N. A. SLADE, AND M. S. GAINES[1]

Natural History Museum and Department of Systematics and Ecology, University of Kansas, Lawrence, Kansas 66045

I. INTRODUCTION

Community ecologists increasingly recognize that interpretations of local community structure must include processes at local, regional, and biogeographical scales (Wiens *et al.*, 1986; Brown, 1987; Ricklefs, 1987; Roughgarden, 1989). Local processes such as predation and competition often play a strong role in shaping communities over short time scales (Paine, 1966; Connell, 1983; Schoener, 1983; Holt and Lawton, 1994). Yet, local communities typically are embedded in larger landscapes; movements of individuals across space can change the influence of local processes on population dynamics (Huffaker,

---

[1] Present address: Department of Biology, University of Miami, Coral Cables, Florida 33133.

Long-Term Studies of Vertebrate Communities

1958; Fahrig and Paloheimo, 1988; Pulliam and Danielson, 1992), and by a variety of mechanisms influence the species richness of local assemblages (Roff, 1974; Vance, 1980; Holt, 1993; McLaughlin and Roughgarden, 1993; Valone and Brown, 1995). Finally, long-term, large-scale processes such as geographical range changes and speciation ultimately determine the pool of possible community members at a given site (Brooks and McLennan, 1993; Ricklefs and Schluter, 1993).

Understanding the interaction of processes at different spatial scales is necessary to understand the effects of habitat fragmentation (Wiens, 1976; Levin, 1992). Fragmentation creates arrays of patches varying in size and distance from one another, arrayed in a matrix of qualitatively different habitats (Wilcox, 1980). The interspersion of different habitats results in spatial heterogeneity, which can alter ecological processes (Saunders *et al.*, 1991). Because species interact with the environment at different spatial scales, species will most likely have different responses to fragmentation (Kareiva, 1986; Gaines *et al.*, 1992a, 1994; Robinson *et al.*, 1992; Diffendorfer *et al.*, 1995a; Margules *et al.*, 1994). For large-bodied species, some patches may be too small for population persistence and such species' disappearances could have indirect effects on community structure, for example, by relaxing competition on smaller bodied species. Because patches become isolated following fragmentation, animal movement patterns will often change, leading to shifts in local population dynamics and in the strength and outcome of interspecific interactions (Holt, 1993). For instance, theoretical and experimental studies have highlighted how dispersal in spatially explicit environments can permit species coexistence that is impossible in closed communities (e.g., of a predator and its prey: Huffaker, 1958; Hassell *et al.*, 1992; or of competitors: McLaughlin and Roughgarden, 1993; Tilman, 1994). Studies of fragmented systems can provide insights into factors influencing community structure in unfragmented habitats and highlight the influence of space on demographic processes (Holt, 1993).

Various investigators at the University of Kansas have been engaged in long-term studies of small mammal populations in the prairie–forest ecotone continually from 1973 to 1996 (with additional though sporadic available data from the 1940s to 1970). These studies have been undertaken mainly by two faculty members—M. Gaines and N. Slade—and their students and associates. Each of the two groups has had a different focus. Gaines and associates initially concentrated on genetics and the biology of dispersal (Gaines and McClenaghan, 1980; Johnson and Gaines, 1987). Since 1984, this group has worked with R. Holt to examine dispersal and population dynamics in the context of an experimental study of habitat fragmentation (Robinson *et al.*, 1992). Slade and associates have worked on a variety of problems with small-mammal ecology, including mass-based demography (Sauer and Slade, 1985, 1986), interference competition (Glass and Slade, 1980), and community structure (Swihart and Slade, 1990).

This chapter and a second paper (Diffendorfer *et al.*, 1995b) represent the first attempts to combine data from the two long-term studies. Here we compare temporal and spatial patterns in old-field small-mammal communities for the period 1984–1992 at study sites separated by just 0.5 km (Fig. 1). One site is a

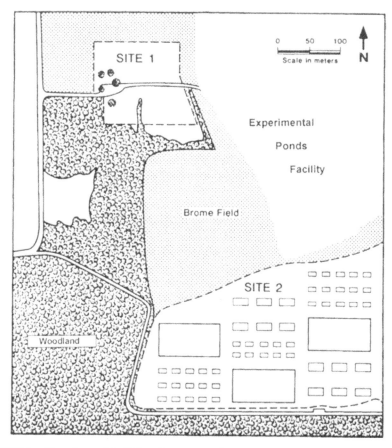

FIGURE 1.   Map of part of the Nelson Environmental Study Area (12 km northeast of Lawrence, Kansas) showing the arrangement of the two study sites. Site 1 is a continuous area of approximately 2.25 ha with 1.90 ha of old-field habitat. Site 2, the experimentally fragmented system, has approximately 1.87 ha of successional old-field habitat contained within 6.9 ha of total area. The 40 small patches in Site 2 are each 4 × 8 m, the 12 medium patches are 12 × 24 m, and the 3 large patches are 50 × 100 m. Note that a single large patch is a large block, a cluster of 6 medium patches is a medium block and a cluster of 10 or 15 small patches is a small block. Blocks are separated by 16–20 m and are numbered from west to east, within a block size (e.g., Large 1 is the western most large block, Large 2 the central large block, and Large 3 the eastern most large block).

tract of continuous old-field habitat; the other an experimentally fragmented old-field. Our intent is not to provide rigorous tests of specific *a priori* theoretical predictions, but rather to compare the continuous and fragmented sites in order to search for potential systematic effects of habitat fragmentation on the long-term spatial and temporal dynamics of a community, and to assess the spatial scale relevant to analyses of community organization.

Because these two sites are close in space, they clearly draw species from the same regional species pool and, moreover, are likely to experience parallel fluctuations in climate. Table I lists the small mammal species in our region and the small mammals that compose the communities found on both sites. Some species in the regional fauna might in principle occur on our sites but never have (e.g., plains harvest mice, pine voles, and meadow jumping mice). Some species are likely absent because local habitat in our sites is inappropriate (e.g., eastern chipmunks, southern flying squirrels, and 13-lined ground squirrels). For other species, there may be local or regional barriers to dispersal. At least one species present, the least shrew, is not accurately censused by our standard trapping techniques. The results reported here consider only comparative patterns for the

TABLE I Small Mammal Species Found in Douglas County, Northeastern Kansas, and Captured More Than 20 Times in the Continuous Site and in the Fragmented Site

Species	Location		
	Northeastern Kansas	Continuous	Fragmented
Least shrew (*Cryptotis parva*)	X	X	X
Prairie vole (*Microtus ochrogaster*)	X	X	X
Deer mouse (*Peromyscus maniculatus*)	X	X	X
Western harvest mouse (*Reithrodontomys megalotis*)	X	X	X
Cotton rat (*Sigmodon hispidus*)	X	X	X
Short-tailed shrew (*Blarina brevicauda*)	X	X	
House mouse (*Mus musculus*)	X	X	
Wood rat (*Neotoma floridana*)	X	X	
White-footed mouse (*Peromyscus leucopus*)	X	X	
Bog lemming (*Synaptomys cooperi*)	X	X	
Meadow jumping mouse (*Zapus hudsonicus*)	X		
Southern flying squirrel (*Glaucomys volans*)	X		
Pine vole (*Microtus pinetorum*)	X		
Plains harvest mouse (*Reithrodontomys montanus*)	X		
13-Lined ground squirrel (*Spermophilis tridecemlineatus*)	X		
Eastern chipmunk (*Tamias striatus*)	X		

species known to be actually present at our sites. Given this focus, we cannot address the question of why some species are missing (though present in the regional pool), even though this is an important dimension of community ecology (Diamond and Case, 1986).

The comparative analyses reported here bear on three distinct issues: (1) the influence of habitat fragmentation on spatial patterns in abundance, (2) the influence of habitat fragmentation on temporal dynamics, and (3) the spatial scale of community processes.

A. The Influence of Habitat Fragmentation on Abundances

We have previously studied the influence of habitat fragmentation on individual movements (Diffendorfer *et al.*, 1995a), population demography (Gaines *et al.*, 1992a,b, 1994; Diffendorfer *et al.*, 1995b), and abundances on different sized patches within the fragmented site (Foster and Gaines, 1991; Robinson *et al.*, 1992). The continuous site studied by N. Slade is in effect a single, very large patch of old-field habitat, four times larger than the largest patch on the fragmented site. This permits us to ask if trends seen in the fragmented site along a gradient of patch size (small to large) extrapolate to a larger patch size.

B. The Influence of Habitat Fragmentation on Temporal Dynamics

Demographic studies of old-field species often show both seasonal and multiannual fluctuations in population size (Gaines and Rose, 1976; Johnson and Gaines, 1988; Swihart and Slade, 1990). Variation in numbers can range over 2–2.5 orders of magnitude and include local extinctions (Swihart and Slade, 1990). Although small-mammal ecologists have traditionally emphasized temporal variability in abundance, few studies have rigorously examined long-term temporal changes in the rank-order of abundances among species, or species turnover (viz., colonization/extinction dynamics) in old-field small-mammal communities. Here we examine the influence of fragmentation on these aspects of temporal variability in communities.

C. The Spatial Scale of Community Processes

Spatial coupling in population dynamics requires individuals to move between different populations. Because the two studies combined in this paper use mark–recapture data, we can directly assess the degree of interchange by individuals between the two sites and the potential for spatial coupling driving local population dynamics over a spatial scale of 0.5 km. Further, because our data come from grids of permanent trap stations, we can examine spatial structuring in the community on a smaller scale within each of our sites. The differences we observe between and within the two long-term studies clarify which aspects of community organization in old-field small-mammal communities might be spatially mediated—effects that would be difficult to gauge otherwise. Finally, the old-field fragments are separated by a road and at least 14 m of mowed turf from nearby woodland. As evident in Fig. 1, the continuous site directly abuts patches of woodland. This difference in surrounding habitat may influence the frequency with which potential community members arrive and use a local site. The results presented in this chapter sharpen our appreciation of the impact of landscape context on local communities, a consideration frequently ignored in comparative community analyses (viz. Holt, 1993).

II. MATERIALS AND METHODS

A. Study Sites

The two long-term study sites each encompass comparable areas of old-field habitat. The continuous area (site 1) is 2.25 ha; it contains 1.90 ha of old-field habitat. The fragmented site (site 2) contains 1.87 ha of successional old-field habitat restricted to patches arrayed within 7 ha of mowed grass (Fig. 1). The two study sites are located at the Nelson Environmental Study Area, 14 km northeast of Lawrence, Kansas (Fig. 1), and are approximately 500 m apart. More detailed descriptions of the sites follow.

Slade and associates have trapped monthly at the continuous site since 1973. This site has a narrow dirt road transecting it east–west and a fence row with small trees running north–south in the southern half. A trailer and storage shed are located near the road on the western half of the site. These landscape details seem relevant in permitting some mammal species to occupy the site (see following discussion). The site is maintained as an old field by sporadic mowing to control the invasion of woody plants. The continuous site was mowed to a height of 30–45 cm (invading saplings and shrubs were cut, but large trees along fencerows and sheds were not) in July 1983, and the northern half of the area was plowed in May 1984.

Gaines and his students trapped small mammals on the fragmented site from 1984 to 1992 (this long-term study is currently being continued by W. Schweiger, R. Pierotti, and others). For comparisons, we analyze data from both sites during this time period. The fragmented site was created in August of 1984 by disking an agricultural field and allowing secondary succession to proceed within rectangular habitat patches. Regular mowing between the habitat patches is used to create these patches of successional habitat, which are organized into clusters of "blocks" (Fig. 1). The interstitial area is closely mowed turf, providing little cover and a substantially distinct plant community from the old field on the patches (Holt et al., 1995). "Large blocks" are single, 50 × 100-m patches; "medium blocks" consist of six 12 × 24-m patches; and "small blocks" consist of either fifteen or ten 4 × 8-m patches. Most blocks span the same amount of total area (5000 m²), but at differing degrees of fragmentation.

Succession was rapid in the first 3 years of the study on the fragmented site (1984–1987) and following the mowing and plowing on the continuous site in 1983 and 1984, as longer-lived perennials replaced annuals. From 1984 to 1987, both sites had few invading woody species and were dominated by grasses and herbaceous plants. The plant communities reached a quasiequilibrium from 1987 to 1992, associated with dominance by perennial herbs (e.g., *Solidago*). Thus, during the time periods for our comparisons, the vegetation on the two study sites was similar. Within the fragmented site, patch size seems to have had little influence on aggregate measures of plant community structure (Foster and Gaines, 1991) or the rate of succession (Holt et al., 1995), although some plant species did show patch-size effects (Robinson et al., 1992). We therefore believe it is appropriate to compare patterns at our two sites. Since 1992, woody plants have begun to invade the fragmented site, again changing the plant community structure. Our current study suggests this is leading to changes in the small-mammal community and the effect is more pronounced as woody plant species invasion proceeds apace. These recent patterns will be addressed in future papers.

Trapping for the two studies has followed typical small-mammal mark–recapture protocols. However, for historical reasons (slightly different techniques were employed by different research teams), the timing and length of the trapping period differed between studies. In the continuous site, 196 traps at 98 trap stations, spaced 15 m apart, were set for three mornings and two afternoons every month. In the fragmented system, 287 traps at 267 stations were set for two mornings and the intervening afternoon twice a month (for more details on trapping protocol and patch layout, see Swihart and Slade, 1990, and Foster and Gaines, 1991). In both studies, when animals were captured their location was noted, and they were individually marked, checked for mass and reproductive condition, and released.

B. Analyses

The analyses reported here consist of three components. First, we analyze how habitat fragmentation modifies the spatial patterning of abundances. Next, we examine temporal stability in the small-mammal communities and compare the degree of variability observed in our studies to studies in other biomes. An issue of particular concern is how fragmentation influences the magnitude and patterning of community variation. Finally, we consider the impact of large- and small-scale movements on local abundances. Combined with our between-site comparison, this permits us to attempt to gauge the spatial scale relevant to community dynamics in our particular system.

1. The Influence of Fragmentation on Abundances

a. Average densities by site and by degree of fragmentation. We calculated monthly minimum number known alive (MNKA) estimates using Fortran programs kindly provided by Dr. Charles Krebs, and corrected for area to achieve density estimates of MNKA per hectare. To correct for the differences in trapping protocols noted previously, we excluded every other week from data collected on the fragmented site and the third day of trapping from data on the continuous site.

The four most common species on the two sites were cotton rats (*Sigmodon hispidus*), prairie voles (*Microtus ochrogaster*), deer mice (*Peromyscus maniculatus*), and harvest mice (*Reithrodontomys megalotis*). These four species all occurred in sufficient numbers to allow statistical analysis. We previously reported abundance and density data for cotton rats, prairie voles, and deer mice (Gaines *et al.*, 1992a,b, 1994; Diffendorfer *et al.*, 1995b), but not for harvest mice. Because harvest mice were not individually marked on the continuous site until April 1989, we estimated MNKA from the number of captures for the time periods prior to 1989. In order to obtain a regression equation relating captures to MNKA for harvest mice, we regressed MNKA against the number of captures for capture periods since 1989 to the present (this equation was MNKA = 1.43 + 0.713(CAPTURES); $R^2 = 0.845$, $t = 15.81$, $df = 46$).

To test for site differences in harvest mouse densities we used general linear models in which site and year-season were entered as independent variables. Treating each 3-month period as a time unit reduced autocorrelation in the data (e.g., the fall year-season was September through November, winter was December through February, spring was March through May, and summer was June through August).

b. Community composition by site and by degree of fragmentation. We analyzed community composition in three ways. First we studied changes in community composition as a function of varying degrees of fragmentation. We calculated monthly community percentage composition estimates by dividing the MNKA for each species by the total MNKA for all species combined. Because this analysis focused on species' relative abundances, we did not correct the raw data for trapping protocol differences between sites (as discussed previously). We used MANOVAs (Manly, 1986) to assess changes in community structure by degree of fragmentation and by year-season. The first MANOVA tested for main effects of habitat type (the entire continuous site vs the entire fragmented site) or year-season on community composition. The second MANOVA tested for main effects of block size within the fragmented site (large vs medium vs small blocks) or year-season on community composition. Because proportions calculated from the MNKA estimates sum to 1, the relative abundance estimates for each species are not independent. We therefore transformed the data by dividing a particular species' proportion by the proportion of the rarest species—in our case, harvest mice (when percentage harvest mice = 0, we added 0.002)—and taking the natural log of this ratio. This transformation provides linearly independent measures (for justification of the technique, see Aebischer *et al.,* 1993).

Second, we compared species richness between the two sites. We investigated the effects of patchiness on community structure by comparing species richness on real medium and small blocks to the same measures calculated on "simulated" small and medium blocks, using data drawn in a structured manner from the large blocks. The simulated blocks were created by sorting raw data by trap locations on large blocks to correspond to the spatial arrangement of traps on medium or small blocks. This procedure permits us to compare a small or medium patch with a similarly spatial-structured sample from within a large block. We calculated the species richness for each trapping period for simulated and real blocks. If a species had a positive MNKA at a sample period, it was included in the richness count, even if it was missing in the actual sample. In this analysis, we included the rarer species, wood rats (*Neotoma floridana*), white-footed mice (*Peromyscus leucopus*), and house mice (*Mus musculus*), and compared average species richness on simulated and real blocks with *t* tests.

Third, we compared Shannon–Weiner diversity estimates (*H*) between the two study areas with *t* tests using only MNKA for the four most common species (cotton rats, prairie voles, deer mice, and harvest mice).

2. The Influence of Fragmentation on Temporal Dynamics

Following Rahel (1990), we analyzed temporal trends in community structure at three levels. First, we examined variation in the relative abundances of species over time. Second, at a higher level of organization, we tested for concordance through time of species' ranked abundances. Third, we assessed variability in community membership by examining patterns in local extinctions.

a. Temporal changes in abundances. We calculated the percentage composition for each species in the community as described previously (Sect. II.B.1.b.), and tested for changes in percentage composition over time using a

MANOVA blocked by unique year-seasons. We also compared coefficients of variation in the average Shannon–Weiner diversity index (H) between the continuous and fragmented study areas. In this case, H was calculated for each trapping period, and the time series of resulting indexes used to obtain the CV. Finally, to compare temporal variation to other communities, following Hanski (1990), we determined the average Spearman's rank correlation among all possible pairwise combinations of the four dominant species in each community and the average of the standard deviation in the log MNKA of each species. We then compared those values to the 14 communities presented in Hanski (1990).

b. Concordance in species ranks over time. We calculated Kendall's W (Rahel, 1990) for both sites and the various degrees of fragmentation as a means of assessing concordance in species ranks over time. Kendall's W assumes as a null hypothesis that ranks among members (in this case species) are randomly assigned. Thus, a significant W indicates consistent patterns in the ranks of community members.

c. Extinctions as a function of site and degree of fragmentation. We looked at trapping periods with MNKAs of zero to determine the influence of habitat patchiness on local extinction. What counts as an "extinction" is, of course, a matter of scale. One problem in some island biogeographic analyses is that they do not correct for sample size (area) when comparing islands differing in size (Holt, 1992). Here we assume the frequency of trapping periods with zero MNKA in a given spatial unit provides a reasonable index of local extinction. MNKA is calculated by keeping track of the times of first and last capture for a marked individual, and then counting that individual as present at all intervening time periods (regardless of whether or not the animal is actually captured). Thus, when MNKA is zero, no marked animals were captured both before and after (or during) the trapping period, making it a more conservative estimate of extinction than just raw captures. We compared real medium and small blocks to medium and small blocks simulated from each of the three large blocks. We created three separate simulated sets of blocks from each of the three large blocks. These data files were then analyzed for MNKA estimates as explained previously. We compared the average number of absences that occurred in the 188 trapping periods between simulated and real blocks using t tests (each real block was replicated at least two times; each simulated block was replicated three times).

Finally, we used trapping periods with zero MNKA as an index of local extinctions over the scale of entire study areas. We compared the overall proportion of 188 trapping periods with zero MNKA between the continuous and fragmented sites for all species using χ^2 tests.

3. The Spatial Scale of Community Processes

Our study also permits us to examine the potential influence of movement on local demography at various spatial scales. By combining our datasets, we have data on movements at three spatial scales: larger scale movements between the two study sites, smaller scale movements between blocks in the fragmented area, and fine scale space-use patterns within each of the sites.

a. Large-scale movements. We examined known immigrants that moved onto the two study sites to assess the potential for spatial coupling over scales of approximately 500 m. All species captured on the fragmented site were given a unique ear tag with an "s" on the back. On the continuous site, only cotton rats were given unique tags with no letters on the back. Smaller species on the continuous site were toe-clipped. Thus, we monitored the movements of cotton rats between the two sites. Unfortunately, toe clips were not checked on the fragmented site, so movements of prairie voles and deer mice from the continuous to the fragmented site were not recorded. However, movements of these smaller species from the fragmented site to the continuous site were determined.

b. Influence of movements on local MNKA. The patchy nature of the fragmented site provides a unique opportunity to characterize how movements influence local demography at smaller scales. To determine the magnitude of immigration and emigration, we calculated the proportion of the MNKA for a given trapping period consisting of marked animals either moving to or leaving a given block relative to local abundances. Thus, for a given block during a given trapping period, we determined the proportion of the MNKA accounted for by animals that entered the block from another block (immigration), or animals that left the block and were captured some place else (emigration). In the case of immigration, animals were continually counted on a given block as long as they remained there after immigrating. We analyzed the data in two ways. First, we compared the proportion of MNKA made up of either immigrants or emigrants across block sizes using all trapping periods, including those when no movement occurred. This method gives an overall picture of the importance of movement to abundance on a block. We also screened the data using only those trapping periods in which movement occurred; then we compared among blocks. We compared these percentages using ANOVAs blocked by year-seasons.

c. The influence of distance on local movements. In addition to determining the overall proportion of the species' abundance on a block composed of immigrants or emigrants (dispersers), we studied the influence of distance on the proportion of immigrants to a block. We calculated the proportion of animals on a block that arrived there from another block. Since blocks are not equidistant to each other on the site, we were able to use t tests to compare the proportion of animals on a block that came from nearby blocks versus blocks farther away. For a given combination of blocks, we calculated the proportion of individuals on the target block that came from the other blocks. For large blocks, there are four possible combinations of moves (Large 1 to Large 2, Large 2 to Large 1, Large 3 to Large 2, and Large 2 to Large 3; See Fig. 1); these are of similar, but shorter, distances than are two other possible moves (Large 1 to Large 3 and vice versa). We analyzed only movements to and from the same size blocks and combined movements between the large blocks with movements between the small blocks. Since very few animals persisted from one season to the next (Gaines *et al.*, 1994), we calculated these proportions for each season and compared the data using t tests.

   d. Within-site spatial heterogeneity in abundances.   Because our grids involve permanent trap stations, the history of captures at particular stations can provide an assessment of within-grid patterns of habitat use at the spatial scale of our studies. Much of this variation is likely to reflect individual habitat selection and permits us to gauge the scale at which small mammals respond to structural features of vegetation. We mapped space use for each species on each site for the entire time period and for periods of low and high densities. We created maps by calculating the percentage of total captures that occurred at a trap location over the given time period, and using this quantity as the radius of a circle centered at the x–y coordinates of the trap location. Visual inspection of patterns in these maps quickly reveals spatial patterns in the data, patterns often not evident when the data are examined in other ways.

III. RESULTS

A. The Influence of Fragmentation on Abundances

1. Average Density by Sites and by Degree of Fragmentation

Summary results for cotton rats, prairie voles, and deer mice have been presented elsewhere (Gaines *et al.*, 1992a,b, 1994; Diffendorfer *et al.*, 1995b; see Fig. 2 for a synopsis of these results). In summary, cotton rats (the largest species) achieve highest densities on the large blocks, prairie voles (the medi-

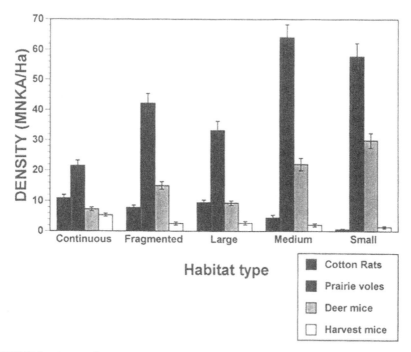

FIGURE 2. Average density (Minimum Number Known to be Alive/ha) with standard error bars by site and by block size for cotton rats, prairie voles, deer mice, and harvest mice. Data come from 7.7 years (August 1984–May 1992) of mark–recapture data.

um-size species) on the medium blocks, and deer mice (the smallest species) on the small blocks. These trends within the fragmented site extrapolate similarly to the continuous site; cotton rats have higher densities on the continuous site, and prairie voles and deer mice have lower densities there.

Harvest mice have higher average densities on the continuous site ($F = 42.21$, $df = 1, 124$, $P < 0.001$, Fig. 2). On average, harvest mice density was higher on large blocks than on small blocks, but not medium blocks ($F = 3.96$, $df = 2, 195$, $P = 0.021$, Fig. 2), consistent with their greater abundance on the continuous site.

2. Community Composition by Site and by Degree of Fragmentation

The continuous site had a number of unique small-mammal species that were captured in low numbers throughout the study. These species were extremely rare or completely absent on the fragmented site. They are house mice

(*Mus musculus*), bog lemmings (*Synaptomys cooperi*), wood rats (*Neotoma floridana*), and meadow jumping mice (*Zapus hudsonicus*). Furthermore, the average H was higher for the continuous site (0.478 ± 0.015) than for the fragmented site (0.369 ± 0.014, $t = 5.380$, $df = 60$, $P < 0.0001$).

Based on the MANOVAs, the average relative frequencies of cotton rats, prairie voles, deer mice, and harvest mice differed by site and by degree of fragmentation (continuous vs fragmented: Wilk's $\lambda = 0.1620$; $df = 3, 122$; $P < 0.001$; between blocks: Wilk's $\lambda = 0.2155$; $df = 6, 386$; $P < 0.001$; Fig. 3). On average, prairie voles were always the most prevalent species at both sites. However, the community in the continuous habitat had relatively fewer prairie voles, and more cotton rats, than did the community in the fragmented site. With increasing fragmentation, the proportion of deer mice in the community increased, while the proportion of harvest mice decreased. Compared to community structure in blocks with greater degrees of fragmentation, overall community structure in the large blocks was thus most similar to that in the continuous area.

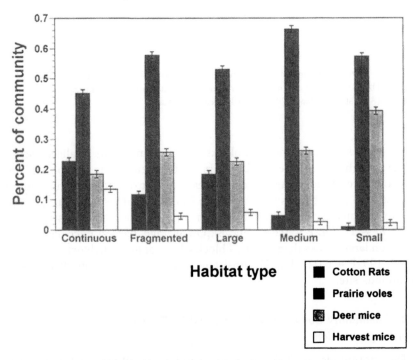

FIGURE 3.   Average proportion of the MNKA with standard error bars for cotton rats, prairie voles, deer mice, and harvest mice by site and by block size. Data come from 7.7 years (August 1984–May 1992) of mark–recapture data.

Patchiness significantly influenced species richness; simulated medium and small blocks (i.e., medium and small blocks simulated from a continuous 5000-m² area) had a mean species richness higher than that of real medium and small blocks (mean ± SE simulated medium = 2.60 ± 0.039; real medium = 2.30 ± 0.038, t = 5.47, df = 371, P < 0.0001; simulated small = 2.64 ± 0.043; real small = 2.20 ± 0.034, t = 8.05, df = 354, P < 0.001). The degree of trap spacing did not influence species richness because there were no differences when comparing simulated medium to simulated small blocks. However, for real blocks, larger blocks (less fragmented) had a greater average species richness (comparing real medium vs real small, t = 1.88, df = 354, P = 0.061).

B. The Influence of Fragmentation on Temporal Dynamics

1. Temporal Changes in Abundances

Species abundances varied through time and in different patterns on each site (Fig. 4). Prairie voles on the fragmented site had a tremendous range in abundance, from nearly zero to ~300 per trapping period. However, on the continuous site, prairie vole populations had a lower range, reaching a maximum abundance of about 140 animals. Cotton rats achieved higher numbers in the continuous area; by contrast, deer mice reached highest abundances in the fragmented area. Local extinctions of harvest mice from the fragmented area after 1989 (Gaines et al., 1992a) are expressed in the high frequency of counts in the zero category in Fig. 4.

In all species, temporal variations in abundances were positively correlated across sites (cotton rats: r = 0.59, P < 0.001; prairie voles: r = 0.72, P = 0.001; deer mice: r = 0.36, P = 0.024; harvest mice: r = 0.79, P < 0.001; in all cases n = 31). In general, a given species was common (or rare) simultaneously at both sites. Cotton rat and prairie vole abundances were positively correlated between species within each site (continuous: r = 0.44, n = 31, P = 0.006; fragmented: r = 0.48, n = 31, P = 0.003). On the continuous site, but not on the fragmented site, prairie voles were negatively correlated with deer mice (r = −0.40, n = 31, P = 0.013) and with harvest mice (r = −0.32, n = 31, P = 0.041), and cotton rats were weakly negatively correlated with deer mice (r = −0.27, n = 31, P = 0.074). No other correlations were significant between species pairs on either site. Pairwise trajectories of abundances through time (Fig. 5) revealed substantial differences between sites in the relationship of species to one another.

Following Hanski (1990), plots of the average correlation coefficient across

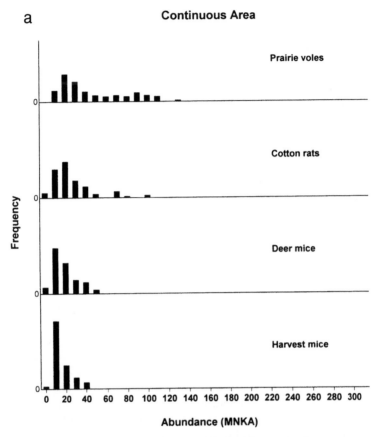

FIGURE 4. Frequency distributions of 92 monthly MNKA estimates for cotton rats, prairie voles, deer mice, and harvest mice on the (a) continuous and (b) fragmented sites.

all pairs of species versus the average standard deviation in MNKA across all species indicates the small-mammal assemblages on each of our sites have both a greater average standard deviation and a lower average correlation than 14 other communities. (Fig. 6).

The relative frequencies of cotton rats, prairie voles, deer mice, and harvest mice changed through time, such that the year-season variable in both MAN-OVAs was significant (continuous vs fragmented: Wilk's $\lambda = 0.0077$; $df = 90$, 366; $P < 0.001$; between blocks with the fragmented site: Wilk's $\lambda = 0.0426$; $df = 90, 578$; $P < 0.001$; Figs. 7a and 7b). Despite similar trends between sites in the relative abundances of both prairie voles and deer mice, the overall temporal changes in community structure were in different patterns across both sites,

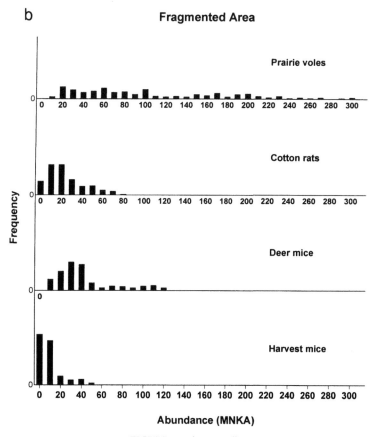

FIGURE 4. (*continued*)

and across block sizes on the fragmented system, leading to significant interactions in the MANOVAs between year-season and either site (Wilk's $\lambda = 0.0953$; $df = 90, 366$; $P < 0.001$) or block size (Wilk's $\lambda = 0.2126$; $df = 180, 580$; $P < 0.001$). Major shifts in community structure occurred multiannually, mainly caused by crashes in the prairie vole population. Cotton rats tended to become more prevalent in the fall, but declined every winter. Harvest mice nearly disappeared from the fragmented site after the fall of 1989, yet remained present on the continuous site. In general, the fragmented site had a greater magnitude of fluctuations in relative abundances than did the continuous site. Coefficients of variation in H were also higher in the fragmented site (21.22) than in the continuous site (16.90).

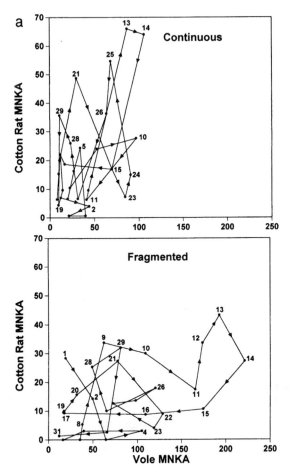

FIGURE 5. Phase plane diagrams of average MNKA for 31 year-seasons for all possible pairwise combinations of the three most common species in the continuous and fragmented communities: (a) prairie voles vs cotton rats, (b) cotton rats vs deer mice, and (c) prairie voles vs deer mice.

2. Concordance in Species Ranks over Time

All five Kendall's W values, for both the continuous and fragmented sites, as well as among the large, medium, and small blocks within the fragmented site, were highly significant, indicating species ranks were positively correlated through time (continuous site: $W = 0.178$, $n = 90$; fragmented site: $W = 0.441$, $n = 96$; large blocks: $W = 0.340$, $n = 96$; medium blocks: $W = 0.515$, $n = 96$; small blocks: $W = 0.597$, $n = 96$; in all cases $P < 0.001$). The value of W calculated for the continuous site was approximately two-fifths that calculated

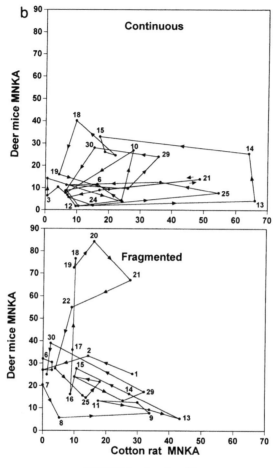

FIGURE 5. (*continued*)

for the fragmented site; this indicates rank abundances were less correlated (changed more often) through time in the continuous site. Relative to other block sizes, the value of W for the large blocks was most similar to that for the continuous site.

3. Extinctions as a Function of Site and Degree of Fragmentation

For cotton rats, there was no difference between simulated and real medium blocks in the average number of absences. However, real small blocks had

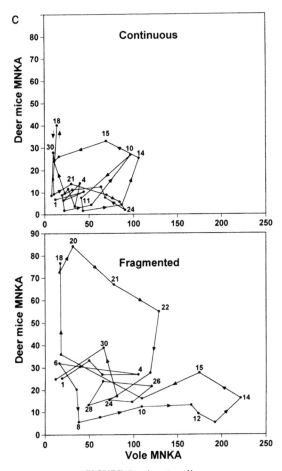

FIGURE 5.   (*continued*)

nearly twice as many cotton rat absences as simulated small blocks (simulated = 91.0 ± 10; real = 176.6 ± 1.8; t = 8.13, one-tailed P = 0.0075, df = 2). For prairie voles, simulated medium blocks had a higher average number of trapping periods with zero MNKA than did real medium blocks (simulated medium = 27.0 ± 6.1; medium = 5.5 ±5.5; t = 2.63, one-tailed P = 0.06, df = 2). Deer mice had a higher number of absences on simulated medium and small blocks than on real blocks (simulated medium = 62.7 ± 16; medium = 22.5 ± 6.5; t = 2.28, one-tailed P = 0.075; simulated small = 82.3 ± 22; small = 15 ± 8.9; t = 2.87, P = 0.05, df = 2 in both cases).

Overall, the two sites had similar and low average numbers of absences

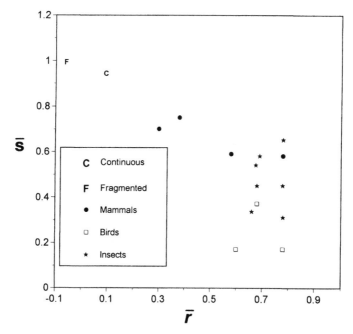

FIGURE 6.   The average Spearman's rank correlation ($\bar{r}$) of all possible pairwise comparisons of species in the community versus the average standard deviation ($\bar{s}$) in the log of abundances of all species. Fourteen communities of boreal insects (stars), birds (open squares), and small mammals (closed circles) are shown from Hanski (1990). The two communities analyzed here are added to the graph. C, continuous site; F, fragmented site.

across sites for both prairie voles and deer mice. Prairie voles never went extinct in either site. Deer mice never went extinct in the fragmented site and had only one absence in the continuous site. In cotton rats, the fragmented site had a higher proportion (0.112) of extinctions than expected and the continuous site a lower proportion (0.045) of extinctions than expected ($\chi^2 = 3.321$, $df = 1$, $P = 0.0684$).

C. The Spatial Scale of Community Processes

1. Large-Scale Movements

Over nearly 8 years of study, only 10 individual cotton rats and 7 individual prairie voles switched from the fragmented to the continuous site (which are ~500 m apart, Fig. 1). During that time a total of 1012 cotton rats and 1702

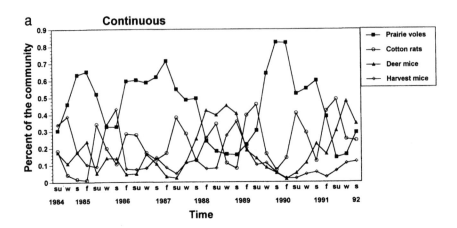

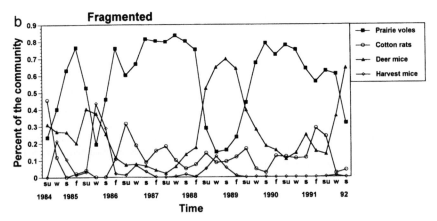

FIGURE 7. Average proportion of the community for 31 year-seasons for cotton rats, prairie voles, deer mice, and harvest mice in (a) the continuous site and (b) the fragmented site.

prairie voles were captured on the continuous site. Since deer mice were not individually marked until April 1989, we have a smaller sample size for this species. From 1989 to 1992, 227 deer mice were captured on the continuous site; during this time 2 deer mice moved from the continuous site to the fragmented site. Thus, for all species combined, fewer than 1% of their total population in the continuous site came from the marked population in the fragmented site.

Similar trends held for cotton rats in the fragmented site. Five cotton rats immigrated from the continuous site to the fragmented site, which contained a total of 1131 marked cotton rats over the study period.

2. The Influence of Movement on Local MNKA

a. Cotton rats. Averaging over all trapping periods, including those in which there was no switching among blocks, the proportion of MNKA accounted for by immigrants changed across year-seasons ($F = 3.491$, $df = 27$, 489, $P < 0.0001$) and in different patterns within each block size, leading to a significant block-size by year-season interaction ($F = 4.876$, $df = 18$, 489, $P < 0.0001$). Despite this temporal variation, large blocks averaged lower proportions of MNKA accounted for by immigrants than did medium blocks ($F = 16.313$, $df = 1$, 489, $P < 0.0001$, Fig. 8). The proportion of MNKA accounted for by emigration also differed with time ($F = 6.340$, $df = 27$, 489, $P < 0.0001$) and in different patterns within blocks across time ($F = 7.814$, $df = 18$, 489, $P < 0.0001$). As with trends in immigration, large blocks had a lower average proportion of their MNKA accounted for by emigrants than did medium blocks ($F = 70.521$, $df = 1$, 489, $P < 0.0001$, Fig. 9). (There were too few cotton rats in small blocks to do this analysis.) Using all blocks simultaneously, there was a strong, negative correlation between log (MNKA) and the log of the proportion of MNKA accounted for by both immigrants ($r = -0.915$, $df = 150$, $P < 0.001$) and emigrants ($r = -0.943$, $df = 86$, $P < 0.001$) into, or from, a block.

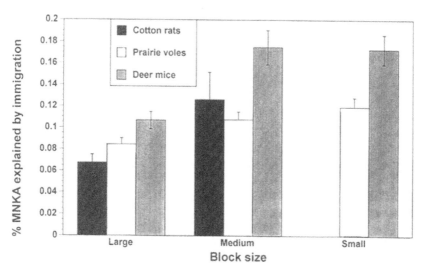

FIGURE 8. The average proportion of the MNKA with standard error bars in large, medium, and small blocks accounted for by marked immigrants from other blocks in the fragmented system for cotton rats, prairie voles, and deer mice.

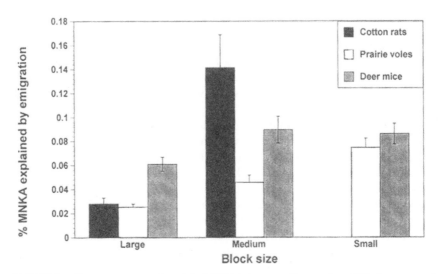

FIGURE 9. The average proportion of the MNKA in large, medium, and small blocks accounted for by marked emigrants that moved to another block in the fragmented system for cotton rats, prairie voles, and deer mice.

b. Prairie voles. Using the entire time series, the proportion of MNKA explained by immigration changed with time ($F = 1.996$, $df = 30, 1305$, $P = 0.029$) and in unique patterns among block sizes ($F = 2.036$, $df = 60, 1305$, $P = 0.049$). However, there were no significant differences among block sizes. The proportion of MNKA lost to emigration changed with time ($F = 2.581$, $df = 30, 1305$, $P < 0.0001$) and in different patterns on each block size ($F = 2.268$, $df = 60, 1305$, $P < 0.0001$). Despite this temporal variation, block size effects on the proportion of MNKA lost to emigration were apparent ($F = 24.483$, $df = 2, 1305$, $P < 0.0001$). Larger block sizes lost a lower proportion of MNKA to emigration than did smaller block sizes (Fig. 9). As with cotton rats, there was a strong and negative correlation between the log of MNKA and both the log of the proportion of MNKA explained by immigration ($r = -0.762$, $df = 760$, $P < 0.001$), and emigration ($r = -0.906$, $df = 444$, $P < 0.001$).

c. Deer mice. Using the entire data set, the proportion of MNKA accounted for by immigrants from other block sizes changed with time ($F = 3.063$, $df = 30, 1276$, $P < 0.001$) and in different patterns on each block size ($F = 1.641$, $df = 60, 1276$, $P < 0.001$). The proportion of MNKA accounted for by immigration was lower on large blocks than on small blocks ($F =$

13.206, df = 2, 1276, P < 0.0001; Fig. 8). As with immigration, the proportion of MNKA lost to emigration changed with time (F = 2.234, df = 30, 1276, P < 0.0001) and in different patterns within each block size (F = 1.674, df = 92, 1276, P < 0.0001). Despite these temporal changes, large blocks lost a lower proportion of MNKA to emigration than did medium and small blocks (F = 9.802, df = 2, 1276, P < 0.0001; Fig. 9). As with the other species, the log of the proportion of MNKA explained by either immigration (r = −0.811, df = 374, P < 0.001) or emigration (r = −0.918, df = 365, P < 0.001) was negatively correlated with log MNKA.

3. The Influence of Distance on Local Movements

The distance between blocks influenced the proportion of the individuals in that block accounted for by immigration. For both prairie voles and deer mice, short movements accounted for a higher proportion of the individuals in a block, relative to longer movements (prairie voles: t = 4.29, df = 44, P < 0.001, short movements = 0.0263 ± 0.0029, long movements = 0.0096 ± 0.0025; deer mice: t = 5.29, df = 32, P < 0.001, short movements = 0.0687 ± 0.011, long movements = 0.0075 ± 0.002). In cotton rats, the trend was the same, but the small sample sizes associated with analyzing movements only between the large blocks precluded a statistically significant result (t = 1.27, df = 15, P = 0.220, short movements = 0.0131 ± 0.0057, long movements = 0.00497 ± 0.0028).

4. Within Site and Spatial Heterogeneity in Abundances

Maps of habitat use (Fig. 10) suggest that at least a few species unique to the continuous site also utilized microhabitats or resources unique to that area, or might be present because the site directly abutted a woodland (Fig. 1). Bog lemmings were captured primarily in areas of brome grass, a forage species which rarely occurs on the fragmented site (Fig. 10a). Wood rats were captured almost exclusively near areas of woody vegetation or young trees in the continuous site (Fig. 10b). Finally, white-footed mice were captured mainly along wooded fencerows or along the southern edge of the plot, immediately adjoining a woodland (Fig. 10c).

A number of species showed nonrandom patterns of space use, including small-scale range expansions with changes in abundances. For instance, cotton rats in the continuous site preferred the northwest corner of the plot (Fig. 11a). In the fragmented site, cotton rats were trapped almost exclusively in the large blocks (Fig. 11b). However, at low abundances (≤10 animals), cotton rats in the fragmented site were restricted to the centermost large block (Fig. 11c). At high densities, prairie voles in the continuous site were scarce in the

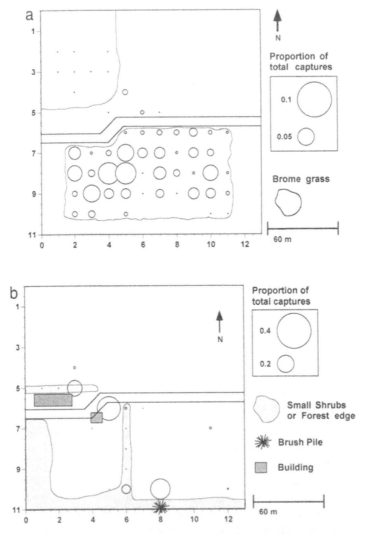

FIGURE 10.   Bubble plots of space use for (a) bog lemmings, (b) wood rats, and (c) white-footed mice in the continuous area. The diameter of a bubble is the proportion of the total captures for a given trap location. The *x* and *y* axes represent the coordinate system for the trapping grid.

southwestern corner (Fig. 12a), whereas at low abundances (≤20 animals), the voles were found almost exclusively in the northeastern portion of the site (Fig. 12b). Deer mice occupied the northern half of the continuous site almost exclusively, regardless of density (Fig. 13a). In the fragmented site, traps in

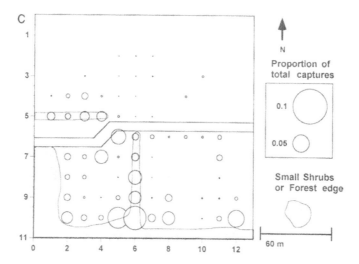

medium and small blocks captured higher proportions of deer mice than traps in large blocks (Fig. 13b). The relative proportions of deer mice captured in the centermost large block (where cotton rats tended to aggregate at low abundances) were lower than in the other large blocks, particularly when deer mice were at high abundances ($\geq$45 animals, Fig. 13b). This pattern is consistent with competitive interactions, though it could also reflect discordant responses to local vegetation. A graduate student (W. Schweiger personal communication) is currently examining our plant data to determine the potential contribution of local heterogeneity in vegetation structure to these spatial patterns.

IV. DISCUSSION

Our comparative analyses have revealed five major patterns: (1) The small-mammal community on the continuous site is more diverse than on the fragmented site. (2) As patch size increases, communities within the fragmented site increasingly resemble communities in the continuous system. Moreover, the species-specific responses to fragmentation extrapolate between sites. (3) Within the fragmented site, patchiness negatively influences species richness. (4) Community structure is highly variable at short time scales and this variability is magnified by fragmentation. (5) Immigration and emigration can have significant impacts on community structure, at a spatial scale of <150 m;

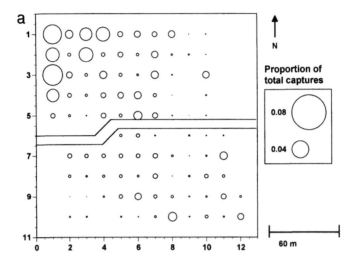

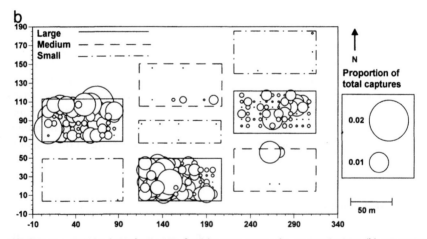

FIGURE 11. Bubble plots of space use for (a) cotton rats in the continuous site, (b) cotton rats at high densities in the fragmented site, and (c) cotton rats at low densities in the fragmented site. The diameter of a bubble is the proportion of the total captures for a given trap location.

however, observed movement rates imply communities may be effectively decoupled at scales >500 m. We discuss each of these patterns in turn and then conclude with some more general observations.

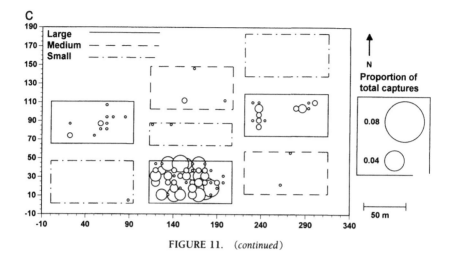

FIGURE 11. (*continued*)

A. The Effects of Fragmentation on Abundances

1. Higher Species Diversity in the Continuous Site

The total area of old field in our two sites is comparable. Nonetheless, the continuous site had a more diverse small mammal community. For instance, the continuous site had three species never captured in the fragmented site. Furthermore, many of the rare species were captured much more frequently in the continuous site than in the fragmented site. For example, white-footed mice were captured 1471 times out of 15,510 total captures in the continuous site, compared to just 38 times out of 23,227 total captures in the fragmented site. House mice were captured 100 times in the continuous site and only 3 times in the fragmented site. The more frequent captures of these rarer species in the continuous site resulted in a higher average species richness. Furthermore, using just the four species common in both sites, Shannon–Weiner diversity indexes were higher in the continuous system, indicating a more equitable distribution of abundances among species in the continuous site than in the fragmented site.

The higher diversity in the continuous site may reflect both patch size effects and the landscape context in which the two sites are placed. The simplest explanation for the unique species in the continuous site is the increased amount of local habitat heterogeneity found there. Unlike the fragmented site, the continuous site contains brome, patches of shrubs and small

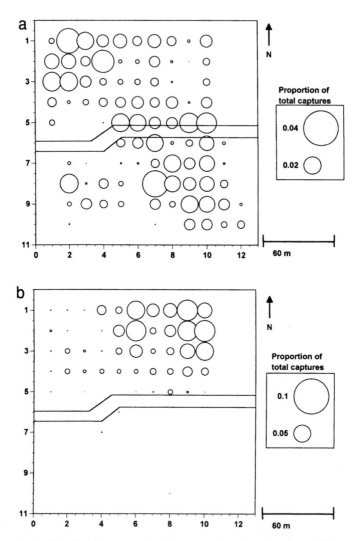

FIGURE 12. Bubble plots of space use for (a) prairie voles at high densities and (b) prairie voles at low densities in the continuous site. The diameter of a bubble is the proportion of the total captures for a given trap location.

trees distributed along fence rows, and small buildings and sheds. The bubble plots indicate that the unique species in the continuous site all preferentially use these habitats. Furthermore, those species (such as white-footed mice) which are rare in the fragmented site but more common in the continuous site also tend to use unique areas of the continuous site.

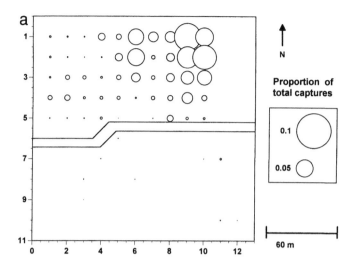

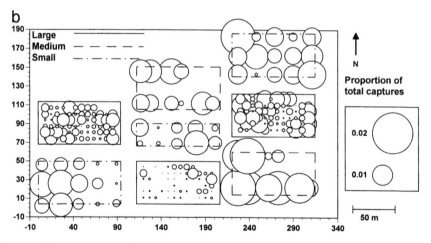

FIGURE 13. Bubble plots of space use for (a) deer mice in the continuous site and (b) deer mice at high densities in the fragmented site. The diameter of a bubble is the proportion of the total captures for a given trap location.

In 1994 the fragmented area was being invaded by woody-stemmed plants and was spotted with many small trees. In that summer wood rats and white-footed mice, species preferring wooded areas, colonized the fragmented site after 10 years of succession (J. E. Diffendorfer and W. Schweiger, unpublished data).

In addition to the unique habitats found in the continuous site, its position in the landscape likely increased the diversity of the small mammal assemblage found there. This is not an effect of fragmentation per se, but rather reflects the fact that the continuous site is juxtaposed to a woodland habitat with a different fauna which can utilize small patches of habitat (i.e., fencerows and small trees) in a way not possible in the fragmented site. There are no obvious barriers to movement from the woodland to the old-field habitat in the continuous site; by contrast, woodland animals must traverse barriers (a road and strips of interstitial habitat) to move into the fragmented site. Thus the landscape context in which a local community is embedded may influence its species richness. Even seemingly minor barriers to dispersal may markedly affect local communities.

2. Species-Specific Responses to Fragmentation Extrapolate between Sites

Figure 2 illustrates that large blocks in the fragmented site have communities that, on average, are relatively more similar to the continuous site than they are to smaller blocks. We interpret this as resulting from, in part, a gradient of decreasing competitive influence by dominant species from continuous habitat through large to small blocks. We are not using the observed patterns in abundance to infer competition, but rather drawing on prior, independent evidence of strong asymmetrical competition to interpret the observed spatial patterns. The negative impact of cotton rats on prairie voles was first inferred by noticing declines in prairie vole populations in the presence of cotton rats (Martin, 1956) and the failure of prairie voles to occupy sites when cotton rats were present (Frydenhall, 1969). Later support for the aggressive dominance of cotton rats over prairie voles came from combined laboratory and field studies (Terman, 1974, 1978; Glass and Slade, 1980; Prochaska and Slade, 1981) which documented prairie vole shifts in space use, diel patterns, and reduced population sizes in the presence of cotton rats. It is reasonable to expect this strong competitive interaction to be present in our system as well.

In the fragmented system, the spatial scale is such that medium and small patches are rarely used by cotton rats and are, in effect, competitor-free habitats for prairie voles. This relaxation of competition leads to (1) prairie voles having the highest densities on the blocks of medium patches when comparing among block sizes (e.g., 43% of the prairie vole population is found in the medium blocks, which make up 23% of the habitat in the fragmented site) and (2) higher overall prairie vole abundances in the fragmented site than in the continuous site.

Previous studies have also documented a negative impact of prairie voles on deer mice populations (Abramsky *et al.*, 1979; Grant, 1971, 1972; Redfield *et al.*, 1977; Doonan, 1993). In the past we have interpreted the higher deer mice densities in the fragmented site as resulting from the ability of deer mice to use resources in the competitor-free interstitial area (Diffendorfer *et al.*, 1995b). This interpretation rests on the assumption that deer mice use the interstitial area, which was initially demonstrated by Foster and Gaines (1991). They showed that for a year (from summer 1986 to summer 1987) more than 50% of the deer mice MNKA in the entire study area came from animals captured in the interstitial area. Early in the fragmentation study, this species apparently viewed the site as a variegated habitat. However, 1994–1995 trapping in the interstitial area captured few deer mice (W. Schweiger, personal communication), indicating deer mice use of the interstitial area had declined. This may reflect a successional shift in the interstitial plant community from annuals to lower productivity perennials (R. D. Holt, unpublished data). Averaged over the years 1984–1992, it seems likely deer mice could utilize the interstitial area as a lower quality habitat, moderating any competitive impact of prairie voles. This could also help explain why deer mice were denser in large blocks than in the continuous site. Animals moving between large blocks and the interstitial area could inflate deer mice densities in patches to levels above those seen in the continuous area, thereby vitiating the effect of resident competitors. The magnitude of this spatial flux should decline with a decreasing perimeter-to-area ratio, leading to the observed decline in density with increasing patch size (small blocks to the continuous site).

Interestingly, prairie voles had higher densities in large blocks than they had in the continuous area. This is somewhat surprising, given our working hypothesis that competition with cotton rats limits their numbers and that cotton rat densities were nearly equal in the continuous area and the large blocks. A significant datum is that a higher percentage of cotton rats in the fragmented site are transients, defined as an animal captured only once (Diffendorfer *et al.*, 1995b). Thus, the actual number of territory-holding cotton rats likely to exhibit strong interference in the fragmented site is almost surely lower. This difference in the social organization of cotton rats could weaken direct competition on prairie voles, and thereby permit prairie voles to reach densities in the large blocks higher than those in the continuous site.

Fragmentation can have at least two possible effects on a community. First, fragmentation could rescale individual species densities among different sized patches, while leaving the dynamical interactions among them unchanged. Second, fragmentation can change the pattern or strength of interspecific processes themselves, which then leads to spatial heterogeneity in abundances. As a metaphor, consider the Lotka–Voltera competition model, which for species i is

$$\frac{dN_i}{dt} = N_i r_i \left[1 - \frac{N_i}{K_i} - \frac{\alpha_{ij} N_i}{K_i} \right],$$

where N_i is the abundance of competitor i where r_i is the intrinsic growth rate, K_i the carrying capacity, and α_{ij} the competition coefficient. Fragmentation could affect r_i and K_i, but leave α_{ij} unchanged. Assuming species 1 is a dominant competitor, such that $\alpha_{12} = 0$, $\alpha_{21} > 0$. In a constant environment, at equilibrium, $N_1 = K_1$, and $N_2 = K_2 - \alpha_{21}K_1$. A given change in the environment could enhance species 2 by either (1) reducing the magnitude of the carrying capacity of the dominant, K_1, or (2) reducing the strength of competition for individuals, measured by α_{21}.

At our particular spatial scale, it is likely that both mechanisms underlie the effects of fragmentation we have observed. In our system, fragmentation changes the local abundance of competitively dominant species, which may, by itself, result in shifts in the mean abundances of subordinate species and changes in community structure.

Beyond this, a number of patterns suggest changes in interaction strength. For instance, negative correlations in abundance in our time series between prairie voles and both deer mice and harvest mice, and between cotton rats and deer mice occurred only in the continuous site. Community ecologists have often tried to use regression techniques on time series of abundances to estimate competition coefficients between species (e.g., Seifert and Seifert, 1976; Hallett and Pimm, 1979; Rosenzweig *et al.*, 1985). There is a variety of problems with this approach (Carnes and Slade, 1988), stemming in part from the absence of an independent estimate for key parameters other than the coefficients (e.g., r, K), and in part from the influence of unmeasured indirect interactions. A perusal of the phase diagrams (Fig. 5), however, provides suggestive evidence that interspecific interactions were likely modified by fragmentation. It is likely that, were we to calculate competition coefficients for a given pair of species, the estimates would differ between the sites. The basic conceptual point here is that a qualitative modification in species interactions by fragmentation may be a common, indirect effect of habitat fragmentation.

3. Species Richness and Patchiness

Patchiness decreased species richness: real medium and small blocks had lower species richness than did similar-size simulated blocks from large, continuous blocks. Furthermore, simulated medium and simulated small blocks (both simulated from the same large blocks) did not differ in species richness, but real medium and small blocks did. Thus, the amount of patchiness, (i.e., the size, number, and spacing of patches) clearly influences species richness. Our analysis of simulated blocks allows us to reject the hypothesis that this is a sampling artifact.

These trends in species richness seem to reflect the influence of patchiness on trapping periods with zero MNKA (absences). For cotton rats and deer mice, real small blocks, when compared to small blocks simulated from large blocks, had twice as many trapping periods with zero MNKA. Furthermore, cotton rats had a higher proportion of absences and deer mice a higher average number of absences in real medium blocks compared to simulated blocks. Thus, because species had more absences in more patchy block sizes, species richness declined.

As one would expect, the trends in periods marked as absences were associated with overall density trends across the patch sizes. For example, cotton rats had the highest densities in large blocks and always had fewer absences in simulated blocks than in real medium and small blocks. Prairie voles had highest densities in the medium blocks and had more absences in simulated medium blocks than in real medium blocks. Finally, deer mice had the highest densities, and the fewest absences, in small blocks.

In our two systems, cotton rats were the only species in which the proportion of overall absences (our measure of site-wide extinction) was influenced by fragmentation. The higher proportion of cotton rat absences in the fragmented site reflects the negative effect of fragmentation on this species, which is essentially confined to three small, partially isolated populations in the large blocks (Diffendorfer et al., 1995a; see Fig. 11b). The total population size of cotton rats is smaller in the fragmented site and more prone to periods of low numbers and even local site-wide extinctions there.

B. The Effects of Fragmentation on Temporal Variation in Communities

1. Old-Field Communities Are Highly Dynamic

Temporal variability is the dominant pattern in our community time series data (Figs. 7a and 7b). As is commonly seen in studies of open communities (Williamson, 1981; Holt, 1993), our rare species are sporadic in their occurrence, whereas common species disappear infrequently, despite large fluctuations in their abundance. The pattern of fluctuations in these persistent species ranges from annual cycles to what appears to be multiannual cycles. These qualitative patterns are observed at both sites, but with substantial quantitative differences. For instance, the fragmented system showed higher amounts of variation in Shannon–Weiner diversity (H), reflecting the high variability in prairie vole abundances there relative to the continuous area. It is tempting to speculate that fragmentation, by creating refuges for prairie voles in medium and small patches, effectively reduces the negative competitive influence of cotton rats on prairie voles and allows the rapid growth and occasional high abundances of this species there.

Relative to other communities, our systems showed quite high levels of variation and lower average interspecific correlations in abundances (Fig. 6). We were rather surprised at how different our systems appear to be from other vertebrate communities. The high average variance in small-mammal numbers in old-field communities may result from the cyclic nature inherent to some of the species' demography, compounded by the strong influence of seasonality on animal abundances (Stokes, 1994). The low average correlation coefficients result because some pairs of species in our systems have negative correlations in abundances through time, whereas others have positive correlations, a phenomenon which must not be as common in other assemblages.

The community in the fragmented site had a higher average standard deviation in abundances and a lower average correlation than did the continuous site. In the fragmented site, we believe a reduction in competitive pressure resulted in higher peaks in both prairie voles and deer mice with lows determined by other factors (e.g., weather). This would increase the average standard deviation. The lower correlation in abundances in the fragmented site is more difficult to interpret. We are tempted to speculate that fragmentation weakens the strength of interspecific interactions. Moreover, because interaction strengths may vary among patch sizes, averaging among patch sizes may make it more difficult to discern correlated changes in abundances.

In terms of rank abundances, the Kendall's *W* test indicates species ranks were positively correlated; hence, by this measure these communities were stable through time. However, based on previously reported values of *W* (Rahel, 1990), the continuous site has a *W* lower than 11 of 12 published values, and the fragmented site has a *W* lower than 10 of 12. Despite being stable in a statistical sense (indicating a degree of order), community structure at both our sites may be unstable relative to other communities. This is particularly so for the continuous site, which has a *W* almost three times lower than that of the fragmented system.

Note that despite the higher average standard deviations in abundances in the fragmented site, rank abundances were more stable in the continuous site. Why is *W* so much lower in the continuous site? We feel the best explanation rests on the restriction of cotton rats to the three large blocks, the competitive release of prairie voles, and the extra resources available to deer mice in the fragmented site. These factors result in high numbers of prairie voles and deer mice and consistently lower numbers of cotton rats, leading to relatively stable rank abundances over time. On the continuous site, cotton rats have more available habitat, prairie voles come under competitive pressure, and deer mice have fewer resources available relative to the fragmented site. Thus, species abundances are more similar permitting ranks to change more frequently.

In terms of presence/absence, the core species in the continuous site and the entire community in the fragmented site rarely went extinct. Thus, the communities in the sense of presence/absence by species were quite stable. There were two clear exceptions. First, harvest mice went extinct in the fragmented site after the first 3 years (Gaines et al., 1992a). Since that initial phase of occupancy they have been sporadic community members, absent in most sampling periods while persistent in the continuous site. Second, the numerous noncore species in the continuous site had frequent extinctions and recolonizations of the area (e.g., possibly reflecting spillover for the adjacent woodland species). Because of the episodic occurrence of these rarer species, one may consider the entire community in the continuous site as somewhat unstable in terms of species presence or absence. The fragmented area did not have the rare species, and the three core species (cotton rats, prairie voles, and deer mice) rarely went extinct.

Our results clearly indicate that different measures of community variability can give quite discordant impressions of stability in comparative analyses of communities at different sites.

C. The Spatial Scale of Community Dynamics

Recent analyses (Diffendorfer et al., 1995a) suggest rates of individual movements may influence spatial patterns in abundances within the fragmented site. We created Markovian matrices of switching probabilities for characterizing movement among blocks, used these to predict the number of animals in a given block, and then compared the predicted abundances to the actual abundances. The matrices contain the probability of an animal switching from one block to another (off-diagonal elements) or remaining in the same block (diagonal elements). Solving the matrices for the dominant eigenvector leads to a predicted steady state of abundances in the various patches, assuming as a null model that observed asymmetries in movement are solely responsible for determining relative abundances (for further details and a sample of the matrices, see Diffendorfer et al., 1995a). We created matrices for prairie voles and deer mice in each season. In all seasons except spring for prairie voles and summer for deer mice, the predicted and observed abundances did not match closely. The results indicated that within the fragmented system (at what we consider a fairly small spatial scale) movements alone cannot account for all the abundance patterns observed over the entire site. However, based on the seasons in which the abundance patterns predicted from the matrices of switching probabilities fit the observed abundances, movements do appear to be capable of influencing the local dynamics in a particular block. In general, these previous findings raised the question of the relative influence of movements on local demography and the spatial scale over which populations are coupled by dispersal.

Our analysis allows us to consider animal movements at three distinct spatial scales. First, we have rough estimates of long-distance dispersal over 0.5 km. Second, we can look at various combinations of movements between blocks in the fragmented site at scales between 15 and 140 m. Third, the bubble plots reflect patterns of space use and are useful in gauging the spatial scale at which small mammals can respond to habitat variation.

1. Large-Scale Movements

For the three dominant community members for which we have data, movements over 0.5 km seem to have little quantitative impact on local populations. If we assume that animals could immigrate to the continuous site from all directions, and that immigrants from the fragmented area account for 60° of the possible 360°, then roughly 5–6% of the total population in the continuous area came from immigrants dispersing from areas outside the site, up to a distance ~0.5 km. This observation is strengthened by considering additional data. In 1984–1985, Foster and Gaines (1991) used 350 traps in brome fields just north of the fragmented site (about 100–200 m away) and used 100 traps in the woodland due south (about 30 m away). They abandoned this additional trapping effort after determining that less than 4% of the total population in the fragmented site over an entire year came from these closely adjacent areas. In addition, our comparisons of the proportion of a given block's population accounted for by short (20–30 m) and long movements (~140 m) within the fragmented site indicate small proportions of local populations (defined as those individuals inhabiting a 50 × 100-m area) are made up of immigrants and further that these proportions decrease with increasing distance between blocks. Thus, even at shorter distances of about 100–200 m, animal movements seem to have little potential to influence local abundances.

2. Small-Scale Movements

Here we discuss the proportion of MNKA that came from or went to any other block on the fragmented site. This procedure measures the cumulative effect of movements on a local population at a scale from 15 to 140 m. For all species, the influence of immigration and emigration on local abundances varies temporally. This temporal variation indicates animal movements may only strongly influence local communities during certain time periods.

Movement of animals explained between 6 to 10% of the MNKA on large blocks across all species. Combining this with our rough estimates of longer scale movements indicates that, in general, 10–20% of a population in a 50 × 100-m block may come from immigrants over the entire disk surrounding the block, up to 0.5 km away. The influence of movements on local abundances would be even less in larger areas, where the perimeter-to-area ratio is smaller. Our data indicate that for all species, larger blocks, with a low perimeter to area ratio, had a lower proportion of MNKA explained by animal movements than did smaller blocks.

However, the strong negative correlations between the proportion of MNKA explained by immigrants and MNKA for all species indicate that at small population sizes, a different picture emerges regarding the impact of animal movements on local abundances and community structure. When populations are small, movements constitute a much higher proportion of the MNKA, for all species. This relationship may partially explain why populations in smaller block sizes (which have lower abundances of animals) were more influenced by movements than were populations on larger block sizes.

Thus, immigration and emigration can have a strong impact on the abundances of species in old-field small-mammal communities, but only sporadically at times of low abundance, and at relatively small spatial scales (<150 m). Correlations between MNKA and the proportion of MNKA explained by movement are strong and negative. Thus, the impact of movements on local MNKA is greatest when populations are low. The familiar scenario of immigrants saving a local population from extinction may be misleading because the correlations between emigration and MNKA (which might result in local extinction) are just as strong.

3. Fine-Scale Patterns in Habitat Use

The bubble plots (Figs. 10–13) dramatically indicate how small mammals respond to small-scale variation in their habitat. The overall impression from the bubble plots is that small mammals may track preferred habitats that are quite localized, cover small areas, and change through time. We anticipate that the close analysis of spatial patterns will be a useful tool in mammal ecology during the upcoming years, given the traditional reliance on permanent trapping grids and the recent growth of GIS and spatial statistics. For instance, the precise locations of captures during times of population collapse to low numbers may be useful in determining attributes of areas where animals are capable of surviving during harsh conditions, or be used to infer the likelihood of source-sink dynamics (Holt, 1985; Pulliam, 1988; Diffendorfer et al., 1995a).

4. An Overall Assessment of the Role of Spatial Dynamics

We should caution that our conclusions regarding spatial coupling result from observational studies, not from direct experimental manipulation. Because temporal fluctuations in abundances were broadly synchronous between our sites (Diffendorfer et al., 1995b), there may be no particular advantage for an animal to move more than a few home ranges from where it was born. By contrast, asynchronous fluctuations between our sites might uncover transient phases when immigration/emigration are much more significant. We should note that we have a documented case of a prairie vole moving about 1.5 km on the Nelson Environmental Study Area and a few more cases of long distance movement (J. E. Diffendorfer and N. A. Slade, unpublished data). This anecdotal evidence demonstrates small mammals can at times move long distances over heterogeneous landscapes. These rare, long distance movements may be crucial to colonization after local or regional extinctions. Finally, in the above analysis, we have only counted the immediate demographic effect of immigrants on a population; we have no information on the longer term reproductive output and hence future contribution to a local population an immigrant might make.

Obviously, when a local population is at low numbers, a female immigrant may make a disproportionate contribution to the future population. Theoretical models suggest even small amounts of dispersal can contribute to synchrony in population dynamics, given weak local density dependence (R. D. Holt, unpublished data). Kareiva and Wennergren (1995) point out that in certain predator–prey models, movements can have stabilizing effects on population dynamics even on extremely large scales, reflecting movements of individuals over multiple generations and leading to characteristic spatial scales in ecological patterns much larger than the scale of individual movement. Thus, we may be missing a more diffuse, longer term impact of movements on small-mammal communities because both our temporal and spatial scales are too small.

Despite all these caveats, our results do strongly suggest to us that the spatial scale characterizing small-mammal populations is on the order of 100–200 m. This is probably a much smaller number than what most field workers might casually believe.

V. CONCLUSIONS

To date, most studies of small mammals in old fields have focused on population dynamics rather than community structure. Our results are congruent with previous studies of old-field small mammals in the central United States with respect to aspects of community structure. We, too, have observed pronounced variability in species richness over time (Swihart and Slade, 1990), indications of interspecific competition (Martin, 1956; Frydenhall, 1969; Glass and Slade, 1980; Terman, 1974, 1978), and habitat selection based on vegetation features (Kaufman and Fleharty, 1974; Grant and Birney, 1979; Seitman *et al.,* 1994). Our work indicates that old-field small-mammal communities are among the most variable vertebrate communities reported to date. Our analyses further suggest that in addition to local processes, patch size, the degree of patchiness of the landscape, and the structure of the surrounding habitat (which influences isolation and defines the source pool) can all strongly influence local community structure. Finally, at short time scales (1–10 generations), and spatial scales exceeding 150 m, movements appear to have little impact on community structure. However, movements at smaller scales can influence community structure, particularly at low abundances. We should caution that our studies concerning movements and local abundances were done on a relatively small scale of 5000-m² blocks, separated by only 16 to 20 m. This small scale may inflate our perceived impact of movements on local abundances and community structure by creating spatial blocks of habitat smaller than typically found in nonexperimental landscapes.

Our data indicate movements have the largest impacts at low abundances. Since the two most common and most competitively dominant species in old fields (cotton rats and prairie voles) have either seasonal, annual, or multiannual fluctuations, movements may influence populations during low periods and, possibly, at somewhat larger spatial scales than our current study.

Given our general conclusions, it is useful to discuss gaps in our knowledge to guide further research directions. First, we and other ecologists studying old-field small mammals have tended to focus on just the rodents and their interactions. Predators and competing species from other taxa may have profound and, as yet, poorly documented impacts on the community. (Indeed, our working hypothesis regarding the effectiveness of our interstitial area habitat as a barrier to dispersal is that small mammals found there incur a serious risk of predation.)

Second, though the circumstantial evidence for habitat selection in our system is compelling, our bubble plots and previous studies barely scratch the surface of the spatial interplay between small mammal herbivores and their resources. Food supplementation experiments have shown that the distribution of old-field small-mammal populations responds to changes in food resources (Doonan, 1994). Desert small-mammal ecologists have successfully quantified the effects of resource heterogeneity on mechanisms of species coexistence using insights from optimal foraging theory (Kotler and Brown, 1988). Attempting such research on old-field small mammals should be fruitful, given the large amounts of spatial and temporal variation in habitat and the diversity of species. Comparing the results of such studies to desert communities could permit more direct engagement with broader issues in community ecology. We find it intriguing that most work on rodent community ecology has been in desert systems; old-field rodent studies have primarily focused on single-species questions. This may be because the mechanisms of resource use and predation avoidance are more sharply delineated and obvious in desert habitats (Kotler and Brown, 1988).

Third, further studies aimed at understanding the interplay among animal movements, local abundances, and interspecific interactions are greatly needed. Communities are open systems and determining if the flux of individuals into or out of a community affects local processes such as predation or competition is essential to a clear understanding of local community dynamics. Our data suggest movements can be influential, but only at much smaller scales than we had expected. Analyzing dispersal in conjunction with colonization and extinction dynamics may help us understand variation in species richness through time and the variation seen among communities in different locations.

We opened the paper by emphasizing the need to consider processes operating at different spatial scales when analyzing the structure of local communities. There is, in general, a rough correspondence between the spatial scale of a natural process and the temporal scale required for the full range of variation in that process to be expressed (Steele, 1995). This implies that long-term time series of data are needed to assay with any precision the influence of mesoscale (Holt, 1993) ecological processes in determining local community structure. In the case of our analyses of small mammal communities in old fields, 8 years of data is barely enough to assess patterns in temporal stability, to provide accurate estimates of local extinction rates, or to detect rare but potentially crucial events, such as bouts of long-distance movement. Short-term studies are likely to miss many longer term and larger scale factors influencing local dynamics. Our results also suggest that long-term studies may be of greatest value when carried out in a comparative spatial context, for instance, along a habitat gradient. We believe analyses of the temporal and spatial dynamics of communities provide interdependent, mutually reinforcing intellectual challenges.

VI. SUMMARY

Local communities are influenced by processes operating at multiple spatial and temporal scales. Understanding the interplay of these processes requires long-term studies conducted at distinct but spatially related sites. In this paper, we compare approximately 8 years of detailed mark–recapture data on small mammals from two sites separated by ~0.5 km. One site is a continuous old field and the other an experimentally fragmented old field. Within the fragmented site, the degree of fragmentation influences species distribution and abundances of animals; these effects appear to reflect a combination of species-specific responses to habitat area and an alteration of competitive interactions. Patterns observed along the gradient from small to large patches in the fragmented site extrapolate to the continuous site.

The continuous site has a higher species richness than the fragmented site. This may reflect the presence of unique microhabitats at the continuous site rather than area effects per se, since the site abuts woodland, and the additional species appear to be present because of these differences in landscape context. Temporal variation for both communities in abundances and rank orders of abundance exceeds that reported in many other vertebrate communities, and community variability increases with fragmentation. However, the continuous site has more sporadic community members and overall more apparent extinction/colonization dynamics than does the fragmented site. Analyses of individual movements between sites indicates old-field small-mammal communities may be effectively decoupled at distances of 0.5 km. Analyses of movements within the fragmented site indicate movements can influence demographic trends at spatial scales ≤ 150 m. In general, our comparative study indicates that both landscape structure and processes occurring outside a local community can influence local community dynamics. However, the spatial scale relevant to governing population dynamics in old-field small-mammal communities may be smaller than many ecologists hitherto have believed.

Acknowledgments

We thank Charles Krebs for developing and giving us the Fortran programs used in some of these analyses. We also appreciate the many previous graduate and undergraduate students whose dogged efforts in the field helped in collecting the data. We thank William Schweiger for last minute help on the figures and updates on the current study. Constructive comments from M. Cody and J. Smallwood added to the chapter's clarity. N. Slade's work was supported by University of Kansas General Research Fund Grants 3349, 3093, 3509, and 3022. Work on the fragmented site was funded by NSF Grants BSR-8718088 and DEB-93-08065, the University of Kansas General Research Funds, and the Experimental and Applied Ecology Program.

References

Abramsky, Z., Dyer, M. I., and Harrison, D. (1979). Competition among small mammals in experimentally perturbed areas of shortgrass prairie. *Ecology* **60**, 530–536.
Aebischer, N. J., Robertson, P. A., and Kenward, R. E. (1993). Compositional analysis of habitat use from animal radio tracking. *Ecology* **74**, 1313–1325.
Brooks, D. R., and McLennan, D. A. (1993). Historical ecology: Examining phylogenetic components of community evolution. *In* "Species Diversity in Ecological Communities" (R. E. Ricklefs and D. Schluter, eds.), pp. 267–280. Univ. of Chicago Press, Chicago.

Brown, J. H. (1987). Variation in desert rodent guilds: Patterns, processes and scales. *In* "Organization of Communities: Past and Present" (J. H. R. Gee and P. S. Giller, eds.), pp. 185–203. Blackwell, Oxford.

Carnes, B. A., and Slade, N. A. (1988). The use of regression for detecting competition with multicollinear data. *Ecology* 69, 1266–1274.

Connell, J. H. (1983). On the prevalence and relative importance of interspecific competition: Evidence from field experiments. *Am. Nat.* 111, 1119–1144.

Diamond, J. M., and Case, T. J., eds. (1986). "Community Ecology." Harper & Row, New York.

Diffendorfer, J. E., Gaines, M. S., and Holt, R. D. (1995a). Habitat fragmentation and the movements of three small mammals species (*Sigmodon, Microtus,* and *Peromyscus*). *Ecology* 76, 827–839.

Diffendorfer, J. E., Slade, N. A., Gaines, M. S., and Holt, R. D. (1995b). Population dynamics of small mammals in fragmented and continuous old-field habitat. *In* "Landscape Approaches in Mammalian Ecology and Conservation" (W. Z. Lidicker, ed.), pp. 175–199. Univ. of Minnesota Press, Minneapolis.

Doonan, T. J. (1994). Effects of an experimental increase in resource abundance on population dynamics and community structure in small mammals. Ph.D Dissertation, University of Kansas, Lawrence.

Fahrig, L., and Paloheimo, J. E. (1988). Determinants of local population size in patchy habitats. *Theor. Popul. Biol.* 34, 194–213.

Foster, J., and Gaines, M. S. (1991). The effects of a successional habitat mosaic on a small mammal community. *Ecology* 72, 1358–1373.

Frydenhall, M. J. (1969). Rodent populations in four habitats in central Kansas. *Trans. Kans. Acad. Sci.* 72, 213–222.

Gaines, M. S., and McClenaghan, L. R., Jr. (1980). Dispersal in small mammals. *Annu. Rev. Ecol. Syst.* 11, 163–196.

Gaines, M. S., and Rose, R. K. (1976). Population dynamics of *Microtus ochrogaster* in eastern Kansas. *Ecology* 57, 1145–1161.

Gaines, M. S., Robinson, G. R., Diffendorfer, J. E., Holt, R. D., and Johnson, M. L. (1992a). The effects of habitat fragmentation on small mammal populations. *In* "Wildlife 2001: Populations" (D. R. McCullough and R. H. Barret, eds.), pp. 875–885. Elsevier Applied Science, London.

Gaines, M. S., Foster, J., Diffendorfer, J. E., Sera, W. E., Holt, R. D., and Robinson, G. R. (1992b). Population processes and biological diversity. *Trans. North Am. Wild. Nat. Resour. Conf.* 57, 252–262.

Gaines, M. S., Diffendorfer, J. E., Foster, J., Wray, F. P., and Holt, R. D. (1994). The effects of habitat fragmentation on three species of small mammals in eastern Kansas. *Pol. Ecol. Stud.* 20, 159–171.

Glass, G. F., and Slade, N. A. (1980). Population structure as a predictor of spatial association between *Sigmodon hispidus* and *Microtus ochrogaster. J. Mammal.* 61, 473–485.

Grant, P. R. (1971). Experimental studies of competitive interaction in a two-species system. III. *Microtus* and *Peromyscus* species in enclosures. *J. Anim. Ecol.* 40, 323–350.

Grant, P. R. (1972). Interspecific competition among rodents. *Annu. Rev. Ecol. Syst.* 3, 79–106.

Grant, W. E., and Birney, E. C. (1979). Small mammal community structure in north American grasslands. *J. Mammal.* 60, 23–36.

Hallett, J. G., and Pimm, S. L. (1979). Direct estimation of competition. *Am. Nat.* 113, 593–600.

Hanski, I. (1990). Density dependence, regulation and variability in animal populations. *Philos. Trans. R. Soc. London* 330, 141–150.

Hassell, M. P., Comins, H. N., and May, R. M. (1992). Spatial structure and chaos in insect population dynamics. *Nature (London)* 353, 255–258.

Holt, R. D. (1985). Population dynamics in two-patch environments: Some anomalous consequences of an optimal habitat distribution. *Theor. Popul. Biol.* **28**, 181–208.

Holt, R. D. (1992). A neglected facet of island biogeography: The role of internal spatial dynamics in area effects. *Theor. Popul. Biol.* **41**, 354–371.

Holt, R. D. (1993). Ecology at the mesoscale: The influence of regional processes on local communities. *In* "Species Diversity in Ecological Communities: Historical and Geographical Perspectives" (R. Ricklefs and D. Schluter, eds.), pp. 77–88. Univ. of Chicago Press, Chicago.

Holt, R. D., and Lawton, J. H. (1994). The ecological consequences of shared natural enemies. *Annu. Rev. Ecol. Syst.* **25**, 495–520.

Holt, R. D., Robinson, G. R., and Gaines, M. S. (1995). Vegetation dynamics in an experimentally fragmented landscape. *Ecology* **76**, 1610–1624.

Huffaker, C. B. (1958). Experimental studies on predation: Dispersion factors and predator-prey oscillations. *Hilgardia* **27**, 343–383.

Johnson, M. L., and Gaines, M. S. (1987). The selective basis for dispersal of the prairie vole, *Microtus ochrogaster. Ecology* **68**, 684–694.

Johnson, M. L., and Gaines, M. S. (1988). Demography of the western harvest mouse, *Reithrodontomys megalotis,* in eastern Kansas. *Oecologia* **75**, 405–411.

Kareiva, P. (1986). Patchiness, dispersal, and species interactions: Consequences for communities of herbivorous insects. *In* "Community Ecology" (J. M. Diamond and T. J. Case, eds.), pp. 192–206. Harper & Row, New York.

Kareiva, P., and Wennergren, U. (1995). Connecting landscape patterns to ecosystem and population processes. *Nature (London)* **373**, 299–302.

Kaufman, D. W., and Fleharty, E. D. (1974). Habitat selection by nine species of rodents in north-central Kansas. *Southwest. Nat.* **18**, 443–452.

Kotler, B. P., and Brown, J. S. (1988). Environmental heterogeneity and the coexistence of desert rodents. *Annu. Rev. Ecol. Syst.* **19**, 281–307.

Levin, S. A. (1992). The problem of pattern and scale in ecology. *Ecology* **73**, 1943–1967.

Manly, B. F. J. (1986). "Multivariate Statistical Methods: A Primer." Chapman & Hall, London.

Margules, C. R., Milkovits, G. A., and Smith, G. T. (1994). Contrasting effects of habitat fragmentation on the scorpion *Cercophonius squama* and an amphipod. *Ecology* **75**, 2033–2042.

Martin, E. P. (1956). A population study of the prairie vole (*Microtus ochrogaster*) in northeastern Kansas. *Univ. Kans. Publ. Mus. Nat. Hist.* **8**, 361–416.

McLaughlin, J. F., and Roughgarden, J. (1993). Species interactions in space. *In* "Species Diversity in Ecological Communities: Historical and Geographical Perspectives" (R. Ricklefs and D. Schluter, eds.), pp. 77–88. Univ. of Chicago Press, Chicago.

Paine, R. T. (1966). Food web complexity and species diversity. *Am. Nat.* **100**, 65–75.

Prochaska M. L., and Slade, N. A. (1981). The effect of *Sigmodon hispidus* on summer diel activity patterns of *Microtus ochrogaster* in Kansas. *Trans. Kans. Acad. Sci.* **84**, 134–138.

Pulliam, R. H. (1988). Sources, sinks and population regulation. *Am. Nat.* **132**, 652–661.

Pulliam, R. H., and Danielson, B. J. (1992). Sources, sinks, and habitat selection: A landscape perspective on population dynamics. *Am. Nat.* **137**, S50–S66.

Rahel, F. J. (1990). The hierarchical nature of community persistence: A problem of scale. *Am. Nat.* **136**, 328–344.

Redfield, J. A., Krebs, C. J., and Taitt, M. J. (1977). Competition between *Peromyscus maniculatus* and *Microtus townsendii* in grasslands of coastal British Columbia. *J. Anim. Ecol.* **46**, 607–616.

Ricklefs, R. E. (1987). Community diversity: Relative roles of local and regional processes. *Science* **235**, 167–171.

Ricklefs, R. E., and Schluter, D. (1993). Species diversity: Regional and historical influences. *In* "Species Diversity in Ecological Communities: Historical and Geographical Perspectives" (R. Ricklefs and D. Schluter, eds.), pp. 350–355. Univ. of Chicago Press, Chicago.

Robinson, G. R., Holt, R. D., Gaines, M. S., Hamburg, S. P., Johnson, M. L., Fitch, H. S., and Martinko, E. A. (1992). Diverse and contrasting effects of habitat fragmentation. *Science* 257, 524–526.

Roff, D. A. (1974). Spatial heterogeneity and the persistence of populations. *Oecologia* 15, 245–258.

Rosenzweig, M. L., Abramsky, Z., Kotler, B., and Mitchell, W. (1985). Can interaction coefficients be determined from census data? *Oecologia* 66, 194–198.

Roughgarden, J. (1989). The structure and assembly of communities. *In* "Perspectives in Ecological Theory" (J. Roughgarden, R. M. May, and S. A. Levin, eds.), pp. 203–266. Princeton Univ. Press, Princeton, NJ.

Sauer, J. R., and Slade, N. A. (1985). Mass-based demography of a hispid cotton rat (*Sigmodon hispidus*) population. *J. Mammal.* 66, 316–328.

Sauer, J. R., and Slade, N. A. (1986). Size-dependent population dynamics of *Microtus ochrogaster.* *Am. Nat.* 127, 902–908.

Saunders, D. A., Hobbs, R. J., and Margules, C. R. (1991). Biological consequences of ecosystem fragmentation: A review. *Conserv. Biol.* 2, 340–347.

Schoener, T. W. (1983). Field experiments on interspecific competition. *Am. Nat.* 122, 240–285.

Seifert, R. P., and Seifert, F. H. (1976). A community matrix analysis of *Heliconia* insect communities. *Am. Nat.* 110, 461–483.

Seitman, B. E., Fothergill, W. B., and Finck, E. (1994). Effects of haying and old-field succession on small mammals in tallgrass prairie. *Am. Midl. Nat.* 13, 1–8.

Steele, J. H. (1995). Can ecological concepts span the land and ocean domains? *In* "Ecological Time Series" (T. M. Powell and J. H. Steele, eds.), pp. 5–16. Chapman & Hall, London.

Stokes, M. K. (1994). Effects of weather and climate on populations of small mammals: Implications for climatic change. Ph.D Dissertation, University of Kansas, Lawrence.

Swihart, R. K., and Slade, N. A. (1990). Long-term dynamics of an early successional small mammal community. *Am. Midl. Nat.* 123, 372–382.

Terman, M. R. (1974). Behavioral interactions between *Microtus* and *Sigmodon:* A model for competitive exclusion. *J. Mammal.* 55, 705–719.

Terman, M. R. (1978). Population dynamics of *Microtus* and *Sigmodon* in central Kansas. *Trans. Kans. Acad. Sci.* 81, 337–351.

Tilman, D. (1994). Competition and biodiversity in spatially structured habitats. *Ecology* 75, 2–16.

Valone, T. J., and Brown, J. H. (1995). Effects of competition, colonization and extinction on rodent species diversity. *Science* 267, 880–883.

Vance, R. R. (1980). The effect of dispersal on population size in a temporally varying environment. *Theor. Popul. Biol.* 18, 342–362.

Wiens, J. A. (1976). Population responses to patchy environments. *Annu. Rev. Ecol. Syst.* 7, 81–120.

Valone, T. J., and Brown, J. H. (1995). Effects of competition, colonization and extinction on rodent species diversity. *Science* 267, 880–883.

Vance, R. R. (1980). The effect of dispersal on population size in a temporally varying environment. *Theor. Popul. Biol.* 18, 342–362.

Wiens, J. A. (1976). Population responses to patchy environments. *Annu. Rev. Ecol. Syst.* 7, 81–120.

Wiens, J. A., Addicott, J. F., Case, T. J., and Diamond, J. (1986). Overview: The importance of spatial and temporal scale in ecological investigations. *In* "Community Ecology" (J. M. Diamond and T. J. Case, eds.), pp. 145–153. Harper & Row, New York.

Wilcox, B. A. (1980). Insular ecology and conservation. *In* "Conservation Biology: An Evolutionary-Ecological Perspective" (M. E. Soulé and B. A. Wilcox, eds.), pp. 95–117. Sinauer Assoc., Sunderland, MA.

Williamson, M. (1981). "Island Populations." Oxford Univ. Press, Oxford.

CHAPTER 15

Long-Term Studies of Small-Mammal Communities from Disturbed Habitats in Eastern Australia

BARRY J. FOX

School of Biological Science, University of New South Wales, Sydney 2052, Australia

I. INTRODUCTION

Long-term studies can do much more than provide a longer run of data than might normally be expected from grant-funded studies that often have a 3-year time limit. In this chapter I begin by introducing four representative studies that I have published (summarized in Table I). These studies demonstrate that monitoring permanent plots for up to 20 years can produce a considerable amount of important information, some unexpected and most not forthcoming by short-term studies. The research involves postfire succession in the small mammal biota and thus addresses temporal changes in the habitat and its occupants. The studies were conducted in a part of Australia where fire is an important and natural event, and all vegetation represents a successional stage following the last fire. The long-term nature of the research is an essential component, allowing validation of the *chronosequence technique* commonly used in studies of succession.

The chronosequence technique is based upon two assumptions: first, postfire succession proceeds in the same way over a sequence of spatially separated sites, and second, the different plots are equivalent to one another in that each can be related, at any point in time, to the successional sequence at any other given site. This equivalence is possible because, although the succession at different sites may be disjunct in time, it proceeds through a similar sequence and with a comparable schedule. Thus each plot has a "regeneration age" that is equivalent among plots but in different calendar or survey years. These were the assumptions upon which Cowles (1899) based his development of the concept of ecological succession, using the initial chronosequence of the Michigan sand dunes; the technique has been followed up on the same dune sequence by Olson (1958). An Australian example is provided by Greenslade and Thompson (1981), who studied ant succession in a coastal dune system in southeastern Queensland.

However, use of the chronosequence technique has not been free of criticism, and two relevant and recent comments are worth mentioning. Austin (1981) emphasized that chronosequence analysis makes no allowance for long-term climatic change or year-to-year variation that may occur at the sites and advocated the use of permanent quadrats. Poissonet *et al.* (1981) have suggested that to verify such studies it may be necessary to monitor temporal changes on several individual sites, rather than recompose the successional sequence from short-term data at many sites in different stages of succession.

To validate the use of the technique in my study of small-mammal succession, I resurveyed in 1987 (5 years after the initial survey) 14 of the original 16 permanent plots reported by Fox and Fox (1984), plus one additional plot. The outcome showed a remarkable degree of congruence between the results from 1982 and 1987 (Twigg *et al.*, 1989). As the waves of colonization for each of two mammal species progressed, the peak abundance for each species in 1987 occurred on plots of regeneration ages that differed from the 1982 peaks by only 1 and 3 months, respectively, and each of the monitored sites supported an equivalent advance of 5 years regeneration age along the successional sequence. Such data provide unequivocal support for the use of chronosequence techniques.

TABLE I   Summary of Information about the Four Long-Term Studies from Myall Lakes National Park, in the Order Published, with an Outline of Procedure Used in Each Study

Habitat/disturbance	Trapping time span	Mammals[a] encountered	Experimental design	Previous publications
		Study 1		
Heathland/sand mining	1976–1977 and 1992	House mouse New Holland mouse Brown antechinus	Chronosequence; 11 1-ha plots, 3 controls; trapped 4 times in 2 years and 12 years later	Fox and Fox (1978); Fox (1990b)
		Study 2		
Open forest/fire	1976–1977, 1982–1983, trapping continued through 1994	House mouse New Holland mouse Common dunnart Brown antechinus Bush rat	Chronosequence; 16 1-ha plots; trapped four or five times over 2 years, 12 years later, then 18 years later	Fox and McKay (1981); Thompson (1983); Fox (1990a)
		Study 3		
Heathland/fire	1974–1980, 1980–1988, trapping continued through 1994	House mouse New Holland mouse Eastern chestnut mouse Common dunnart Brown antechinus Bush rat Swamp rat	One 7-ha plot; diverse habitats—swamp, wet heath to woodland, and forest, trapped 20 years (1974–1994) through two fire cycles	Fox (1982); Fox (1990a); Fox (1990b)
		Study 4		
Open forest/sand mining	1982, 1987, 1992 at 5-year intervals	House mouse New Holland mouse Common dunnart Brown antechinus Bush rat Swamp rat	Chronosequence; 17 1-ha plots on two mining paths, simultaneouslyy trapped at 5-year intervals, using Study 2 as control	Fox and Fox (1984); Twigg et al. (1989); Fox (1990b)

[a] The species are house mouse, *Mus domesticus*; New Holland mouse, *Pseudomys novaehollandiae*; eastern chestnut mouse, *P. gracilicaudatus*; common dunnart, *Sminthopsis murina*; brown antechinus, *Antechinus stuartii*; bush rat, *Rattus fuscipes*; swamp rat, *R. lutreolus*.

The studies listed in Table 1 are arranged in the order in which the results were published. First (Study 1) is a study of the small-mammal succession on heathland following disturbance by mineral sand mining (Fox and Fox, 1978). It is a chronosequence study that was resampled 16 years later, after the oldest sites had been rehabilitated for 26 years. Second, in Study 2 I review the succession in open eucalypt forest following fire, initially a chronosequence study covering regeneration from 0.1 to 9 years (Fox and McKay, 1981), in which some of the plots were monitored for 18 years, and the oldest plot was unburned for 26 years. Study 3 originally reported (Fox, 1982) 5 years of succession data from one 7-ha plot that had postfire regeneration monitored for 20 years, encompassing two fire cycles. The final example, Study 4, began as another chronosequence of small-mammal succession on forest plots that had been rehabilitated from 0 to 11 years (Fox and Fox, 1984); these plots were resampled twice more at 5-year intervals, with the first resampling forming the basis for a validation of the chronosequence technique (Twigg *et al.*, 1989). One review of this body of research, covering the longer term effects of mammalian succession in both forest and heath, was published previously, but it dealt mainly with describing changes in species composition as a function of postdisturbance time and showing the trajectories in "species space" followed by each community after fires (Fox, 1990a). A second review dealt with disturbance effects on a wider range of taxa than just small mammals (Fox, 1990b).

One focus in this overview is on 20 years of data from the small mammal community on the 7-ha site (SL1) first studied in 1974 (Study 3; Fox, 1981, 1982). This community comprises seven mammal species: two insectivorous marsupials, the common dunnart (*Sminthopsis murina*) and the brown antechinus (*Antechinus stuartii*); two native mice, the New Holland mouse (*Pseudomys novaehollandiae*) and the eastern chestnut mouse (*Pseudomys gracilicaudatus*); two native rats, the bush rat (*Rattus fuscipes*) and the swamp rat (*Rattus lutreolus*); and one introduced species, the house mouse (*Mus domesticus*). Fire is a common occurrence in this area, and the 7-ha site has been through two cycles of regeneration over the study period. The first followed a fire in August 1974 and provided 6 years of the small-mammal succession (Fox, 1982); the second cycle, begun after a fire in August 1980, was first analyzed after 8 years (Fox, 1990a) and provided 14 years of available data.

At this heathland site (SL1), the first period of regeneration demonstrated changes in relative abundance of species, but monitoring over the longer time period (14 years) produced additional species changes and the loss of early successional species over 3 years. These changes became apparent only with the extended monitoring period and demonstrate the importance of longer term studies; when I originally reported on the first 8 years of the second cycle, I concluded that it was quite similar to the first cycle (Fox, 1990a).

A second focus here is on the mammal community occupying an extensive (8000 ha) open eucalypt forest which abuts the heathland (over 3000 ha, and containing site SL1; see Fig. 1) in Myall Lakes National Park. This community commonly contains five of the same mammal species, although two species from wetter heath habitats, the eastern chestnut mouse and the swamp rat, are encountered only rarely. An initial 2-year study was based on 16 replicated 1-ha

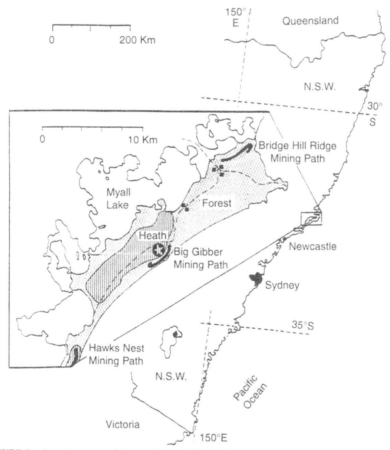

FIGURE 1. Location map of the study area showing forest and heath vegetation in the Euru-deree Embayment of Myall Lakes National Park; the mining path, to which control sites are adjacent, in heath at Hawks Nest (Study 1 in Table I); the location of the study plots (square symbols) in forest with heath understory (Study 2 in Table I); the main 7-ha heath site (SL1), shown as a star symbol (Study 3 in Table I); and the sand mining paths in forest at Bridge Hill Ridge and Big Gibber (Study 4 in Table I).

trapping grids in a chronosequence from areas of the forest that had regenerated after fire for five different lengths of time. Information produced from this chronosequence in the forest small-mammal community (Study 2) extended from 1 month to 9 years of succession (Fox and McKay, 1981). This eucalypt forest has been disturbed further by sand mining for heavy minerals, and a similar study has illustrated the small-mammal community response (Study 4) to this additional form of disturbance (Fox and Fox, 1984). It employs the same technique of chronosequence analysis, validated by retrapping the same grids 5 years later (Twigg et al., 1989). However, a considerable amount of information was collected from this forest from 1976 to 1994, and the oldest patch of forest had not burned from 1968 to 1994. I have information from the chronosequence analyses supplemented with data from long-term monitoring of these same sites.

II. STUDY AREA

Myall Lakes National Park is approximately 300 km north of Sydney on the east coast of Australia. Figure 1 shows the location of the park together with the distribution of major habitats and location of the study sites and mining activities within the park. A detailed description of the area follows.

A. Description

The area of the park used for the studies reported here is part of an embayment comprising an inner barrier of Pleistocene dunes and an outer barrier of Holocene high dunes situated between the beach and an extensive system of coastal lakes. The geomorphology of this embayment is typical of many such systems along the east coast and has been described by Thom et al. (1992). The older Pleistocene sand formations occur as parallel dunes, with a subdued topographic relief of 4 to 5 m, and are very heavily leached with currently low nutrient levels. These dunes support shrubby vegetation varying in height from 50 cm to 1 m in wet heaths and swamps, and up to 3 or 4 m in the drier heaths found on the tops of low dune ridges. By contrast, the sand from the Holocene dunes, although still relatively low in nutrients, may be an order of magnitude higher in nitrogen and phosphorus than the Pleistocene dunes. These younger dunes support an open eucalypt forest with trees 25 m high and 65% foliage projected cover (M. D. Fox, 1988). This open forest is relatively uniform, comprising a mixture of two tree species (*Eucalyptus pilularis* and *Angophora costata*), and has a history of frequent fires of either accidental or deliberate origins, as well as some prescribed burning intended for hazard reduction. The heath habitat is more heterogeneous, as changes in the topography over the parallel dune system and interbarrier lagoons produce a complex mosaic of dry heath, wet heath, and swamp vegetation. Maps detailing the pattern of these vegetation types have been produced by Myerscough and Carolin (1986), while the vegetation types on my main heath site have also been described in detail (Fox, 1981, 1982). The eucalypt forest may have either a heath understory or a forest understory as described and mapped by Fox and McKay (1981); I restrict the current analysis to that part of the forest with heath understory in order to emphasize comparisons with the mammal community in adjacent heathlands on the Pleistocene sands.

B. Mammal Trapping

A similar trapping protocol has been used in all of the studies reported here. A single Elliott aluminum small-mammal trap ($33 \times 10 \times 10$ cm) was placed at each trap station, with stations arranged on a 20-m grid spacing. Plots generally consisted of 25 traps in a 5×5 grid covering an effective trapping area of 1 ha (including a 10-m boundary strip). During the studies of regeneration following mining activities, the narrowness of the mining path at some places meant that a few plots had to be configured as 6×4 grids with 24 traps. The main heath study plot, monitored over 20 years, covered 7 ha comprising a complex mosaic of different habitat types (Fox, 1981, 1982). In the earlier studies the trapping period encompassed 4 consecutive nights, while the later studies utilized a 3-night trapping period since the 4th night was shown to provide only a small amount of additional data. In effect, the results from 3-night trapping periods differed little from those from 4-night trapping periods. A bait of peanut butter and oatmeal was used in all traps, and these were checked as early as possible each morning. Trapped animals were identified to species, individually marked, weighed, and standard reproductive information recorded from each capture. Most sites were trapped at the same time of year over the time period reviewed here; a few trapping episodes were conducted as a one-time effort covering all sites in a chronosequence. Hence I use the simplest measure of abundance, the number of individuals caught on the trapping plot, as this requires no assumptions or estimations. When chronosequence analyses were used, all (or as many as logistically possible) of the sites were trapped on the same nights. For analyses using data pooled from several sites, pooling was based on plot regeneration age class, and as such may cover a range of different years and sometimes different times of year (e.g., Fox and McKay, 1981; Twigg et al., 1989). When regeneration age classes were used, species abundance values were plotted at the center of the time span representing regeneration age class (age of habitat since disturbance).

III. CHANGES IN THE STRUCTURE OF FOREST UNDERSTORY WITH VEGETATION SUCCESSION

A. Post-Fire Succession

Fires in eucalypt woodland remove the understory vegetation and the smaller tree saplings but do not kill the overstory trees. As regeneration time advances, the general pattern of change in the structure of the understory is for the bulk of the leaves to move up into higher layers as the understory regrows, eventually exceeding the 2-m level. With understory regrowth, there is an increasing shading effect that reduces the amount of vegetation in each lower layer, with a subsequent drop in their contribution to each layer.

B. Post-Mining Succession

Because wave action concentrated many heavy minerals such as rutile, zircon, ilmenite, monazite, magnetite, and others along the strand lines of both present and ancient beaches, sand mining for heavy minerals has occurred on most of the beach and dune systems along the east coast of Australia. Mining activities are concentrated along narrow ribbons of highly concentrated ore bodies, and the heavy minerals are extracted from dunes up to 100 m high (in the case of Bridge Hill Ridge) using centrifugal and magnetic separation. First the vegetation is removed, and the topsoil cleared and stockpiled; then the mineral-bearing sand is processed as a slurry through a floating dredge in an artificial pond that moves with the mineface along the dune system. After the passage of the dredge, the homogenized sand tailings are recontoured behind it, the topsoil is respread, and a cover crop is sown with a small amount of fertilizer. A mix of seed intended to approximate the premining vegetation, along with hand-planted eucalypt trees and banksia shrubs from tube stock, begins the rehabilitation process. Because of the severity of the disturbance, involving a reconstitution of the entire soil strata, the succession following disturbance is more akin to a primary succession than to the secondary succession following fire. The vegetation and animal succession continues along the ribbon-like mining path with a documented history, thus offering an ideal outdoor laboratory for the study of the successional process using chronosequence techniques.

On regenerating mined paths, a cover crop of *Sorgum halepense* is used to reduce topsoil loss through erosion; subsequently, an initial flush of vegetation, almost a monoculture of *Acacia longifolia*, occurs as this early colonizer rapidly sequesters most of the nutrients available from the fertilizing of the cover crop. However, native species that have evolved on the low-nutrient soils of Australia may suffer phosphorus toxicity; *Acacia longifolia* is one such species, and it usually dies out after 4 years. This leads to a substantial drop in the biomass of vegetation before other native species take over the plant succession on sites where nutrient excess is no longer a problem (Fox and Fox, 1978, 1984).

Fox and Fox (1984) studied understory regeneration and mammal recolonization on sand-mined areas previously supporting open forest in Myall Lakes National Park; this study illustrated the way in which the structure of the understory vegetation changes with the vegetation succession. They used a measure of understory vegetation termed "horizontal cover index," determined as the percentage of a 50 × 20-cm coverboard that is obscured by vegetation at a distance of 5 m, and also presented data for a "vegetation index" which is measured using a light meter to assess the attenuation of light passing vertically through the vegetation (see Fox, 1979). Both techniques measure vegetation indexes for each of the layers at heights of 0–20, 20–50, 50–100, 100–150, and 150–200 cm as well as total values for the vegetation below 2 m. The vertical method is used by Fox and McKay (1981) for the postburn response of forest understory (used as controls by Fox and Fox, 1984) and by Fox and Fox (1978) for mined heathland. The coverboard method, although less objective, became more favored because it does not require the uniform sunshine conditions needed for the light meter method.

Evaluation of comparable data for burned forest (Fox and McKay, 1981) and mined forest (Fox and Fox, 1984) reveals less understory on mined forest sites (total index = 0.30 at 5 to 7 years) than on burned forest sites (total index ranging from 0.65 at 5 years to 0.45 at 7 years as understory subcanopy grows above 2 m). Both methodologies indicate that the total amount of understory vegetation on mined forest sites remains relatively constant or increases only slightly from 4 to 9 years, with the slight hint of a decrease from 9 to 10 years (Fox and Fox, 1984). This contrasts with the marked decrease from 4 to 8 years for understory on burned forest sites (Fox and McKay, 1981) and the increase from 5 to 11 years observed for mined heath sites (Fox and Fox, 1978).

IV. MAMMALIAN RESPONSES TO DISTURBANCE BY FIRE

In Australia it is easy to understand why fire is the type of disturbance most often studied. Fire is of major importance in most Australian ecosystems, and both plants and animals must cope with fire to occupy successfully their respective environments. Fire must be considered an endogenous disturbance, a natural part of the system (B. J. Fox and Fox, 1986). Fire regimes (Gill, 1975) are described by a number of component variables, such as frequency, intensity, seasonality, areal extent, and time interval from the previous fire (M. D. Fox and

Fox, 1987). Frequency is a measure of how many fires occur in a specified period, and different fire frequencies can have an important impact on mammal species by producing marked changes in the structure and floristics of the understory vegetation (M. D. Fox and Fox, 1986). Fire frequency is generally related to a second component, that of fire interval, since a high frequency usually means short fire intervals on average. More important than the average fire interval are its maximum and minimum values, as these have the greatest biological impact, particularly for plant species that require a minimum time for seed production. On the other hand, maximum times are a real factor in the viability of seeds in the soil seed bank, and in the longevity of resprouting species that senesce and can die out. In particular the penultimate fire interval seems to be of considerable importance to regeneration pattern and may be a key variable in understanding how fire interacts with plant and animal species (B. J. Fox, unpublished data).

Fire intensity is determined by the amount of available fuel, the heat yield of the fuel, and the rate of spread of the fire and is an important aspect to the impact of fire on plants and animals. Most obvious in this respect is the seed hardiness of species of *Acacia* (such as *A. suaveolens*) and other genera that will germinate only after the passage of a fire (Auld, 1986). The seasonality of fire is another component of the fire regime with confounding effects; clearly fires in summer will be more intense than those in winter, because higher temperatures and lower humidities greatly increase the heat yield of fuel and the rate of spread of fire. There are also additional effects relating to the breeding seasons of both plants and animals, so that timing of a fire relative to reproductive phenology will influence the severity of its impact on the biota. Further, the area burned by a fire is important because of problems associated with dispersal and recolonization, and it seems reasonable that the more extensive a fire the greater will be its impact (M. D. Fox and Fox, 1986).

This brief review of the components of a fire regime indicates that a considerable amount of interaction among them is expected and that different components are often confounded one with another such that it may be difficult to isolate single effects. The term *fire regime* encompasses a variety of component variables, and its use conveys the complexity involved with fire and its consequences in natural systems.

Fire produces substantial effects on the scleromorphic vegetation of forests and heathlands of Australia. The recovery of the plant and animal communities from this disturbance has been documented in *Fire and the Australian Biota,* a key reference to fire research in Australia (Gill et al., 1981). Here papers on fire history, the physical phenomenon of fire, responses of the Australian biota and of selected ecosystems, and the role of fire in ecosystem management are presented. Most of the published research is on plant species or communities, but there has been some work on animal responses, and some of this has been long term. Andersen and Yen (1985) studied the immediate effects that fires have on ant communities in semiarid mallee vegetation, demonstrating the great resilience of ants to disturbance by fire. York (1994) has reported on a long-term study of ant communities that demonstrated gradual reductions in species richness in forests that remained long unburned; he used both chronosequence techniques for sites regenerating from 2 to 14 years as well as the monitoring of permanent control plots over a 7-year period.

The ground parrot (*Pezoporus wallicus*) is a fire specialist species that reaches maximum abundance at 5 to 10 years postfire in shrub heath and 8 to 17 years in graminoid heath and is completely absent on sites unburned for 28 years (Meredith *et al.*, 1984). Animal responses to fire have been reported from the Northern Territory for small mammals (Begg, 1981) and birds (Braithwaite and Estbergs, 1985, 1987). The results of long-term studies on the northern quoll (*Dasyurus hallucatus*) reported by Braithwaite and Griffiths (1994) indicate that range contraction and demographic variation are among the consequences of fires. Masters (1993) surveyed sites in spinifex grassland at Uluru from 1987 to 1990 and found a greater species richness of small mammals in areas that had burned in 1976 compared to more recently burned (1986) sites, and she emphasized the importance of fire as a management tool.

Long-term research on the effects of fire on small-mammal communities has been conducted in the forests and heathland of Nadgee Nature Reserve (Newsome *et al.*, 1975; Newsome and Catling, 1979, 1983; Catling and Newsome, 1981; Catling, 1986). This area lacks *Pseudomys* species, which have been identified as fire specialists (Cockburn, 1981; Fox, 1982, 1983; M. D. Fox and Fox, 1987). Catling (1986) found a substantially different succession of small mammals following fire when *Pseudomys* was absent and the swamp rat (*Rattus lutreolus*) colonized much earlier; in a nearby forest Lunney *et al.* (1991) investigated the combined effects of fire and logging disturbance on small mammals.

Elsewhere in the world fire is generally less important, except in some specific ecosystems. A review of the role of fire in ecological systems (Trabaud, 1987) contains papers on European, North American, and Australian research. Most fire research is conducted in fire-prone regions such as those with summer drought and Mediterranean-type climates. The Mediterranean basin has had a long history of fire affecting its biota (Naveh, 1975; Trabaud, 1987). Long-term studies of fire effects on arthropods and birds as well as small mammals have been undertaken from the field station at Banyuls near the border between France and Spain (Prodon *et al.*, 1987). In this study, large areas have been burned according to an experimental timetable and the outcomes have been monitored; similar work on postfire succession for small mammals has been conducted in other areas nearby (Arrizabala *et al.*, 1993; Fons *et al.*, 1993). In a similar habitat in California, Mills (1986) studied the impact of insect and mammal herbivores on the postfire plant succession. Small-mammal succession following chaparral fires has been studied, with several papers identifying the sequence of species in the regenerating vegetation (Quinn, 1979; Wirtz, 1982; Wirtz et al., 1988).

In South Africa the importance of fire may be similar to its importance in Australia, as attested by a growing literature. A major review of this corpus (Booysen and Tainton, 1984) is comparable in scope to that of the Australian work (Gill et al., 1981). Willan and Bigalke (1982) have studied the small-mammal succession in montane fynbos, and in a short-term study Midgeley and Clayton (1990) observed only a small change in species composition over the first 8 months following a fire. A comparison of the small-mammal succession following fire in Australia, South Africa, and California indicated that differences were related largely to differences among continents in productivity and the nutrient levels of the soils and vegetation (Fox et al., 1985).

A. Forest

Our initial study of the mammalian succession following fire in open eucalypt forest at Myall Lakes National Park reported changes in relative abundance of species over a 9-year chronosequence in sites that were trapped in 1976 and 1977 (Fox and McKay, 1981). This study documented a replacement sequence of species' dominance relative to regeneration age, with introduced house mouse and native New Holland mouse as the early successional species, common dunnart increasing in abundance in mid seral stages, and bush rat and brown antechinus most abundant in the late seral stage. Most information was available for the bush rat; Fox and McKay (1981) used stepwise multiple regression and path diagrams to show that leaf litter explained 43% of the variance in bush rat biomass, and this figure increased to 50% when analysis was restricted to sites of the same regeneration age, with an additional 48% explained by variables quantifying vegetation structure. Based on the information then available, it was concluded that Bush rat biomass reached a plateau of 0.68 kg ha^{-1} (Fox and McKay, 1981) in parallel with leaf litter reaching a steady-state value asymptotically. Additional information from some of the same sites was presented by Thompson (1983), who trapped these and some new sites 6 years later. His data indicate that, to the contrary, the abundance of the bush rat had decreased (Fox, 1990a). Further monitoring has confirmed a continued decrease (Fig. 2a) up to a 20-year regeneration age, but there is an indication of an increasing trend in bush rat abundance in the oldest unburned plots monitored through 25.5 years of regeneration.

The abundance of the New Holland mouse also has shown marked changes as the forest understory has aged since the fire. Its maximum abundance was

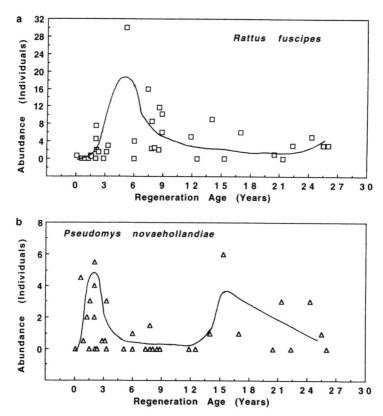

FIGURE 2. Abundance (individuals) as a function of regeneration age for all 1-ha study plots surveyed in open forest with heath understory in Myall Lakes National Park from 1976 to 1994 (curves fitted by eye). (a) Bush rat (*Rattus fuscipes*); (b) New Holland mouse (*Pseudomys novaehollandiae*).

reported for early seral stage (1.5-year age class), whence it decreased toward the mid seral stage (4-year age class), with a possible increase at the 8-year age class (see Fig. 6b in Fox and McKay, 1981). However, additional data from extended monitoring support the trend of an increasing abundance of the New Holland mouse in later seral stages with an apparent peak at 16 years (Fig. 2b), perhaps followed by another decrease in the latest age class (see Fig. 2b and Fig. 3b), somewhat unexpected in a species previously regarded as an early successional opportunist. The changes in abundance for these two species appears to be out of phase, most likely due to their different habitat requirements. Population density in the New Holland mouse has been shown to be correlated positively with bare ground and vegetation cover at 50 cm, negatively with cover at 10 cm and with total vegetation (Posamentier, 1976), and positively with floristic variables (Fox and Fox, 1978) and again with both structural and floristic variables (Fox and Fox, 1981). All of this points to a species which appears to select bare ground and vegetation open in the lower layers (up to 20 cm), typical of early seral stages. On the other hand bush rat abundance has been shown to correlate positively with the amount of leaf litter and with dense vegetation both below 20 cm and also in the shrub layer from 50 cm to 1 m (Fox and Fox, 1981). The structure of the forest understory has changed markedly over the monitoring period (1976 to 1994), and these changes appear to be driving variations in abundance and habitat use by these species. In the oldest plot (ML6) the understory was dense at 9 years, but as regeneration age increased shrubs grew taller, some even reaching into the canopy, while other shrubs began senescing, thereby opening up the vegetation to more light. In effect, the understory vegetation succession from ground level to 2 m, with the exception of the amount of leaf litter present, progressed through dense to more open stages similar to the earliest postfire physiognomy, so that it is not surprising to see the mammalian succession also "regressing." This very interesting outcome would not be recognized in short-term studies and could be identified only through long-term monitoring at permanently established study sites.

The changes in mammal abundances have been traced on three individual plots (ML2, ML5, and ML6), shown as solid lines in Fig. 3, while a chronosequence of three sites monitored in November 1982 is shown as a dashed line. For the bush rat (Fig. 3a) this approach demonstrates that data from chronosequences are compatible with data obtained from the continued monitoring of specific sites over time. Identification of successional trajectories for the same three sites and the successional trajectory for a chronosequence monitored in November 1982 for New Holland mouse confirm a marked bimodality in abundances, at least for this species (Fig. 3b).

An overview of the successional changes in populations of the five species of small mammals encountered in the forest regeneration is gained from pooling the data presented in Figs. 2 and 3 into regeneration age classes. For ages up to 9 years I have used 1.5-year classes (as did Fox and McKay, 1981), whereas 3-year classes were used over the later period of succession (Fig. 4a). I have flagged one point in Fig. 4a where only a single value is available for that age class, and the value is markedly greater than those from immediately before and after. In the absence of a corroborative value in this age class I have plotted the value as a three-point smoothed average. The dominant position of the bush rat in this community is apparent for all classes except the earliest class and a few of the later classes. The abundance of the other species is much lower, and to counterbalance the effects of differing relative abundance on the graphical presentation of the data I have redrawn the figure (as Fig. 4b), where

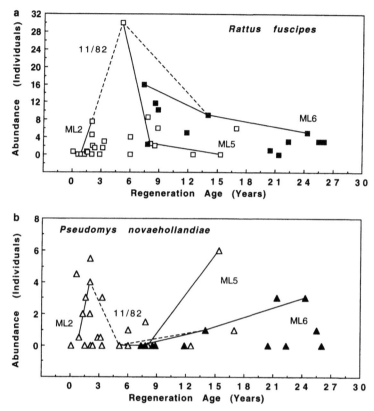

FIGURE 3. The same data as in Fig. 2, but (1) all values for the oldest study plot (ML6) are shown with solid symbols; (2) plots sampled at the same time of year but in different years are joined by solid lines (ML2, ML5, and ML6); and (3) the chronosequence for these same three sites (all sampled in November 1982) is indicated with a dashed line. (a) Bush rat (*Rattus fuscipes*); (b) New Holland mouse (*Pseudomys novaehollandiae.*).

the abundance of each species is expressed as a percentage of its maximum value. This method of presentation clearly illustrates multiple peak abundances for the New Holland mouse and brown antechinus, and indicates also that the bush rat appears to be increasing toward a second peak in late succession. The house mouse clearly has only one peak and although an opportunist species it does not appear to be able to reinvade the forest succession when the understory opens up after 2 decades. However, this interpretation should be tempered with caution as, from my experience in this area and many others on the east coast, the presence of *Mus domesticus* at any site is closely linked to its abundance in the region as a whole, rather than being determined largely by site-specific features. Such variations in regional abundances are controlled

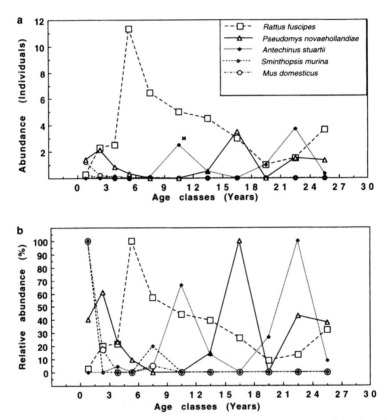

FIGURE 4. (a) Data from the mammal surveys of 1-ha plots in open forest with heath understory are pooled into age classes and positioned at the center of each age class for each of the five species encountered. One value (indicated with a cross) represents a single value for that age class; as it was markedly greater than those immediately before and after, a three-point smoothed average was included. (b) Data from (a) are transformed to relative abundance by scoring each species as a percentage of its maximum value.

apparently by climatic and weather factors, which trigger plague behaviour in the wheatbelt and are reflected also in high *Mus* abundance in other regions from which the species may be all but absent at other times. Saunders and Giles (1977) have suggested that for three agricultural regions of southeastern Australia over the period from 1900 to 1970, the cause of *Mus* plagues has been drought conditions which preceded each outbreak. This is contradicted by the later findings of Chapman (1981) who suggested that abundant rainfall in the preceding two years was the cause of the plague he observed in 1975. His conclusion is consistent with the findings of Newsome and Corbett (1975), that successive years of good rainfall result in *Mus* plagues in central

Australian desert areas, and also provides further support for Newsome's (1969) hypothesis that *Mus* plagues occur when summer rains ameliorate seasonal drought in the Mediterranean-type climate region of South Australia, so that both food and shelter become available simultaneously. Redhead (1988) suggested that good autumn rainfall triggers mouse plagues by extending the breeding season and increasing the rate of survival of mice over winter, leading to a peak abundance in the following autumn so that the recommencement of breeding in the following summer builds into a plague. However, Singleton (1989) concluded that one rainfall event was unlikely to be a good predictor and observed that, for the plague considered by Redhead (1988), the autumn rain was preceded by severe drought and then higher than average winter and summer rains. Singleton (1989) succeeded in incorporating the proposed mechanisms of both Saunders and Giles (1977) and Redhead (1988) into one hypothesis, as reviewed by Caughley *et al.* (1994). Such studies demonstrate not only the need to take into account the regional abundance of species when interpreting local patterns but also the importance of long-term studies which can provide the necessary information to test hypotheses and models.

Despite its name, common dunnart is rarely common, and my experience indicates that it has been quite rare in this east coast region at times over the past two decades. It was reasonably abundant in Myall Lakes National Park from 1975 to 1981, reaching densities of up to 6 ha^{-1} in at least two areas of the forest during 1979–1980 (B. J. Fox, unpublished data). It was recorded on sand-mined areas in 1983–1984 (Fox and Pople, 1984) and was again caught in reasonable numbers in the heathland in 1989; however, an extensive trapping program specifically for this species in 1990–1991 revealed it to be at extremely low numbers, and it was not until 1994 that it became reasonably abundant again on heathland sites (V. Monamy, personal communication). Only in the last year or two has it appeared in reasonable numbers in the suitable early to mid seral stages of its preferred habitat. Thus in considering its absence during the 10- to 25-year regeneration age classes in this forest one must take its low regional abundance into account before concluding that this species also is not destined to reappear in the succession. Extensive trapping at Tomago (another study area farther south), on a similar embayment just north of Newcastle (see Fox *et al.*, 1993) during 1990–1995, indicated similar results.

Climatic and regional effects can also have important bearing on the breeding phenology, as well as on the presence or absence, particularly with reference to the length of the breeding season. This is exemplified by our long-term studies at Myall Lakes and in the Tomago region. The observation in early May 1992 of breeding in 63% of female New Holland mice on the Hawks Nest mining path (between Myall Lakes and Tomago; Study 1 in Table I) came as somewhat of a surprise for a species that is normally expected to breed in spring and summer from August to early January, with an occasional extension to March. Examination of trapping records from 1972 to 1992 revealed that extended breeding occurred only when rainfall exceeded a threshold level with a particular distribution in the early and later parts of the normal breeding season. Our results demonstrated unequivocally that under appropriate, specific climatic conditions this species could breed for 10 months of the year (Fox et al., 1993). They have a direct bearing on the hypotheses put forward by Redhead (1988, see above), and could be drawn only because of two long-term studies from replicated sites in two separate embayments, each with an independent weather station.

B. Heath

Studies began in the complex mosaic of wet and dry habitats that constitute the 7-ha coastal heathland plot (SL1) following an intense wildfire. Four years of monthly or bimonthly mammal trapping data in the heath were collected and published together with 3 months of prefire data (Fox, 1982). A single trapping period from the 5th year of regeneration at the same time of year as the prefire data was also reported (Fox, 1982). These trapping results were compared to those from a 4-ha unburned control plot (SL2), with a similar range of vegetation but separated from SL1 by a gravel road. The density of mammals on the burned plot reached levels similar to those in the unburned control toward the end of the 2nd year postfire (Fox, 1982, p. 1335), supporting the view that this small-mammal community was able to respond quickly to habitat perturbation by wildfire. As the time of the previous fire was not known, the site was arbitrarily aged at 6 years postfire when it was burned. Diagrams of species' relative abundance as a function of regeneration time were presented for wet and dry heath habitats separately (Fox, 1982, p. 1339) and discussed in relation to a "habitat accommodation" model for animal succession. The observed succession in both major habitat types (wet and dry heath) followed a replacement sequence in dominance by the small-mammal species. Both species of Pseudomys were seen clearly as early successional and could be regarded as fire or disturbance specialists because their postfire abundances were substantially increased over the prefire values. In contrast, the house mouse was seen as more of an opportunist species than a fire specialist. Of the two species of insectivorous marsupials, common dunnart appeared to be a mid successional species and brown antechinus a late seral species, while both species of Rattus were classified as late successional (Fox, 1982).

The postfire mammal succession in this heath shrubland has been compared to equivalent shrublands in California and South Africa (Fox et al.,

1985). In the South African postfire fynbos, the small mammal fauna involved in the succession was markedly richer in species (a total of 12 in all, with a maximum of 9 in the 2nd year) than those in either Australian heath (a total of 7 species, all present most of the time) or Californian chaparral (a total of 8 species with a maximum of 7 species most of the time). A second comparison questions whether postfire changes involve species turnover (the actual replacement of species over time) or less drastically a replacement sequence of relative dominance. Using 1974–1980 data, the Australian data appeared to differ markedly from those of the other two regions and to illustrate the latter rather than the former scenario (Fox et al., 1985). In contrast, the South African and California results demonstrated a clear species replacement, more marked for South Africa where data from 25-year-old sites were available. When considered in a multivariate analysis, the Australian heathland sites supported fewer total species in the succession with low densities but high equitability, while the California sites had both high total species counts and high density with low equitability, and the South African sites had low total species and density, like California, and a range in equitability values from low to high. The similarities and contrasts were illuminated by discriminant function analysis showing that Australia differed significantly from both California ($P = 0.037$) and South Africa ($P = 0.002$), but the latter two were not significantly different ($P = 0.356$; Fox et al., 1985). Nutrient limitation, ex hypothesi, likely plays an important role in these observed differences.

Twenty years of data for the Australian site cover a second fire event and 14 years of its postfire regeneration. Mammal abundance data are presented in Fig. 5, in which marsupial species and the rodents of wet and dry habitats are graphed separately. Several interesting points emerge from this figure. First, the early successional species characteristic of wet heath habitats, eastern chestnut mouse, appears to be heading toward local extinction on the SL1 site (Fig. 5a). This interpretation is supported by observations from other heathland sites of advanced regeneration age from which the eastern chestnut mouse is absent or present at very low abundance (Higgs and Fox, 1993; V. Monamy, personal communication). The mouse apparently loses competitive dominance as succession progresses; reciprocal manipulation experiments (Higgs and Fox, 1993; Thompson and Fox, 1993) have demonstrated that its decline is directly attributable to asymmetric interspecific competition with the swamp rat. Furthermore, the interaction explains a continued increase in abundance of swamp rats (Fig. 5a).

Second, the presence of common dunnarts was recorded intermittently throughout the census period, but dunnarts were absent from the site the last 3 years (Fig. 5b). While I have already indicated that regional rarity of common dunnart over the last decade is a cautionary consideration, I am confident that its later absence here is directly associated with site regeneration age, an

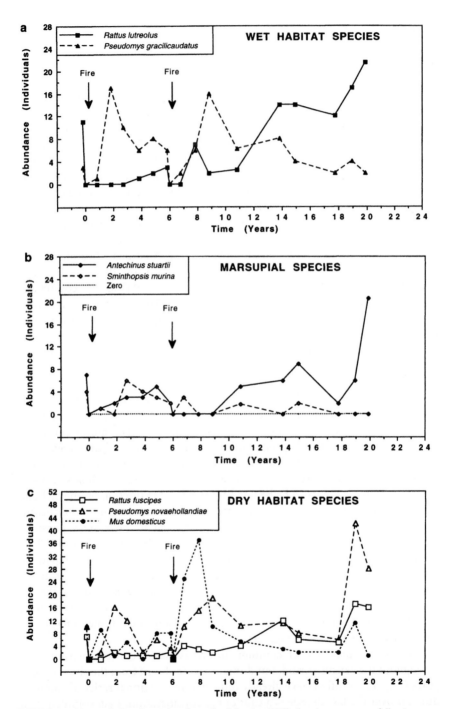

FIGURE 5. Mammal abundance data from the 7-ha site (SL1) showing responses following two fires (August 1974 and August 1980), with the zero time scale set at the date of the first fire. The mammals are divided into (a) wet habitat species, (b) marsupial species, and (c) dry habitat species.

interpretation supported by trapping results over the last 5 years at several heathland sites, 50 m to 2 km distant, that are similar in all respects except regeneration age. Common dunnart has been recorded from several of these sites; specifically, six common dunnarts were caught in July 1994 at one site (5 years postfire) of eight separate 0.5-ha plots trapped over the same 3 nights as the nearby 14-year site (V. Monamy, personal communication). These observations strengthen the inference that the common dunnart is now locally extinct at the older site, and its demise there most likely results from changes in the vegetation structure associated with advanced regeneration age. Earlier Fox and Fox (1981) used stepwise multiple regression to show that dunnart abundance at SL1 was sensitive to vegetation structure, with 46% of the variance in common dunnart abundance explained by habitat variables, and floristic variables improving the figure only a further 1%.

Third, the abundance peaks of New Holland mouse in the dry heath habitats are shown in Fig. 5c. During the second fire cycle, New Holland mouse showed an early peak 3 years after the 1980 fire and another peak 13 years postfire. House mouse also showed two peaks in the second cycle, the first in 1982, 2 years postfire and 1 year ahead of the peak in New Holland mouse (as in the first fire cycle), although the two mice are in phase at their second peaks. Reciprocal manipulation experiments have shown that these two mice are competitors (Fox and Pople, 1984; Fox and Gullick, 1989). New Holland mouse enters the succession at Year 2 when the habitat meets its requirements (as has been demonstrated by transplant experiments; Fox and Twigg, 1991), and thence it competitively excludes the house mouse. The two species follow well the habitat accommodation model for animal succession proposed originally on the basis of postfire heath studies (Fox, 1982) and extended with the later work described above (Fox, 1990b).

The relationship between *Pseudomys* and *Rattus* species in the dry heath habitats is less clear than that of their congeners in wet habitats. The New Holland mouse, with an adult body mass of 18 g, is an order of magnitude smaller than the 150–200 g bush rat, so that direct competition would seem unlikely. The bush rat is more abundant in forest than heath, while the New Holland mouse reaches its greatest abundance in heath (see the previous discussion about their different habitat requirements). Note that, as in the eucalypt woodland, both dry heath species tend to increase in the later stages of postfire regeneration.

One advantage from the long-term monitoring of the SL1 heathland is the ability to replicate over successive fires the responses of each species relative to the regeneration age of the habitat. Figure 6 plots on a single graph species' abundances as a function of time for both fire cycles, enabling direct comparisons of responses for two 5-year postfire periods, 1974–1979 and 1980–1985. Both *Pseudomys* species show reasonable agreement between fire cycles (Figs. 6a and 6b), although in each case the peak for the first cycle occurs at 2 years

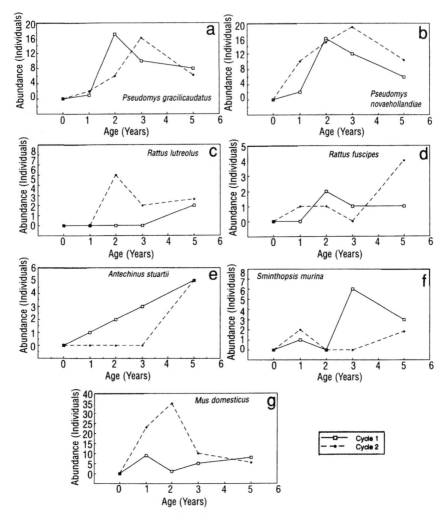

FIGURE 6.   The response of each small mammal species over the first 5 years of two consecutive fire cycles as shown in Fig. 5. Each species is identified on the graph.

and for the second cycle at 3 years. This offset in peak abundance may reflect differences in weather, primarily rainfall and temperature, that influence the rate at which the vegetation regenerates and which, in turn, determines the pace of the mammal succession. In the habitat accommodation model for animal succession, emphasis is placed on the response of animal succession to the changing vegetation, and not to time per se (Fox, 1990b).

    For later successional *Rattus* species, differences in responses to the two fire cycles are more apparent (Figs. 6c and 6d) and include earlier increases during the second cycle. In fact there was more rain in the 2nd and 3rd years of the second regeneration cycle, producing a much more rapid vegetation regrowth than in the first cycle and denser vegetation in the first meter above ground level, particularly in the area of swamp forest on one edge of SL1 (vegetation type 7 of Fox, 1982). The trap stations in the swamp forest are excluded from this analysis of heath succession, but the two individuals of swamp rat caught in swamp forest would increase the species' abundance in Year 2 from five to seven.

The dasyurid marsupials also showed considerable differences between fire cycles, but overall regeneration rates were generally slower, not faster, in the second cycle (Figs. 6e and 6f). Common dunnarts recovered marginally faster in the 1st year of the second cycle, but at a markedly slower rate after that. The differences between fire cycles in the marsupial responses and in house mouse (Fig. 6g) were already noted by Fox (1990a, p. 324); as discussed above, the abundance of both house mouse and common dunnart at this site are likely to be influenced by regional in addition to local factors related to the two fire cycles.

V. RESPONSES TO DISTURBANCE BY SAND MINING

A. Nonmammalian Fauna

Bradshaw and Chadwick (1980) reviewed the way in which areas of land have been rehabilitated following mining, while Majer (1989a) focused particularly on the role of fauna. In Australia, Greenslade and Thompson (1981) investigated the way in which ant communities recolonize sand-mined areas in coastal dunes in Queensland. Jasper et al. (1989) found that vesicular arbuscular mycorrhizal infectivity was greatly reduced during bauxite mining and that this may limit the growth of some species. This would then impact animal species recolonization as well. The return of both vertebrate and invertebrate animals in these bauxite mined areas of Western Australia have been studied by Nichols et al. (1989), while Greenslade and Majer (1993) looked specifically at the recolonization of Collembola on rehabilitated bauxite mines. Long-term studies on such recolonization by fauna have been undertaken by Majer (1989b, 1990).

B. Mammalian Fauna: Forest

Although both the Bridge Hill Ridge and Big Gibber mining paths pass through vegetation that was originally open forest, for the first 10 to 12 years they regenerate as shrubland or heath. Eventually hand-planted eucalypt saplings begin to form a canopy, after which the regeneration proceeds as a forest. The mammal succession on these regenerating mining paths was studied, and the chronosequence assumptions were tested and validated. Mammal abundances from 15 plots were pooled into eight age class categories following the protocol of Twigg et al. (1989) and graphed for each species separately. These data, from trapping sessions in 1982, 1987, and 1992, are shown in Figs. 7a (house mouse), 7b (New Holland mouse), and 7c (other mammal species).

For the house mouse, both the 1982 and 1987 data show a similar timing of early peak abundance, at 3.7 years and 3.4 years, respectively, and a minor, secondary peak at 13.8 years in 1987 (similar to the later, secondary peaks seen during postfire regeneration in New Holland mouse, see Figs. 2 and 3). The 1992 data, reflecting later regeneration ages, show a peak value at 15 years (Fig. 7a), quite close to the 1987 secondary peak. For New Holland mouse, peak abundances in the 1982 and 1987 surveys occur in shrubby vegetation at regeneration ages of 8.8 years and 8.7 years, respectively (Fig. 7b). The 1992 survey was conducted when an emerging eucalypt canopy had developed at most sites, and the peak abundance was recorded at 19 years postdisturbance (Fig. 7b).

Three other species were recorded during the succession. Common dunnart was the most abundant of these; although the 1982, 1987, and 1992 surveys recorded 0, 3, and 1 animal, respectively, a total of 16 animals were caught over six trapping periods from November 1983 to February 1984 as part of a study of competition between the house mouse and the New Holland mouse in the succession (Fox and Pople, 1984). These supplementary dunnart data are added to Fig. 7c as a separate line. In addition, bush rat occurred on just one site, at 16.6 years postdisturbance, and brown antechinus was trapped only on sites from 15 to 20 years postdisturbance and appeared to be increasing in abundance (Fig. 7c). Thus all five of the species encountered in open forest following fire now have been at least recorded on the regenerating mining paths, in locations that imply they are resident and breeding and in vegetation of regeneration ages consistent with what has been found in open forest, postfire succession.

FIGURE 7.   Mammal abundance from a chronosequence of 1-ha plots along a regenerating sand mining path in open forest pooled into age classes and shown for three separate trapping periods (1982, 1987, and 1992). (a) House mouse, *Mus domesticus;* (b) New Holland mouse, *Pseudomys novaehollandiae;* and (c) for 1992 trapping period with dashed lines: bush rat, *Rattus fuscipes* (open squares); brown antechinus, *Antechinus stuartii* (solid diamonds); and common dunnart, *Sminthopsis murina* (open circles). A separate set of data for common dunnart collected during 1983–1984 is shown with a separate line.

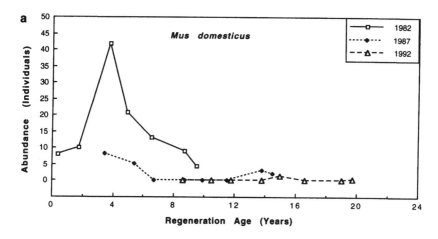

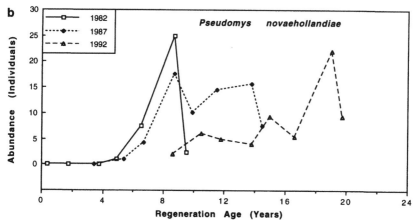

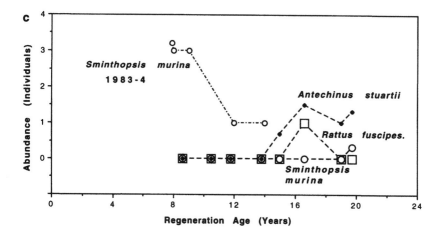

With three sets of chronosequence surveys available, it is possible to construct trajectories for individual plots; this has been done for five plots in Figs. 8a–8e. Inspection of these sites in order clearly shows the successive waves of abundance in house mouse and New Holland mouse, as the species sweep along the mining path. For the youngest site (BHR1, Fig. 8a) the house mouse followed a convex decreasing trajectory over the three sampling periods (1982, 1987, and 1992), while on the same site over the same period we see the New Holland mouse following a concave increasing trajectory. At an older site (BHR7, Fig. 8b) we see that the house mouse follows a concave decreasing trajectory over the three sampling periods and that the New Holland mouse follows a peaking convex trajectory, increasing and then decreasing. At the first of the Big Gibber sites (BG4, Fig. 8c) the house mouse has low abundance while the New Holland mouse decreases markedly from a plateau. By the next

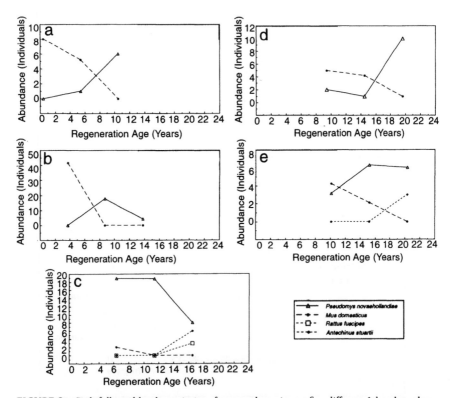

FIGURE 8. Path followed by the majority of mammal species on five different 1-ha plots along the Bridge Hill Ridge and Big Gibber mining paths (see Figs. 1 and 7) with values shown for the three trapping periods: 1982, 1987, and 1992. (a) Plot BHR1, (b) Plot BHR7, (c) Plot BG4, (d) Plot BG7, and (e) Plot BG8.

stage (BG7, Fig. 8d) New Holland mouse is again following a concave increasing trajectory, and the house mouse a convex decreasing trajectory. The final example (BG8, Fig. 8e) shows a convex increasing trajectory for the New Holland mouse and a linear decreasing trajectory for the house mouse, while the brown antechinus exhibits the beginnings of an increase.

These observations provide additional support for the consistency of the succession and roles different mammal species play in it. The role of the house mouse is that of the earliest species in the mammalian succession, with high dispersal ability and a high reproductive rate. The New Holland mouse follows it in every case, with reduced dispersal and reproductive rates but superior competitive abilities (as demonstrated by reciprocal removal experiments, Fox and Pople, 1984; Fox and Gullick, 1989). Note, however, that its role would have been that of the earliest successional species before the arrival of the introduced house mouse. At this stage, the role of mid successional species is beginning to be undertaken by common dunnart, although there are signs that brown antechinus is just starting to increase. The role of late successional species is taken by the bush rat in forest regenerating after fire, but is not yet defined on the mined sites since the vegetation there does not yet provide the necessary habitat requirements for the bush rat.

C. Mammalian Fauna: Heath

Chronosequence analysis of the Hawks Nest mining path (Fox and Fox, 1978) was the earliest of the studies forming the basis of this chapter. In the 1978 paper we predicted it would take 20 years for the abundance of New Holland mouse at this site to reach those of the control plots. When regenerating sites along this path were resampled in 1992, the youngest of the sites was 20 years and the oldest site was 26 years; the recent data, placed in the same age class categories used by Fox and Fox (1978), are shown in Fig. 9a. The abundance values of New Holland mouse in undisturbed control plots are included, and the intervening period between sampling times is shown as a dashed line. Results from individual sites (rather than regeneration age classes) are shown in Fig. 9b. The regression line of the earlier (1978) equation predicting population increases in New Holland mouse with regeneration age is shown (as Abundance = $-2.52 + 0.85 \pm 0.37 \times$ Regeneration age; $n = 20$, $r^2 = 0.22$, $P = 0.04$), and can be compared to the regression line incorporating the 1992 data (Abundance = $-0.087 + 0.56 \pm 0.185 \times$ Regeneration age; $n = 25$, $r^2 = 0.28$, $P = 0.006$). There is no significant difference between the slopes of the two regression lines. Trajectories through time at two specific sites (HN3 and HN7) demonstrate that the sites actually did move over time in the directions indicated by the regressions. Equivalent results for the house mouse are shown

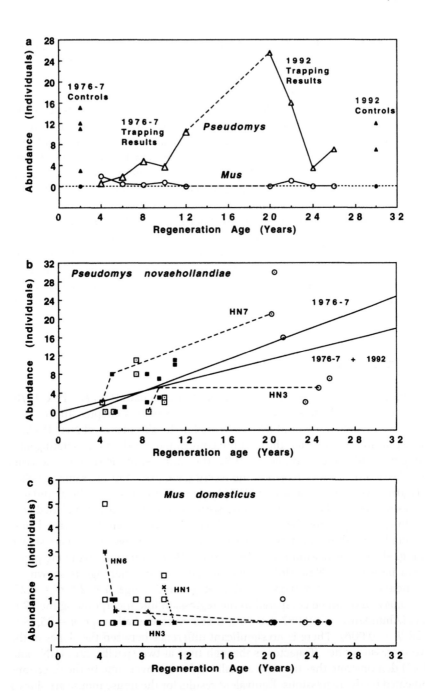

in Fig. 9c, in which the declining abundance levels suggested by the earlier data are confirmed by the later trapping results. In this case trajectories are shown for three sites where there were two replicates at each site and the crosses shown are mean values for these sites (HN1, HN3, and HN6).

VI. THE RELATIONSHIP BETWEEN MAMMALIAN AND VEGETATION SUCCESSION

Although these successional studies relate mammalian responses to a time scale, I emphasize that the mammal populations are responding not to time per se but rather to successional changes in the vegetation. The mammalian succession is appropriately regarded as an externally induced succession and is so represented in the habitat accommodation model developed for postfire mammal succession (Fox, 1982), extended to postmining succession and further applied to ant succession (Fox, 1990b). In this model, species enter the succession when local habitats meet their specific requirements, but as conditions move beyond species' optimal ranges competitive ability falls, their dominance in the succession is lost, and they are replaced by species whose competitive abilities have been enhanced by the changes in habitat corresponding to the ongoing vegetation succession.

The relationship between the biomass of each mammal species and habitat variables, particularly those associated with the structure of the understory, have been well documented. In forest regenerating after fire, the variance in bush rat biomass can be explained by the accumulated dry mass of leaf litter (43%) and variability in the height of the understory (a further 23%; Fox and McKay, 1981, Figs. 8 and 9). In turn, 85% of the variance in the leaf litter was attributable to the time elapsed since the site last burned, so that 37% of the variance in bush rat biomass was attributed to the regeneration age of the site. In a similar way the diversity of plant species at the site, the amount of vegetation in the 20- to 50-cm layer, and the variability in the amount of

FIGURE 9.   Mammal abundance from a chronosequence of 1-ha plots along a regenerating heath following sand mining at Hawks Nest, shown for three separate trapping periods: 1976 and 1977, pooled, and 1992. (a) Data for New Holland mouse (*Pseudomys novaehollandiae*) and house mouse (*Mus domesticus*) are pooled into age classes, with control values for each trapping period from unmined sites adjacent to the mining path. (b) Individual plots for New Holland mouse with values from two sites (HN3 and HN7) for the three trapping periods joined by dashed lines. Regression lines (1976–1977 and 1976–1977 + 1992) for all plots are also shown (see text for explanation and discussion). (c) Individual plots for house mouse with values from three sites (HN1, HN3, and HN6) for the three trapping periods joined by dashed lines.

vegetation in the lowest 20 cm of the understory ultimately contribute 7, 7, and 5%, respectively, to predictions of bush rat biomass. When Fox and McKay (1981) selected a set of sites with the same regeneration age, the direct contributions to variance in bush rat biomass from habitat variables were: litter, 50%; vegetation from 0 to 20 cm, 22%; vegetation from 50 to 100 cm, 12%; and variability in the height of the understory, 14%, for a total of 98%. It is clear that structural components of the understory and accumulated litter play key roles in determining the abundance of bush rats, emphasizing how tightly abundance is related to the changes in the vegetation succession and also why regeneration age can be used as a surrogate variable.

A similar picture has been presented for the biomass variations of New Holland mice on sand-mined areas (Fox and Fox, 1978; Fig. 6), where the contributing variables are: plant species diversity (52%), the proportion of plant species that are "heath" species (35%), and the hardness of the soil in the first 15 cm (9%); 96% of the total variance can be explained by these variables. Of the plant species diversity component, 96% of its variance in plant could be partitioned among regeneration age (67%), vegetation from the 20- to 50-cm layer (18%), and the lowest layer of vegetation from 0 to 20 cm (11%), so that these three variables ultimately contribute 35, 9, and 6%, respectively, to the variance in New Holland mouse biomass. For the mined forest sites a regression equation for New Holland mouse biomass explains 69% of the variance and the variables included are the proportion of "heath" species, the amount of vegetation in the layers from 20 to 100 cm, the amount of dead shrub cover, and the top soil depth (Fox and Fox, 1984). In this case there is no direct relationship with regeneration age, as changes in mammal abundance with time are not linear, nor in some cases even unimodal (see Figs. 7 and 8). These observations reinforce the view that the animal succession proceeds in response to the vegetation succession rather than in response to time itself.

The understanding that I have gained of how small-mammal communities inhabiting the heathlands and forests of this area respond to disturbance could have been obtained only from a commitment to maintain long-term studies. One of the greatest difficulties in such work is to ensure a continuity of funding over the, e.g., 20-year period I have worked in the Myall Lakes area, with grant funding coming in 3-year blocks that only sometimes can be extended to 6. However, despite the difficulties I would strongly support the establishment of studies with long-term aims and encourage the continuation of studies that are already in or are verging on the long-term category. I can confidently say that long-term studies such as those reported here not only repay the effort put into them, with the enhanced opportunities for cross-referencing and linkages among sites as the data accumulate, but in a number of cases information becomes available that could not be obtained in any other way.

VII. SUMMARY

Long-term monitoring of small mammals in these four specific projects has provided a substantial amount of information that would not have been available in shorter term surveys. A forest postfire study has identified repeating mammalian successions within the single 26-year period since the last fire. The repeated mammalian successions in this interval appear to be driven by changes in the structure of the vegetation. Monitoring individual sites throughout the extended time period provides data supporting published chronosequence analyses.

The results of the extended heath fire study demonstrate that, given a sufficient amount of time, the Australian mammalian postfire succession does become a species replacement sequence, rather than merely a change in the relative dominance of species as had previously been suggested (Fox, 1982). Close examination of the two heathland fire cycles shows that, while some species follow similar patterns in each cycle, others show markedly different responses that in turn may be due to regional factors and broad-scale differences in the abundance of species.

Sand-mined forest regeneration has proved to be a valuable system for study and validation of the chronosequence analysis technique, where individual sites are followed through time and reach certain regeneration ages in different calendar years. In heathland regeneration after sand mining, data collected 18 years after disturbance confirms original predictions made about the growth rates of New Holland mouse populations, specifically that it would take some 20 years for abundances to reach those of control levels.

Few or none of these results would have been apparent from shorter term datasets, or even from chronosequences covering longer time spans; the long-term monitoring of sites has delivered more interpretable information than either of these alternative approaches. In particular, a second wave of mammalian succession in response to vegetation senescence in later succession could not have been predicted from shortcut methods. However, regional and local (on-site) effects still remain to be disentangled, perhaps by the use of longitudinal studies, and wider regional surveys might serve to partition local weather factors from those associated with successional trends following local disturbances. It is clear that a strong case can be made for the value of long-term studies, which can produce superior returns per time and effort relative to the investment of the same time in a number of shorter projects.

Acknowledgments

The scope of the research reviewed in this chapter, covering the last two decades, could not have been achieved without the valued assistance of friends and colleagues, including my graduate students, many of whom appear in the joint publications providing the database for this review. However, I especially thank Marilyn Fox, who has provided ever-present support ranging from trapping assistance to editing skills and critical analysis; she has been a most valued colleague. I have mentioned the difficulties of maintaining continuity of funding; these projects could not have continued long-term without ongoing grant support from the Australian Research Council and its predecessors.

References

Andersen, A. N., and Yen, A. L. (1985). Immediate effects of fire on ants in the semi-arid mallee region of north western Victoria. *Aust. J. Ecol.* 10, 25–30.

Arrizabala, A., Montagud, E., and Fons, R. (1993). Post-fire succession in small mammal communities in the Montserrat Massif (Cartalonia, Spain). *In* "Fire in Mediterranean Ecosystems" (L. Trabaud and R. Prodon, eds.), Ecosyst. Res. Rep. No. 5, pp. 281–291. Commission of the European Communities, Brussels.

Auld, T. D. (1986). Dormancy and viability in *Acacia suaveolens* (Sm.) Willd. *Aust. J. Bot.* 34, 463–472.

Austin, M. P. (1981). Permanent quadrats: An interface for theory and practice. *Vegetatio* 46, 1–10.

Begg, R. J. (1981). The small mammals of Little Nourlangie Rock, N.T. V. The effects of fire. *Aust. Wildl. Res.* 8, 515–527.

Booyson, P. de V., and Tainton, N. M., eds. (1984). "Ecological Effects of Fire in South African Ecosystems." Springer-Verlag, Berlin.

Bradshaw, A. D., and Chadwick, M. J. (1980). "The Restoration of Land: The Ecology and Reclamation of Derelict and Degraded Land." Blackwell, Oxford.

Braithwaite, R. W., and Estbergs, S. A. (1985). Fire patterns and woody vegetation trends in the Alligator River Region of the Northern Territory. *In* "Ecology and Management of the World's Savannas" (J. C. Tothill and J. J. Mott, eds.), pp. 359–364. Aust. Acad. Sci., Canberra.

Braithwaite, R. W., and Estbergs, S. A. (1987). Fire-birds of the Top End. *Aust. Nat. Hist.* 22, 299–302.

Braithwaite, R. W., and Griffiths, A. D. (1994). Demographic variation and range contraction in the Northern quoll, *Dasyurus hallucatus* (Marsupialia: Dasyuridae). *Wildl. Res.* 21, 203–217.

Catling, P. C. (1986). *Rattus lutreolus,* colonizer of heathland after fire in the absence of *Pseudomys* species? *Aust. Wildl. Res.* 13, 127–139.

Catling, P. C., and Newsome, A. E. (1981). Responses of the Australian vertebrate fauna to fire: An evolutionary approach. *In* "Fire and the Australian Biota" (A. M. Gill, R. H. Groves, and I. R. Noble, eds.), pp. 273–310. Aust. Acad. Sci., Canberra.

Caughley, J., Monamy, V., and Heiden, K. (1994). "Impact of the 1993 Mouse Plague," *Occas. Pap. No. 7.* GRDC, Bureau of Resource Sciences, Canberra.

Chapman, A. (1981). Habitat preference and reproduction of feral house mice, *Mus musculus,* during plague and non-plague situations in Western Australia. *Aust. Wildl. Res.* 8, 567–580.

Cockburn, A. (1981). The response of the heath rat *Pseudomys shortridgei* to pyric succession: A temporally dynamic life-history strategy. *J. Anim. Ecol.* 50, 649–666.

Cowles, H. C. (1899). The ecological relationships of vegetation on the sand dunes of Lake Michigan. *Bot. Gaz. (Chicago)* 27, 95–117, 167–202, 281–308, 361–391.

Fons, R., Grabulosa, I., Feliu, C., Mas-Coma, S., Galan-Puchades, M. T., and Comes, A. M. (1993). Postfire dynamics of a small mammal community in a Mediterranean forest. *In* "Fire in Mediterranean Ecosystems" (L. Trabaud and R. Prodon, eds.), Ecosyst. Res. Rep. No. 5, pp. 259–270. Commission of the European Communities, Brussels.

Fox, B. J. (1979). An objective method of measuring the vegetation structure of animal habitats. *Aust. Wildl. Res.* 6, 297–303.

Fox, B. J. (1981). Niche parameters and species richness. *Ecology* 62, 1415–1425.

Fox, B. J. (1982). Fire and mammalian secondary succession in an Australian coastal heath. *Ecology* 63, 1332–1341.

Fox, B. J. (1983). Mammal species diversity in Australian Heathlands: the importance of pyric succession and habitat diversity. *In* "Mediterranean-type Ecosystems: The Role of Nutrients" (F. J. Kruger, D. T. Mitchell, and J. U. M. Jarvis, eds.), pp. 473–489. Springer-Verlag, Berlin.

Fox, B. J. (1990a). Changes in the structure of mammal communities over successional time scales. *Oikos* 59, 321–329.

Fox, B. J. (1990b). Two hundred years of disturbance: How has it aided our understanding of "succession" in Australia? *Proc. Ecol. Soc. Aust.* 16, 521–529.

Fox, B. J., and Fox, M. D. (1978). Recolonization of coastal heath by *Pseudomys novaehollandiae* following sand mining. *Aust. J. Ecol.* 3, 447–465.

Fox, B. J., and Fox, M. D. (1981). A comparison of vegetation classifications as descriptors of small mammal habitat preference. *In* "Vegetation Classification in the Australian Region" (A. N. Gillison and D. J. Anderson, eds.), pp. 66–80. CSIRO/Aust. Nat. Univ. Press, Canberra.

Fox, B. J., and Fox, M. D. (1984). Small-mammal recolonization of open-forest following sand-mining. *Aust. J. Ecol.* 9, 241–252.

Fox, B. J., and Fox, M. D. (1986). Resilience of animal and plant communities to human disturbance. *In* "Resilience of Mediterranean-Type Ecosystems" (B. Dell, A. J. M. Hopkins, and B. Lamont, eds.), Tasks Veg. Sci. Ser., pp. 39–64. Dr. W. Junk, The Hague, The Netherlands.

Fox, B. J., and Gullick, G. (1989). Interspecific competition between mice: A reciprocal field manipulation experiment. *Aust. J. Ecol.* 14, 357–366.

Fox, B. J., and McKay, G. M. (1981). Small mammal response to pyric successional changes in eucalypt forest. *Aust. J. Ecol.* 6, 29–42.

Fox, B. J., and Pople, A. R. (1984). Experimental confirmation of interspecific competition between native and introduced mice. *Aust. J. Ecol.* 9, 323–334.

Fox, B. J., and Twigg, L. E. (1991). An experimental transplant of *Pseudomys novaehollandiae* onto young stages of post-mining regeneration. *Aust. J. Ecol.* 16, 281–287.

Fox, B. J., Quinn, R. D., and Breytenbach, G. J. (1985). A comparison of small-mammal succession following fire in shrublands of Australia, California and South Africa. *Proc. Ecol. Soc. Aust.* 14, 179–197.

Fox, B. J., Higgs, P., and Luo, J. (1993). Extension of the breeding season of the New Holland mouse: A response to above average rainfall. *Wildl. Res.* 20, 599–605.

Fox, M. D. (1988). Community changes in coastal eucalypt forest in response to fire. *Cunninghamia* 2, 85–96.

Fox, M. D., and Fox, B. J. (1986). The effect of fire frequency on the structure and floristic composition of a woodland understorey. *Aust. J. Ecol.* 11, 77–85.

Fox, M. D., and Fox, B. J. (1987). The role of fire in the scleromorphic forests and shrublands of eastern Australia. *In* "The Role of Fire in Ecological Systems" (L. Trabaud, ed.), pp. 23–48. SPB Academic Publishing, The Hague, The Netherlands.

Gill, A. M. (1975). Fire and the Australian flora: A review. *Aust. For.* 38, 4–25.

Gill, A. M., Groves, R. H., and Noble, I. R., eds. (1981). "Fire and the Australian Biota." Aust. Acad. Sci., Canberra.

Greenslade, P., and Majer, J. D. (1993). Recolonization by Collembola of rehabilitated bauxite mines in Western Australia. *Aust. J. Ecol.* 18, 385–394.

Greenslade, P. J. M., and Thompson, C. H. (1981). Ant distribution, vegetation and soil relationships in the Cooloola-Noosa River area, Queensland. *In* "Vegetation Classification in the Australian Region" (A. N. Gillison and D. J. Anderson, eds.), pp. 192–207. CSIRO/Aust. Nat. Univ. Press, Canberra.

Higgs, P., and Fox, B. J. (1993). Interspecific competition: A mechanism for rodent succession after fire in wet heathland. *Aust. J. Ecol.* 18, 193–201.

Jasper, D. A., Abbott, L. K., and Robson, A. D. (1989). The loss of VA mycorrhizal infectivity during bauxite mining may limit the growth of *Acacia pulchella* R. Br. *Aust. J. Bot.* 37, 33–42.

Lunney, D., Eby, P., and O'Connell, M. (1991). Effects of logging and fire on small mammals in Mumbulla State Forest, near Bega, NSW. *Aust. J. Ecol.* 16, 33–46.

Majer, J. D., ed. (1989a). "Animals in Primary Succession: The Role of Fauna in Reclaimed Lands." Cambridge Univ. Press, Cambridge, UK.

Majer, J. D. (1989b). Long-term colonization of fauna in reclaimed land. *In* "Animals in Primary Succession: The Role of Fauna in Reclaimed Lands" (J. D. Majer, ed.), pp. 143–174. Cambridge Univ. Press, Cambridge, UK.

Majer, J. D. (1990). Rehabilitation of disturbed land: Long-term prospects for the recolonisation of fauna. *Proc. Ecol. Soc. Aust.* 16, 509–519.

Masters, P. (1993). The effects of fire driven succession and rainfall on small mammals in spinifex grassland at Uluru National Park, Northern Territory. *Wildl. Res.* 20, 803–813.

Meredith, C. W., Gilmore, A. M., and Isles, A. C. (1984). The ground parrot (*Pezoporus wallicus* Kerr) in south-eastern Australia: A fire-adapted species? *Aust. J. Ecol.* 9, 367–380.

Midgely, J. J., and Clayton, P. (1990). Short-term effects of an autumn fire on small mammal populations in Southern Cape coastal mountain fynbos. *S. Afr. For. J.* 153, 27–30.

Mills, J. N. (1986). Herbivores and early postfire succession in southern California chaparral. *Ecology* 67, 1637–1649.

Myerscough, P. J., and Carolin, R. C. (1986). The vegetation of the Eurunderee sand mass, headlands and previous islands in the Myall Lakes area, New South Wales. *Cunninghamia* 1(4), 399–466.

Naveh, Z. (1975). The evolutionary significance of fire in the Mediterranean region. *Vegetatio* 29, 199–208.

Newsome, A. E. (1969). A population study of house-mice permanently inhabiting a reed-bed in South Australia. *J. Anim. Ecol.* 38, 361–377.

Newsome, A. E., and Catling, P. C. (1979). Habitat preferences of mammals inhabiting heathlands of warm temperate coastal, montane and alpine regions of south-eastern Australia. *In* "Ecosystems of the World" (R. L. Specht, ed.), Vol. 9A, pp. 301–316. Elsevier, Amsterdam.

Newsome, A. E., and Corbett, L. K. (1975). Outbreaks of rodents in semi-arid and arid Australia: Causes, prevention and evolutionary considerations. *In* "Rodents in Desert Environments" (I. Prakash and P. G. Ghosh, eds.), pp. 117–153. Dr. W. Junk, The Hague, The Netherlands.

Newsome, A. E., and Catling, P. C. (1983). Animal demography in relation to fire and shortage of food: Some indicative models. *In:* "Mediterranean-Type Ecosystems: The Role of Nutrients" (F. J. Kruger, D. T. Mitchell, and J. V. M. Jarvis, eds.), pp. 490–505. Springer-Verlag, Berlin.

Newsome, A. E., McIroy, J., and Catling, P. C. (1975). The effects of an extensive wildfire on populations of twenty ground-dwelling vertebrates in south-east Australia. *Proc. Ecol. Soc. Aust.* 9, 107–123.

Nichols, O., Wykes, B. J., and Majer, J. D. (1989). The return of vertebrate and invertebrate fauna to bauxite mined areas in south-western Australia. *In* "Animals in Primary Succession: The Role of Fauna in Reclaimed Lands" (J. D. Majer, ed.), pp. 397–422. Cambridge Univ. Press, Cambridge, UK.

Olson, J. S. (1958). Rates of succession and soil changes on southern Michigan sand dunes. *Bot. Gaz. (Chicago)* 199, 125–170.

Poissonet, J., Poissonet, P., and Thialt, M. (1981). Development of flora, vegetation and grazing value in an experimental plots of a *Quercus coccifera* garrigue. *Vegetatio* 46, 93–104.

Posamentier, H. G. (1976). Habitat requirements of small mammals in coastal heathlands of New South Wales. M.Sc. Thesis, University of Sydney, Sydney, Australia.

Prodon, R., Fons, R., and Athias-Binche, F. (1987). Effects of fire on animal communities in the Mediterranean area. *In* "The Role of Fire in Ecological Systems" (L. Trabaud, ed.), pp. 121–157. SPB Academic Publishing, The Hague, The Netherlands.

Quinn, R. D. (1979). Effects of fire on small mammals in the chaparral. *Cal-Neva Wildl. Trans.*, pp. 125–133.

Redhead, T. D. (1988). Prevention of plagues of house mice in rural Australia. In "Rodent Pest Management" (I. Prakash, ed.), pp. 191–205. CRC Press, Baton Rouge, LA.

Saunders, G. R., and Giles, J. K. (1977). A relationship between plagues of the house mouse *Mus musculus* (Rodentia: Muridae) and prolonged periods of dry weather in south-eastern Australia. *Aust. Wildl. Res.* **4**, 241–248.

Singleton, G. R. (1989). Population dynamics of an outbreak of house mice (Mus domesticus) in mallee wheatfields of Australia_hypothesis of plague formation. *J. Zool.* **219**, 495–515.

Thom, B. G., Shepard, M., Ly, C. K., Bowman, G. M., and Hesp, P. A. (1992). "Coastal Geomorphology and Quaternary Geology of the Port Stephens-Myall Lakes Area." Aust. Natl. Univ. Press, Canberra.

Thompson, J. (1983). The effects of fire frequency on small mammal habitats in open forest. B.Sc. (Honours) Thesis, University of New South Wales, Sydney, Australia.

Thompson, P. T., and Fox, B. J. (1993). Asymmetric competition in Australian heathland rodents: A reciprocal removal experiment demonstrating the influence of size-class structure. *Oikos* **67**, 264–278.

Trabaud, L., ed. (1987). "The Role of Fire in Ecological Systems." SPB Academic Publishing, The Hague, The Netherlands.

Twigg, L. E., Fox, B. J., and Luo, J. (1989). The modified primary succession following sand mining: A validation of the use of chronosequence analysis. *Aust. J. Ecol.* **14**, 441–447.

Willan, K., and Bigalke, R. C. (1982). The effects of fire on small mammals in S. W. Cape montane fynbos (Cape Macchia). In "Dynamics and Management of Mediterranean-type Ecosystems" (C. E. Conrad and W. C. Oechel, eds.), Gen. Tech. Rep. PSW-58, pp. 207–212. Pacific Southwest Forest and Range Experiment Station, Forest Service, U.S. Department of Agriculture, Berkeley, CA.

Wirtz, W. O., II (1982). Post-fire community structure of birds and rodents in southern California chaparral. In "Dynamics and Management of Mediterranean-type Ecosystems" (C. E. Conrad and W. C. Oechel, eds.), Gen. Tech. Rep. PSW-58, pp. 241–246. Pacific Southwest Forest and Range Experiment Station, Forest Service, U.S. Department of Agriculture, Berkeley, CA.

Wirtz, W. O., II, Hoekman, D., Muhm, J. R., and Souza, S. L. (1988). Postfire rodent succession following prescribed fire in southern California chaparral. In "Proceedings of the Symposium on the Management of Amphibians, Reptiles, and Small Mammals in North America," Tech. Rep. RM-166, pp. 333–339. Rocky Mountain Forest and Range Experiment Station, Forest Service, U.S. Department of Agriculture, Berkeley, CA.

York, A. (1994). The long-term effects of fire on forest ant communities: Management implications for the conservation of biodiversity. *Mem. Queensl. Mus.* **36**, 231–239.

CHAPTER 16

Organization, Diversity, and Long-Term Dynamics of a Neotropical Bat Community

Elisabeth K. V. Kalko,*,†,‡ Charles O. Handley, Jr.,‡
and Darelyn Handley‡

*University of Tübingen, Tierphysiology, 72076 Tübingen, Germany; †Smithsonian
Tropical Research Institute, Balboa, Republic of Panama; and ‡Division of Mammals,
National Museum of Natural History, Smithsonian Institution, Washington, DC 20560

I. INTRODUCTION

The study of processes that influence the distribution, diversity, and abundance of species is one of the most challenging and complex fields in all of biology. Many basic questions are still controversial. Is community composition primarily the result of deterministic processes, with interactions among species profoundly influencing community organization, or is it best understood as a fundamentally random process? Further, what relative importance do these processes have in promoting and maintaining diversity of species, particularly in species-rich tropical communities? As the results of an increasing number of studies clearly demonstrate, good data on temporal and spatial variability in community composition and species abundance are essential for formulating and evaluating hypotheses about the processes determining organization of communities and diversity of species.

With respect to communities of terrestrial, warm-blooded vertebrates, long-term sampling has been limited mostly to communities of birds and desert rodents (e.g., Holmes *et al.*, 1986; Wiens, 1986, 1989; Karr *et al.*, 1989; Brown and Heske, 1990; Terborgh *et al.*, 1990; Thiollay, 1994). Because they are difficult to sample and study, the abundant and speciose bats, an important target group for such long-term approaches, have been neglected. To appreciate the importance of bats, one needs only consider how they influence forest diversity and regeneration by means of insectivory, seed dispersal, and pollination (Dobat and Peikert-Holle, 1985; Marshall, 1985; Charles-Dominique, 1986; Foster *et al.*, 1986; Fleming, 1988, 1991; Cox *et al.*, 1992). This is particularly true in the tropics, where the true magnitude of bat biomass and its effect on ecosystems is only beginning to be recorded and understood.

Unfortunately, our knowledge of organization and dynamics of chiropteran communities, particularly those of species-rich tropical ecosystems, is very poor. There are few exceptions (see Fleming *et al.*, 1972; Handley, 1976; Bonaccorso, 1979; Willig, 1986; Crome and Richards, 1988; Fleming, 1988; Handley *et al.*, 1991; Willig *et al.*, 1993). Otherwise most studies are limited to lists describing communities in terms of presence or absence of species. Moreover, most such compilations have been made in a single season without standardized protocols. The few long-term studies of bats have mostly described communities in the temperate zone, where bats are monitored in hibernacula and maternity roosts or, on rare occasions, sampled with mist nets in their foraging areas (Glass, 1956; Humphrey, 1975; Gardner *et al.*, 1989; Gaisler *et al.*, 1990).

The difficulty of sampling and studying bats has lead to reliance on eco-morphological characters for studying patterns in bat communities (McNab, 1971; Fenton, 1972; Findley, 1973, 1976, 1993; Schum, 1984; Fleming, 1986; Willig and Moulton, 1989; Heller and Volleth, 1995). In place of detailed field studies, morphological characters can be measured easily on museum specimens and broadly reflect some aspects of ecological preferences of species (Fenton, 1972; Findley and Wilson, 1982; Findley, 1993). In particular, this method has been used to address questions about the role and extent of deterministic and stochastic processes in structuring bat communities. To date, all studies (except McKenzie and Rolfe, 1986) addressing this question and using primarily ecomorphological characters have concluded that species assemblages of bats cannot readily be distinguished from communities formed largely by stochastic events (Fleming, 1986; Willig and Moulton, 1989; Findley, 1993; Willig *et al.*, 1993).

However, recent studies indicate that ecomorphologically similar species may be clearly differentiated by behavioral traits such as spatial segregation in habitat use, roost site selection, foraging strategies, and diet (Saunders and Barclay, 1992; Kalko, 1996). Ecomorphological predictions alone would have placed these species closer together in the community than they actually are.

Here, we begin an in-depth study and analysis of tropical bat communities, using the longest-studied and best-known bat community in the Neotropics as our data source. Barro Colorado Island (BCI), Panamá, with a known fauna of at least 66 species of bats (Table I), has been the site of bat studies for about 85 years. We will focus our preliminary analysis on the extensive mark–recapture netting studies of the BCI Bat Project (Handley et al., 1991) to describe community organization and temporal variability in guild composition and species abundance.

II. METHODS

A. Study Site

The BCI Bat Project, starting in 1975 and continuing to the present, has been conducted on Barro Colorado Island and on nearby peninsulas and small islands within a 10×10-km square. BCI (15.6 km^2) is located in central Panamá ($9°09'$ N, $79°51'$ W) in Gatún Lake, part of the Panama Canal. It is a field station of the Smithsonian Tropical Research Institute (STRI).

Prior to 1914 BCI was called West Hill, rising from the south bank of the Río Chagres about 20 km upstream from its mouth on the Caribbean coast of Panamá. The hill became BCI soon after 1914 when the Chagres valley was flooded to form Gatún Lake and the Panama Canal. Before canal construction much of the dry ground near the river and its tributaries, including most of the north and east slopes of West Hill, was cleared periodically for agriculture. On the back side of West Hill, away from the river, old forest remains.

Now, except for the small Laboratory Clearing and a long, narrow, grassy strip between a pair of canal range lights, BCI is entirely covered with tropical semi-deciduous lowland forest (Foster and Brokaw, 1982). The last deforestation occurred during the era of French canal construction, so forest regrowth on the north and east half of the island now is about 100 years old. Forest on the back side of the island, facing away from the canal, is estimated to be 400–600 years old (Piperno, 1992). Since the data of the BCI Bat Project have been collected almost exclusively in the tall secondary forest of BCI, we distinguish only a single more or less uniform habitat type as characteristic of our study site.

Rainfall on BCI averages about 260 cm annually, about 90% of which falls during a May–December rainy season (Windsor, 1990). The January–April dry season is characterized by very little rainfall and high trade winds from the north and northeast.

B. Sampling Period and Project Organization

The original BCI Bat Project studied bats on and near BCI for 63 months between 1975 and 1985. In this interval we censused four complete 12-month cycles in the field (1977–1980 and 1984–1985) and 3–4 months in each of the other years. No data were collected in 1983. The project was directed by M. A. Bogan, A. L. Gardner, C. O. Handley, and D. E. Wilson as principal investigators from 1977 to 1980, and by Handley alone in 1975–1976 and 1981–1985. After a 6-year respite Handley joined E. K. V. Kalko, H.-U. Schnitzler, and their group from the University of Tübingen to resume the BCI Bat Project in 1991. It continues as a program involving conventional mist net sampling together with a variety of other emphases including behavioral studies, dietary analyses, food choice experiments, radio-tracking, and acoustic monitoring.

C. Sampling Protocol and Data Analysis

1. Mist Netting

During the reconnaissance phase of the project in 1975 and 1976 we randomly set a varying number of mist nets and a Tuttle trap (Tuttle, 1974). From 1977 until 1985 our sampling protocol used only mist nets placed whenever possible close to ripe fruit trees. Most often we netted near *Ficus* sp., occasionally near *Spondias* sp., *Quararibea asterolepis, Poulsenia armata,* and *Dipteryx panamensis,* and rarely near other species (Handley *et al.,* 1991; Kalko *et al.,* 1996).

Our standard netting station consisted of 10 12-m mist nets set at ground level where the shrub layer was open enough for easy passage of bats. We occasionally used 6-m nets, which we recorded as "half nets" in our analysis. In all we used 105 netting stations on BCI and 15 more on nearby peninsulas and islands, all within a 10 × 10-km square. Depending on local conditions we set nets end to end, in a T pattern, or scattered without regard to positioning of nearby nets.

We routinely netted five nights per week. Netting nights usually began at dusk, regardless of moon phase, and continued as long as bats were present. Rain time was discounted from net time because bats were not flying. Net hours were calculated to the nearest quarter hour for each net for the time the net was actually open. Frequently we used distress calls of bagged bats or squeaks of

Audubon bird calls to attract passing bats into the nets. Bats were identified, examined for reproductive condition, measured, marked with individually numbered stainless steel ball chain necklaces (punch marks in wing membranes were used prior to 1977), and released at the capture site (for details, see Handley et al., 1991).

The placement of the mist nets near ripe fruit trees tended to oversample fruit-eating phyllostomids, particularly fig-eating species. However, for several reasons we are convinced that our database is useful not only for describing patterns of community organization of fig-eating bats but also for demonstrating patterns in the whole community:

a. We used the same sampling protocol consistently throughout the project.

b. Although bats that catch insects in the air are undersampled or missed altogether with mist-netting, this is a good sampling method not only for fruit-eaters but also for other phyllostomids that glean insects from the vegetation and ground or feed on nectar. Nevertheless, our consistent capture bias must be taken into account particularly when comparing relative abundances of these bats and the aerial insectivores.

c. In addition to fig-eating bats, our catches contained high numbers of nonfig fruit-eaters such as *Carollia perspicillata* and *C. castanea*. Since the sampling was not targeted toward nonfig fruit-eaters or insect gleaners, we assume that the samples of these species were mostly random.

d. Results of random mist-netting on BCI (1975–1976) in the reconnaissance phase of the BCI Bat Project and studies elsewhere in the Neotropics also documented high numbers of fruit-eating phyllostomids, suggesting that this phenomenon was not only related to our capture protocol but also reflected general patterns in this community most likely linked to availability and distribution of resources.

e. In our preliminary data analysis we focused mostly on the well-sampled phyllostomids, particularly on gleaning insectivores and gleaning frugivores because they were the best sampled groups.

f. Since we marked the bats we captured, we could exclude recaptures (27.7% of captures) from our analyses. Thus we avoided artificially high counts due to individual bats caught more than once.

g. To account for possible differences between random and fruit-targeted positioning of the mist nets, we compared data from 1975–1976 (random mist-netting) with data from 1977–1985 (fruit-targeted mist-netting).

2. Acoustic Monitoring

Usually mist-netting at ground level is ineffective for sampling bats that avoid nets or catch insects in midair in open space, in forest gaps, or at forest edges.

Thus, aerial insectivores are underrepresented in mist-netting. To compensate for this, beginning in 1991 we supplemented mist-netting with acoustic monitoring for identifying high-flying bats. Most of these bats can be identified to genus and many to species, based on their species-specific signal types (e.g., Ahlén, 1981; Fenton and Bell, 1981; Fenton *et al.*, 1983, 1987; Kalko, 1995b). Currently we are developing a key for identifying echolocation calls of Neotropical bats.

3. Analysis of Data

During the marking phase of the project, 1975–1985, we recorded 48,222 captures. Since part of the data was collected with random mist-netting (1975–1976) and part with fruit-targeted mist-netting (1977–1985), we reconfigured this database for analysis:

a. To determine species richness, we combined all data from the fruit-targeted mist-netting (1977–1985) and from the reconnaissance phase (1975–1976), excluding recaptures, in a single pool together with records before 1975 and after 1985.

b. For rank abundance diagrams we combined the records, minus recaptures, of all bats we caught with fruit-targeted mist-netting (1977–1985) in one pool and all bats from random netting (1975–1976) in another pool, and then we ranked the species according to their proportional abundance on a logarithmic scale. We excluded species infrequently caught, such as those catching insects in the air as well as fish-eating and blood-drinking bats.

c. For comparisons of long-term population trends and to exclude seasonality, we selected all data collected in the late rainy season (September to December) from each of seven years (1977–1984). Each year throughout the project this rainy season segment was the best sampled period. We did not include 1975 or 1976 because we used a different capture protocol in those years, and 1983 and 1985 are lacking because we did no mist-netting in the late rainy season in those years. This sample totaled 348 netting nights, 24,266 net hours, and 13,408 captures of 46 species. In each of these periods Handley was the principal investigator. We excluded species infrequently caught, such as vampires, fishing bats, and most of the insect-eating bats that catch flying prey.

d. To estimate long-term population trends, we used data set *c* (described in the preceding paragraph), calculated species abundance as captures per 100 mist net hours, and transformed them to log abundances to reflect the exponential growth potential of natural populations (Berthold *et al.*, 1986; Böhning-Gaese *et al.*, 1993; Böhning-Gaese, 1994). We added 1 to the values to keep the log abundances positive. Long-term population trends were calculated by regressing log abundance as a function of time. The long-term population trend was the slope of the regression line.

III. RESULTS

A. Species Richness of the BCI Bat Community

Our knowledge of bats of BCI and vicinity began in 1908 with the precanal inventories of E. A. Goldman (1920) and others in the much disturbed Río Chagres corridor. They found 11 species of bats in roosts in old buildings and abandoned canal construction machinery at the foot of the hill that would become BCI (Table I). Enders (1930, 1935) listed 12 species of bats he collected or saw on BCI during visits between 1929 and 1932 when the island was less than 20 years old. With Enders' additions, the list of bats known from BCI in 1932 stood at 19 species (Table I). It included those species most easily observed in flight or captured by hand, usually in roosts.

Twenty years after Enders' observations, in 1952, the first mist net was set on BCI (Hall and Jackson, 1953). With one mist net open on 32 evenings, in or very near the Laboratory Clearing, Hall and Jackson added six common, easily netted phyllostomid frugivores. These additions raised to 25 species the bat fauna known on BCI in 1952 (Table I). Small-scale, sporadic mist-netting and opportunistic captures in and near the Laboratory Clearing in 1956–1957 enabled Carl B. Koford (unpublished checklist, 1957) to add 9 more species to the known fauna of bats of BCI, bringing the island total to 34 species (Table I). Ten to 15 years later, small-scale collecting by Don E. Wilson, A. L. Gardner, and David Klingener added 5 more species to the island list, raising the BCI total to 39 species (see Table I for specimen references).

Long-term intense mist-netting and trapping for bats on BCI began with Frank Bonaccorso in 1971–1974. During monthly capture sessions, January to December 1973, he caught 2324 bats (Bonaccorso, 1979). In that year he captured or observed 35 species of bats, 9 of them new for BCI, increasing the island total to 48 species (Table I).

During 10 years of sampling (1975–1985; no records in 1983), the marking phase of the BCI Bat Project documented 48,222 mist net captures of bats in 35,750 mist net hours. Of these captures, 34,896 (72.3%) were first captures and 13,326 (27.7%) were recaptures of previously marked bats. Overall, this phase of the BCI Bat Project added 13 species to the known fauna, raising the total to 61 species (Table I). In the listening phase of the project, begun in 1991, Kalko used ultrasound recording equipment and a night vision scope to tentatively identify 5 more species new for the island, bringing the BCI total to 66 species (Table I).

As of 1995, the known bat fauna of BCI included eight families composed of 40 genera and 66 species (Table I). The family Phyllostomidae (New World leaf-nosed bats) contributed most to species richness, making up about three-fifths of the community (39 species). Sheath-tailed bats (Emballonuridae) were

TABLE I Summary Data for the Bat Community on BCI and Vicinity, Assembled by the BCI Bat Project and Other Major Sources, 1908–1995

FAM/ Subfam.[a]	No. marked[b]	Guild[c]	Mass (g)[d]	Diet[e]	Sources of BCI records[f]							
					GO	EN	HJ	KO	MI	BO	HA	KA
Superabundant (>10,000)												
Artibeus jamaicensis Leach PHY/Ste	17,564	8f	48.3 ± 1.8 (n = 10)	99.2% FR, 0.7% PO, 0.1% IN (N = 2990)		x^	x	x		x	x	x
Abundant (1,000–10,000)												
Artibeus lituratus (Olfers) PHY/Ste	4,093	8f	67.3 ± 3.0 (n = 10)	93.8% FR, 6.2% PO (N = 486)			x	x		x	x	x
Uroderma bilobatum Peters PHY/Ste	1,957	8f	16.8 ± 1.2 (n = 10)	96.6% FR, 3.4% PO (N = 205)				x			x	x
Carollia perspicillata (Linmaeus) PHY/Car	1,692	8p	19.7 ± 1.6 (n = 10)	96.8% FR, 3.2% PO (N = 372)	x^		x	x		x	x	x
Vampyrodes caraccioli (Thomas) PHY/Ste	1,174	8f	36.2 ± 3.3 (n = 10)	99% FR, 1% PO (N = 103)	x^		x	x		x	x	x
Common (100–999)												
Chiroderma villosum Peters PHY/Ste	722	8f	22.0 ± 1.6 (n = 10)	100% FR (N = 117)			x	x		x	x	x
Artibeus phaeotis (Miller) PHY/Ste	636	8f	11.7 ± 1.0 (n = 10)	96.4% FR, 3.6% PO (N = 28)				x		x	x	x
Carollia castanea H. Allen PHY/Car	476	8p	13.3 ± 2.0 (n = 10)	91.4% FR, 4.3% PO, 4.3% IN (N = 187)			x	x		x	x	x
Tonatia silvicola (D'Orbigny) PHY/Phy	448	4	34.3 ± 2.2 (n = 10)	2.8% FR, 97.2% IN (N = 106)				x		x	x	x
Artibeus watsoni Thomas PHY/Ste	440	8f	12.0 ± 1.1 (n = 10)	94.4% FR, 5.6% PO (N = 18)		x^	x	x		x	x	x
Vampyressa pusilla (Wagner) PHY/Ste	417	8f	8.5 ± 0.8 (n = 10)	100% FR (N = 46)			x	x		x	x	x
Pteronotus parnellii (Gray) MOR	362	3	23.4 ± 1.4 (n = 10)	100% IN (N = 65)						x	x	x

Species	Code			Mass (n)	Diet (N)						
Phyllostomus discolor Wagner	PHY/Phy	357	9	42.1 ± 3.1 (n = 10)	17.9% FR, 82.1% PO (N = 28)	x̂			x	x	x x
Micronycteris hirsuta (Peters)	PHY/Phy	307	4	15.5 ± 0.8 (n = 10)	100% IN (N = 44)	x̂			x	x	x x
Tonatia bidens (Spix)	PHY/Phy	284	4	36.8 ± 2.0 (n = 10)	2.1% FR, 97.9% IN (N = 48)					x	x x
Trachops cirrhosus (Spix)	PHY/Phy	208	4	34.9 ± 2.0 (n = 9)	82.8% IN, 17.2% VE (N = 29)					x	x x
Phyllostomus hastatus (Pallas)	PHY/Phy	180	10	125.6 ± 7.9 (n = 10)	41.2% FR, 44.1% PO, 11.8% IN, 2.9% VE (N = 34)	x			x	x	x x
Vampyressa nymphaea Thomas	PHY/Ste	180	8f	14.0 ± 0.9 (n = 10)	100% FR (N = 22)		x		x	x	x x
Glossophaga soricina (Pallas)	PHY/Glo	163	9	11.3 ± 0.8 (n = 10)	31.3% FR, 68.7% PO (N = 16)	x̂	x̂	x	x	x	x x
Micronycteris megalotis (Gray)	PHY/Phy	154	4	7.2 ± 0.9 (n = 10)	100% IN (N = 19)	x̂			x	x	x x
Mimon crenulatum (E. Geoffroy)	PHY/Phy	138	4	15.0 ± 1.3 (n = 8)	100% IN (N = 21)					x	x x
Platyrrhinus helleri (Peters)	PHY/Ste	133	8f	15.8 ± 2.0 (n = 8)	100% FR (N = 19)		x		x	x	x x
Uncommon (10–99)											
Micronycteris nicefori Sanborn	PHY/Phy	99	4	11.1 ± 1.2 (n = 9)	22.2% FR, 77.8% IN (N = 9)					x	
Micronycteris brachyotis (Dobson)	PHY/Phy	66	10	14.3 ± 1.9 (n = 9)	21.4% FR, 7.2% PO, 71.4% IN (N = 14)				x	x	x x
Desmodus rotundus (E. Geoffroy)	PHY/Des	44	7	34.4 ± 2.2 (n = 10)		x			x	x	x x
Micronycteris schmidtorum Sanborn	PHY/Phy	42	4	7.1 ± 0.5 (n = 10)	100% IN (N = 12)					x	x

(continues)

TABLE 1 (Continued)

FAM/Sublam.[a]	No. marked[b]	Guild[c]	Mass (g)[d]	Diet[e]	Sources of BCI records[f]							
					GO	EN	HJ	KO	MI	BO	HA	KA
Saccopteryx bilineata (Temminck) EMB	41	2	6.8 ± 1.1 (n = 10)		x`	x`	x`	x`		x	x	x*
Glossophaga commissarisi Gardner PHY/Glo	36	9	7.2 ± 0.8 (n = 10)	100% PO (N = 2)					1		x	
Macrophyllum macrophyllum (Schinz) PHY/Phy	28	2	8.4 ± 1.3 (n = 10)	100% IN (N = 5)							x	x
Cormura brevirostris (Wagner) EMB	26	2	9.1 ± 1.5 (n = 10)							x	x	x*
Myotis nigricans (Schinz) VES	25	2	4.3 ± 0.3 (n = 10)		x`	x`	x	x		x	x	x*
Noctilio albiventris Desmarest NOC	22	2	30.7 ± 2.0 (n = 6)	100% IN (N = 4)				x		x	x	x*
Centurio senex Gray PHY/Ste	17	8o	20.5 ± 3.9 (n = 8)	100% FR (N = 5)				x		x	x	x
Phylloderma stenops Peters PHY/Phy	17	8o	61.4 ± 2.7 (n = 8)	100% FR (N = 3)						x	x	x
Rhogeessa tumida H. Allen VES	13	2	4.7 ± 0.6 (n = 3)		x`					x	x	x
Molossus molossus (Pallas) MOL	11	1	12.0 ± 1.6 (n = 3)		x`			x`			x`	
Rare (<10)												
Carollia brevicauda (Schinz) PHY/Car	9	8p	14.7 ± 1.3 (n = 10)	60% FR, 20% PO, 20% IN (N = 5)							x	
Uroderma magnirostrum Davis PHY/Ste	7	8f	16.5 ± 1.3 (n = 10)								x	x

Species	Family			Mass	50% IN, 50% VE (N = 2)									
Chrotopterus auritus (Peters)	PHY/Phy	6	5	84.4 ± 5.2 (n = 10)									x	x
Molossus bondae J. A. Allen	MOL	4	1	21.9 ± 1.8 (n = 9)									xˆ	
Saccopteryx leptura (Schreber)	EMB	4	2	4.1 ± 0.5 (n = 10)				x		x			x	x*
Vampyrum spectrum (Linnaeus)	PHY/Phy	4	5	151.3 ± 20.3 (n = 5)					2	x			x	
Lonchophylla robusta Miller	PHY/Lon	3	9	15.3 ± 1.0 (n = 10)				x					x	
Thyroptera tricolor Spix	THY	3	2	4.0 ± 0.6 (n = 10)		xˆ	xˆ	xˆ					x	
Centronycteris maximiliani (Fischer)	EMB	2	2	4.8 ± 0.2 (n = 2)				x		x			x	
Myotis albescens (E. Geoffroy)	VES	2	2	6.5 ± 0.7 (n = 10)							xˆ		xˆ	
Noctilio leporinus (Linnaeus)	NOC	2	6	55.3 ± 3.9 (n = 4)		x		x		x			x	x*
Ametrida centurio Gray	PHY/Ste	1	80	7.8 ± 1.4 (n = 3)									x	
Artibeus hartii Thomas	PHY/Ste	1	80	16.7 ± 1.2 (n = 10)									x	
Myotis riparius Handley	VES	1	2	5.2 ± 0.7 (n = 6)									x	
Rhynchonycteris naso (Wied)	EMB	1	2	3.4 ± 0.8 (n = 10)		xˆ		x					x	x
Sturnira luisi Davis	PHY/Stu	1	80	20.5 ± 2.4 (n = 7)									x	
Thyroptera discifera (Lichtenstein and Peters)	THY	1	2	3.2 (n = 1 female)									x	

(continues)

TABLE I (Continued)

	FAM/ Subfam.[a]	No. marked[b]	Guild[c]	Mass (g)[d]	Diet[e]	Sources of BCI records[f]							
						GO	EN	HJ	KO	MI	BO	HA	KA
1975–1976 BCI Bat Project captures													
Molossus coibensis J. A. Allen	MOL	13	1	15.7 ± 2.7 (n = 5)		xˆ	xˆ	xˆ	xˆ		xˆ	xˆ	x*
Mesophylla macconnelli Thomas	PHY/Ste	4	8o	7.0 ± 0.9 (n = 7)								x	
Pteronotus gymnonotus Natterer in Wagner	MOR	1	2	14.8 ± 0.9 (n = 10)							x	x	
Not found by BCI Bat Project, 1975–1985													
Diclidurus albus Wied	EMB		1	15.5 ± 1.7 (n = 6)									x*
Eptesicus furinalis (D'Orbigny)	VES		2	6.7 ± 1.5 (n = 5)		xˆ							x*
Eumops auripendulus (Shaw)	MOL		1	33.1 ± 1.4 (n = 4)		x				3			
Lasiurus (blossevillii?) (Lesson and Garnot)	VES		2	8.2 ± 1.0 (n = 6)									x*
Molossus sinaloae J.A. Allen	MOL		1	27.3 ± 2.8 (n = 6)						4			
Natalus stramineus Gray	NAT		2	6.1 ± 0.8 (n = 3)									x
Nyctinomops laticaudatus (E. Geoffroy)	MOL		1	12.6 ± 1.5 (n = 8)						5			

Peropteryx (kappleri?) Peters	EMB	1	7.5 ± 1.1 (n = 10)							x*
Pteronotus (personatus?) (Wagner)	MOR	2	8.0 ± 0.7 (n = 4)							x*
Tonatia brasiliensis (Peters)	PHY/Phy	4	8.6 ± 0.8 (n = 8)			6				
Individual source total species:				11	12	15	29	6	35	56	44
Species new for BCI:				11	8	6	9	5	9	13	5
BCI total:				11	19	25	34	39	48	61	66

a Family/subfamily affiliation is given (EMB, Emballonuridae; NOC, Noctilionidae; MOR, Mormoopidae; PHY, Phyllostomidae, with subfamilies Phy, Phyllostominae; Glo, Glossophaginae; Lon, Lonchophyllinae; Car, Carolliinae; Stu, Sturnirinae; Ste, Stenodermatinae; Des, Desmodontinae; NAT, Natalidae; THY, Thyropteridae; VES, Vespertilionidae; and MOL, Molossidae).

b Bats mist-netted and necklaced during the fruit-targeted capture phase of the Bat Project, 1977–1985, are ranked in order of number of individuals marked.

c See Fig. 2 and text for guild designation and description. N.B. Guild 8, gleaning frugivores of hightly-cluttered space, has three subdivisions: 8f, mostly fig. diet; 8p, mostly piper diet; and 8o, other frugivores.

d Mass of males (mean ± standard deviation, in grams); sample size (n =).

e Diet categories are fruit (FR), pollen and nectar (PO), insects (IN), and vertebrates (VE). Sample size (N =).

f Sources include GO, Goldman (1920); EN, Enders (1930, 1935); HJ, Hall and Jackson (1953); KO, Koford (unpublished checklist, 1957); MI, miscellaneous (1—Wilson, 1973, Buena Vista Peninsula, USNM 503470; 2—Gardner and Wilson, 1973, BCI, USNM 503833; 3—Koford, 1956, Frijoles, USNM 304937-40; 4—Klingener, 1975, Buena Vista Peninsula, Univ. Mass. Mus. Zool. 2642; 5—Wilson, 1969, BCI, Univ. New Mexico 29058; 6—Wilson, 1972, Buena Vista Peninsula, USNM 503446); BO, Bonaccorso (1979); HA, BCI Bat Project, Handley *et al.* (1991); KA, Kalko (1991–1995). Superscripts (ˆ) from roost, (*) acoustic identification.

second with 7 species, followed by evening bats (Vespertilionidae) and free-tailed bats (Molossidae) with 6 species each, leaf-chinned bats (Mormoopidae) with 3 species, bulldog bats (Noctilionidae) and disk-winged bats (Thyropteridae) with 2 species each, and funnel-eared bats (Natalidae) with 1 species.

B. Cumulative Species Curves

Cumulative species curves of bats reflect species richness of a given area and convey a first impression of relative abundances of species. We compared cumulative numbers of species captured in 10 mist nets in the first 30 consecutive netting nights in each of our 7 study periods in the late rainy season (1977–1985). This revealed to us how much capture effort had to be expended to sample a large fraction of the community and whether the species accumulation curves differed between the yearly sampling periods. Although netting success varied dramatically from night to night in numbers of individuals caught, all cumulative species curves showed remarkably similar trends in terms of numbers of species and rate of accumulation (Figs. 1a and 1b). The 1977 curve was somewhat lower than the curves of the other 6 study periods overall, but the similarity of the curves showed that we were drawing our samples from a stable community. On average it took two netting nights to catch 10 species, seven nights to net 20 species, and 22 nights to capture 30 species. The species total for the seven 30-night intervals was 43. Typically, more than 90% of these species were phyllostomids. They included almost all of the frequently netted species of BCI.

C. Guild Structure and Species Composition

In discussions with H.-U. Schnitzler we categorized the bats of BCI and vicinity in guilds following the concept of Root (1967), who characterized guilds by habitat, foraging mode, and diet (Fig. 2). To allow for standardized comparisons, we defined categories in terms of tasks which must be accomplished by all foraging bats anywhere in the world (Schnitzler and Kalko, 1996). Most notably there are obstacles such as foliage, ground, or water surfaces that represent mechanical as well as perceptual problems for bats (Fenton, 1990). For each guild, we found that echolocation signal design and wing shape indicate adaptations of bats to specific ecological requirements that are the guild determinants (Norberg and Rayner, 1987; Neuweiler, 1989, 1990; Fenton, 1990; Schnitzler and Kalko, 1996). Mechanically, wing morphology determines the habitats in which a bat can forage. For example, long, pointed wings adapt bats for fast, agile flight in open space whereas short, broad wings are best suited to slow,

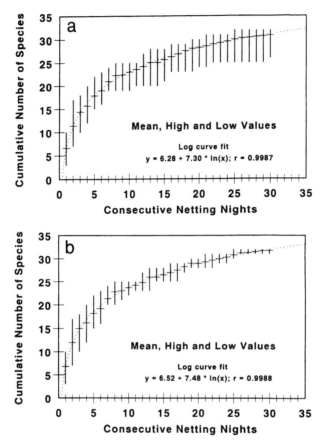

FIGURE 1.   Mean, high, and low values of cumulative species curves for seven sampling periods in the late rainy season (September–December), with fitted log curve. (a) Years 1977–1984; (b) years 1978–1984 (N.B., no records for 1983).

maneuverable flight in an obstacle-rich environment (Norberg and Rayner, 1987; Fenton, 1990). Perceptually, bats are constrained by their sensory capabilities (echolocation, vision, olfaction, and passive listening) to habitats in which they can detect and locate food (Neuweiler, 1989; Fenton, 1990; Schnitzler and Kalko, 1996).

Habitat in our proposed guild concept refers to the main foraging area of a bat in terms of its proximity to obstacles which we hereafter refer to as *clutter* (Fig. 2). In our categorization of habitats we discriminate between uncluttered, background cluttered, and highly cluttered space.

Uncluttered space is the open space far from obstacles, high above the

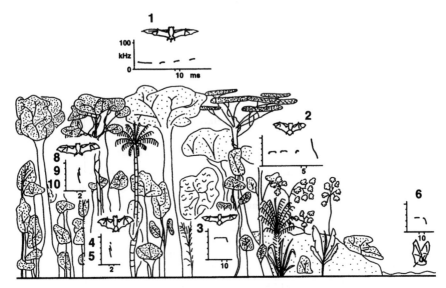

FIGURE 2.   Schematic representation of the BCI bat community guilds. For each guild (num-bered 1 through 10), foraging locations, characteristic wing shape, and echolocation signals (sonagrams: frequency versus time) of representative species are shown. Guilds are further charac-terized by diet and labeled as follows: (1) Uncluttered space/aerial insectivore; (2) back-ground cluttered space/aerial insectivore; (3) highly cluttered space/aerial insectivore; (4) highly cluttered space/gleaning insectivore; (5) highly cluttered space/gleaning carnivore; (6) highly cluttered space/gleaning piscivore; (8) highly cluttered space/gleaning frugivore; (9) highly clut-tered space/gleaning nectarivore; and (10) highly cluttered space/gleaning omnivore. (Guild 7, highly cluttered space/gleaning sanguivore, is not shown here.)

canopy, or high above the ground in open areas (Fig. 2). All bats foraging in uncluttered space rely almost exclusively on echolocation to detect, classify, and localize food and for orientation in space. Design of search phase echo-location signals in uncluttered space typically includes narrowband, low-frequency, shallowly modulated pulses of long duration and long pulse interval.

Background cluttered space includes forest edges and larger gaps or trails surrounded by vegetation or other clutter, the space within the forest between subcanopy and canopy, and, depending on the position of the targeted food, foraging areas over water surfaces (Fig. 2). In this habitat, in contrast to uncluttered space, bats must discriminate between target echoes (food) and clutter-producing background (see Kalko and Schnitzler, 1993; Schnitzler and Kalko, 1996). Some bats are constrained by their echolocation system in how close to vegetation they can feed. Most bats detect and react to prey only when their search signals do not overlap the returning echo from the prey and, moreover, when the echo from the prey does not overlap echoes reflected from the clutter-producing background (Kalko and Schnitzler, 1989, 1993; Kalko,

1995a,b). Search phase echolocation signals in background cluttered space typically have reduced pulse duration and interval; they are wideband, steep frequency modulated signals or they are signals composed of shallow frequency modulated and steep frequency modulated components.

Highly cluttered space is very close to or within vegetation or close to ground or water (Fig. 2). Bats that forage in such clutter-rich environments face the problem of echoes from potential food frequently being buried in echoes from the clutter-producing background. To detect and locate flying prey in highly cluttered space, the aerial insectivores possess a highly specialized echolocation system composed of a long constant frequency component followed by a steep frequency modulated component. Such a search phase signal allows these bats to discriminate between echo patterns produced by rhythmical movements of the prey (e.g., wingbeats of insects) and the clutter-producing background, even when returning echoes overlap the emitted signals (Schnitzler, 1987; Neuweiler, 1990). Gleaning insectivores and frugivores, on the other hand, use other sensory cues such as olfaction, vision, passive listening, or touch to detect, classify, and find food (Tuttle and Ryan, 1981; Laska, 1990; Kalko and Condon, 1993; Schnitzler and Kalko, 1996). They use echolocation mostly for orientation in space. Typically, search phase signals of these bats consist of short, steep, wideband, frequency modulated, often multiharmonic, components.

Foraging mode describes the method and place bats obtain their food. *Aerial* bats catch flying insects on the wing (Griffin *et al.*, 1960; Kalko, 1995a). *Gleaning* bats take more or less stationary food such as fruit, nectar, pollen, nonflying insects, small vertebrates, and blood from clutter-rich surfaces such as ground, vegetation, and water.

Diet is a bat's major food. For BCI we have based our assignments to guilds on feces and food carried into nets by bats (Table I) and, when our samples were small, on detailed field studies reported in literature (Wilson, 1971, 1973; Fleming *et al.*, 1972; Heithaus *et al.*, 1975; Bradbury and Vehrencamp, 1976; Gardner, 1977; Bonaccorso, 1979; LaVal and LaVal, 1980; Whitaker and Findley, 1980; Dunkle and Belwood, 1982; Humphrey *et al.*, 1983; Belwood, 1988a,b; Alonso-Mejia and Medellín, 1991; Willig *et al.*, 1993).

Guilds are useful because they group local associations of taxa living under similar ecological conditions. If, as we suppose, a community is structured in a largely deterministic way, we would expect strong interactions among species within guilds and weak or no interactions between guilds (unless roosts are critical resources). Guilds set standards for describing the structure of communities and provide a standard way of comparing communities, even ones that are geographically remote, with totally different assemblages of species.

The usefulness of the guild concept is often disputed, since many organisms are highly flexible in their behavior and make it difficult to clearly define boundaries between guilds and to assign individual species to a particular guild. Birds, for example, often alter their diets dramatically between seasons or even between habitats (Poulin *et al.*, 1994). However, bat guilds appear to be generally more discrete than guilds of other major taxa. For assignment of species to our guilds, we used choices of habitat, foraging mode, and diet that appear to be the norm for each species at a particular locality.

Since behavior varies geographically with some bats, particularly in terms of diet, occasionally a species may not occupy the same guild throughout its range. In the BCI bat community, diets of most species are rather consistent (Table I), but compare those diets with ones found by Willig *et al.* (1993) for the same species in the Brazilian cerrados; *Phyllostomus discolor, P. hastatus,* and *Glossophaga soricina* require different guild assignments there.

Our scheme of guilds provides a category for omnivores, but on BCI only *P. hastatus* and *Micronycteris brachyotis* qualify as true omnivores. Their diets include both animal and plant material on a regular basis. *Phyllostomus discolor,* which we tentatively list as a nectar-drinking bat based on our field observations, may also qualify as an omnivore. We have seen it avidly feed on fruit and insects in captivity. Several other species are marginally omnivorous. For example, some frugivorous phyllostomids (e.g., *Artibeus jamaicensis* and *A. lituratus*) occasionally ingest insects and leaves (Gardner, 1977; Zortéa and Mendes, 1993; Kunz and Diaz, 1995), but fruit is the overwhelmingly dominant component of their diets on BCI (Table I and Morrison, 1978; Handley *et al.*, 1991). Gleaning insectivores such as *Tonatia bidens* and *T. silvicola* occasionally eat fruit, and *Micronycteris nicefori* often eats fruit (Table I and Gardner, 1977), but the major fraction of their diets is insects.

Variations in choice of the other guild determinants are much less frequent than variation in diet. In some instances aerial insectivorous bats may switch foraging habitat and foraging mode (Fenton, 1990), but such changes seem to be limited. Bats cannot access all habitats and resources equally well, since they are constrained by wing morphology and sensory capabilities to a limited range in which they can operate efficiently over long periods of time (Norberg and Rayner, 1987; Fenton, 1990; Schnitzler and Kalko, 1996).

Based on our definitions, we recognize 10 guilds of bats. Representatives of all 10 occur in the BCI bat community (Table I; Fig. 2). Natural history observations in the guild accounts are our own, made on BCI or nearby in Panamá, unless otherwise noted. To describe capture frequency of bats within the guilds on BCI, we assigned abundance status based on the pooled capture records (or marks) from 1977 to 1985: "superabundant"—more than 10,000 individuals marked; "abundant"—1000–10,000 marked; "common"—100–999 marked; "uncommon"—10–99 marked; and "rare"—fewer than 10 individuals marked (Table I). To categorize component species in BCI guilds by size we recognized six classes of body mass: "very small"—less than 10 g; "small"—10–19 g; "medium"—20–39 g; "large"—40–59 g; "very large"—

60–90 g; and "giant"— >90 g. Mass was obtained from bats in the field on BCI and from museum specimen labels. Several species in the BCI bat fauna weigh as little as 3 g, and *Vampyrum spectrum,* the largest New World bat, reaches 190 g (Table I).

1. Uncluttered Space/Aerial Insectivores

Bats of this guild forage for flying insects in open space far from obstacles and frequently high above the ground or canopy (Fig. 2). For this reason we did not capture any of them with our standard mist-netting procedure. They eat their prey on the wing and thus do not use dining roosts. We know of their presence by day roosts, echolocation signals, and visual observation in flight. By these means, we found six species of at least two families representing this guild on BCI (Table I). Undoubtedly there are others we have not detected.

Three kinds of free-tailed bats (Molossidae) roost in attics of buildings in the Laboratory Clearing. They have narrow wings and fast, direct flight. The small *Molossus coibensis* has colonies in most of the attics, and it is a common sight over the clearing at twilight. We found two other free-tails (medium-size *M. bondae* and small *M. molossus*) only in the attic of the tallest building (three stories) in the lab complex.

We have recorded the echolocation signals of two kinds of sheath-tailed bats (Emballonuridae). On two occasions we observed and recorded the small white bat, *Diclidurus albus,* high above the Lab Clearing. Even though it forages only after dark, sometimes it can be detected by its audible calls, and its white fur and wings make it easy to spot with a night vision scope or strong light beam. On the Bohio Peninsula across the canal from BCI, we have recorded the ultrasounds of a *Peropteryx* that we presume, on the basis of habitat, to be the very small *P. kappleri.* We think scarcity of open ground for foraging and lack of caves for roosts may exclude the very small *P. macrotis* from Bohio.

We have also recorded the ultrasounds of a bat that we have previously misidentified as *P. kappleri* (Kalko, 1995b). Some of the echolocation calls of this species are audible to human ears. We observed this bat foraging in evening and morning twilight, 10–15 m or sometimes even higher above the ground, in open areas adjacent to forest, in large forest gaps, above the canopy, and among buildings in the BCI Lab Clearing. We have not included it in any of the tables because we are uncertain of its familial affiliation.

Except for the molossids, we have not found any of these bats roosting on BCI. Elsewhere we have found *Diclidurus albus* roosting under palm fronds and under a bridge and *P. kappleri* in caves and tree holes. *Molossus coibensis* appears to be the only common member of this guild on BCI. The other species may be rare, or their numbers may be small. Intensive acoustic monitoring is needed to determine their true abundance. In time, additional species may be detected on BCI by this method.

2. *Background Cluttered Space/Aerial Insectivores*

Typically these bats hunt small insects on the wing in open spaces in the forest, in forest gaps, and at forest edges (Fig. 2). This guild is taxonomically more diverse than any other on BCI in terms of families represented and second in terms of species. It includes 18 species of 7 families, representing 27.3% of all species of bats documented on the island. With the exception of *Noctilio albiventris*, a medium-size bat, the members of this guild are small or very small (Table I).

We know very little about the specific diet of these bats on BCI, but we do know that foraging and roosting behavior vary widely among the species and in most cases are not correlated with family affiliation. Many species begin to forage during twilight or even earlier (most notably *Rhynchonycteris naso*, *Cormura brevirostris*, and *Noctilio albiventris*). Four or five species characteristically forage over water (*R. naso*, *N. albiventris*, *Macrophyllum macrophyllum*, *Myotis albescens*, and maybe *Myotis riparius*). Seven species usually forage in the forest interior (*Saccopteryx bilineata* along trails, low above the shrub layer, and in gaps; *S. leptura* and *C. brevirostris* in the space between canopy and subcanopy and high in gaps; *Centronycteris maximiliani*, *Thyroptera discifera*, and *T. tricolor* probably at low levels, but we are not sure exactly where; and *Rhogeessa tumida* around tops of shrubs). All of these sometimes come out of the forest to forage at forest edges. The other six species usually forage in large gaps or in the open along forest edges (*Pteronotus gymnonotus*, *P. personatus*, *Natalus stramineus*, *M. nigricans*, *Eptesicus furinalis*, and *Lasiurus blossevillii*). Bats of this guild eat their prey on the wing.

Roosting behavior of these species also varies without respect to family:

a. Associated with water: *R. naso* under tree trunks and logs leaning over water; *M. albescens* and *N. albiventris* in tree holes and artificial structures over or near water; *M. macrophyllum* in hollow logs near water, in tunnels through which water flows (Brazil), or in houses (Venezuela).

b. Tree trunks: *S. bilineata* low on tree trunks, in large tree hollows, under eaves of houses, and in houses; *S. leptura* high on tree trunks, near the canopy; *M. nigricans*, *E. furinalis*, and *R. tumida* in tree holes and in buildings; *C. brevirostris* under buttresses or trunks of fallen trees.

c. Foliage: *C. maximiliani* and *L. blossevillii* under leaves; *T. discifera* and *T. tricolor* in large furled leaves.

d. Caves: *P. gymnonotus*, *P. personatus*, and *N. stramineus* in caves or cave-like structures (thus accidental or rare on BCI where there are no caves).

Allowing for our inadequate sampling of this guild, it appears that most of its species are rare or uncommon on BCI. Based on acoustic monitoring and visual observation, four species can be classed as common to abundant: *C. brevirostris, S. bilineata, S. leptura,* and *M. nigricans.*

3. Highly Cluttered Space/Aerial Insectivores

These bats forage for flying insects in more or less thick vegetation (Fig. 2). On BCI this guild is represented by a single species (Table I), the medium-size mustached bat, *Pteronotus parnellii* (Mormoopidae). A highly specialized echo-location and hearing system adapts this bat to hunt on the wing for fluttering insects in dense vegetation at low levels in the forest (Schnitzler, 1987; Neu-weiler, 1990). It roosts in tree holes and is a common species on BCI.

4. Highly Cluttered Space/Gleaning Insectivores

The gleaning insectivores are forest bats, perhaps good indicators of forest quality. They glean prey from surfaces. The nine species found on BCI are all in the family Phyllostomidae. Three species (*Micronycteris megalotis, M. schmid-torum,* and *Tonatia brasiliensis*) are very small, three are small (*M. hirsuta, M. nicefori,* and *Mimon crenulatum*), and three are medium-sized (*Tonatia bidens, T. silvicola,* and *Trachops cirrhosus*) (Table I).

Bats of this guild glean prey from foliage, tree trunks, logs, the ground, and water surfaces (Fig. 2). They may specialize in particular species or particular size classes of insects, including beetles, roaches, katydids, cicadas, moths, butterflies, and dragonflies. We observed a tendency for bigger bats to eat bigger insects, but small and very small gleaning insectivores such as *M. megalotis* frequently handle prey with linear dimensions approximating those of the bat itself. Although bats of this guild are primarily or exclusively insectivorous, *T. cirrhosus* sometimes catches frogs. However, most individuals we handled on BCI had eaten insects. We also found *T. bidens* eating insects, although Martuscelli (1995) reported birds in its diet in Brazil. All gleaning insectivores have large ears, which they use with a passive hunting strategy. Our preliminary radio-tracking studies of *T. silvicola* and *T. cirrhosus* indicated that these species mostly employ a sit-and-wait foraging strategy. They hang from perches and wait for prey, listening for calls or sounds of movements (Krull and Kalko, 1994). Bats of this guild usually carry prey to a dining roost different from the day roost. There, the inedible parts of the prey are discarded.

Roosting behavior is diverse in this guild. *M. hirsuta* and *M. megalotis* roost in hollow logs, tree holes, and buildings; *M. crenulatum* in tree snags open at the top; and *T. silvicola* in active termite nests hollowed-out at the bottom. We do not know where the other species roost on BCI. Except for *T. brasiliensis,* which was caught only once, prior to the Bat Project, this guild does not include any rare or infrequently caught species (Table I).

5. Highly Cluttered Space/Gleaning Carnivores

These are forest bats that glean prey from surfaces (Fig. 2). The guild is represented on BCI, as elsewhere in Central and South America, by two species of phyllostomids, the very large *Chrotopterus auritus* and the giant *Vampyrum spectrum* (Table I). *Chrotopterus auritus* has been reported to capture rodents, birds, and lizards (Medellín, 1988), but those we have handled on BCI and elsewhere in Panamá had eaten only beetles. The *V. spectrum* we caught in Panamá had preyed on a spiny rat (*Proechimys semispinosus*) and a tent-building bat (*Uroderma bilobatum*). Captive individuals on BCI expertly ate rats, bats, birds, and lizards they were offered. There is at least one permanent group of *C. auritus* on BCI which we suppose roosts in a hollow log. Probably the *V. spectrum* we infrequently caught on BCI were transients from the mainland (Table II).

6. Highly Cluttered Space/Gleaning Piscivore

Foraging over water, these bats mostly glean prey from the surface (Fig. 2). The fishing bat, *Noctilio leporinus* (Noctilionidae), a large bat, is the only member of this guild on BCI as well as in all of Central and South America. It uses its long, curved, fishhook-like claws to gaff small fish while randomly dragging or repeatedly dipping its claws in the water, or by sporadically dipping for specific prey. It also apparently feeds on small shrimp and occasionally catches insects in the air (Brooke, 1994; Schnitzler *et al.*, 1994). Since this bat forages primarily over water, it is inherently underrepresented in our samples (Table II), but acoustic monitoring and observations with the night vision scope showed it to be fairly common around BCI. We occasionally caught it in mist nets over small streams, where it probably was en route to or from a roost in a hollow tree.

7. Highly Cluttered Space/Gleaning Sanguivores

Bats of this guild forage in the forest and in the open to lap blood of vertebrate prey from wounds they inflict with their razor-sharp incisors. The common vampire (*Desmodus rotundus;* Phyllostomidae), a medium-size bat, is the only representative of this guild on BCI (Table I), but two other vampires occur elsewhere in Panamá and could eventually be found on the island. On BCI, where it must attack wild prey, we caught few *D. rotundus* in each sampling period. Its low capture rate on the island contrasts with mainland areas in central Panamá where it is a common bat. Absence of livestock on and near the island can account for its rarity there. On BCI it roosts in hollow trees.

8. Highly Cluttered Space/Gleaning Frugivores

Bats of this guild use mostly olfaction and to a lesser extent echolocation to glean fruit from foliage of trees and shrubs (Kalko and Condon, 1993; Kalko, 1994; Fig. 2). With a total of 20 species, all in the family Phyllostomidae, this is the most speciose guild on BCI (Table I). Size is reflected in diet. Larger species tend to eat larger fruit; smaller species tend to eat smaller fruit (Kalko *et al.,* 1996). These frugivores vary in size from very small to very large. Except for the small bats, with seven common to abundant species, each size category is represented by only one or two common to very abundant species, thus presumably spreading available fruit resources rather evenly among many species—very small: *Vampyressa pusilla* (rare: *Ametrida centurio* and *Mesophylla macconnelli*); small: *Artibeus phaeotis, Artibeus watsoni, Platyrrhinus helleri, Uroderma bilobatum, Vampyressa nymphaea, Carollia castanea,* and *Carollia perspicillata* (rare: *Uroderma magnirostrum, Carollia brevicauda,* and *Artibeus hartii*); medium: *Chiroderma villosum,* and *Vampyrodes caraccioli* (rare: *Centurio senex* and *Sturnira luisi*); large: *Artibeus jamaicensis;* very large: *Artibeus lituratus* (rare: *Phylloderma stenops*).

The distribution of these bats apparently coincides with availability of suitable fruit, whether in the open or in the depths of the forest. Bats seem to fly directly to a fruit source (following odor?); then, if it is a tree, they circle inside or outside the tree, and soon approach an individual fruit (Kalko and Condon, 1993; Kalko, 1994; Kalko *et al.,* 1996). The bats bite into the fruit and pull it free in flight or with a brief landing. They then fly to a temporary dining roost, often under a palm frond, to eat the fruit. Large seeds are discarded and small seeds swallowed (Handley *et al.,* 1991). At all times flight is swift and darting.

In terms of major diet items, frugivorous bats form three distinct groups (Table I). The first group, equivalent to the "canopy frugivores" of Bonaccorso (1979), consists of 11 species, which on BCI subsist mostly on fruits of figs (*Ficus* sp.) (Bonaccorso, 1979; Handley *et al.,* 1991; Kalko *et al.,* 1996). The second group, equivalent to the "understory frugivores" of Bonaccorso (1979), includes the three species of *Carollia,* which on BCI feed mostly on the fruits of Piperaceae (Bonaccorso, 1979; Fleming, 1988; our unpublished data). The remaining species, the third group, either are not fig or piper specialists or their diets are poorly known.

The *Carollia* roost in tree holes, under tree roots and overhanging banks, and occasionally in hollow logs and houses. Most of the very small and small species of fig bats are tent-builders. They modify leaves to form shelters. The medium, large, and very large species roost under sheltering foliage. Nursery colonies of *Artibeus jamaicensis* are in tree holes. Elsewhere in Panamá we have

TABLE II Temporal Variability in Capture Rate of Infrequently Caught Species (Not Caught Every Year) in the BCI Bat Project

	1975	1976	1977	1978	1979	1980	1981	1982	1984	1985	1995
Guild 1: Uncluttered space/aerial insectivores											
Molossus bondae									4˜		
Molossus coibensis	5˜	13˜									
Molossus molossus	1˜								11˜		
Guild 2: Background cluttered space/aerial insectivores											
Centronycteris maximiliani				2							
Myotis albescens	1	1					2				
Myotis nigricans	33˜	11˜	4	1	7	4	2	4		3	
Myotis riparius									1		
Natalus stramineus											1
Noctilio albiventris	15˜	21˜	11˜			5	1	2	2	2	
Pteronotus gymnonotus	1										
Rhogeessa tumida	4	10	6	1	3	1		2	1	1	
Rhynchoncteris naso		4					1				
Saccopteryx leptura			2	2							
Thyroptera discifera									1		
Thyroptera tricolor		2		3							

Guild / Species	3	5	4	3	4	9	14
Guild 4: Highly cluttered space/gleaning insectivores							
Micronycteris schmidtorum				3	4	9	14
Guild 5: Highly cluttered space/gleaning carnivores							
Chrotopterus auritus				2	3		
Vampyrum spectrum	1			1		1	2
Guild 6: Highly cluttered space/gleaning piscivores							
Noctilio leporinus	2				1	1	
Guild 8: Highly cluttered space/gleaning frugivores							
Ametrida centurio		1					
Artibeus hartii		1		1			
Carollia brevicauda	1		2	3	1	1	
Centurio senex	4		2	2	8		3
Mesophylla maccomelli	4						
Phylloderma stenops	3	5	6		1		
Sturnira luisi							4
Uroderma magnirostrum	1	2		1		1	1
Guild 9: Highly cluttered space/gleaning nectarivores							
Glossophaga commissarisi	1	2	13		2		
Lonchophylla robusta	2	1	1				4

Note. Within guilds, species are sorted alphabetically. Symbol (ˆ) from roost.

527

found it roosting also in caves and houses. We do not know where *Chiroderma villosum*, *Sturnira luisi*, and *Phylloderma stenops* roost on BCI.

The frugivores dominated our capture records in both numbers of species and frequency of capture (Table I). The first eight species in rank abundance in the BCI bat community are in this guild. Twelve species are regarded as super-abundant, abundant, or common. On the other hand, eight are rare or uncommon, caught less than 20 times by the project (Table I).

9. Highly Cluttered Space/Gleaning Nectarivores

Bats of this guild forage from flowers of trees, shrubs, or vines (Fig. 2). On BCI the guild includes four species of Phyllostomidae (Table I). *Glossophaga commissarisi* is very small, *G. soricina* and *Lonchophylla robusta* are small, and *Phyllostomus discolor* is large. All of these bats may be found in the open or in the forest, depending on the availability of suitable flowers. The three smaller species have long extensible tongues tipped with feathery papillae for lapping nectar. They also eat pollen and sometimes insects they find in the flowers. *Glossophaga soricina* occasionally eats fruit. The tongue of *P. discolor* is not extensible, so it feeds on larger flowers such as those of *Ochroma pyramidale* and *Lafoensia punicifolia* into which it can stick its head if necessary to reach the nectar. The small species roost in hollow logs and tree holes, and *G. soricina* sometimes roosts in houses. *Phyllostomus discolor* roosts in tree holes. *Glossophaga soricina* and *P. discolor* were common and frequently caught, but *P. discolor* appeared to decline significantly from 1977 to 1985. *Glossophaga commissarisi* and *L. robusta* were uncommon or rare.

10. Highly Cluttered Space/Gleaning Omnivores

These bats glean a wide variety of animal and vegetable foods (Fig. 2). We assign only two phyllostomids, *Phyllostomus hastatus* and *Micronycteris brachyotis*, to this guild (Tables I and III). The small *M. brachyotis* is a forest bat probably foraging at low levels, while the giant *P. hastatus* seems not to be restricted to any particular habitat and forages at all levels. In both species we consistently found a mixed diet of nectar, pollen, fruit, insects, and, in *P. hastatus*, also small vertebrates (Table I; Gardner, 1977; our unpublished data). *Phyllostomus hastatus* roosts both in forest and in open areas in hollow trees and in canal markers around BCI. It is a common species. We do not know where the uncommon *M. brachyotis* roosts on BCI.

D. Patterns of Relative Abundance

We pooled the 1977–1985 records of all bats mist-netted with our standard (fruit-targeted) capture protocol, excluded recaptures, and ranked the species according to their numerical abundance (Table I). The rank abundance curve of the BCI bat community (Fig. 3a) provides a familiar picture, characteristic of many other groups of organisms. A few abundant and superabundant species dominated the community, followed by a number of common species, and many uncommon and rare species forming the tail-end of the distribution.

Numerically, the common fruit bat (*Artibeus jamaicensis*) was the overwhelmingly dominant species. With almost 18,000 captures it made up 53.8% of all marks (Table I). It was followed by the four abundant gleaning frugivorous species, which contributed 27.3% of the marks, and 17 common species of five guilds accounting for 17.2% of the marks. Fourteen uncommon species of six guilds and 16 rare species of seven guilds translated to only 1.5 and 0.2% of all marks, very small fractions of the nettable bat community.

The 22 most numerous species (superabundant, abundant, and common in our classification) amounted to 98.3% of all marks and a third of the species documented for BCI. More than half of the most numerous species (12 of 22) were frugivores, far exceeding the rest of the community in numerical abundance. For instance, the five most abundant frugivores amounted to less than 10% of all species found on BCI, but accounted for more than 80% of individual bats mist-netted. In contrast, the 30 species that were labeled as uncommon or rare, and which represented half the species documented by mist-netting, contributed only 1.7% to the total catch of individuals.

To account for the effects that our fruit-targeted capture protocol may have had on occurrence and relative abundance of species, we compared the data obtained from 1977 to 1985 with the data collected with a random sampling protocol in 1975 and 1976. We found the rank abundance curves for all species to be similar in both data sets (Fig. 3a). Comparing the 20 highest ranking species in the two data sets reveals the dominance of frugivorous bats in each (Fig. 3b). These included "fig-eating" bats (e.g., *Artibeus* sp., *Vampyressa* sp., *Vampyrodes caraccioli*, and *Uroderma bilobatum*) as well as "mostly piper-eating" bats (*Carollia* sp.). *Artibeus jamaicensis*, of course, was by far the most sampled species. Both curves also included gleaning insectivores, gleaning nectarivores, and one highly cluttered space/aerial insectivore (*Pteronotus parnellii*).

With few exceptions, species composition was similar regardless of sampling method. The most striking difference was the rank position of two fig-eating bats, *Uroderma bilobatum* and *Chiroderma villosum*. They occupied ranks 3 and 6 respectively in the fruit-targeted sampling, but ranks 19 and 20 in the random sampling (Fig. 3b). These species clearly were oversampled with the fruit-targeted protocol. We concluded nevertheless that fruit-targeting provides a rather accurate picture of actual occurrence and relative abundance of frugivorous and insectivorous gleaning bats on BCI.

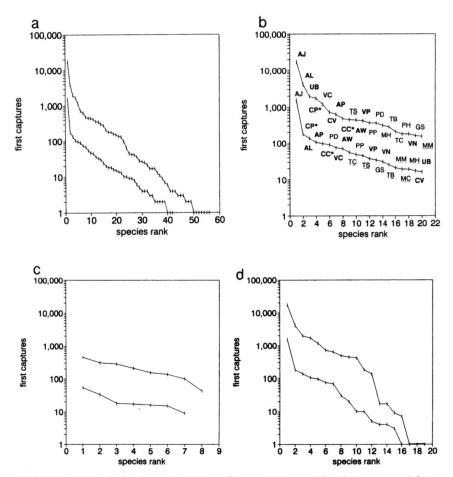

FIGURE 3. (a) Rank abundance distribution of bats on BCI sampled with (upper curve) fruit-targeted mist-netting, 1977–1985, and (lower curve) random mist-netting, 1975–1976. (b) Rank abundance of the 20 most common species of bats on BCI sampled with (upper curve) fruit-targeted mist-netting, 1977–1985, and (lower curve) random mist-netting, 1975–1976. Guild 8 species (gleaning frugivores) abbreviations are in bold caps, with an asterisk for "mostly piper-eating" species; guild 4 species (gleaning insectivores) are in underlined caps, and species of other guilds are in plain caps: **AJ,** *Artibeus jamaicensis;* **AL,** *A. lituratus;* **AP,** *A. phaeotis;* **AW,** *A. watsoni;* **CC*,** *Carollia castanea;* **CP*,** *C. perspicillata;* **CV,** *Chiroderma villosum;* GS, *Glossophaga soricina* (guild 9); M̲C̲, *Mimon crenulatum;* M̲H̲, *Micronycteris hirsuta;* **M̲M̲,** *M. megalotis;* PD, *Phyllostomus discolor* (guild 9); PH, *P. hastatus* (guild 10); PP, *Pteronotus parnellii* (guild 3); T̲B̲, *Tonatia bidens;* T̲C̲, *Trachops cirrhosus;* T̲S̲, *Tonatia silvicola;* **UB,** *Uroderma bilobatum;* **VC,** *Vampyrodes caraccioli;* **VN,** *Vampyressa nymphaea;* **VP,** *V. pusilla.* (c) Rank abundance distribution of guild 4 species, highly cluttered space/gleaning insectivores sampled with (upper curve) fruit-targeted mist-netting, 1977–1985, and (lower curve) random mist-netting, 1975–1976. (d) Rank abundance distribution of guild 8 species, highly cluttered space/gleaning frugivores, sampled with (upper curve) fruit-targeted mist-netting, 1977–1985, and (lower curve) random mist-netting, 1975–1976.

E. Rank Abundance within Guilds

We selected for further analysis the two best sampled guilds, gleaning fru-givores and gleaning insectivores. Comparing rank abundance curves of these guilds, sampled with our standard fruit-targeted protocol (1977–1985), re-veals striking differences in species richness and relative abundance (Figs. 3c and 3d).

In the frugivorous guild, with 19 species (*Mesophylla macconnelli* was caught only in 1975–1976), species richness was more than twice that of the insectivorous guild, which had 8 species (Table I; Figs. 3c and 3d). Whereas species abundance in the frugivorous guild covered the full range from super-abundant (1), abundant (4), common (7), uncommon (2), to rare (5), most species of the insectivorous guild were common (6) and 2 were uncommon. This pattern is well reflected in the rank abundance diagrams, which reveal a higher degree of evenness for the insectivorous guild (Fig. 3c) than for the frugivorous guild (Fig. 3d) with extremes at both ends of the distribution. The frugivores span a broad spectrum of sizes, from very small (*Vampyressa pusilla*) to small and medium-sized, to large (*Artibeus jamaicensis*), to very large *Arti-beus lituratus* (Table I). The gleaning insectivores likewise include a very small species (*Micronycteris megalotis*), but maximum mass in this guild, with 3 species weighing about 35 g, is much less than that in the frugivorous guild. Adding masses of species within a guild and including differences in relative abundances lead to a mean mass of 44.4 g for the frugivorous guild and a mean mass of 25.5 g for the gleaning insectivorous guild on BCI.

Since the frugivores, particularly the larger species, are numerically domi-nant species in the community, this guild is likely to contribute dispropor-tionally to the total biomass of the bat community on BCI. However, to ade-quately estimate biomass, data on actual population densities are necessary. Ongoing analysis of the 1975–1985 capture–recapture records will facilitate community biomass estimates.

To ensure that the observed differences between the two guilds were not just an artifact of our sampling protocol, we compared the results obtained by fruit-targeted mist-netting (1977–1985) with the data obtained by random sampling (1975–1976). Generally, the shape of the curves and differences in species richness and abundance patterns between the two guilds were similar in each sample (Figs. 3c and 3d). The curves differed in the number of individuals per species and in the number of species in the random sampling (Table II). These discrepancies may be attributable to the large difference in capture effort between the sampling periods.

F. Long-Term Trends in Guild Structure

All 10 guilds of the BCI community were present throughout the study period, 1977–1985. However, species composition of the guilds varied. A basic group of 27 species (48.3% of all species mist-netted) was present in each of the yearly samples, and 29 species (51.7%) were caught only infrequently (Table II). Among the bats infrequently caught, 16 species belonged to the guilds conspicuously undersampled in our mist net monitoring. In general, species that contributed to annual increase or decrease of species richness within guilds were never sampled in large numbers. Of those, 19 species were represented by fewer than 10 individuals, and 7 of them were caught only once during the study (Tables I and II).

G. Long-Term Population Trends

For analysis of long-term population trends we used annual samples from the late rainy season (see Methods). We chose regression analysis to give a first impression of population trends in the BCI bat community. However, we are aware that estimation of temporal variability of populations requires complex methodology (Stewart-Oaten *et al.*, 1995) to minimize analysis artifacts. A more refined analysis including population estimates based on capture–recapture data is in progress.

Among the 24 species analyzed in detail, 19 (79%) showed positive and five (21%) showed negative trends (Table III); positive trends outweighed negative trends by a ratio of 4:1. After a standard Bonferroni adjustment one nectarivorous bat (*Phyllostomus discolor*) showed a significantly negative trend at the $P < 0.001$ level. The trends of one gleaning insectivorous bat (*Tonatia bidens*) and two frugivorous bats (*Artibeus jamaicensis* and *Vampyressa nymphaea*) were marginally positive at $P < 0.02$ and $P < 0.05$ levels, respectively. Population trends, as indicated by the slopes of the regression, were remarkably similar for most species, regardless of guild (Table III). Long-term trends for the BCI bat community indicated an almost uniform, slightly positive increase for most species.

In contrast, our preliminary analysis indicated considerable differences in short-term (annual) fluctuations among species (Figs. 4a–4x). Most species among the gleaning insectivores and gleaning frugivores showed a synchronous increase in relative abundance from 1978 to 1979 followed by a decrease from 1979 to 1980 (Figs. 4b–4t). On the other hand, most species showed dissimilar short-term fluctuations in other years. During the study period some species pairs, such as *Carollia castanea* and *C. perspicillata* (Figs. 4s and 4t), sometimes fluctuated inversely (i.e., increase in one species appeared to be associated with simultaneous decrease in the other species).

TABLE III   Long-Term Population Trends of the Superabundant, Abundant, and Common Bats plus an Uncommon Omnivore (*Micronycteris brachyotis*) in the BCI Bat Community during the Late Rainy Season over 7 Years (1977–1984)

	Slope	t	r^2 (%)	P	Standard error of estimate
Guild 3: Highly cluttered space/aerial insectivores					
Pteronotus parnellii	−0.0124	−0.72	9.5	0.501	0.017
Guild 4: Highly cluttered space/gleaning insectivores					
Micronycteris hirsuta	+0.0119	0.64	4.5	0.644	0.024
Micronycteris megalotis	+0.0478	1.76	38.3	0.138	0.027
Micronycteris nicefori	+0.0493	1.48	30.4	0.198	0.033
Mimon crenulatum	+0.0375	0.19	30.3	0.199	0.025
Tonatia bidens	+0.0447	3.41	69.9	0.019	0.013
Tonatia silvicola	+0.0422	1.66	35.4	0.158	0.025
Trachops cirrhosus	+0.0159	1.10	19.3	0.322	0.014
Guild 8: Highly cluttered space/gleaning frugivores					
Species with a "mostly fig diet"					
Artibeus jamaicensis	+0.0519	3.19	67.1	0.024	0.016
Artibeus lituratus	−0.0364	−1.01	17.0	0.357	0.035
Artibeus phaeotis	−0.0097	−0.20	0.7	0.851	0.049
Artibeus watsoni	+0.0504	2.42	53.9	0.059	0.020
Chiroderma villosum	+0.0510	0.50	4.7	0.639	0.102
Ptalyrrhinus helleri	+0.0532	1.74	37.8	0.141	0.030
Uroderma bilobatum	+0.0615	1.26	24.0	0.263	0.048
Vampyressa nymphaea	+0.0935	2.86	62.1	0.035	0.030
Vampyressa pusilla	+0.0675	2.41	53.7	0.060	0.028
Vampyrodes caraccioli	+0.0483	1.93	42.7	0.111	0.025
Species with a "mostly piper diet"					
Carollia castanea	+0.0000	0.00	0.0	0.997	0.026
Carollia perspicillata	+0.0375	1.41	28.4	0.217	0.026
Guild 9: Highly cluttered space/gleaning nectarivores					
Glossophaga soricina	+0.0949	2.28	50.9	0.071	0.041
Phyllostomus discolor	−0.1803	−7.57	91.9	0.0006*	0.023
Guild 10: Highly cluttered space/gleaning omnivores					
Micronycteris brachyotis	+0.0048	0.26	13.4	0.804	0.018
Phyllostomus hastatus	−0.0730	−1.81	39.7	0.129	0.040

Note. Relative abundance is calculated as log of capture frequency per 100 mist net hours. Within guilds, species are sorted alphabetically.
*$P = 0.001$.

Furthermore, whereas gleaning insectivorous bats, such as *Tonatia bidens*, *T. silvicola*, and *Trachops cirrhosus* (Figs. 4f–4h) and frugivores, such as *Artibeus jamaicensis*, *Vampyrodes caraccioli*, *Carollia perspicillata*, and *C. castanea* (Figs. 4i and 4r–4t), showed only small year-to-year population fluctuations,

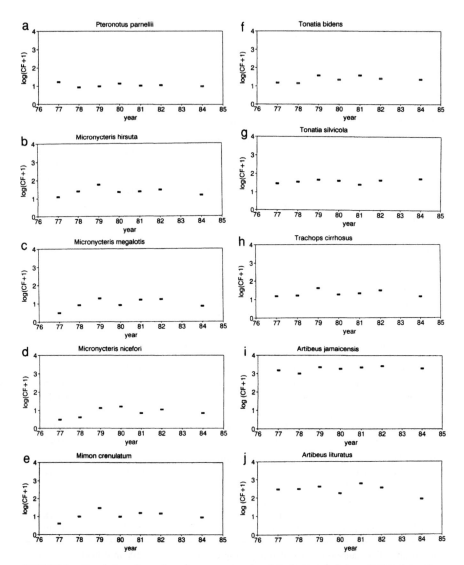

FIGURE 4. Population dynamics of common to abundant bats sampled on BCI in late rainy season (September–December) with fruit-targeted mist-netting, 1977–1984. For each species the ordinate measures annual average relative abundance as log of capture frequency (CF) per 100 mist net hours plus one: log (CF+1). The bats are ordered according to guild membership and, within guilds, in alphabetic order. (a) guild 3: highly cluttered space/aerial insectivores; (b–h) guild 4: highly cluttered space/gleaning insectivores; (i–t) guild 8: highly cluttered space/gleaning frugivores, with "mostly fig-eating" bats; (i–r) and "mostly piper-eating" bats (s–t); (u–v) guild 9: highly cluttered space/gleaning nectarivores; (w–x) guild 10: highly cluttered space/gleaning omnivores. For 7-year summary see Table III.

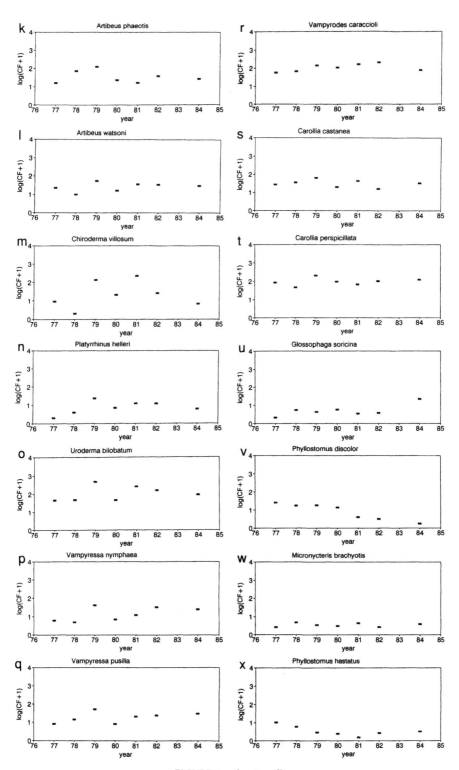

FIGURE 4. *(continued)*

other frugivorous bats, particularly *Chiroderma villosum* and *Uroderma bilobatum* (Figs. 4m and 4o), displayed high year-to-year variability.

IV. DISCUSSION

This study represents the first quantitative treatment of any bat community for which data on relative abundance and long-term population trends are available for a large fraction of the community. One of the strengths of the study is the combination of data and observations gathered by many techniques: mist-netting, acoustic monitoring, observations with a night vision scope, radio-tracking, and dietary analysis. If we consider only the mark–recapture data of the mist-netting (48,222 records), our sample is very large, even by the standards of large-scale mist-netting studies of tropical birds (Bierregaard, 1989; Karr *et al.*, 1989; Terborgh *et al.*, 1990).

By using many techniques we can appreciate their individual limitations. For instance, the use of mist nets is the method of choice for sampling a large subset of bats, the Neotropical leaf-nosed bats (Phyllostomidae). Although phyllostomids continuously echolocate and theoretically could avoid nets, they seem to depend mostly on spatial memory when commuting between roosts and feeding sites. Thus, they can be captured easily with mist nets set in their flyways and around fruit trees. So mist-netting provides excellent presence–absence data for the phyllostomids and, combined with mark–recapture studies, allows estimates of relative abundances.

However, mist-netting grossly underestimates or totally misses aerial insectivores such as sheath-tailed bats (Emballonuridae), free-tailed bats (Molossidae), leaf-chinned bats (Mormoopidae), and evening bats (Vespertilionidae). This is because these bats rely almost exclusively on echolocation for orientation in space and when foraging, and thus mostly avoid mist nets. In addition, many of them routinely forage far too high to be reached by any mist net.

Fortunately, recent advances in acoustic monitoring are likely to bring a solution to this problem because many of the aerial insectivores can be identified by the signature of their unique echolocation signals (e.g., Ahlén, 1981; Fenton and Bell, 1981; Fenton *et al.*, 1983, 1987; Aldridge and Rautenbach, 1987; Crome and Richards, 1988; Kalko, 1995a,b). In addition, it will be possible not only to assess presence or absence with acoustic monitoring, but also to develop standardized measuring techniques such as circular plots and transects to assess relative abundance.

Unfortunately, like mist-netting, acoustic monitoring also has its limitations. It misses phyllostomids and other bats that glean prey, since their echolocation signals usually are very faint and thus difficult to pick up with ultrasound receivers. Furthermore, so far as is known, the structure of phyllostomid signals is so uniform that differentiation of species is difficult if not impossible.

Thus, to determine presence–absence and relative abundance of species in a whole community we must employ both mist-netting and acoustic monitoring. Furthermore, we must also have abundant mark–recapture and radio-tracking data to determine home ranges if we are to convert relative abundance into density. Usually the size of the area from which individuals of a sample have been drawn is unknown. Preliminary analyses of mark–recapture data and radio-tracking studies on BCI have revealed large differences in home ranges within and between guilds (Handley et al., 1991; Kalko and Handley, 1994; Kalko et al., 1996).

While BCI is so close to the mainland (as little as 250 m in some places) that it would seem that bats could commute regularly, this seems not to be the case with many phyllostomids. For these species, reluctance to cross open spaces, including water, outside the forest is likely to be an isolating factor. Furthermore, our studies show that small frugivores and most gleaning insectivores have home ranges from no more than a few hundred meters to less than 2 km² in area. The water barrier combined with small home ranges may severely limit exchange between island and mainland. For many phyllostomids BCI may represent their maximum home range. On the other hand, we also know that some species, particularly some gleaning frugivores, vary dramatically in abundance on BCI, occurring in large numbers periodically during the year (Handley et al., 1991). This strongly suggests (seasonal?) migration onto and off the island. With these caveats in mind, we will consider some of the implications of our preliminary analyses of the BCI bat community.

A. Patterns of Species Richness

Patterns of species richness found in the BCI community, particularly the dominance of phyllostomids in terms of numbers of individuals and species and the lower number of aerial insectivores in the mist net samples, are consistent with those found in other mist net studies in tropical areas of Central and South America (see Dalquest, 1954; Handley, 1967, 1976; Fleming et al., 1972; Heithaus et al., 1975; Fleming, 1986; Brosset and Charles-Dominique, 1990; Handley et al., 1991; Reis and Muller, 1995). After 85 years of observations using a variety of techniques, very few bats known to occur in central Panamá remain to be discovered on BCI (Handley, 1966). In fact, several species are known in Panamá only on BCI or only on the island and a few other localities (Handley, 1966). Composition and number of species in the BCI bat fauna are consistent with faunas of other lowland semi-evergreen sites in Panamá and elsewhere in lower Central America. In this region the total bat fauna will average 65–70 species per site. In contrast, in the Amazon Basin bat faunas of about 100 species are possible.

B. Patterns of Rarity

Five frugivorous phyllostomids (*Artibeus jamaicensis, A. lituratus, Uroderma bilobatum, Carollia perspicillata,* and *Vampyrodes caraccioli*), with 81.1% of all individuals marked, are the dominant fraction of the BCI bat community. The remaining part, although species-rich, contributes little to the overall biomass. However, to interpret the significance of capture frequency, particularly of species labeled as rare in our sample, we must consider a number of ecological and methodological factors that contribute to the observed commonness and rarity of species.

1. Trophic Specialization

For species such as *Vampyrum spectrum* and *Chrotopterus auritus* body size and trophic category are likely to be responsible for their rarity. Large carnivorous species are intrinsically rare because they require extensive areas for foraging and must have large home ranges. We assume that individuals of the top predator *V. spectrum* have their roosts on the mainland and occasionally wander on and off BCI. Since a few *C. auritus* have been caught repeatedly on BCI since 1980, two of them as recently as 1993–1995, a small population may have become resident on the island after inception of the Bat Project.

Some noncarnivorous bats, such as *Ametrida centurio, Centurio senex,* and *Phylloderma stenops,* also have large geographic ranges and appear to maintain low population densities everywhere. Since our knowledge of the ecology and natural history of these species is limited, explaining the causes of their rarity must be speculative at best. It may be that they are extreme food specialists whose populations are limited by low frequency of particular food sources. For instance, although on one occasion *P. stenops* has been found feeding on larvae and pupae of social wasps (Jeanne, 1970), evidence from feces suggests that at least in Panamá and Venezuela it is largely frugivorous, highly specialized for fruits of Cucurbitaceae (our unpublished data). *Ametrida centurio* is a special case. Not only may it be a trophic specialist, but also we attribute its rarity on BCI to probable disjunct distribution.

2. Ecological Requirements

A number of species that are not particularly difficult to net and are abundant elsewhere in central Panamá have been documented only rarely or not at all on BCI. Some of these species may be rare on BCI, or they may not be residents of the island because of the absence of critical resources such as particular foods or roost sites. For instance, flowering shrubs, vines, and early successional trees that bear the flowers upon which nectarivorous bats depend are declining on BCI, where the forest is maturing and natural openings are limited to temporary and small tree-fall gaps. *Glossophaga commissarisi*, which may prefer more mesic conditions, is uncommon and was infrequently caught. *Lonchophylla robusta*, a species of open shrubby habitats, is rare on BCI and has not been caught since 1980 (Table II).

Lack of caves to provide roost sites is likely the major constraint for *Pteronotus gymnonotus*, which has been captured only twice and is heard rarely on BCI; *Natalus stramineus*, which has been recorded only once; and *Peropteryx macrotis* and *Lonchorhina aurita*, which have not been found on BCI. All of these species are known to roost in caves and cave-like structures within 20–25 km of BCI, frequently in large numbers.

Other environmental conditions may have contributed to the rarity of *Artibeus hartii* and *Sturnira luisi*, each of which has been netted only once on BCI. Both are common to abundant in evergreen forests at medium elevations elsewhere in Panamá, and *S. luisi* is common in lowland evergreen forests in Bocas del Toro. They may require more mesic conditions than are found on BCI.

3. Disjunct Distribution

Occurrence of a species outside its normal contiguous range may account for some rarities. *Ametrida centurio*, netted once on BCI, seems to be in this category. This capture is remarkable, since the known range of the species is across the Caribbean at least 1500 km east of BCI, east of Lake Maracaibo in Venezuela, and entirely east of the Andes elsewhere in South America (Handley, 1976; Wilson and Reeder, 1993). Most likely it may have an undiscovered isolated relictual distribution somewhere in Panamá near BCI, or this may have been a wandering individual that accidentally came overland to the vicinity of the island. On the other hand, this individual may have gotten accidentally onto a ship, survived the sea passage, and escaped as the ship passed through the Panama Canal. However, this is unlikely since stenodermatines, especially small ones such as *Ametrida centurio*, quickly starve to death under conditions of stress and food deprivation (Handley *et al.*, 1991).

There are striking examples of apparently disjunct distributions of bats widespread in Central America and South America but unknown in Panamá (*Cyttarops alecto, Pteronotus davyi, Mormoops megalophylla, Leptonycteris curasoae,* and *Lasiurus cinereus*). In addition to these there are several species with similar distributions but including marginal Panamanian localities (*Mimon bennetti, Glyphonycteris daviesi,* and *Lasiurus egregius*) (Hall, 1981; Wilson and

Reeder, 1993). Species of this latter group are all recent discoveries in Panamá (within the past 25–30 years), so it would not be too surprising to find another, *Ametrida centurio.* When Handley discovered *Lasiurus egregius* in northeastern Panamá in 1963, it had been known for almost 100 years by a single specimen from southeastern Brazil.

4. Sampling Techniques

The largest class of rare species in our study consists of bats that forage for insects on the wing, in forest gaps, at forest edges, above the canopy, or high above the ground in open space. Since these bats are difficult or impossible to capture with mist nets, the rarity suggested by our data is at least in part an artifact of inadequate sampling methods and not of actual low numbers. For example, *Cormura brevirostris, Myotis nigricans, Molossus coibensis, Saccopteryx bilineata,* and *Saccopteryx leptura,* all categorized as uncommon or rare on BCI by mist net data, can be scored visually, as abundant in roosts and during crepuscular foraging (Tannenbaum, 1975; our unpublished data), and acoustically, by their characteristic echolocation signals (Kalko, 1995b).

In addition, we know from analysis of sound recordings that the actual number of species of aerial insectivorous bats on BCI is greater than the five species listed here, and some of them are common. However, since our handbook of ultrasounds of Neotropical bats is still incomplete, we can only guess that the unidentified recordings may include molossids known from BCI and the adjacent mainland but not found by the Bat Project (*Nyctinomops laticaudatus, Molossus sinaloae,* and *Eumops auripendulus*) or other species such as *Molossops paranus* and *Promops centralis* that are known from more distant localities in central Panamá.

Type and placement of traps and nets also influence monitoring results. For instance, Bonaccorso (1979) documented mustached bats (*Pteronotus parnellii*) in larger numbers on BCI than the BCI Bat Project found. His use of Tuttle traps and inadvertent placement of traps near roosts could have contributed to his high capture rates. Furthermore, mist nets set at various heights in the forest have demonstrated apparent vertical stratification in bat communities (Handley, 1967; Bonaccorso, 1979; Brosset and Charles-Dominique, 1990). Whereas species composition, with few exceptions, did not differ much between heights, there were striking differences in capture frequencies of species.

In conclusion, to interpret the significance of rarity of species of a community, it is necessary to closely examine the probable causes of rarity. It is important to differentiate between bats such as top predators or extreme food specialists that are intrinsically rare, bats that are rare because of locally inadequate resources such as food, habitat, and roost sites, and bats that only appear rare because of inappropriate sampling techniques.

C. Rank Abundance

The patterns of abundance in the BCI bat community, characterized by a few numerically dominant species and many uncommon and rare species (Table I and Fig. 3a), are similar to those reported in other communities of bats and birds (see Fleming et al., 1972; Bierregaard, 1989; Terborgh et al., 1990). Interpretation of such curves is controversial. Especially troublesome are the kind and extent of biological factors that may influence the shape of the curve. A pattern such as that shown by the BCI community is expected for species-rich systems influenced by a variety of more or less independent ecological factors (MacArthur, 1957; May, 1975; Putman, 1994). Since the BCI bat community is composed of ecologically diverse species, availability, variety and abundance of food, habitat types, and roost sites are likely to influence community organization. Relationships within a community similar to those seen in the rank abundance distribution of the BCI bat community characterize rather undisturbed systems (Ugland and Gray, 1982; Putman, 1994). The relative stability of guild and species composition and the rather uniform population trends of the BCI community during our 10-year study, in spite of numerous ecologically diverse species, support this characterization. This, however, does not imply that the community is at a static equilibrium; relative abundance of species showed continual although limited change (Figs. 4a–4x).

D. Rank Abundance within Guilds

Compared with the rank abundance curve of the BCI bat community as a whole (Fig. 3a), curves of the well-sampled guilds of gleaning insectivores (Fig. 3c) and gleaning frugivores (Fig. 3d) differed. Whereas the guild of gleaning frugivores resembled the community as a whole, with a few numerically abundant species followed by less common species, the guild of gleaning insectivores was characterized by a higher degree of evenness. Part of the observed differences between gleaning frugivores and gleaning insectivores was undoubtedly based on the sampling method which was aimed toward frugivores. Nonetheless, comparison with data obtained by random mist-netting clearly demonstrates that the observed pattern was present in both samples (Figs. 3c and 3d).

We conclude that differences in availability, abundance, and distribution of food are likely to be key factors in the divergence of the rank abundance curves. Whereas many distinct dietary classes are present in the community as a whole, each of the guilds is characterized by a rather uniform diet.

Abundance and distribution of figs (Moraceae) on BCI are likely to be major reasons for the large populations of fig-eating frugivorous bats. Eleven species of the BCI bat community (Table I) are known to feed heavily on fruits of 14 species of figs (Bonaccorso, 1979; Handley *et al.*, 1991; Kalko *et al.*, 1996). Our continuing demographic studies reveal that the free-standing *Ficus insipida* and *Ficus yoponensis* (subgenus *Pharmacosyce*) with more than 1,200 individuals are the numerically dominant figs on BCI, and their fruits are favored by the most abundant frugivores. Large free-standing figs are characteristic of tall secondary forest. In addition to a few other species of *Pharmacosyce*, there are at least 400 large strangler figs (subgenus *Urostigma*) on the island. Figs are an ideal food source for frugivores since they fruit year-round. Furthermore, fruiting in free-standing figs actually peaks on BCI during a seasonal bottleneck when other fruit is scarce (Howe, 1982; Milton *et al.*, 1982; Milton, 1991). When we compare the results of our random versus fruit-targeted mist-netting, the aseasonal fruiting pattern of figs and high fig abundance on BCI are likely to account for the many similarities in the two samples, particularly the dominance of fig-eating bats. Even with a random sampling protocol there is likely to be a nearby fruiting fig tree wherever nets are set.

However, our studies on BCI show that the population of free-standing figs is rapidly declining due to aging and lack of recruitment in the protected forest on the island. Since free-standing figs are a major food source (Handley *et al.*, 1991; Kalko *et al.*, 1996) and possibly a determinant of population size of frugivorous phyllostomids on BCI, we anticipate a decline in relative abundance of fig-eating bats in the future. This supposition must be tempered, however, by the fact that phyllostomid frugivores vary their diets to compensate for seasonal changes in resource abundance, and most bats that we have labeled as "fig-eating" on BCI have different diets in places where figs are rare or absent (Fleming *et al.*, 1972; Gardner, 1977; Handley *et al.*, 1991; Willig *et al.*, 1993; our unpublished data). So, the consequences of the disappearance of free-standing figs may be mediated somewhat for fig-eating bats by the possibility of fallback on alternative fruits. Our continuing studies on BCI will document this predicament faced by the bats. In addition, comparative studies at sites elsewhere are needed to better define the whole spectrum of dietary niche breadths.

Differences in abundance and distribution of resources may be an important reason for the apparently great difference in relative abundance of gleaning frugivorous and gleaning insectivorous bats. Whereas BCI has many trees producing large crops of fruit preferred by bats, insects—particularly larger insects such as katydids, cicadas, cockroaches, dragonflies, and scarab beetles, the preferred prey of gleaning insectivores (LaVal and Fitch, 1977; Belwood, 1988a; Kalko, 1996)—may be less abundant, more dispersed, and probably not as easily obtained as fruit. Hence, a given area on BCI might support more gleaning frugivorous bats than gleaning insectivorous bats.

Furthermore, distribution and abundance of resources may be associated with the striking differences in mobility and feeding strategies reflected in size of home ranges of gleaning frugivores and gleaning insectivores. While gleaning frugivores, particularly larger species, have large home ranges and thus may exploit fruiting trees in a large area, gleaning insectivores appear to be more sedentary, with smaller home ranges. This assumption is supported by radio-tracking studies of the gleaning insectivore *Tonatia silvicola*, which exploits foraging areas no more than a few hundred meters in diameter (Krull and Kalko, 1994; our unpublished data). The use of small foraging areas is linked to the sit-and-wait strategy employed by this bat. Our radio-tracking studies indicated that *Tonatia silvicola* does not fly continuously while foraging but rather hangs from low perches, continuously scanning its surroundings by turning its body, listening for rustling noises or mating calls produced by potential prey. As soon as the bat detects a sound of interest, it dives at the site of the prey.

In contrast, most frugivorous bats track resources that are highly variable in space and time. Home ranges of some frugivores average larger than those of gleaning insectivores on BCI. For example, we postulate nightly commuting distances averaging 1.5–3 km for the large frugivore, *Artibeus jamaicensis,* on BCI (Handley *et al.,* 1991; our unpublished data), and it has been radio-tracked 8 ± 2 km (one way) between day roost and feeding sites in Mexico (Morrison, 1978). However, preliminary analysis of mark–recapture data of the BCI project indicates smaller home ranges for some of the smaller frugivores such as *Artibeus watsoni,* most individuals of which were recaptured no more than 200–1000 m from their original mark sites (Kalko and Handley, 1994).

The greater evenness in rank abundance of gleaning insectivores compared with the gleaning frugivores may reflect to some extent patterns of resource partitioning. More even distribution of resources in combination with largely nonoverlapping diets in gleaning insectivores may lead to the greater evenness observed in this guild. Preliminary analysis of the diets of some gleaning insectivores on BCI revealed little overlap (Belwood, 1988b; our unpublished data), whereas the gleaning frugivores on BCI, particularly the fig-eating species, overlapped substantially in their diets. Possibly, competition between groups of gleaning frugivores with overlapping diets led to the rather large differences in rank abundance within this guild.

We tentatively conclude that differences in distribution, abundance, and availability of resources leading to differences in feeding strategies and resource partitioning are likely to account for the lower numbers, lower biomass, and more even distribution of gleaning insectivores, compared with gleaning frugivores. However, to substantiate these impressions we must learn much more about spatial and temporal distribution and abundance of resources, and we must conduct behavioral and dietary studies of the bats. For now, our interpretations must remain speculative.

E. Rank Order of Species Pairs

Pairs of congeneric and ecologically similar species such as *Micronycteris hirsuta* and *M. megalotis, Carollia perspicillata* and *C. castanea,* and *Artibeus jamaicensis* and *A. lituratus* are composed of a numerically dominant species and a much less abundant species. Although species within pairs frequently showed dissimilar short-term fluctuations, the numerical dominant in each pair remained so throughout the study (1977–1985). Does interspecific competition within the species pairs contribute to the observed pattern or does it largely reflect individual responses of the species to changes in the physical environment, independent of competitive interactions? Since controlled field experiments are difficult if not impossible, comparative studies of bat communities that differ in species composition suggest a promising approach. For instance, in some parts of Central America up to four species of *Carollia* coexist (Hall, 1981), and in southern Venezuela four large species of *Artibeus* are sympatric (Handley, 1987, 1990). We have evidence that the dominance hierarchy of species may shift between localities in Panamá. For example, at higher elevations and on the Caribbean coast, *Carollia brevicauda* is the dominant *Carollia,* but on BCI and on the dry Pacific slope, *C. perspicillata* is the dominant species. On BCI, *Artibeus phaeotis* outnumbers *A. watsoni,* but in Bocas del Toro, *A. watsoni* is by far the more abundant of the two (our unpublished data). Ultimately, differences in species abundance need to be quantified, compared, and related to differences in species composition and ecological conditions such as diet, habitat, and roost sites at various localities.

F. Long-Term Population Trends and Short-Term Variability

Some guilds were well sampled by mist-netting and provided bountiful information on species composition and on relative abundance of species; for other guilds little information is available, due to inadequate sampling techniques. Their examination reveals a clear pattern of community composition during the study period. The well-sampled guilds were characterized by a group of species that was consistently present and ranged from superabundant to common and another group of uncommon to rare species that was caught only sporadically during the study period. Species that were consistently present showed rather uniform long-term, usually positive, population trends. With the exception of *Phyllostomus discolor,* which declined significantly, we did not observe any drastic shifts in the abundance or dominance hierarchy among these species. Rare species were persistently rare, and abundant species remained abundant.

The decrease and disappearance of plants with bat-pollinated flowers on BCI due to forest regrowth during the past 80–100 years may be an underlying cause of low numbers of nectar-drinking bats in general and for the significant decline of the nectarivorous *P. discolor* during our study. This contention is supported by the observation that a post-1985 increase of balsa (*Ochroma pyramidale*), a fast-growing pioneer species, on construction sites on BCI has led to a simultaneous increase in capture frequency of *P. discolor*. Flowers of the balsa seem to be a dietary favorite of this species.

The causes of the population trends we have observed in many species of the BCI community as well as the conspicuous peak and subsequent decline of the relative abundance of most bats during the 1978–1980 interval have not been identified. Since trends include several guilds, ranging from gleaning insectivores to gleaning frugivores and gleaning nectarivores, it is likely that a single key factor affected all of these species in a similar way. Possibly variability in resource abundance related to changes in physical conditions was one of the causes. Weather may have been a factor. On BCI wet and dry years alternated in the interval between 1975 and 1981, beginning with rainfall just above the yearly average (2616 mm) in 1975, followed by the driest year on record (1707 mm) in 1976, average alternating high and low rainfall years, 1977–1980, and finishing with the wettest year of the century (4133 mm), 1981. What may have been more important to forest phenology than amount of rainfall was its seasonal distribution. Months with less than 100 mm declined progressively from 6 in the unusually dry 1976 to only 2 in the very wet 1981 (Windsor, 1990). Several bat populations increased as dry months decreased, possibly allowing increasing flower, fruit, and insect production.

In contrast to the rather uniform long-term trends in the community, we also noticed more or less pronounced short-term, year-to-year, variations in abundance of species. The short-term fluctuations of *Uroderma bilobatum* and *Chiroderma villosum* seemed to be seasonal (Handley *et al.*, 1991; our unpublished data), suggesting migratory behavior in these species. They sometimes appeared at fruiting trees in large numbers that may have constituted flocks.

Within species pairs such as *Carollia perspicillata* and *C. castanea* and *Artibeus jamaicensis* and *A. lituratus* which share ecological preferences, relative abundance of the species may fluctuate at least in part inversely; as one species increases, the other decreases. This may indicate competition for similar resources among these species. However, to better understand the significance and the underlying causes of the short-term variations and long-term trends in populations observed in the BCI bat community, additional in-depth study is needed. Further analyses of our data, examining similarities and differences in population variations among species and searching for links to annual variations in physical conditions such as precipitation and their effect on resource abundance, are in progress.

G. Community Dynamics

The recurrent patterns in species composition and species abundance as well as the rather uniform population trends in guilds and in the community as a whole point toward predictable patterns in organization of the BCI bat community. During the main study period, 1977–1985, no drastic hierarchical shifts occurred. Abundant species remained abundant and rare species remained rare. None of the superabundant, abundant, or common species were extirpated, nor did any of the rare or uncommon species become more common. Short-term fluctuations of individual species might be interpreted as stochastic events. However, to evaluate the extent to which this variability may reflect a deterministic response of species to the availability of essential resources, more knowledge of the ecology of species and a better understanding of the dynamics of the physical and ecological environment are needed.

It is important to keep in mind that the conclusions we have reached are scale-dependent. For instance, had we compared results of individual mist-netting nights, rather than an 8-year pool of 3-month segments, we would have found a much higher variability in species composition and relative abundance. Alternatively, even longer studies might reveal that the observed short- and long-term population fluctuations are insignificant, or they might reveal drastic changes in relative abundance and species composition due to changes in physical conditions and/or species interactions which could not be detected in the eight-year period.

V. FUTURE DIRECTIONS

This study is the first to document the long-term dynamics of organization and population trends of a Neotropical bat community. There are no similar long-term studies of bat communities anywhere in tropical or temperate zones. In contrast to community studies of birds, lack of standardized methodology and extensive databases prevents comparisons among bat communities. Field studies of bat communities at many places, using standardized sampling protocols and on sufficiently large spatial and temporal scales to include most or all resident species, are needed for intercommunity comparisons. Long-term inventories employing multiple techniques are critical to avoid the biases of inadequate sampling (Fleming *et al.*, 1972; Wilson, 1989). Only intense, long-term sampling efforts can lead to accurate estimates of species richness and abundance.

Although the ecomorphological approach has given valuable insights to the basic organization of bat communities, only the combination of such approaches and long-term studies that accumulate field data on a large scale promise meaningful descriptions of community dynamics and interpretation of observed patterns. The data gained from such approaches are essential for generating hypotheses about processes influencing community organization, diversity, and dynamics. Our own continuing studies of morphological characters such as wing shape, size, and skull morphology in the BCI bat community are a step toward filling the gap between field data and ecomorphological predictions.

Another promising approach to better understanding the factors shaping community organization in bats is comparison with the community organization of ecologically similar organisms such as birds. Since both birds and bats fly and are highly diverse ecologically in the tropics, comparisons of patterns and organization found in bird and bat communities promise to yield useful results. In the tropics bats exceed all other mammals in ecological diversity and in species richness, but in the Neotropics local species richness in bats is generally two to four times lower than that in birds (Karr *et al.*, 1989; Terborgh *et al.*, 1990; Thiollay, 1994). Why aren't there more bats? One striking difference is the larger number of guilds in bird communities than in bat communities. For instance, there are nocturnal as well as diurnal birds; terrestrial, aquatic, wading, and climbing birds; and seed-eating and wood-excavating birds, to name a few that have no bat equivalents. Morphological, behavioral, and sensory constraints as well as the phylogenetic background of birds and bats play important roles in this scenario. Long-term studies focusing on comparison of community and guild structure, dynamics of relative abundance of species and biomass, and study of morphological, behavioral, and sensory adaptations are promising approaches to identifying the factors underlying the disparity in species richness of bats and birds.

VI. SUMMARY

Bats (Chiroptera) are distributed globally and are more diverse ecologically than any other group of mammals. In many terrestrial ecosystems they are important pollinators, seed dispersers, and consumers of insects. However, detailed information on composition and dynamics of bat communities and relative abundance of species is essentially nonexistent. Here, we report results of preliminary analyses in an ongoing study of community organization and long-term dynamics of bats in a seasonal tropical lowland forest on and near Barro Colorado Island (BCI) in central Panamá. Using a variety of techniques (mist-netting, capturing at roosts, acoustic monitoring, radio-tracking, and surveying literature) we have documented the occurrence of 66 species in eight families. These bats can be sorted into 10 guilds based on habitat, foraging mode, and diet. We found that rank order abundance in the community as a whole, characterized by a few abundant species followed by a large number of common, uncommon, and rare species, was similar to that of other vertebrate communities in the tropics. The best sampled guilds, gleaning frugivores and gleaning insectivores, had different individual rank abundance patterns. The rank abundance curve of the gleaning frugivores mirrored the pattern of the community as a whole, but gleaning insectivores showed a more even distribution. Part of the observed differences between the guilds may be related to our sampling protocol, but most appear to be linked to variations in resource abundance and distribution.

Furthermore, relative abundance of well-sampled species showed individual year-to-year variation. Some frugivorous species with large seasonal variation in numbers may be migratory but, except for one species that declined, long-term population trends remained rather constant. Our preliminary analyses of long-term trends indicated that composition of the BCI bat community and relative abundances of species remained rather stable throughout the study period. Our data support the idea that deterministic processes, especially the availability, abundance, and distribution of resources, are important factors influencing organization and dynamics of this bat community. Long-term monitoring of bats at many localities, using a standard methodology, is urgently needed to build databases for intercommunity comparisons.

Acknowledgments

We thank the many people and institutions, in particular the Smithsonian Tropical Research Institute (STRI, Panamá) and the National Museum of Natural History (NMNH, Washington), that with logistic support and good working conditions helped us accomplish this project. We especially thank Allen Herre for discussions of conceptual issues and presentation of data, Uli Schnitzler for encouragement and support throughout, and Egbert Leigh and Joe Wright for many helpful suggestions and comments. We are indebted to Katrin Böhning-Gaese for her generous advice and help on data analysis, and Ralph Chapman and George Venable for advice and assistance on graphic presentation. We thank Don E. Wilson, A. L. Gardner, and E. G. Leigh for participation in initial data analysis, and Ingrid Kaipf for compilation of data. Of course, we will always be indebted to the many individuals who handled the bats and gathered the data in the field. This research and data analysis were supported by a NATO postdoctoral fellowship to Kalko; by grants of the Deutsche Forschungsgemeinschaft, Schwerpunktprogramm "Mechanismen der Aufrechterhaltung tropischer Diversität" to Schnitzler and Kalko; and by ESP funds from the Smithsonian Institution to Handley and Kalko.

References

Ahlén, I. (1981). "Identification of Scandinavian Bats by their Sounds," Rep. No. 6. Swedish University of Agricultural Sciences, Department of Wildlife Ecology, Upsala.

Aldridge, H. D. J. N., and Rautenbach, I. R. (1987). Morphology, echolocation and resource partitioning in insectivorous bats. *J. Exp. Biol.* **128**, 419–425.

Alonso-Mejia, A., and Medellín, R. A. (1991). *Micronycteris megalotis. Mamm. Spec.* 376, 1–6.

Belwood, J. J. (1988a). The influence of bat predation on calling behavior in Neotropical forest katydids (Insecta: Orthoptera: Tettigoniidae). Unpublished Ph.D. Thesis, University of Florida, Gainesville.

Belwood, J. J. (1988b). Foraging behavior, prey selection, and echolocation in phyllostomine bats (Phyllostomidae). *In* "Animal Sonar" (P. E. Nachtigall and P. W. B. Moore, eds.), pp. 601–605. Plenum, New York.

Berthold, P. G., Fliege, U., Querner, U., and Winkler, H. (1986). Die Bestandsentwicklung von Kleinvögeln in Mitteleuropa: Analyse von Fangzahlen. *J. Ornithol.* 127, 397–437.

Bierregaard, R. O. (1989). Species composition and trophic organization of the understory bird community in a central Amazonian Terra Firme forest. *In* "Four Neotropical Rainforests" (A. H. Gentry, ed.), pp. 217–236. Yale Univ. Press, New Haven, CT.

Böhning-Gaese, K. (1994). Avian community dynamics are discordant in space and time. *Oikos* 70, 121–126.

Böhning-Gaese, K., Taper, M. L., and Brown, J. H. (1993). Are declines in North American insectivorous songbirds due to causes on the breeding range? *Conserv. Biol.* 1, 76–86.

Bonaccorso, F. J. (1979). Foraging and reproductive ecology in a Panamanian bat community. *Bull. Fl. State Mus., Biol. Sci.* 24, 359–408.

Bradbury, J. W., and Vehrencamp, S. L. (1976). Social organization and foraging in emballonurid bats. I. Field studies. *Behav. Ecol. Sociobiol.* 1, 337–381.

Brooke, A. P. (1994). Diet of the fishing bat, *Noctilio leporinus* (Chiroptera: Noctilionidae). *J. Mammal.* 75, 212–218.

Brosset, A., and Charles-Dominique, P. (1990). The bats from French Guiana: A taxonomic, faunistic, and ecological approach. *Mammalia* 54(4), 509–560.

Brown, J. H., and Heske, E. J. (1990). Temporal changes in a Chihuahuan desert rodent community. *Oikos* 59, 290–302.

Charles-Dominique, P. (1986). Inter-relations between frugivorous vertebrates and pioneer plants: *Cecropia*, birds and bats in French Guyana. *In* "Frugivores and Seed Dispersal" (A. Estrada and T. H. Fleming, eds.), pp. 119–136. Dr. W. Junk, The Hague, The Netherlands.

Cox, P. A., Elmqvist, T. E., Pierson, E. D., and Rainey, W. E. (1992). Flying foxes as pollinators and seed dispersers in Pacific Island ecosystems. *In* "Pacific Island Flying Foxes: Proceedings of an International Conservation Conference" (D. E. Wilson and G. L. Graham, eds.), pp. 18–23. U. S. Fish Wildl. Serv., Washington, DC.

Crome, F. H., and Richards, G. C. (1988). Bats and gaps: Microchiropteran community structure in a Queensland rain forest. *Ecology* 69, 1960–1969.

Dahlquest, W. W. (1954). Netting bats in tropical Mexico. *Trans. Kans. Acad. Sci.* 57, 1–10.

Dobat, K., and Peikert-Holle, T. (1985). "Blüten und Blumenfledermäuse. Bestäubung durch Fledermäuse und Flughunde (Chiropterphilie)." Waldemar Kramer, Frankfurt am Main.

Dunkle, S. W., and Belwood, J. J. (1982). Bat predation on Odonata. *Odonatologica* 11, 225–229.

Enders, R. K. (1930). Notes on some mammals from Barro Colorado Island, the Canal Zone. *J. Mammal.* 11, 280–292.

Enders, R. K. (1935). Mammalian life histories from Barro Colorado Island, Panama. *Bull. Mus. Comp. Zool.* 78(4), 385–502.

Fenton, M. B. (1972). The structure of aerial-feeding bat faunas as indicated by ears and wing elements. *Can. J. Zool.* 50, 287–296.

Fenton, M. B. (1990). The foraging behavior and ecology of animal-eating bats. *Can. J. Zool.* 68, 411–422.

Fenton, M. B., and Bell, P. G. (1981). Recognition of species of insectivorous bats by their echolocation calls. *J. Mammal.* 62, 233–243.

Fenton, M. B., Merriam, H. G., and Holroyd, G. L. (1983). Bats of Kootenay, Glacier, and Mount

Revelstroke national parks in Canada: Identification by echolocation calls, distribution, and biology. *Can. J. Zool.* **61**, 2506–2508.

Fenton, M. B., Tannant, D. C., and Wyszecki, J. (1987). Using echolocation calls to measure the distribution of bats: The case of *Euderma maculatum. J. Mammal.* **68**(1), 142–144.

Findley, J. S. (1973). Phenetic packing as a measure of faunal diversity. *Am. Nat.* **107**, 580–584.

Findley, J. S. (1976). The structure of bat communities. *Am. Nat.* **110**, 129–139.

Findley, J. S. (1993). "Bats: A Community Perspective." Cambridge Univ. Press, Cambridge, UK.

Findley, J. S., and Wilson, D. E. (1982). Ecological significance of chiropteran morphology. *In* "Ecology of Bats" (T. H. Kunz, ed.), pp. 243–260. Plenum, New York.

Fleming, T. H. (1986). The structure of neotropical bat communities: A preliminary analysis. *Rev. Chil. Hist. Nat.* **59**, 135–150.

Fleming, T. H. (1988). "The Short-tailed Fruit Bat: A Study in Plant-animal Interactions." Univ. of Chicago Press, Chicago.

Fleming, T. H. (1991). Fruiting plant-frugivore mutualism: The evolutionary theater and the ecological play. *In* "Plant-animal Interactions" (P. W. Price, T. M. Lewinsohn, G. W. Fernandes, and W. W. Benson, eds.), pp. 119–144. Wiley, New York.

Fleming, T. H., Hooper, E. T., and Wilson, D. E. (1972). Three Central American bat communities: Structure, reproductive cycle, and movement patterns. *Ecology* **53**, 556–569.

Foster, R. B., and Brokaw, N. V. L. (1982). Structure and history of the vegetation of Barro Colorado Island. *In* "The Ecology of a Tropical Forest" (E. G. Leigh, Jr., A. S. Rand, and D. M. Windsor, eds.), pp. 151–172. Smithsonian Inst. Press, Washington, DC.

Foster, R. B., Arce B., J., and Wachter, T. S. (1986). Dispersal and the sequential plant communities in Amazonian Peru floodplain. *In* "Frugivores and Seed Dispersal" (A. Estrada and T. H. Fleming, eds.), pp. 357–370. Dr. W. Junk, The Hague, The Netherlands.

Gaisler, J., Chytil, J., and Vlasin, M. (1990). The bats of S-moravian lowlands (Czechoslovakia) over thirty years. *Acta Sci. Nat. Brno* **24**(9), 1–50.

Gardner, A. L. (1977). Feeding habits. *Spec. Publ.—Mus., Tex. Tech. Univ.* **13**, 293–350.

Gardner, J. E., Garner, J. D., and Hoffmann, J. E. (1989). A portable mist netting system for capturing bats, with emphasis on *Myotis sodalis* (Indiana bat). *Bat Res. News* **30**, 1–8.

Glass, B. P. (1956). Effectiveness of Japanese mist nets for securing bats in temperate latitudes. *Southwest. Nat.* **1**, 136–138.

Goldman, E. A. (1920). Mammals of Panama. *Smithson. Misc. Collect.* **69**(5), 1–309.

Griffin, D. R., Webster, F. A., and Michael, C. R. (1960). The echolocation of flying insects by bats. *Anim. Behav.* **8**, 141–154.

Hall, E. R. (1981). "The Mammals of North America," Vols. 1 and 2. Wiley, New York.

Hall, E. R., and Jackson, W. B. (1953). Seventeen species of bats recorded from Barro Colorado Island, Panama Canal Zone. *Univ. Kans. Publ., Mus. Nat. Hist.* **5**(37), 641–646.

Handley, C. O., Jr. (1966). Checklist of the mammals of Panamá. *In* "Ectoparasites of Panama" (R. L. Wenzel and V. J. Tipton, eds.), pp. 753–795. Field Mus. Nat. Hist., Chicago.

Handley, C. O., Jr. (1967). Bats of the canopy of an Amazonian forest. *Atas Simp. Sôbre Biota Amazônica* **5**, 211–215.

Handley, C. O., Jr. (1976). Mammals of the Smithsonian Venezuelan Project. *Brigham Young Univ. Sci. Bull., Biol. Ser.* **20**, 1–91.

Handley, C. O., Jr. (1987). New species of mammals from Northern South America: Fruit-eating bats, genus *Artibeus* Leach. *In* "Studies in Neotropical Mammalogy: Essays in Honor of Philip Hershkovitz" (B. D. Patterson and R. M. Timm, eds.), *Fieldiana*, 2 vol., Num Ser. No. 39, Field Mus. Nat. Hist., Chicago.

Handley, C. O., Jr. (1990). The *Artibeus* of Gray 1838. *In* "Advances in Neotropical Mammalogy" (K. H. Redford and J. F. Eisenberg, eds.), pp. 443–468. Sandhill Crane Press, Gainesville, FL.

Handley, C. O., Jr., Gardner, A. L., and Wilson, D. E., eds. (1991). "Demography and Natural

History of the Common Fruit Bat, *Artibeus jamaicensis*, on Barro Colorado Island, Panamá." *Smithson. Contrib. Zool.*, No. 511. Smithsonian Inst. Press, Washington, DC.

Heithaus, E. R., Fleming, T. H., and Opler, P. A. (1975). Foraging patterns and resource utilization in seven species of bats in a seasonal tropical forest. *Ecology* 56, 841–854.

Heller, K.-G., and Volleth, M. (1995). Community structure and evolution of insectivorous bats in the Palaeotropics and Neotropics. *J. Trop. Ecol.* 11, 429–442.

Holmes, R. T., Sherry, T. W., and Sturges, F. W. (1986). Bird community dynamics in a temporal deciduous forest: Long-term trends at Hubbard Brook. *Ecol. Monogr.* 56(3), 201–220.

Howe, H. F. (1982). Fruit production and animal activity in two tropical trees. *In* "The Ecology of a Tropical Forest" (E. G. Leigh, A. S. Rand, and D. M. Windsor, eds.), pp. 189–199. Smithsonian Inst. Press, Washington, DC.

Humphrey, S. R. (1975). Nursery roosts and community diversity of Nearctic bats. *J. Mammal.* 56, 321–346.

Humphrey, S. R., Bonaccorso, F. J., and Zinn, T. L. (1983). Guild structure of surface-gleaning bats in Panamá. *Ecology* 64, 284–294.

Jeanne, R. L. (1970). Note on a bat (*Phylloderma stenops*) preying upon the brood of a social wasp. *J. Mammal.* 51, 624–625.

Kalko, E. K. V. (1994). The use of echolocation and other sensory cues in the frugivorous bat *Artibeus jamaicensis*, while foraging. *Bat Res. News* 35(1), 28.

Kalko, E. K. V. (1995a). Insect pursuit, prey capture and echolocation in pipistrelle bats (Microchiroptera). *Anim. Behav.* 50, 861–880.

Kalko, E. K. V. (1995b). Echolocation signal design, foraging habitats and guild structure in six Neotropical sheath-tailed bats (Emballonuridae). *Symp. Zool. Soc. London* 67, 259–273.

Kalko, E. K. V. (1996). "Diversity in Tropical Bats. Systematics and Biodiversity in Tropical Ecosystems." Alexander König, Research Institute and Zoological Museum, Bonn, Germany. (in press).

Kalko, E. K. V., and Condon, M. (1993). Bat-plant interactions: How frugivorous leaf-nosed bats find their food. *Bat Res. News* 34(4), 115.

Kalko, E. K. V., and Handley, C. O., Jr. (1994). Evolution, biogeography, and description of a new species of fruit-eating bat, genus *Artibeus* Leach (1821), from Panamá. *Z. Säugetierkd.* 59, 257–273.

Kalko, E. K. V., and Schnitzler, H.-U. (1989). The echolocation and hunting behavior of Daubenton's bat, *Myotis daubentoni. Behav. Ecol. Sociobiol.* 24, 225–238.

Kalko, E. K. V., and Schnitzler, H.-U. (1993). Plasticity in echolocation signals of European pipistrelle bats in search flight: Implications for habitat use and prey detection. *Behav. Ecol. Sociobiol.* 33, 415–428.

Kalko, E. K. V., Herre, A. E., and Handley, C. O., Jr. (1996). Relation of fig fruit characteristics to fruit-eating bats in the New and Old World tropics. *J. Biogeogr.* (in press).

Karr, J. R., Robinson, S., Blake, J., and Bierregaard, R. O., Jr. (1989). Birds of four neotropical forests. *In* "Four Neotropical Rainforests" (A. H. Gentry, ed.), pp. 237–272. Yale Univ. Press, New Haven, CT.

Krull, D., and Kalko, E. K. V. (1994). Foraging and roosting behavior in the two phyllostomid gleaning bats, *Tonatia silvicola* and *Trachops cirrhosus. Bat Res. News* 35(1), 31.

Kunz, T. H., and Diaz, C. A. (1995). Folivory in fruit-eating bats, with new evidence from *Artibeus jamaicensis* (Chiroptera: Phyllostomidae). *Biotropica* 27, 106–120.

Laska, M. (1990). Olfactory sensitivity to food odor components in the short-tailed fruit bat, *Carollia perspicillata* (Phyllostomatidae, Chiroptera). *J. Comp. Physiol. A* 166A, 395–399.

LaVal, R. K., and Fitch, H. S. (1977). Structure, movements, and reproduction in three Costa Rican bat communities. *Occas. Pap. Mus. Nat. Hist., Univ. Kans.* 69, 1–28.

LaVal, R. K., and LaVal, M. L. (1980). Prey selection by a Neotropical foliage-gleaning bat, *Micronycteris megalotis. J. Mammal.* 61(2), 327–330.

MacArthur, J. W. (1957). On the relative abundance of bird species. *Proc. Natl. Acad. Sci. U.S.A.* **43**, 293–295.

Marshall, A. J. (1985). Old world phytophageous bats (Megachiroptera) and their food plants: A survey. *Zool. J. Linn. Soc.* **83**, 351–369.

Martuscelli, P. (1995). Avian predation by the round-eared bat (*Tonatia bidens,* Phyllostomidae) in the Brazilian Atlantic forest. *J. Trop. Ecol.* **11**, 461–464.

May, R. M. (1975). Patterns of species abundance and diversity. *In* "Ecology and Evolution of Communities" (M. L. Cody and J. M. Diamond, eds.), pp. 81–120. Belknap, Cambridge, MA.

McKenzie, N. L., and Rolfe, J. K. (1986). Structure of bat guilds in the Kimberley mangroves, Australia. *J. Anim. Ecol.* **55**, 401–420.

McNab, B. K. (1971). The structure of tropical bat faunas. *Ecology* **52**, 352–358.

Medellín, R. (1988). Prey of *Chrotopterus auritus,* with notes on feeding behavior. *J. Mammal.* **69**, 841–844.

Milton, K. (1991). Leaf change and fruit production in six neotropical Moraceae species. *J. Ecol.* **79**, 1–26.

Milton, K., Windsor, D. M., Morrison, D. W., and Estribi, M. (1982). Fruiting phenologies of two Neotropical *Ficus* species. *Ecology* **63**, 752–762.

Morrison, D. W. (1978). Foraging ecology and energetics of the frugivorous bat *Artibeus jamaicensis. Ecology* **59**, 716–723.

Neuweiler, G. (1989). Foraging ecology and audition in echolocation bats. *Trends Ecol. Evol.* **6**, 160–166.

Neuweiler, G. (1990). Auditory adaptations for prey capture in echolocating bats. *Physiol. Rev.* **70**, 615–641.

Norberg, U. M., and Rayner, J. M. V. (1987). Ecological morphology and flight in bats (Mammalia: Chiroptera): Wing adaptations, flight performance, foraging strategy and echolocation. *Philos. Trans. R. Soc. London* **316**, 335–427.

Piperno, D. R. (1992). Fitolitos, arqueología y cambias prehistoricos de la vegetación en un lote de cincuenta hectareas de la isla de Barro Colorado. *In* "Ecología de un Bosque Tropical" (E. G. Leigh, Jr., A. S. Rand, and D. M. Windsor, eds.), pp. 153–162. Smithsonian Inst. Press, Washington, DC.

Poulin, B., Lefebvre, G., and McNeil, R. (1994). Characteristics of feeding guilds and variation in diets of bird species of three adjacent tropical sites. *Biotropica* **26**(2), 187–197.

Putman, R. J. (1994). "Community Ecology." Chapman & Hall, London.

Reis, N. R., and Muller, M. F. (1995). Bat diversity of forests and open areas in a subtropical region of South Brazil. *Ecol. Aust.* **5**, 31–36.

Root, R. B. (1967). The niche exploitation pattern of the blue-gray gnatcatcher. *Ecol. Monogr.* **37**, 317–350.

Saunders, M. B., and Barclay, R. M. R. (1992). Ecomorphology of insectivorous bats: A test of predictions using two morphologically similar species. *Ecology* **73**, 1335–1345.

Schnitzler, H.-U. (1987). Echoes of fluttering insects: Information for echolocating bats. *In* "Recent Advances in the Study of Bats" (M. B. Fenton, R. Racey, and O. W. Henson, Jr., eds.), pp. 226–243. Cambridge Univ. Press, Cambridge, UK.

Schnitzler, H.-U., and Kalko, E. K. V. (1996). Echolocation behavior in foraging bats. *BioScience.* (in press).

Schnitzler, H.-U., Kalko, E. K. V., Kaipf, I., and Grinnell, A. D. (1994). Fishing and echolocation behavior of the greater bulldog bat, *Noctilio leporinus* in the field. *Behav. Ecol. Sociobiol.* **35**, 327–345.

Schum, M. (1984). Phenetic structure and species richness in North and Central American bat faunas. *Ecology* **65**, 1315–1324.

Stewart-Oaten, A., Murdoch, W. W., and Walde, S. J. (1995). Estimation of temporal variability in populations. *Am. Nat.* **146,** 519–535.

Tannenbaum, B. R. (1975). Reproductive strategies in the white-lined bat. Unpublished Ph.D. Thesis, Cornell University, Ithaca, NY.

Terborgh, J., Robinson, S. K., Parker, T. A., III, Munn, C. A., and Pierpont, N. (1990). Structure and organization of an Amazonian forest bird community. *Ecol. Monogr.* **60,** 213–238.

Thiollay, J.-M. (1994). Structure, density and rarity in an Amazonian rainforest bird community. *J. Trop. Ecol.* **10,** 449–481.

Tuttle, M. D. (1974). An improved trap for bats. *J. Mammal.* **55**(2), 475–477.

Tuttle, M. D., and Ryan, M. J. (1981). Bat predation and the evolution of frog vocalizations in the Neotropics. *Science* **214,** 677–678.

Ugland, K. I., and Gray, J. S. (1982). Log normal distributions and the concept of community equilibrium. *Oikos* **39,** 171–178.

Whitaker, J. O., and Findley, J. S. (1980). Foods eaten by some bats from Costa Rica and Panama. *J. Mammal.* **61,** 540–544.

Wiens, J. A. (1986). Spatial scale and temporal variation in studies of shrubsteppe birds. *In* "Community Ecology" (J. M. Diamond and T. J. Case, eds.), pp. 154–172. Harper & Row, New York.

Wiens, J. A. (1989). "The Ecology of Bird Communities," Vol. 2. Cambridge Univ. Press, Cambridge, UK.

Willig, M. R. (1986). Bat community structure in South America: A tenacious chimera. *Rev. Chil. Hist. Nat.* **59,** 151–168.

Willig, M. R., and Moulton, M. P. (1989). The role of stochastic and deterministic processes in structuring neotropical bat communities. *J. Mammal.* **70,** 323–329.

Willig, M. R., Camilo, G. R., and Noble, S. J. (1993). Dietary overlap in frugivorous and insectivorous bats from edaphic cerrado habitats of Brazil. *J. Mammal.* **74,** 117–128.

Wilson, D. E. (1971). Food habits of *Micronycteris hirsuta* (Chiroptera: Phyllostomidae). *Mammal.* **35,** 107–110.

Wilson, D. E. (1973). Bat faunas: A trophic comparison. *Syst. Zool.* **22,** 14–29.

Wilson, D. E. (1989). Mammals of La Selva, Costa Rica. *In* "Four Neotropical Rainforests" (A. H. Gentry, ed.), pp. 273–286. Yale Univ. Press, New Haven, CT.

Wilson, D. E., and Reeder, D. M. (1993). "Mammal Species of the World, a Taxonomic and Geographic Reference," 2nd ed. Smithsonian Inst. Press, Washington, DC.

Windsor, D. M. (1990). Climate and moisture variability in a tropical forest: Long-term records from Barro Colorado Island, Panama. *Smithson. Contribib. Earth Sci.* **9,** 1–145.

Zortéa, M., and Mendes, S. L. (1993). Folivory in the big fruit-eating bat, *Artibeus lituratus* (Chiroptera: Phyllostomidae) in eastern Brazil. *J. Trop. Ecol.* **9,** 117–120.

CHAPTER 17

Desert Rodents

Long-Term Responses to Natural Changes and Experimental Manipulations

THOMAS J. VALONE* AND JAMES H. BROWN†

*Department of Biology, California State University at Northridge, Northridge, California 91330; and †Department of Biology, University of New Mexico, Albuquerque, New Mexico 87131

I. INTRODUCTION

Although ecological communities are the natural unit of study for community ecologists, they can be difficult to characterize precisely. In the broadest sense, a community can be defined as an entire set of interacting species populations, or it can be specified as a subset of species that share taxonomic or ecological

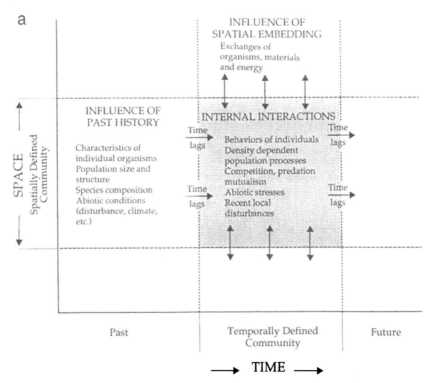

FIGURE 1. (a) Representation of factors that affect the structure and dynamics of all communities. (b) The spatial–temporal scales of our study. Note the very small region of space occupied by a typical ecological study.

affinities (Ricklefs, 1979; Begon *et al.*, 1986; Diamond and Case, 1986). Ecologists usually also stipulate that the populations in a community occur together within a specified area over a defined period, although such boundary specifications are almost always arbitrary (Ricklefs, 1979).

All communities, no matter how defined, consist of populations of multiple species, and so they can be characterized by the number, abundances, and characteristics of the coexisting species. Such characterizations have led to the development of indices of community structure including species diversity, size ratios, food webs, connectivity, and dynamics such as measures of stability and turnover (e.g., Pimm, 1982; Pianka, 1983). Numerous processes are known to influence the structure and dynamics of ecological communities. Intrinsic processes include the local biotic interactions among coexisting species, and a great deal of recent evolutionary and community ecology has been devoted to understanding the processes of competition, predation, and mutualism and their effects on community organization (Begon *et al.*, 1986; Gee and Giller, 1987).

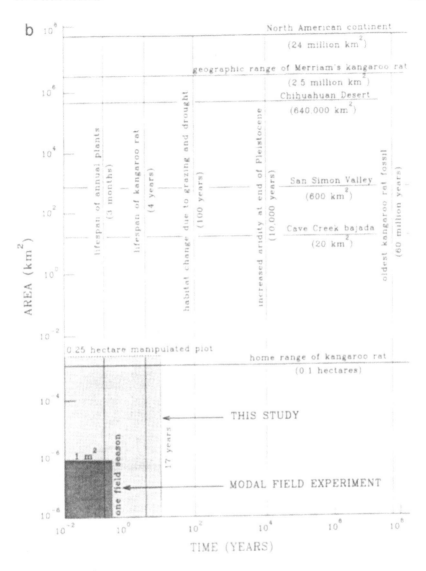

Community structure and dynamics are also affected by extrinsic abiotic conditions. For example, climatic variables such as temperature and precipitation, as well as physical disturbance, have been shown to influence strongly the abundances of component species and thus the composition of many terrestrial communities (Harper, 1977; Pickett and White, 1985; Levey, 1988).

And, finally, because ecological communities are open systems, in both space and time, they are influenced by regional and historical factors (Ricklefs and Schluter, 1993). While one can define arbitrarily and operationally the limits of a community (i.e., the physical area and duration of a study), few ecological systems are so closed. In reality, ecological communities are embedded in larger systems such that energy, materials, and organisms move across the arbitrarily defined boundaries of the community (Fig. 1). Many mathematical models have assumed that communities are closed systems, and most ecological experiments artificially close communities by preventing immigration and emigration of species. Only recently have ecologists begun to appreciate the openness of ecological systems (Roughgarden et al., 1987, 1988; Brown and Heske, 1990a; Gilpin and Hanski, 1991; Heske et al., 1994). Indeed, the increasing attention to regional and historical processes is a direct outgrowth of the explicit appreciation of the openness of ecological systems (Ricklefs and Schluter, 1993).

In this chapter, we use 17 years of data from long-term monitoring and experimentation to examine how intrinsic biotic, extrinsic abiotic, and regional factors have affected the structure of a Chihuahuan Desert rodent community. This community is defined by the nocturnally active rodents that occurred together on a 20-ha site near Portal, Arizona. We begin by showing how the structure of this community and the population dynamics of component species have been influenced by both interspecific competitive interactions and climatic fluctuations. Then we examine the importance of a regional factor, colonization, on rodent community diversity. Finally, we discuss the relative importance of each of these factors in structuring this community, and consider the implications of our findings for future work in community ecology.

II. THE CHIHUAHUAN DESERT RODENT SYSTEM

In 1977, 24 experimental plots (50 × 50 m) were established on a 20-ha site of relatively homogeneous Chihuahuan Desert scrub vegetation (Heske et al., 1993) on the Cave Creek Bajada in the San Simon valley of southeastern Arizona. Each plot was surrounded by wire mesh topped with aluminum flashing to control access by rodents. Plots were assigned to various experimental treatments that included the exclusion of different rodent species based on their body size. Equal access (or control) plots had 16 large (3.7 × 3.7 cm) gates in the wire mesh, which allowed access to all rodents. Kangaroo rat (Dipodomys spp.) removal plots had 16 small (1.9 × 1.9 cm) gates in the wire mesh, which allowed access by all small-bodied rodents but prevented the larger-bodied kangaroo rats

(*D. spectabilis, D. ordii,* and *D. merriami*) from entering (Brown and Munger, 1985). From 1977 through 1987, there were 14 equal access plots and 4 kangaroo rat removal plots. In 1988, some equal access plots were converted into new kangaroo rat removal plots by changing their gate size. Thus, from 1988 to present, there were 8 equal access plots and 8 kangaroo rat removal plots: 4 "old" removal plots maintained continuously since 1977, and 4 "new" removal plots maintained since 1988 (Heske *et al.,* 1994).

Rodents have been censused on all plots at approximately monthly intervals since 1977. For one night, 49 Sherman live-traps were placed on each plot and baited with millet. During each census, all gates were closed so that only resident individuals on the experimental plots were captured. Note that on all other nights, all gates remained open so that rodents of appropriate size from the surrounding habitat could freely enter and exit the plots. All individuals captured on the plots were identified, measured, uniquely marked, and released. See Brown and Munger (1985) and Heske *et al.* (1994) for further details.

The rodent community at the study site is very diverse: So far, we have captured 17 species of nocturnal rodents. These range in body size from less than 10 g (*Perognathus flavus* and *Baiomys taylori*) to well over 100 g (*Dipodomys spectabilis* and *Neotoma albigula*). The species can be divided into two classes based on trophic considerations. There is a well-defined guild of granivores, composed of species that feed primarily on seeds, and a set of other species whose diets are composed primarily of other food types, such as green vegetation or insects (Table I).

TABLE I Average Body Size and Dietary Classification of the 11 Common Species of Rodents at the Chihuahuan Desert Study Site

Species	Body size (g)	Dietary classification
Dipodomys spectabilis	123	Granivore
D. ordii	48	Granivore
D. merriami	43	Granivore
Chaetodipus penicillatus	16	Granivore
Perognathus flavus	7	Granivore
Peromyscus eremicus	21	Granivore
P. maniculatus	21	Granivore
Reithrodontomys megalotis	10	Granivore
Neotoma albigula	174	Omnivore
Onychomys leucogaster	35	Carnivore
O. torridus	25	Carnivore

III. THE INFLUENCE OF BIOTIC PROCESSES
ON COMMUNITY STRUCTURE

A. Direct Competitive Interactions

One of the original goals of the study was to examine the role of interspecific competition in organizing this rodent community (Munger and Brown, 1981; Brown and Munger, 1985). Comparison of rodent densities on equal access and kangaroo rat removal plots allowed direct testing of the hypothesis that competition from larger-bodied kangaroo rats negatively affected the abundance of small granivorous rodents.

Initial results (using data from 1977 through 1982) clearly demonstrated interspecific competition: five species of small granivores (*Chaetodipus penicillatus, Perognathus flavus, Peromyscus eremicus, P. maniculatus,* and *Reithrodontomys megalotis*) were significantly more abundant on kangaroo rat removal plots than on equal access plots, although there was a considerable time lag in the initial increase of these species (Munger and Brown, 1981; Brown and Munger, 1985). More recent results (using data from 1977 to 1991) yielded identical results: the above five species of small granivores were two to four times more abundant on both old and new kangaroo rat removal plots than on equal access plots (Heske *et al.,* 1994).

Such consistency in response by small granivorous rodents to the exclusion of kangaroo rats raises an important question: Is the degree of competition between kangaroo rats and smaller granivores relatively constant through time, or does it vary with fluctuations in environmental conditions or population densities as suggested by Wiens (1977, 1986, 1989)? If the former condition holds, it would suggest that competition may often have an important effect on the structure and dynamics of communities, while the latter condition would relegate competition to only a minor, perhaps episodic, role as an ecological and evolutionary force affecting communities.

To address this question, Brown *et al.* (1996) examined the differences in density of the above five species of small granivores between equal access and kangaroo rat removal plots over time. They calculated the per capita effect of kangaroo rats on the small granivorous rodents as an index of competition, α. Specifically they asked whether variation in α was periodic or constant and whether it was related to any of several measurements of environmental variation.

Brown *et al.* (1996) found that competition was surprisingly constant over time. Over the 12 years of their study, they found that small granivore densities were consistently higher on plots where kangaroo rats had been removed (Fig. 2). Further, they found that their index of competition was not correlated with annual plant density, total rodent population density, energy use, or population

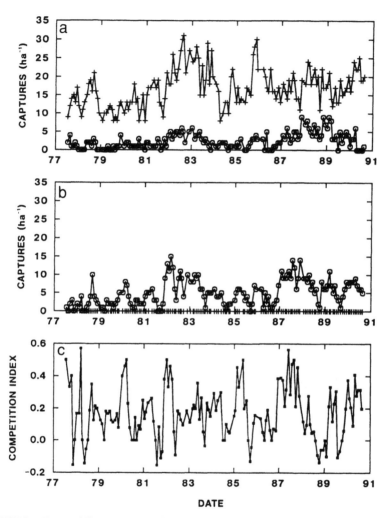

FIGURE 2. Temporal fluctuations in rodent populations and per capita competition. (a) Kangaroo rat (crosses) and small granivore (open circles) captures on control plots; (b) kangaroo rat (crosses) and small granivore (open circles) captures on kangaroo rat removal plots; and (c) the computed value of the competition coefficient, α. Values for rodent populations and total captures are from four replicate plots that have a combined area of 1 ha. Statistical tests were conducted on the last 12 years of data (Brown *et al.*, 1996).

growth rate. Thus, there was no evidence that competition between kangaroo rats and small granivores was limited to periods when resources were scarce or when populations were either high or low. The results suggest that competition is not confined to infrequent environmental "crunches" (*sensu* Wiens, 1977), but is a consistent, pervasive force structuring this rodent community.

B. Indirect Interactions

While direct interactions such as interspecific competition have long been known to play a role in the structure and dynamics of communities, equally important effects of indirect interactions have been documented increasingly both by our experiments (Brown *et al.*, 1986; Davidson *et al.*, 1984, 1985; Thompson *et al.*, 1991; Heske *et al.*, 1993) and by those of other investigators (e.g., Schmidt, 1987; Carpenter, 1988; Wootton, 1993). Indirect interactions are chains of two or more interactions involving three or more links, with each link being a species or a functional group of organisms.

Our long-term "press" experiments (i.e., continuously sustained manipulations) (Bender *et al.*, 1984) have demonstrated a complex pattern of responses by both animals and plants to the selective removal of species or functional groups of granivores. As the manipulations have continued, there has been a continual increase in the number of species, higher taxa, and functional groups that have changed in abundance and/or distribution. During the first few years, large changes in the abundances of certain species of granivorous rodents and annual plants reflected interactions through direct pathways of competition and predation, respectively (Munger and Brown, 1981; Brown and Munger, 1985; Brown *et al.*, 1986; Samson *et al.*, 1992). While most of these direct responses are still evident, an increasing number of indirect responses have become apparent (Brown *et al.*, 1986; Davidson *et al.*, 1984, 1985). Now, 17 years after the initiation of the first cohort of treatments, we can document dramatic changes in species composition and habitat structure, most notably on plots where either kangaroo rats or all rodents have been removed: in particular, such plots have significantly more grass cover than plots where kangaroo rats (or all rodents) have been present (Brown and Heske, 1990a; Heske *et al.*, 1993, 1994).

In addition to demonstrating the existence of indirect interactions, it is important to begin to quantify the relative magnitudes of direct versus indirect pathways of interactions within communities (Strauss, 1991). One potential method is path analysis, a technique of decomposing correlations along hypothesized paths of causation (Kingsolver and Schemske, 1991). Developed by Sewell Wright (1934), path analysis has been used to measure the strengths of interactions in networks of complex evolutionary, physiological, and genetic relationships, and increasingly, to quantify ecological interactions (Strauss, 1991; Wootton, 1994).

Interactions between the western harvest mouse, *R. megalotis*, kangaroo rats, *Dipodomys* spp., and grass cover in our system are ideal for using path analysis to quantify the relative magnitudes of direct versus indirect effects. Our work has shown that harvest mouse densities increase shortly after the removal of kangaroo rats from experimental plots, indicating a negative (i.e., competitive) interaction (Heske *et al.*, 1994). We also know that the long-term absence of kangaroo rats from experimental plots results in an increase in grass cover: kangaroo rats inhibit the abundance of grasses (Brown and Heske, 1990a). Furthermore, studies have demonstrated that harvest mouse densities attain their highest level on long-term kangaroo rat removal plots (Heske *et al.*, 1994), presumably because they respond positively not only to the absence of kangaroo rats, but also to the increased grassiness, a preferred habitat (Hoffmeister, 1986). Thus, kangaroo rats inhibit harvest mouse densities directly via competition and indirectly through their negative effects on grass cover.

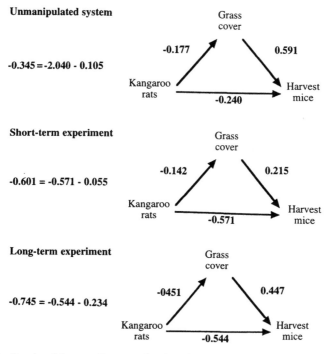

FIGURE 3. Results of three applications of path analysis to the relationships between kangaroo rat density, harvest mouse density, and density of grass on experimental plots. Numbers next to each arrow indicate strength of the correlation coefficient. The path equation (overall effect = direct effect + indirect effect) is located to the left of each diagram. The unmanipulated system includes data from non-kangaroo rat manipulated plots only. The short-term experiment results were obtained by combining data from the short-term kangaroo rat removal plots with data from the unmanipulated system. The long-term experiment results were obtained by combining data from long-term kangaroo rat removal plots with the unmanipulated system.

We used path analysis to partition the effect of kangaroo rats on *R. megalotis* density both directly via competition and indirectly via grass cover (Smith *et al.*, 1996). In addition, our experimental design allowed an examination of the consistency of the results given by path analysis when applied to nonexperimental data as well as to the results of manipulations of varying duration.

We found that path analysis gave very different results when applied to different sets of data (Fig. 3; Smith *et al.*, 1996). Specifically, the method failed to reveal an appreciable indirect interaction between kangaroo rats and harvest mice when data from only unmanipulated or short-term experimental plots were used. When data from long-term manipulations were included in the analysis, however, a strong indirect interaction via grass cover was detected. Furthermore, we also obtained varying results from the long-term data when we analyzed two different 2-year periods, suggesting that the strength of indirect interactions in our system varies over time, perhaps in response to environmental fluctuations (Smith *et al.*, 1996). Given the variety of results that we obtained from path analysis, we conclude that the technique can be a useful tool with which to quantify the magnitude of direct and indirect effects in ecological systems, but that it should be used cautiously and perhaps only with long-term experimental data.

IV. THE INFLUENCE OF ABIOTIC FACTORS ON POPULATION DYNAMICS AND COMMUNITY STRUCTURE

A by-product of our long-term monitoring project has been the accumulation of data on the population dynamics of the rodents in our community. Figures 4 and 5 summarize fluctuations in the rodent community as a whole (total number of rodents and biomass) and show the population dynamics of the 11 common rodent species at our site, respectively. Two features of the 17-year records of population densities stand out: (1) the peaks and valleys of the densities of many species often seem to coincide; and (2) there are several striking long-term trends: *D. spectabilis* declined while *R. megalotis*, *C. penicillatus*, and *D. ordii* exhibited a general increase in density during most of the study. In the next three sections we discuss how abiotic factors may have contributed to these patterns.

A. Population Dynamics of Species Pairs

Most empirical work on small mammals has focused on the population dynamics of single species (but see O'Connell, 1989; Brown and Heske, 1990b). Such studies often have been concerned with determining whether the dynamics of a

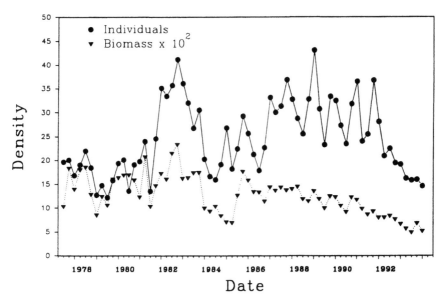

FIGURE 4. Temporal fluctuations in the density of rodents (number/ha) and community biomass density (g/ha). Data are plotted as 3-month averages for clarity.

certain species could be categorized as being stable, cyclical, or chaotic (e.g., Hansson, 1991; Turchin, 1992). Unfortunately, while our study has been ongoing for more than 17 years, we still do not have sufficiently long time series to characterize confidently the dynamics of our rodent populations.

Instead, we focus on a different question: how tightly linked are the population dynamics of coexisting species? The answer has important implications for community organization. If the populations of coexisting competing species are limited by common resources whose availability fluctuates over time, one might expect many species to exhibit positively correlated population dynamics. Alternatively, a postulated mechanism of coexistence for competing species is different responses to environmental variation, which should be reflected in asynchronous or negatively correlated changes in abundance (Abrams, 1984; J. S. Brown, 1989). Indeed, there is evidence in some desert rodent communities of negative correlations in the seasonal or interannual patterns of abundance of species (MacMillen, 1964; Reichman and Van de Graaf, 1973; J. S. Brown, 1989).

To determine whether species pairs exhibited positive or negative population dynamics, we calculated cross-correlations for the time series of abundances for all pairwise combinations of the 11 most common rodents at our site. A previous analysis (Brown and Heske, 1990b), performed on the first 10 years of data, was

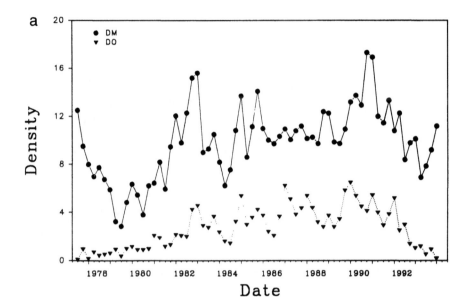

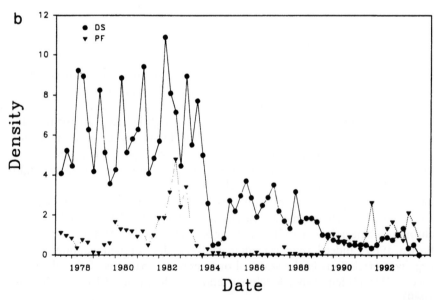

FIGURE 5.   Temporal fluctuations in the density (individuals/ha) of 11 common rodent species. Data are plotted as 3-month averages for clarity. (a) DM—*Dipodomys merriami*; DO—*D. ordii*; (b) DS—*D. spectabilis*; PF—*Perognathus flavus*; (c) CP—*Chaetodipus penicillatus*; NA—*Neotoma albigula*; (d) OL—*Onychomys leucogaster*; OT—*O. torridus*; (e) PE—*Peromyscus eremicus*; PM—*Peromyscus maniculatus*; RM—*Reithrodontomys megalotis*.

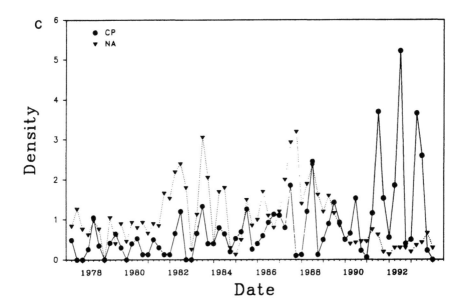

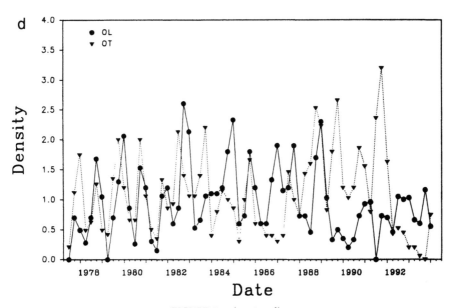

FIGURE 5. (continued)

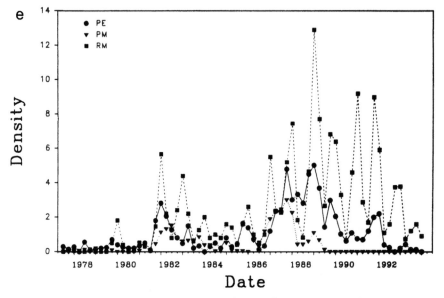

FIGURE 5.   (*continued*)

conducted using information from all plots at the site. This was done to maximize the sample size (number of rodents captured) and thus increase statistical power (Brown and Heske, 1990b). Combining data from both equal access and kangaroo rat removal plots, however, may mask important features of the dynamics of coexisting species.

Thus, we conducted several separate analyses: first, we repeated the earlier analysis of Brown and Heske (1990b) but included six additional years of data. We then performed a second analysis but used rodent densities estimated only from equal access plots: 14 such plots were used to estimate species densities from 1977 through 1988, and 8 plots were used to estimate densities from 1988 through 1994. Finally, we compared the population densities of individual species on equal access plots with their densities on kangaroo rat removal plots to determine the influence of kangaroo rats on the population dynamics of the small granivorous rodents.

Tables II and III show that the results of all three analyses are remarkably similar. Of the 55 pairwise cross-correlations, 37 were positive for rodents on equal access plots, while 35 were positive for data from all plots (Table II). Further, 16 were significantly positive for equal access plots compared to 18 for all plots. Only 2 and 4 pairwise correlations were significantly negative for the equal access and all plot analyses, respectively (Table II). Table III shows that the densities of all non-*Dipodomys* rodents on equal access plots were

TABLE II Cross-Correlation Coefficients at Lag 0 for Time Series of Monthly Rodent Census Date ($N = 192$ Trapping Sessions) for Pairwise Comparisons among the 11 Most Common Species

	D.s.	D.o.	D.m.	C.p.	Pg.f.	Pm.e.	Pm.m.	R.m.	N.a.	O.l.	O.t.
D. spectabilis		−0.24	−0.28	−0.03	0.41	−0.14	0.07	−0.19	0.28	0.01	−0.05
D. ordii	−0.36		0.59	−0.12	−0.19	0.29	0.21	0.32	0.10	0.25	0.18
D. merriami	−0.35	0.61		0.01	0.24	0.12	0.33	0.04	0.27	0.29	
C. penicillatus	−0.26	0.04	−0.02		0.15	−0.15	−0.05	−0.19	0.07	−0.21	−0.01
Pg. flavus	0.32	−0.13	0.12	0.05		−0.15	0.01	−0.06	0.02	0.08	0.04
Pm. eremicus	−0.16	0.44	0.32	−0.03	−0.11		0.35	0.59	0.22	0.29	0.13
Pm. maniculatus	0.13	0.32	0.15	−0.05	−0.07	0.48		0.14	0.36	0.29	−0.11
R. megalotis	−0.33	0.55	0.48	−0.13	0.03	0.63	0.23		−0.04	0.30	0.15
N. albigula	0.26	0.07	0.01	−0.02	−0.06	0.39	0.57	−0.04		0.11	0.02
O. leucogaster	−0.08	0.16	0.17	−0.20	−0.02	0.20	0.23	0.23	0.04		0.14
O. torridus	−0.10	0.19	0.22	−0.01	0.08	0.32	−0.03	0.22	0.13	0.10	

Note. Results above the diagonal were obtained from the mean density of rodents on equal access plots only; results below the diagonal were obtained from the mean density of rodents captured on all plots. Statistically significant correlations are indicated by boldface (negative) and underline (positive). Table-wide rejection level of 0.05 for each analysis was maintained by the sequential Bonferroni method (Rice, 1989). Column heading abbreviations correspond to species names listed as row labels.

TABLE III Correlation Coefficients for Species Densities on Equal Access Plots versus Their Densities on Kangaroo Rat Removal Plots at Time Lag 0, for N = 192 Trapping Sessions

Species	Correlation coefficient
Chaetodipus penicillatus	0.60***
Perognathus flavus	0.63***
Peromyscus eremicus	0.57***
P. maniculatus	0.52***
Reithrodontomys megalotis	0.79***
Onychomys leucogaster	−0.04
O. torridus	0.28***
Neotoma albigula	0.37***

***P <0.001.

significantly positively correlated with their densities on kangaroo rat removal plots, with the exception of *O. leucogaster.* Thus, the presence or absence of kangaroo rats had little effect on the temporal pattern of population fluctuations of other rodents.

The present analyses support the general conclusions of Brown and Heske (1990b), who found 26 significantly positive and 7 significantly negative correlations between these same species pairs. Most species shared similar population dynamics, likely indicating that general environmental conditions were of overriding importance in determining the population dynamics of the species that constitute this assemblage (cf. Dunning and Brown, 1982).

B. The Effects of Infrequent Climatic Events on Species Abundances

One of the advantages of long-term studies is that only these can shed light on the importance of highly sporadic or infrequently periodic events (Likens, 1989). In the desert southwest, at least two kinds of such climatic events occur: El Niño Southern Oscillations and flooding due to dissipating tropical storms (Webb and Betancourt, 1992). Because these climatic events are widespread, they potentially affect large areas and many biological systems. Below, we examine the effects that such events have had on our desert rodent community.

1. El Niño Events

The El Niño Southern Oscillation (ENSO) phenomenon is characterized by anomalous sea level barometric pressures and sea surface temperatures in the

Pacific Ocean (Wallace, 1985), and these are associated with changes in global weather patterns (Hansen, 1990). Marine biologists have long documented the effects of ENSO events on numerous marine organisms, including marine mammals, seabirds, fishes, corals, invertebrates, algae, and plankton (Barber and Chavez, 1983; Wooster and Fluharty, 1985; Tegner and Dayton, 1987; Glynn, 1988, 1990). While dramatic biological impacts of ENSO events may be most readily observed in marine systems, there is growing consensus that these events also influence terrestrial systems (e.g., Lough and Fritts, 1990).

In the desert regions of the southwestern United States, ENSO events usually result in unusually heavy winter and spring precipitation during El Niño years (Rasmussen, 1985; Ropelewski and Halpert, 1986). Given that precipitation is often a limiting resource in this arid region, ENSO events should also affect the biota of the region.

Brown and Heske (1990b) showed that three peaks in the abundance and biomass of rodents at our site corresponded to the ENSO events of 1977–1978, 1982–1983, and 1986–1987 (Fig. 4). Inspection of Fig. 5 shows that these events were associated with exceptionally high densities of four species, *P. eremicus, P. maniculatus, N. albigula,* and *R. megalotis.* Brown and Heske suggested that these peaks reflected the increased availability of seeds following the heavy winter rains. In addition, Meserve *et al.* (1995) showed that some populations of South American rodents and their predators reach exceptionally high densities following El Niño events.

The above data suggest that climatic fluctuations associated with ENSO phenomenon can strongly influence local populations of rodents in the arid regions of western North and South America. Given that rodent populations at our site increased following three consecutive ENSO events, one might conclude that a strong link exists between ENSO and rodent population dynamics in arid North America.

The strength of the presumed association between ENSO and rodent population dynamics, however, is brought into question when the response of the rodents at our site to the El Niño events of 1991–1992 and 1992–1993 is examined. Figure 5 reveals that none of the species in our community, with the possible exception of *P. eremicus,* exhibited the expected increase in abundance. One explanation for the varied response of the rodents at our site may lie in ENSO events themselves: they vary in both strength and pattern of development (Cannon *et al.,* 1985; Hansen, 1990; Diaz and Kiladis, 1992). Given that numerous marine systems exhibit variation in response to ENSO events, it is not surprising that our terrestrial system exhibited similar variation in response. Clearly, future work is required to further elucidate the link between ENSO events and rodent population dynamics in arid regions of the Western Hemisphere.

2. Dissipating Tropical Storms

Occasionally, in late summer or early autumn, dissipating tropical storms affect the southwestern portion of North America, producing widespread and intense precipitation. While such storms can affect the region several times per decade, they are sufficiently severe to cause heavy flooding approximately once every 20 years (Webb and Betancourt, 1992). In late September 1983, tropical storm Octave generated record flooding throughout southeastern Arizona and southwestern New Mexico (Roeske *et al.*, 1989). During a period of less than a week, our study site received 129 mm of precipitation (equal to approximately 50% of the yearly average).

Such rare, severe climatic events often can have dramatic effects on biological systems (e.g., Dayton and Tegner, 1984). While no obvious effect was detected immediately following the storm, Fig. 5b reveals that populations of *D. spectabilis*, the second most abundant species in the community, declined dramatically over the winter of 1983–1984. Valone *et al.* (1995) postulated that damaged seed stores, caused by the unusually heavy precipitation, may have caused the observed decline of *D. spectabilis* at that time.

Dipodomys spectabilis is the largest granivore in the community. It is known to influence strongly both vegetation structure and the use of space by other granivorous rodents (Bowers and Brown, 1992). Thus, the dramatic decline of this keystone species provided a unique opportunity to examine the response of the rodent community to a natural experiment. Because *D. spectabilis* is known to compete directly with other granivorous rodents (Frye, 1983; Brown and Munger, 1985), one might anticipate an increase in the abundances of these species following the decline in *D. spectabilis* density but little or no concomitant increase by nongranivorous species.

To test this hypothesis, Valone *et al.* (1995) examined the response of 10 rodent species to the change in abundance of *D. spectabilis* over the winter of 1983–1984 by comparing the mean density of each species on plots before and after the decline. To reduce the effects of year-to-year fluctuations in densities, they used data from 12 plots over two 3-year periods: 1981–1983, when *D. spectabilis* was abundant; and 1985–1987, when *D. spectabilis* was scarce (Fig. 5b). Table IV shows that the densities of three species (in addition to *D. spectabilis*) in these two periods differed significantly. *Dipodomys ordii* and *P. eremicus* were significantly more abundant on the plots when *D. spectabilis* densities were low, but *P. flavus* was more abundant on the plots when *D. spectabilis* densities were high. As expected, the nongranivorous *Onychomys* spp. and *N. albigula* showed no response to the change in *D. spectabilis* density. But, surprisingly, the most abundant species and most likely competitor, *D. merriami* (Brown and Munger, 1985), exhibited essentially no difference in density in the two periods.

TABLE IV   Mean Density of Rodent Taxa before (1981–1983) and after (1985–1987) the
Decline in Density of *D. spectabilis*

| Taxon | Mean ± SD (individuals/ha) | | t | P |
	Before	After		
Dipodomys spectabilis	7.28 ± 2.88	1.60 ± 1.84	−8.9	0.001
D. merriami	10.84 ± 4.36	11.04 ± 3.44	0.2	0.857
D. ordii	2.08 ± 1.28	4.20 ± 3.72	2.2	0.048
Chaetodipus penicillatus	0.52 ± 0.56	0.36 ± 0.40	−0.8	0.438
Peromyscus eremicus	0.60 ± 0.72	1.20 ± 0.92	2.5	0.023
P. maniculatus	0.20 ± 0.24	0.44 ± 0.48	2.0	0.076
Reithrodontomys megalotis	0.88 ± 0.68	0.68 ± 0.56	−1.4	0.192
Perognathus flavus	1.04 ± 0.76	0.02 ± 0.08	−4.9	0.001
Onychomys spp.	2.52 ± 1.16	2.72 ± 1.60	0.5	0.645
Neotoma albigula	1.60 ± 0.88	1.72 ± 1.20	0.4	0.678

Note. Statistical analyses were conducted on actual numbers of individuals on the plots, but we
have converted these numbers to density (individuals/ha) for consistency. Differences in mean
density for each species were evaluated using a paired *t*-test ($N = 12$ plots). *Onychomys* spp.
includes the ecologically similar *O. leucogaster* and *O. torridus*.

The results in Table IV are only partially consistent with other information
indicating the importance of *D. spectabilis* in this rodent community. Because
D. spectabilis is known to be competitively dominant to all granivorous rodents
(e.g., Frye, 1983; Brown and Munger, 1985), it is not surprising that its decline
was associated with increases in the abundance of some species. It is far from
clear, however, why *D. ordii* and not the more common *D. merriami* increased
in abundance and why the only small granivore to increase in abundance was
P. eremicus.

3. Evidence of a Long-Term Climate Change?

The above discussion brings us back to the other striking features in the
population dynamics of Fig. 5: the long-term population trends of several
species. The response by *D. ordii* to the decline and lack of recovery of
D. spectabilis may explain, at least in part, its long-term population increase.
But recall that *C. penicillatus* and *R. megalotis* also have increased dramatically
over time. Table IV reveals that neither of these species appears to have increased in response to the decline in *D. spectabilis* density, however, so we have
no ready explanation of their increase. Furthermore, we have no explanation
for the lack of recovery of *D. spectabilis* after its dramatic reduction in 1984.

Long-term population trends such as these often result from changes in
habitat quality. In the mid-1990s P. Waser suggested that the vegetation at our

study site (and his nearby site) changed during our study, with the density of small shrubs (especially *Gutierrezia sarothrae* and *Haplopappus* sp.) exhibiting dramatic increases since the mid-1980s. If true, different habitat preferences of the species may explain at least some of the long-term trends. *Dipodomys spectabilis* prefers more open desert habitats (Hoffmeister, 1986), and so increased vegetative cover would represent a general deterioration of its habitat and might explain its lack of recovery from the mid-1980s to the mid-1990s. *Reithrodontomys megalotis* and *C. penicillatus,* on the other hand, prefer more densely vegetated habitats than *D. spectabilis*. Thus, denser shrubs might represent a general improvement in habitat quality for these species, perhaps accounting for their increased densities over time. We are currently investigating the hypothesis that long-term vegetation changes can account for the long-term population trends we have observed.

V. THE INFLUENCE OF REGIONAL FACTORS ON COMMUNITY STRUCTURE: COLONIZATION EVENTS

Up to this point we could have viewed the rodent community on our plots as a "closed" system. We have seen that kangaroo rats compete directly with several species of small granivores. Now, we examine more closely the effects of kangaroo rats on the structure of the rodent community. Specifically, we ask whether kangaroo rats have affected the species diversity of other rodents in this community and, if so, how. The answer will demonstrate the unique insights that our "open" system has yielded.

Figure 6 shows the mean number of non-*Dipodomys* species captured over time on equal access and kangaroo rat removal plots. Note that kangaroo rat removal plots usually supported more species and that the magnitude of this difference in diversity increased over time. To assess the significance of these differences, we analyzed the data in Fig. 6 for three roughly equal time periods. We found no differences in number of non-*Dipodomys* species between equal access and kangaroo rat removal plots during the first years of the experiments, from 1977 through 1981. From 1982 through 1987, and again from 1988 through 1994, however, we found that there were significantly more species of small granivorous rodents on kangaroo rat removal than on equal access plots: kangaroo rats significantly suppressed the diversity of other rodents on our experimental plots (Valone and Brown, 1995).

To determine precisely how kangaroo rats affected species diversity, we focused on the colonization and extinction probabilities of rodents on our plots. We found that differences in species diversity could be accounted for by significant differences in local colonization and extinction probabilities of several rodents; six species of small granivores had a higher colonization probability, a lower extinction probability, or both on kangaroo rat removal plots compared to plots that contained kangaroo rats (Table V) (Valone and Brown, 1995). Neither of the insectivorous *Onychomys* species nor the herbivorous *Neotoma albigula* differed in colonization or extinction probabilities across plot types. *Sigmodon fulviventer,* a species characteristic of grasslands (Hoffmeister, 1986), had a higher colonization probability on kangaroo rat removal plots—apparently a response to the increased grass cover owing to the long-term exclusion of kangaroo rats (Brown and Heske, 1990a; Heske *et al.,* 1993).

These results support the hypothesis that interspecific competition affects species diversity in this desert rodent community. Granivorous kangaroo rats directly affected the colonization and extinction probabilities of small granivorous rodents, but had essentially no effect on these probabilities for nongranivorous species. Such results demonstrate the importance of the openness of the system. Clearly, we could not have obtained such results had our plots been closed to prevent movement of individuals of small granivorous species across the fences. Our open experimental design enabled us to show that the competitively subordinate species differentially colonized plots from which kangaroo rats had been removed.

The above results have important implications for studies of metapopulation dynamics (Gilpin and Hanski, 1991). Many populations are composed of subunits, metapopulations, between which individuals migrate. Theoretical treatments of metapopulation dynamics often assume that colonization and

TABLE V   Responses of Rodent Species to the Absence of Kangaroo Rats from Experimental Plots

Increased colonization probability
  Perognathus flavus
  Peromyscus maniculatus
  Reithrodontomys fulvescens
  R. megalotis
  Sigmodon fulviventer

Decreased extinction probability
  Chaetodipus penicillatus
  Perognathus flavus
  Peromyscus eremicus
  Reithrodontomys megalotis

No differences in colonization or extinction probability
  Onychomys leucogaster
  O. torridus
  Neotoma albigula

extinction events in these metapopulations are stochastic (Gilpin and Hanski, 1991), but few empirical data can address this important assumption. Our results demonstrate that these probabilities are not stochastic, but rather are significantly influenced by the competitive environment.

VI. GENERAL DISCUSSION

We have shown how intrinsic biotic interactions, extrinsic climatic events, and exchanges of individuals with the regional species pool have influenced the population dynamics and community structure of a Chihuahuan Desert rodent community over 17 years. Clearly, our understanding of the nature of this community would be less complete had we not been able to assess the effects of each of these processes.

It may appear at first that the results of our studies are contradictory: kangaroo rats have little influence on the population fluctuations of small granivores (Table III), but at the same time they strongly influence their local abundance (Fig. 2) and overall species diversity (Fig. 6). The apparent paradox of these findings underscores the importance of multiple levels of investigation in community ecology and provides insight into the relative importance of intrinsic, extrinsic, and regional factors in structuring communities.

Clearly, abiotic climatic conditions are the primary force influencing the population dynamics of the rodents in our community. In desert systems such as ours, the amount and timing of precipitation strongly determine the availability of plants, seeds, and insects which are consumed by animals (e.g., Tevis, 1958; Beatley, 1974; Dunning and Brown, 1982). Thus, it is not surprising that most species exhibited positive population dynamics: at our site there were "good" and "bad" years for the rodents, regardless of diet, depending on the pattern of precipitation.

Underneath the ultimate influence of precipitation on food resources and population densities, however, interspecific interactions have strong proximate effects on the local abundances of species. Dominant kangaroo rats suppress the local densities of several small granivores, and this competitive effect is largely independent of the seasonal and interannual fluctuations in abiotic conditions. The effect of such competition ultimately leads to reduced species diversity of other rodents in the presence of kangaroo rats.

Thus, while kangaroo rats have little effect on the overall seasonal and year-to-year population fluctuations of small granivores, at the same time they strongly influence their local abundance and, ultimately, their diversity. The apparent paradox results from responses assessed at different spatial and temporal scales and from the different abiotic and biotic factors that combine to influence the structure and dynamics of communities (Fig. 1).

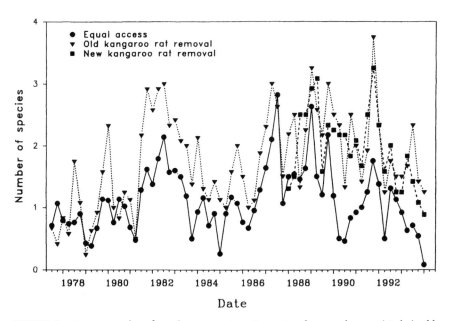

FIGURE 6. Average number of non-kangaroo rat species captured on equal access (circles), old kangaroo rat removal (triangles), and new kangaroo rat removal (squares) plots over time. Data are plotted as 3-month averages for clarity. [Reprinted with permission from T. J. Valone and J. H. Brown (1995). Copyright 1995 American Association for the Advancement of Science.]

Desert rodent communities around the world have been the focus of much ecological research on competitive coexistence (e.g., Price, 1986; Kotler and Brown, 1988). Typically, such work has emphasized short-term empirical studies of habitat selection (Rosenzweig, 1973, 1987; Price, 1978a), resource utilization (Brown and Lieberman, 1973; Reichman and Oberstein, 1977; Price, 1978b), or anti-predator adaptations (Price and Brown, 1983; Kotler, 1984) of presumed close competitors; often this work has been structured around theoretical models of coexistence that involve some degree of temporal or spatial variation in environmental parameters (Kotler and Brown, 1988; J. S. Brown, 1989). Our long-term work complements this body of literature both by emphasizing the profound effects of environmental variation on rodent population dynamics and by illustrating the pervasive nature of competition in one particular community.

Prior empirical work on long-term rodent population dynamics often has focused on single species, especially of microtenes in mesic or boreal habitats, and attempted to determine whether the observed dynamics were stable, cyclical, or chaotic (e.g., Turchin, 1992). Our work demonstrates that desert rodent populations fluctuate widely, sometimes by more than an order of magnitude

(Fig. 5). More importantly, our data implicate abiotic conditions as an important ultimate influence on rodent populations because the dynamics of most pairs of species in the community were positively correlated. This was true for species-pairs that differed both in diet (e.g., granivorous *D. merriami* and insectivorous *O. torridus*) and by an order of magnitude in body size (e.g., *D. spectabilis* and *P. flavus;* Tables I and II). Of course, the effects of temporal (and spatial) variation in climate and other abiotic variables on rodent populations can be mediated through biotic factors such as food supply, competition, and predation. Whether similar patterns exist in rodent communities of different habitats remains to be seen.

Our rodent system is hardly unique. As indicated in Fig. 1, all communities are influenced by a combination of abiotic, biotic, and regional or historical factors. Other excellent examples are afforded by experimental studies of communities in rocky intertidal habitats. These have shown how the outcomes of intrinsic competitive and predator–prey interactions among the locally coexisting species are influenced by the extrinsic abiotic environments in which the local community is embedded (e.g., Connell, 1961, 1972, 1975; Dayton, 1971; Lubchenco and Menge, 1978; Underwood, 1984, 1985; Wootton, 1992; Menge *et al.,* 1994). Most of these studies have focused on situations in which variation in abiotic conditions is caused by spatial gradients of exposure to wave action and desiccation between tides, however, rather than by temporal variation in climate as in our study. Many of the above and other (e.g., Paine, 1966, 1974; Schmidt, 1987) intertidal studies have also documented the importance of indirect interactions: how effects of certain predators or competitors can alter the outcomes of interactions between other species and thus affect the organization of communities. Work of Roughgarden *et al.* (1987, 1988) has shown how the organization of local intertidal communities depends on their openness and hence on regional processes, i.e., how colonization of planktonic larvae, which is influenced by offshore currents and ENSO events, influences the initial composition of communities and thus the outcomes of subsequent interactions.

One unusual aspect of our study is its design: the combination of long-term monitoring, controlled experimental perturbations, and an "open" system that permits free exchange of individuals with the regional environment. Most experimental studies of communities not only manipulate small patches of space for short periods of time (Fig. 1), but also study systems effectively closed to movements of individuals. Either the experimental plots are surrounded by fences or other exclosures to actively prevent immigration and emigration or the organisms are relatively sedentary and the study is conducted for a sufficiently brief period that dispersal is drastically curtailed.

Our design, which uses selectively permeable fences to exclude certain species but which does not enclose any populations, together with our long-term measurements, allows us to assess the roles of dispersal and local colonization—extinction dynamics in the responses of the community to both natural environmental variation and experimental perturbations. Such movements are important. The openness of communities and their resulting sensitivity to regional and historical influences profoundly affect their structure and dynamics (see Ricklefs and Schluter, 1993; Valone and Brown, 1995).

VII. SUMMARY

We have examined how intrinsic biotic, extrinsic abiotic, and regional factors have affected the structure and dynamics of a Chihuahuan Desert rodent community. Several lines of evidence suggest that precipitation is the primary force influencing the population dynamics of the rodents in this community: (1) most species pairs exhibited positive population dynamics, even those that shared little dietary overlap; (2) population densities of a few species increased following several El Niño Southern Oscillation events; and (3) a dissipating tropical weather system may have caused the dramatic decline of the largest rodent at the site.

Underneath this ultimate abiotic influence on population densities, however, interspecific interactions had strong proximate effects on the local abundances of species. Kangaroo rats competitively suppressed the densities of several species of small granivorous rodents, and this competitive effect was relatively independent of seasonal and interannual fluctuations in abiotic conditions. Further, we showed that such competition reduced the diversity of rodents on experimental plots by affecting the regional colonization and local extinction probabilities of species; most granivorous rodent species exhibited a higher colonization probability, lower extinction probability, or both where kangaroo rats were absent.

All systems will be influenced by a combination of biotic, abiotic, and regional factors. Most experimental work in community ecology manipulates small, closed patches for brief periods of time and focuses on one of these factors. Our work demonstrates how long-term studies of open systems can yield information on the relative importance of biotic, abiotic, and regional processes on community structure and dynamics.

Acknowledgments

We thank the numerous individuals who have contributed to the "Portal project" over the years. Research at this site has been supported by grants from the National Science Foundation, most recently DEB-9221238.

References

Abrams, P. (1984). Variability in resource consumption rates and the coexistence of competing species. *Theor. Popul. Biol.* 25, 106–124.

Barber, R. T., and Chavez, F. P. (1983). Biological consequences of El Niño. *Science* 222, 1203–1210.

Beatley, J. C. (1974). Phenological events and their environmental triggers in Mojave Desert ecosystems. *Ecology* 55, 856–863.

Begon, M., Harper, J. L., and Townsend, C. R. (1986). "Ecology," 2nd ed. Blackwell, Boston.

Bender, E. A., Case, T. J., and Gilpin, M. E. (1984). Perturbation experiments in community ecology: Theory and practice. *Ecology* 65, 1–13.

Bowers, M. A., and Brown, J. H. (1992). Structure in a desert rodent community: Use of space around *Dipodomys spectabilis* mounds. *Oecologia* 92, 242–249.

Brown, J. H., and Heske, E. J. (1990a). Control of a desert-grassland transition by a keystone rodent guild. *Science* 250, 1705–1707.

Brown, J. H., and Heske, E. J. (1990b). Temporal changes in a Chihuahuan Desert rodent community. *Oikos* 59, 290–302.

Brown, J. H., and Lieberman, G. (1973). Resource utilization and coexistence of seed-eating rodents in sand dune habitats. *Ecology* 54, 788–797.

Brown, J. H., and Munger, J. C. (1985). Experimental manipulation of a desert rodent community: Food addition and species removal. *Ecology* 66, 1545–1563.

Brown, J. H., Davidson, D. W., Munger, J. C., and Inouye, R. C. (1986). Experimental community ecology. *In* "Community Ecology" (J. M. Diamond and T. J. Case, eds.), pp. 41–61. Harper & Row, New York.

Brown, J. H., Taper, M. L., and Heske, E. J. (1996). Constant competition in a variable environment: Experiments with desert rodents. In preparation.

Brown, J. S. (1989). Desert rodent community structure: A test of four mechanisms of coexistence. *Ecol. Monogr.* 59, 1–20.

Cannon, G. A., Reed, R. K., and Pullen, P. E. (1985). Comparison of El Niño events off the pacific northwest. *In* "El Niño North. Niño Effects in the Eastern Subarctic Pacific Ocean" (W. S. Wooster and D. L. Fluharty, eds.), pp. 75–84. Washington Sea Grant Program, University of Washington, Seattle.

Carpenter, S. R. (1988). "Complex Interactions in Lake Communities." Springer-Verlag, New York.

Connell, J. H. (1961). The influence of interspecific competition and other factors on the distribution of the barnacle *Chthamalus stellatus*. *Ecology* 42, 710–723.

Connell, J. H. (1972). Community interactions on marine rocky intertidal shores. *Annu. Rev. Ecol. Syst.* 3, 169–192.

Connell, J. H. (1975). Some mechanisms producing structure in natural communities: A model and evidence from field experiments. *In* "Ecology and Evolution of Communities" (M. L. Cody and J. Diamond, eds.), pp. 460–490. Harvard Univ. Press, Cambridge, MA.

Davidson, D. W., Inouye, R. S., and Brown, J. H. (1984). Granivory in a desert ecosystem: Experimental evidence for indirect facilitation of ants by rodents. *Ecology* 65, 1780–1786.

Davidson, D. W., Samson, D. A., and Inouye, R. S. (1985). Granivory in the Chihuahuan Desert: Interactions within and between trophic levels. *Ecology* 66, 486–502.

Dayton, P. K. (1971). Competition, disturbance and community organization: The provision and subsequent utilization of space in a rocky intertidal community. *Ecol. Monogr.* 41, 351–389.

Dayton, P. K., and Tegner, M. J. (1984). Catastrophic storms, El Niño, and patch stability in a southern California kelp community. *Science* 224, 283–285.

Diamond, J. M., and Case, T. J., eds. (1986). "Community Ecology." Harper & Row, New York.

Diaz, H. F., and Kiladis, G. N. (1992). Atmospheric teleconnections associated with extreme phases of the Southern Oscillation. In "El Niño: Historical and Paleoclimatic Aspects of the Southern Oscillation" (H. F. Diaz and V. Markgraf, eds.), pp. 7–28. Cambridge Univ. Press, Cambridge, UK.

Dunning, J. B., and Brown, J. H. (1982). Summer rainfall and winter sparrow densities: A test of the food limitation hypothesis. Auk 99, 123–129.

Frye, R. J. (1983). Experimental field evidence on interspecific aggression between two species of kangaroo rat (Dipodomys). Oecologia 59, 74–78.

Gee, J. H. R., and Giller, P. S., eds. (1987). "Organization of Communities: Past and Present." Blackwell, Oxford.

Gilpin, M. E., and Hanski, I. (1991). "Metapopulation Dynamics: Empirical and Theoretical Investigations." Cambridge Univ. Press, Cambridge, UK.

Glynn, P. W. (1988). El Niño-Southern Oscillation 1982–83: Nearshore population, community and ecosystem responses. Annu. Rev. Ecol. Syst. 19, 309–345.

Glynn, P. W., ed. (1990). "Global Ecological Consequences of the 1982–83 El Niño-Southern Oscillation." Elsevier, Amsterdam.

Hansen, D. V. (1990). Physical aspects of the El Niño event of 1982–83. In "Global Ecological Consequences of the 1982–83 El Niño-Southern Oscillation" (P. W. Glynn, ed.), pp. 1–20. Elsevier, Amsterdam.

Hansson, L. (1991). Levels of density variation: the adequacy of indices and chaos. Oikos 61, 285–287.

Harper, J. L. (1977). "The Population Biology of Plants." Academic Press, London.

Heske, E. J., Brown, J. H., and Guo, Q. (1993). Effects of kangaroo rat exclusion on vegetation structure and plant species diversity in the Chihuahuan Desert. Oecologia 95, 520–524.

Heske, E. J., Brown, J. H., and Mistry, S. (1994). Long-term experimental study of a Chihuahuan Desert rodent community: 13 years of competition. Ecology 75, 438–445.

Hoffmeister, D. F. (1986). "Mammals of Arizona." Univ. of Arizona Press, Tucson.

Kingsolver, J. G., and Schemske, D. W. (1991). Path analysis of selection. Trends Ecol. Evol. 6, 276–280.

Kotler, B. P. (1984). Predation risk and the structure of desert rodent communities. Ecology 65, 689–701.

Kotler, B. P., and Brown, J. S. (1988). Environmental heterogeneity and the coexistence of desert rodents. Annu. Rev. Ecol. Syst. 19, 281–307.

Levey, D. J. (1988). Tropical wet forest treefall gaps and distributions of understory birds and plants. Ecology 69, 1076–1089.

Likens, G. E. (1989). "Long-term Studies in Ecology." Springer-Verlag, New York.

Lough, J. M., and Fritts, H. C. (1990). Historical aspects of El Niño/Southern Oscillation—information from tree rings. In "Global Ecological Consequences of the 1982–83 El Niño-Southern Oscillation" (P. W. Glynn, ed.), pp. 1–20. Elsevier, Amsterdam.

Lubchenco, J., and Menge, A. (1978). Community development and persistence in a low rocky intertidal zone. Ecol. Monogr. 48, 67–94.

MacMillen, R. E. (1964). Population ecology, water relations, and social behavior of a southern California semidesert fauna. Univ. Calif., Berkeley, Publ. Zool. 71, 1–66.

Menge, B. A., Berlow, E. L., Blanchette, C. A., Navarrete, S. A., and Yamada, S. B. (1994). The keystone species concept: Variation in interaction strength in a rocky intertidal habitat. Ecol. Monogr. 64, 249–286.

Meserve, P. L., Yunger, J. A., Gutierrez, J. R., Contreras, L. C., Milstead, W. B., Lang, B. K., Cramer, K. L., Herrera, S., Lagos, V. O., Silva, S. I., Tabilo, E. L., Torrealba, M.-A., and Jaksic, F. M. (1995). Heterogeneous responses of small mammals to an El Niño (ENSO) event in north-central semiarid Chile and the importance of ecological scale. J. Mammal. 76, 580–595.

Munger, J. C., and Brown, J. H. (1981). Competition in desert rodents: An experiment with semipermeable exclosures. *Science* 211, 510–512.

O'Connell, M. A. (1989). Population dynamics of neotropical rodents in seasonal environments. *J. Mammal.* 70, 532–548.

Paine, R. T. (1966). Food web complexity and species diversity. *Am. Nat.* 100, 65–75.

Paine, R. T. (1974). Intertidal community structure: Experimental studies on the relationship between a dominant competitor and its principle predator. *Oecologia* 15, 93–120.

Pianka, E. R. (1983). "Evolutionary Ecology." Harper & Row, New York.

Pickett, S. T. A., and White, P. S. (1985). "The Ecology of Natural Disturbance as Patch Dynamics." Academic Press, Orlando, FL.

Pimm, S. L. (1982). "Food Webs." Chapman & Hall, London.

Price, M. V. (1978a). The role of microhabitat specialization in structuring desert rodent communities. *Ecology* 58, 1393–1399.

Price, M. V. (1978b). Seed dispersion preferences in coexisting desert rodent species. *J. Mammal.* 59, 624–626.

Price, M. V. (1986). Structure of desert rodent communities: A critical review of questions and approaches. *Am. Zool.* 26, 39–49.

Price, M. V., and Brown, J. H. (1983). Patterns of morphology and resource use in North American desert rodent communities. *Great Basin Nat. Mem.* 7, 117–134.

Rasmussen, E. M. (1985). El Niño and variations in climate. *Am. Sci.* 73, 168–177.

Reichman, O. J., and Oberstein, D. (1977). Selection of seed distribution types by *Dipodomys merriami* and *Perognathus amplus. Ecology* 58, 636–643.

Reichman, O. J., and Van de Graaf, K. M. (1973). Seasonal activity and reproductive patterns of five species of Sonoran Desert rodents. *Am. Midl. Nat.* 90, 118–126.

Rice, W. R. (1989). Analyzing tables of statistical tests. *Evolution (Lawrence, Kans.)* 43, 223–225.

Ricklefs, R. E. (1979). "Ecology," 2nd ed. Chiron Press, New York.

Ricklefs, R. E., and Schluter, D. (1993). "Species Diversity in Ecological Communities." Univ. of Chicago Press, Chicago.

Roeske, R. H., Garrett, J. M., and Eychaner, J. H. (1989). Floods of October 1983 in southeastern Arizona. *Water Resour. Invest. Rep. (U.S. Geol. Surv.)* 85-4225-C, 1–77.

Ropelewski, C. F., and Halpert, M. S. (1986). North American precipitation and temperature patterns associated with El Niño/Southern Oscillation (ENSO). *Mon. Weather Rev.* 114, 2352–2362.

Rosenzweig, M. L. (1973). Habitat selection experiments with a pair of coexisting heteromyid rodent species. *Ecology* 62, 327–335.

Rosenzweig, M. L. (1987). Community organization from the point of view of habitat selectors. *In* "Organization of Communities: Past and Present" (J. H. R. Gee and P. S. Giller, eds.), pp. 469–490. Blackwell, Oxford.

Roughgarden, J., Gains, S., and Pacala, S. (1987). Supply side ecology: The role of physical transport processes. *In* "Organization of Communities: Past and Present" (J. H. R. Gee and P. S. Giller, eds.), pp. 491–518. Blackwell, Oxford.

Roughgarden, J., Gains, S., and Possingham, H. (1988). Recruitment dynamics in complex life cycles. *Science* 241, 1460–1466.

Samson, D. A., Philippi, T. E., and Davidson, D. W. (1992). Granivory and competition as determinants of annual plant diversity in the Chihuahuan Desert. *Oikos* 65, 61–80.

Schmidt, R. J. (1987). Indirect interactions between prey: Apparent competition, predator aggregation, and habitat segregation. *Ecology* 68, 1887–1897.

Smith, F. A., Brown, J. H., and Valone, T. J. (1996). Path analysis: A critical evaluation using long-term experimental data. *Am. Nat.*, in press.

Strauss, S. Y. (1991). Indirect effects in community ecology: Their definition, study, and importance. *Trends Ecol. Evol.* 6, 206–210.

Tegner, M. J., and Dayton, P. K. (1987). El Niño effects on southern California kelp forest communities. *Adv. Ecol. Res.* 17, 243–281.

Tevis, L. (1958). Germination and growth of ephemerals induced by sprinkling a sandy desert. *Ecology* 39, 681–688.

Thompson, D. B., Brown, J. H., and Spencer, W. D. (1991). Indirect facilitation of granivorous birds by desert rodents: Experimental evidence from foraging patterns. *Ecology* 72, 852–863.

Turchin, P. (1992). Chaos and stability in rodent population dynamics: Evidence from non-linear time-series analysis. *Oikos* 68, 167–172.

Underwood, A. J. (1984). Vertical and seasonal patterns in competition for microalgae between gastropods. *Oecologia* 64, 211–222.

Underwood, A. J. (1985). Physical factors and biological interactions: The necessity and nature of ecological experiments. *In* "The Ecology of Rocky Coasts" (P. G. Moore and R. Seed, eds.), pp. 371–390. Hodder & Stoughton, London.

Valone, T. J., and Brown, J. H. (1995). Effects of competition, colonization, and extinction on rodent species diversity. *Science* 267, 880–883.

Valone, T. J., Brown, J. H., and Jacobi, C. L. (1995). Catastrophic decline of a desert rodent, *Dipodomys spectabilis*: Insights from a long-term study. *J. Mammal.* 76, 428–436.

Wallace, J. M. (1985). Atmospheric responses to equatorial sea-surface temperature anomalies. *In* "El Niño North. Niño Effects in the Eastern Subarctic Pacific Ocean" (W. S. Wooster and D. L. Fluharty, eds.), pp. 9–21. Washington Sea Grant Program, University of Washington, Seattle.

Webb, R. H., and Betancourt, J. L. (1992). Climatic variability and flood frequency of the Santa Cruz River, Pima county, Arizona. *Geol. Surv. Water-Supply Pap. (U.S.)* 2379, 1–40.

Wiens, J. A. (1977). On competition and variable environments. *Am. Sci.* 65, 590–597.

Wiens, J. A. (1986). Spatial scale and temporal variation in studies of shrubsteppe birds. *In* "Community Ecology" (J. M. Diamond and T. J. Case, eds.), pp. 154–172. Harper & Row, New York.

Wiens, J. A. (1989). "The Ecology of Bird Communities," Vols. 1 and 2. Cambridge Univ. Press, Cambridge, UK.

Wooster, W. S., and Fluharty, D. L., eds. (1985). "El Niño North. Niño Effects in the Eastern Subarctic Pacific Ocean." Washington Sea Grant Program, University of Washington, Seattle.

Wootton, J. T. (1992). Indirect effects, prey susceptibility and habitat selection: Impacts of birds on limpets and algae. *Ecology* 73, 981–991.

Wootton, J. T. (1993). Indirect effects and habitat use in an intertidal community: Interaction chain and interaction modifications. *Am. Nat.* 141, 71–89.

Wootton, J. T. (1994). Predicting direct and indirect effects: An integrated approach using experiments and path analysis. *Ecology* 75, 151–165.

Wright, S. (1934). The method of patch coefficients. *Ann. Math. Stat.* 5, 161–215.

INDEX

Underlined page numbers indicate figures or tables.

Printed and bound by CPI Group (UK) Ltd, Croydon, CR0 4YY

08/05/2025

01864885-0002